GREEK LETTERS USED IN THE TEXT

α	alpha	ν	nu
β	beta	Ξ	xi
γ	gamma	π	pi
Δ	delta	ρ	rho
ε	epsilon	σ, Σ	sigma
η	eta	τ	tau
θ	theta	ϕ	phi
λ, Λ	lamda	ψ	psi
μ	mu	ω, Ω	omega

ASTRONOMICAL DATA

1 Light year (L.Y.)	$= 9.46 \times 10^{17}$ cm
1 Parsec (pc)	$= 3.08 \times 10^{18}$ cm
1 Astronomical unit (A.U.) (Earth-Sun distance)	$= 1.50 \times 10^{13}$ cm
Radius of Sun	$= 6.96 \times 10^{10}$ cm
Earth-moon distance	$= 3.84 \times 10^{10}$ cm
Radius of Earth	$= 6.38 \times 10^{8}$ cm
Radius of moon	$= 1.74 \times 10^{8}$ cm
Mass of Sun	$= 1.99 \times 10^{33}$ g
Mass of Earth	$= 5.98 \times 10^{27}$ g
Mass of moon	$= 7.35 \times 10^{25}$ g
Average orbital speed of Earth	$= 2.98 \times 10^{6}$ cm/sec

NUMERICAL QUANTITIES

$\sqrt{2} = 1.414$

$\sqrt{3} = 1.732$

$\sqrt{5} = 2.236$

$\sqrt{10} = 3.162$

$\pi = 3.14159$

1 rad $= 57.296$ degrees

1 degree $= 0.01745$ rad

PHYSICAL CONSTANTS

Velocity of light in vacuum	$c = 2.998 \times 10^{10}$ cm/sec
Charge of electron	$e = 4.80 \times 10^{-10}$ statC
	$= 1.60 \times 10^{-19}$ C
	$e^2 = 1.44 \times 10^{-13}$ MeV-cm
Planck's constant	$h = 6.63 \times 10^{-27}$ erg-sec
	$= 4.14 \times 10^{-15}$ eV-sec
	$\hbar = 6.58 \times 10^{-16}$ eV-sec
	$hc = 1.24 \times 10^{-4}$ eV-cm
Boltzmann's constant	$k = 1.38 \times 10^{-16}$ erg/°K
	$= 0.862 \times 10^{-4}$ eV/°K
Avogadro's number	$N_o = 6.022 \times 10^{23}$ mole^{-1}
Electron mass	$m_e = 9.11 \times 10^{-28}$ g
	$m_e c = 0.511$ MeV
Proton mass	$m_p = 1.6726 \times 10^{-24}$ g
	$= 1836.11\, m_e$
	$m_p c^2 = 938.259$ MeV
Neutron mass	$m_n = 1.6749 \times 10^{-24}$ g
	$m_n c^2 = 939.553$ MeV
Atomic mass unit	1AMU $= 1.6605 \times 10^{-24}$ g
	$(1\text{ AMU}) \times c^2 = 931.481$ MeV
Gravitational constant	$G = 6.673 \times 10^{-8}$ dyne-cm^2/g^2
Gas constant	$R = 1.986$ cal/mole-°K
	$= 8.314 \times 10^{7}$ erg/mole-°K

Physics and the Physical Universe

Physics and the Physical Universe

JERRY B. MARION

Department of Physics and Astronomy
University of Maryland
College Park, Maryland

John Wiley & Sons, Inc.
New York   London   Sydney   Toronto

The illustration on the cover is adapted from a photograph courtesy of Martin Marietta Corporation.

Library of Congress Catalogue Card Number: 73-125274

ISBN 0-471-56915-1

Printed in the United States of America

10 9 8 7 6 5 4 3 2 1

Preface

We live today in a scientific world. Wherever we look we readily see the evidence of the scientific and technological underpinning of our society. Science is a *real* part of the *real* world—and it will remain so. Consequently, no one who hopes to understand or to influence the world he lives in can do so without some appreciation of science and the scientific basis of modern technology.

In the world of the 1970's we must reassert science as a fundamental channel for inquiry into the many basic and unanswered questions of our time. Some of these questions lie not in physics, at least not in the strict sense, but in related fields: astrophysics, geoscience, and the life sciences. Today's student of physics must be made aware of these implications and attention must be given to the interfaces between the various sciences.

It is my hope that, in this introductory physics text, I have succeeded in making it possible to present the subject in this comprehensive light. This book is designed to guide the student to the basic ideas of physics and to explore some of the modern scientific concepts that have emerged in the mid-twentieth century. The emphasis is on contemporary thinking in physics—ideas concerning the structure and the constituents of matter (the *micro*world) as well as modern notions of astronomy, astrophysics and cosmology (the *macro*world). These subjects and the connection between the physical microworld and the life sciences are the topics that are exciting and challenging today. They are, as well, the topics that will shape our future science and technology. No longer is it sufficient for the educated citizen to be aware of Newton's laws and the elementary ideas of gravity and electricity; he must also be acquainted with the new concepts that are emerging from the science of today.

It is necessarily impossible in a relatively short and non-mathematical survey to discuss more than a small fraction of the immense quantity of important ideas in physics. This book represents a synthesis of compromises, in that the choice has been made to treat only briefly or to eliminate entirely many topics in classical physics that seem not directly related to modern physics. In their place has been included a proportionately larger amount of material from areas of contemporary physics, astrophysics, and cosmology. One feature of this book which I hope is both evident and useful is the logical and gradual progression of the level of sophistication required to understand and to appreciate the series of concepts presented.

This text is a part of a larger whole which, taken together, will serve to present a more complete picture of physics and the physical universe. There is available a book of selected readings from the literature of physics (*A Universe of Physics*) that generally follows the organization of the text; it contains historical, biographical, and descriptive articles from a wide variety of sources. There is also available a more-elaborate-than-usual instructor's manual and a student study guide of comparable quality.

I wish to take this opportunity to acknowledge the contribution of several persons to the successful completion of this book. Professors G. J. Stephenson, Jr., P. DiLavore, N. S. Wall, and F. C. Young provided extremely helpful criticism of the manuscript in various stages of development. Donald Deneck of John Wiley and Sons gave enthusiastic support to the project from its inception. Several secretaries, particularly Mrs. Elizabeth Lee, typed the several drafts with great efficiency. Thanks are also due the numerous students who used this material in its various mimeographed forms and made many helpful suggestions.

Jerry B. Marion
College Park, Maryland

Table of Contents

Physics and the Physical Universe

The Chemist

1 *The Structure and the Language of Physics*

What is physics about? What questions regarding the nature of things have physicists been able to answer? What questions are they trying to answer *today?* Why is physics important? How does physics affect mankind?

In this book we shall try to give some answers to these questions. Our goal is to present an elementary account of the current status of physical theory and to describe the character of physics research as it is being conducted today.

The phases of development in physics can be divided (somewhat arbitrarily) into the *classical, modern,* and *contemporary* periods. Before the end of the 19th century, a detailed knowledge had been acquired of such subdisciplines of physics as mechanics, thermodynamics, electromagnetism, optics, and hydrodynamics. By about 1900, the theoretical descriptions of these areas seemed to be essentially complete and it appeared that there were no more basic discoveries to be made. Collectively, this subject matter is referred to as *classical physics.*

The last few years of the 19th century and the first three decades of the 20th century produced a series of startling new ideas in physics. During this period *radioactivity* was discovered and then used to probe the core of the atom. The development of the theory of *relativity* forced us to examine carefully and to modify our previous views of space and time. And *quantum theory* was formulated from our attempts to describe the inner workings of atomic systems. These decisive years, during which the entire complexion of physics research was changed, we call the era of *modern physics.*

The 1930s witnessed the first observation of radio emissions from stars, the discoveries of the neutron and fission, and the identification of the first elementary particle not found naturally in atoms. Discoveries such as these led to a tremendous outpouring of results in all of the new fields of physics—a growth that continues to the present time. These developments and the new ideas and discoveries to which they have given birth constitute *contemporary physics.*

In the first part of this book we concentrate on equipping ourselves with the necessary background to be able to appreciate a discussion of the problems of modern and contemporary physics that have been solved and those for which we still seek solutions today.

In this chapter we begin our preparation by considering what physics is about, why experiments are crucial to physics, what we mean by a physical theory, and how we apply such theories to the real physical world. We must also recognize that physics is *quantitative* in character and therefore uses mathematics as its natural language. Thus, before we can commence the business of discussing physics, we must mention some of the essentials of mathematics. Here, as throughout the book, we introduce only the minimum of mathematical detail necessary for a clear understanding of the subject.

1.1 *What Is Physics?*

THE STUDY OF THE LAWS OF NATURE

A few hundred years ago the entire body of scientific knowledge was sparse enough that one person could be familiar with most of the essential ideas. Indeed, the scientist of that day was termed a *natural philosopher,* one who studied all facets of natural science. The accumulation of scientific information since those Renaissance days has been so rapid that the natural philosopher (the *compleat* man) has long since ceased to exist. Instead we have physicists, chemists, biologists, zoologists, geologists, and several dozen other designations for the working scientists of today. However, it is still our goal to unify the separate disciplines of science. When our understanding of Nature is more complete, we shall be more able to appreciate the connection between physics and biology, between chemistry and geology, and between meteorology and oceanography.

The *physicist* seeks, first, to understand the way in which the most elementary systems in Nature really operate. The discoveries made by the physicist not only broaden our view of fundamental processes, but frequently are of crucial importance in the advancement of other sciences. The development of quantum theory, for example, permitted the chemist to understand the wide variety of facts that he had gathered about chemical structures and chemical reactions. The rules that the physicist formulated concerning the propagation of sound waves in solid materials allowed the geologist to use seismological techniques for the investigation of the interior of the Earth. The theory of fluid flow is of great importance to the meteorologist and to the oceanographer. The laws of physics determine all physical processes. We have discovered some of these laws—others still elude us.

PHYSICS BEGINS WITH SIMPLE SYSTEMS

Sciences such as geology, meteorology, or physical oceanography attempt to describe the general behavior of very complex systems. Physics, on the other hand, first examines the most elementary systems but in great detail. Thus, while a geologist might be concerned with the description of the process of forming a mountain of rock, the physicist must approach problems of matter in the solid state with a thorough understanding of the inner workings of a simple atom, such as hydrogen. Only then can he progress to the study of the more complicated hydrogen molecule and then to systems of greater complexity, such as matter in the solid state. At each stage of this procedure, the physicist encounters new fundamental problems that must be solved before proceeding to the next step. Frequently the solution to a problem will elude him. He must then be aware of the effect of this lack of knowledge on the answer to his next question. If he is fortunate, he may find that his next problem or two will close the circle and bring him back to his unsolved problem with a different approach that permits him, this time, to find a solution. It is by such methods—asking the proper question, by-passing unsolved problems only to return with a new approach, sharpening the answers to both old and new questions—that progress in physics is made.

SERENDIPITY

Sometimes physics is pure luck. There is always the chance that while studying a certain problem there will come, quite by accident, some important new discovery. When Galileo was using his newly invented telescope to study the planets, there suddenly appeared, viewed by man for the first time, four moons circling the planet Jupiter. And Becquerel *just happened* to discover radioactivity when he developed some photographic film on which he had placed some pitchblende (uranium ore) some weeks earlier. Of course, we cannot rely on luck to provide the answers to all of our questions, but science is an endeavor of discovery and some discoveries do happen by accident. The scientist must always be alert to appreciate and take advantage of a stroke of good fortune.

1.2 *Why Is Physics Important?*

PHYSICS AND TECHNOLOGY

Having generally described what physics is about, we now ask "Why is physics important; what *good* is physics?" Certainly, the physicist does not create new buildings or construct new modes of transportation. He does not cure our ills nor provide greater comforts in our homes. Physics deals with the pursuit of knowledge about our Universe, its constituents, and their behavior. However, it *is* true that the architects and engineers who construct our buildings and aircraft make constant use of the laws of mechanics and dynamics as formulated by physicists. Many of the diagnostic and thera-peutic techniques used in modern medicine were developed in the physics laboratory. Refrigeration, radio, and television are outgrowths of discoveries by physicists. The discovery of the transistor in a solid-state physics labora-tory has led to a new age of miniaturized electronics and also to an increasing reliance on the computer in research activities and in our everyday lives. Without the injection of new ideas that have been produced by physicists our great technological industries would not exist and the level of our society would be stark and primitive. Physics is therefore intimately connected with technology and it is the impact of this association that is the most apparent effect of physics on society.

PHYSICS AND THE PURSUIT OF KNOWLEDGE

Although its contribution to technology is obvious, there is an equally important *why* to physics. Man does not live by technology alone. The fruits of technology influence him *physically,* but it is decisive to the continuing development of the stature of man that he have *intellectual* stimulation. Physics—indeed, *any* science—is therefore a legitimate pursuit of the mind, just as is history, or philosophy, or music.

Man has always had a never satisfied curiosity of the unknown. Physics provides him a special kind of adventure into this unknown. It offers the challenge of a new problem, the excitement of the development of a new idea, and the intellectual satisfaction of at last seeing a problem solved. One can imagine the excitement of Isaac Newton when the idea of universal gravitation first occurred to him. Or the thrill that must have been Enrico Fermi's when he finally observed his instruments showing that a fission chain-reaction was taking place for the first time. What satisfaction it must have been for Hans Bethe to have looked at the stars and to be the first to know *why* they were shining, to know that these distant suns were pouring out energy generated by converting hydrogen to helium.

Explorers have climbed mountains and penetrated jungles, divers have probed the depths of the seas, and we have now begun an era of exploration in space. It is no less of an adventure to search out the secrets of Nature in the laboratory.

1.3   *Physics as an Experimental Science*

THE FINAL TEST IS IN THE LABORATORY

The scientist seeks to learn the "truth" about Nature. In physics we can never learn "absolute truth" because physics is basically an experimental science; experiments are never perfect and, therefore, our knowledge of Nature must always be imperfect. We can only state at a certain epoch in time the extent and the precision of our knowledge of Nature, with the full realization that both the extent and the precision will increase in the next epoch. Our understanding of the physical world has as its foundation experimental measurements and observations; on these are based our theories that organize our facts and deepen our understanding.

Physics is not an armchair activity. The ancient Greek philosophers debated the nature of the physical world, but they would not test their conclusions, they would not experiment. Real progress was made only centuries later, when man finally realized that the key to scientific knowledge lay in observation and experiment, coupled with reason. The Greeks argued that the heavier of two objects would fall the faster. A simple experiment would have tested this conclusion and shown that it was in error. But it remained for Galileo to resolve the point with his careful measurements and well-constructed logic. Of course, the generation of ideas in physics involves a certain amount of just plain *thinking*, but when the final analysis is made, the crucial questions can only be answered in the laboratory.

PHYSICS DEALS WITH CONTROLLED SITUATIONS

In what way does physics differ from a subject such as history? In both fields we try to analyze events and situations by considering causes and effects. However, a given set of historical circumstances presents itself only once. By using the known facts we deduce the chain of occurrences that led to an important event, and we can sometimes establish motives for the actions. But it makes no sense to attempt to reconstruct history as if a certain event did not happen. One can make the idle speculation that if the Spanish had not been turned back by Sir Francis Drake's fleet in 1588, this book might have been written in Spanish instead of English. But such is sheer fantasy—history did not develop that way and it is fruitless to suppose that it might have.

In physics, on the other hand, if a given set of circumstances produces a specific result, we can indeed ask the question "What would have happened if A had been absent?" To answer such a question we simply set up our apparatus again, with A absent, and perform the experiment once more. Ideally, then, physics deals with precise conditions that can be altered and controlled while the effects on the object under investigation are studied. Our results can be unambiguously stated, based on logical deductions from experimental facts.

PHYSICS AND NUMBERS

The goal of physics is to provide an orderly and precise description of all physical phenomena in such a manner that the description can ultimately be reduced to relationships among *numbers*. The results of all experiments can be stated with the use of numbers. Thus, we might measure a certain physical quantity and find the result to be 2.17 units. But this is not sufficient. The statement of the result of a measurement is never complete unless an indication of the reliability of the result is also given. The customary way of doing this is to state the result as, for example, 2.17 ± 0.06 units. This expression means that the experimenter has measured the quantity many times and that the average or mean value is 2.17. Furthermore, he has carefully assessed his equipment, in terms of calibration against standards and in terms of reproducibility, and from these considerations has assigned a *probable error* of ± 0.06 to the result. In this case "probable error" means that if additional measurements are made (either by the same experimenter or by others), the chance that the result will lie in the range from $2.17 + 0.06 = 2.23$ to $2.17 - 0.06 = 2.11$ is equal to the chance that the result will lie outside this range. If another experimenter (with better equipment) remeasures the quantity and finds 2.1746 ± 0.0004, this result is compatible with the previous value because it lies within the range of the probable error of the first measurement.

By improving the result from 2.17 ± 0.06 to 2.1746 ± 0.0004, we have sharpened our knowledge of this particular phenomenon. The improvement of precision plays an important role in the progress of physics, but this is not the only method by which we can increase our understanding. Ideally, we always try to design an experiment to answer a question with a "yes" or "no." That is, we might arrange an experimental test of a theory in such a way that if we observe a certain type of event take place we can state "Yes, the theory is verified on this point," but if we observe a different type of event instead, we can state "No, the theory is not correct." The probable error in such an experiment would be expressed in terms of the degree of confidence that our "yes" or "no" was the proper result. Sometimes a "yes-no" experiment must be extremely elaborate, but there are situations (when we are presented with fortunate circumstances or because we are particularly clever) in which even a crude experiment will suffice. (Galileo made such a "yes-no" test of Aristotle's theory when he dropped objects of unequal mass from the Tower of Pisa.) But whether it is a precision measurement or only the determination of an approximate result, the crucial point is that we must ultimately rely on an *experiment*.

1.4   Is Physics an "Exact" Science?

WHAT IS THE ULTIMATE PRECISION?

How far can we push the precision of a measurement? What is required to determine the *exact* value of a certain quantity? Suppose that we make better and better equipment until finally it is *perfect*. Can we then obtain an *exact* value for the quantity? A first response to this question is: "No, because we can never actually make a perfect instrument." We next ask: "Disregard the fact that we can never make a perfect instrument and *imagine* that we have such an ideal instrument. Can we then, *in principle*, make an exact measurement?" This is an entirely different question from that originally asked. The first question was a practical one and it had an immediate and obvious answer. The second question refers to a *thought experiment*, a device frequently used by physicists to examine problems that cannot be attacked directly because of the inadequacy of existing apparatus.

The thought experiment technique permits us to ask fundamental questions—questions that we can no longer avoid by blaming the quality of our instruments. In this case, the thought experiment with a perfect instrument requires that we answer the fundamental question: "Does there actually *exist* an exact value for a physical quantity?" This question carries us into a realm of smallness that is difficult to appreciate in terms of everyday experience. Nevertheless, the question can be answered within the framework of *quantum theory*, the theory that must be applied to all problems that arise in the domain of the ultimate smallness of things. This theory specifically states that there are no *exact* physical quantities. Moreover, according to quantum theory, it is meaningless to inquire about the exact value of a physical quantity because all measurements are subject to uncertainties—not just uncertainties associated with flaws in equipment but fundamental uncertainties that are the result of the nature of things. Thus, no imaginable experiment can ever be performed to measure the exact size of an atom because an atom has no exact size! Consequently an exact measurement, for example, of the length of a rod, can never be made because all material things are composed of atoms.

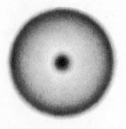

Fig. 1.1 *An atom is a fuzzy thing; it has no "exact" size.*

THE UNCERTAINTY PRINCIPLE

Quantum theory also tells us that we can never predict the outcome of *any particular measurement*. However, if we make a large number of measurements of a certain physical quantity, there will be a well-defined *average value*. Associated with this average value will be a certain "probable error," which is a property of the quantity measured and not of the apparatus used. This is the essence of the famous *uncertainty principle*, about which we shall learn more in Chapter 12. On this point, quantum theory is firmly established. No one has ever been able to devise an experiment (even a thought experiment) that is not limited by the uncertainty principle. The magnitude of the uncertainty involved in the measurement of everyday objects is exceedingly small—so small, in fact, that we have no instruments capable of detecting the effect of the uncertainty principle in the measurement, for example, of the length of a rod. The effect is manifest only when problems of extreme smallness arise (as, for example, in the question of atomic sizes). But this is just the domain that is important when we inquire about *exactness*.

We can now answer the question posed in the heading of this section. Physics is not an "exact" science; indeed, there can be no "exact" experimental science. But physics is a *precise science*. Experiments can be performed that produce results that are uncertain to only 1 part in 100 billion. This, certainly, is high precision, but it is not exactitude.

A QUESTION OF SEMANTICS

The reader will note that nowhere in the preceding discussion was the term *accuracy* used. At the risk of being decried for splitting hairs, the following explanation is offered. As used in science and mathematics, the term *accurate* means no deviation from the truth or from exactitude. For example, the product 2.14 × 3.76 is *exactly* equal to 8.0464. If we were to use a slide-rule to obtain the result, we might read the value 8.05. This result is therefore *inaccurate* because it deviates from the exact value. In physical measurements, on the other hand, we do not know *a priori* what the *truth* is; indeed, we have already argued that the question of the exact value of a physical quantity has no meaning. Under such circumstances, we do not know the *accuracy* of the result; we can only state the uncertainty in the measurement, that is, the *precision* of the result. Although it is well to keep in mind the distinction between *accuracy* and *precision*, we are sometimes not as careful as in the discussion above and the two terms may be used interchangeably on occasion.

1.5 *The Scientific Method*

THE PHILOSOPHY OF DISCOVERY

The mere accumulation of facts does not constitute good science. Certainly, facts are a necessary ingredient in any science, but facts alone are of limited value. In order to fully utilize our facts, we must understand the relationships among them; we must systematize our information and discover how one event produces or influences another event. In doing this, we follow the *scientific method:* observation, reason, and experiment.

The scientific method is not a formal procedure or a detailed map for the exploration of the unknown. In science we must always be alert to a new idea and prepared to take advantage of an unexpected opportunity. Progress in science occurs only as the result of the symbiotic relationship that exists between observational information and the formulation of ideas that correlate the facts and allow us to appreciate the interrelationships among the facts. The scientific method is actually not a "method" at all; instead, it is an attitude or philosophy concerning the way in which we approach the real physical world and attempt to gain an understanding of the way Nature works.

Johannes Kepler (Fig. 1.3) followed the scientific method when he analyzed an incredible number of observations of the positions of planets in the sky. From these facts he was able to deduce the correct description of planetary motion: the planets move in elliptical orbits around the Sun.

Fig. 1.2 *Tycho Brahe (1546–1601) and one of his quadrant sextants. Brahe, a Prussian astronomer, used instruments such as this to make extensive and precise measurements of the positions of stars and planets. He was unable to formulate a consistent description of planetary motion, but his meticulous observations were of great value to later astronomers, particularly Kepler (see Fig. 1.3).*

Kepler's procedure—amassing facts and trying various hypotheses until he found one that accounted for all the information—is not the only way to utilize the scientific method. When Erwin Schrödinger was working on the problems associated with the new experiments in atomic physics in the 1920s, he set out to find a description of atomic events that could be formulated in a mathematically beautiful way. Schrödinger deviated from the "normal" procedure of the scientific method. Instead of closely following the experimental facts and attempting to relate them, he sought only to find an aesthetically pleasing mathematical description of the general trend of the results. This pursuit of mathematical beauty led Schrödinger to develop modern quantum theory. In the realm of atoms, where quantum theory applies, Nature does indeed operate in a beautiful way. At essentially the same time, Werner Heisenberg followed the more conventional approach and formulated an alternative version of quantum theory, which is equivalent in every respect that can be experimentally tested to the theory constructed by Schrödinger.

Therefore, we cannot impose any rigid constraints on the development of science. Different individuals work in different ways. As long as we couple logic and experiment we follow the scientific method.

Fig. 1.3 *Johannes Kepler (1571–1630) and the Copernican description of the solar system which Kepler's calculations firmly established in concept and improved in detail (by showing that the orbits are ellipses).*

1.6 *Physical Theories*

THE EVOLUTION OF IDEAS

If we are confronted with a set of facts, it is our task to find the simplest possible way to relate these facts one to another. A successful relationship is called a *theory*. An acceptable theory must account for all the empirical information accumulated about the particular subject and, furthermore, it must be capable of predicting the results of any new experiments that can be performed. (Frequently, a theory will predict effects that are beyond our capabilities to detect at present; in such cases, the tests must await the development of more sensitive techniques.) If any disagreement is found between theory and experiment, then the theory must be modified to account for the new information. Thus, theories evolve by successive refinements.

Physical theories are meant to be tested by confrontations with new facts. Indeed, perhaps the greatest value of a theory is the way in which it can sharply delineate the point at which it fails. No theory can ever be *proved* to be correct; it can only be proved to be *incorrect*. Suppose that certain facts are assembled and a theory is developed which accounts for these facts. With this theory we then make a number of predictions that can be tested by new experiments. If we had made 20 predictions and 19 were verified by experiment, the theory would not be proved correct because it could fail on the twentieth prediction. But such a failure would not be without value, for then we would have a key clue as to where in the theory there was a flaw. Hopefully, this information would enable us to eliminate the fault by modifying the theory and thereby to bring it into agreement with all of the experimental facts.

WHAT CONSTITUTES A "GOOD" THEORY?

The requirements of a good physical theory are:

1. *The theory must be concise.* It is almost always possible to construct a theory that is so complicated that it can account for any given number of facts. But such a theory is highly artificial and offers nothing in the way of intellectual satisfaction. Given two theories that explain a certain body of facts, one complicated and one concise, our preference is always for the concise theory. It is a great triumph to be able to formulate a theory based on only a few postulates and from this theory to extract a wealth of verifiable predictions. The theories of relativity and quantum mechanics are models of brevity, but we are still working out the consequences and putting them to the test of experiment.

2. *The theory must be general.* A theory that is constructed to explain only one or a few facts and is incapable of relating to other facts is useless. A simple (and ludicrous) example of such an *ad hoc* theory is the following. Suppose that we wish to formulate a theory to explain why apples fall downward (the proverbial problem of Newton). A possible theory is: "All apples fall downward." This "theory" explains the observed fact but does nothing else; it is therefore without value. We might attempt to generalize the theory: "All objects fall downward." But now we have gone too far. Not all objects fall downward; helium-filled balloons do not fall nor do airplanes (usually). Newton did not frame an empty *ad hoc* theory. His statement was that all objects are *attracted* toward the Earth. Some objects fall, but for other objects, such as the helium-filled balloon, there is an additional agency that overcomes the Earth's attraction and causes the balloon to rise rather than fall. A good theory must not be overly restrictive; it must attempt to explain the greatest possible number of facts.

3. *The theory must be precise.* A theory is of little value if it cannot make precise and unambiguous predictions. Otherwise, the theory cannot be verified by subjecting it to the test of experiment. Because the results of experiments can be described by numbers with known precision, theories must also be capable of producing numbers with comparable precision. An experimenter will naturally be discouraged from making a measurement to test a theory if the theory predicts a wide range of possible results of the experiment.

4. *The theory must be capable of modification.* Physical theories must be organic—growing and changing as new information becomes available. A theory that cannot be modified will collapse completely with the first failure, however slight. In ancient times, man believed the Earth to be flat. This is, in fact, a primitive theory: "The Earth is flat." This theory can be tested by elementary experiments. A person standing on the sea shore watching

Courtesy of NASA

Fig. 1.4 *Direct photographic evidence that the Earth is round. This television picture of the Earth as viewed from 22,400 mi in space was radioed from the ATS-III communications and meteorological satellite. The outline of South America is clearly distinguishable in the lower half of the picture.*

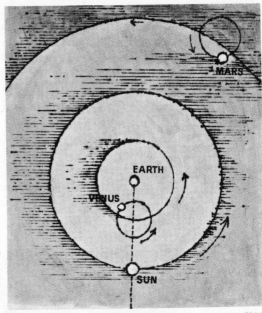

Fig. 1.5 *Ptolemy's model of the planetary system. The Sun makes a simple circular orbit around the Earth. Venus and Mars make circular orbits on which are superimposed additional circular motions (epicycles) to explain the "back and forth" motion of these planets in the sky.*

ESCP

a ship sailing out to sea will observe the apparent size of the ship grow smaller but, if the Earth is flat, he should never lose sight of the ship. But, of course, the ship seems to disappear. Since the ship has not sunk (it returns and the sailors report it was always afloat), the theory of a flat Earth is disproved. Because the theory is rigid and cannot be modified (*flat is flat*, there are no two ways about it), it must be completely discarded. Similarly, a theory that is characterized by the statement "the Earth is the center of the solar system" can be refuted and no modification can be made to render it tractable. On the other hand, a theory that states that "the Earth moves around the Sun in a circular orbit" can be modified by replacing "circular" with "elliptical" to bring it into agreement with observations.

1.7   *Models, Theories, and Laws*

POLISHED THEORIES FROM CRUDE MODELS GROW

In its embryonic stages, a theory is often called a "model." A physical model is constructed for essentially the same reasons that an architect constructs a model of the design of a group of buildings—to visualize better the relationships among the buildings, the open areas between, and the walkways or streets connecting them. A model in physics can be either mathematical or basically mechanical. The advocates of the geocentric theory of the planetary system constructed fantastic mechanical models in their attempts to describe the motion of the Sun and the planets. The "flat Earth theory" that we referred to earlier was a primitive model, sometimes made more elaborate by placing the flat disc on an elephant which stood on the back of a tortoise which, in turn, swam in a bottomless sea. These frills added a certain degree of mysticism but did little to overcome the basic difficulty of the model. Bohr's model of the atom (Fig. 1.6), in which he pictured electrons executing orbits about the nuclear center (a miniature planetary system), was the forerunner of today's quantum mechanical theory of atomic structure.

Most people think more clearly in concrete terms than in the abstract. Models are therefore a great aid to the physicist in the early stages of the development of a theory. When facts are few, the model is necessarily a crude one. But as more information becomes available, the model becomes more sophisticated and eventually (at least *sometimes*) blossoms into a full-fledged theory.

THEORIES THAT BECOME LAWS

It has become traditional that after a theory has been tested and verified by numerous experiments, it is accorded the status of a *law* of physics. The term is used loosely, however, and many "laws" are now recognized to be inaccurate. Ohm's law of electricity, for example, is obeyed only by a limited class of materials and only under certain conditions. Newton's laws of dynamics are now known to require modifications in certain situations; the theory of relativity has shown us how these modifications are to be made. (There is some inconsistency here: Newton's theory, which is not completely accurate, is called a "law," while relativity, which is able to correct the failings of Newton's laws, is still termed a "theory." Historical tradition is often slow to be overturned by new facts.)

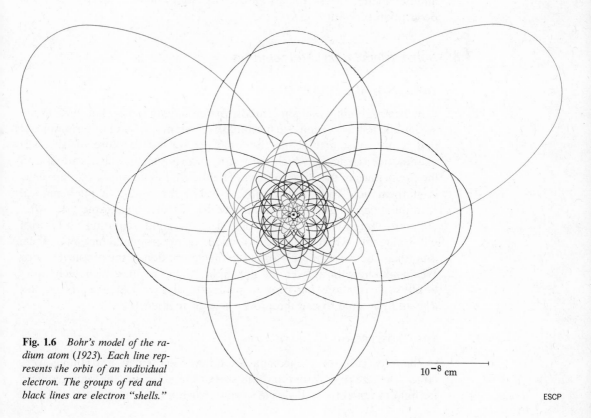

Fig. 1.6 *Bohr's model of the radium atom (1923). Each line represents the orbit of an individual electron. The groups of red and black lines are electron "shells."*

10^{-8} cm

ESCP

THE CONTINUING SEARCH FOR NEW THEORIES

We are now evolving theories (or models) relating to almost every aspect of physics—the structure of nuclei; the way elementary particles behave and interact; the evolution of stars and the formation of the elements; the behavior of matter in bulk; and even space, time, and matter. These theories have enjoyed many remarkable successes but they have also been confronted, on occasion, with dismal failure. Progress continues; we have no "complete" or "perfect" theories, but many new ideas are being formulated and tested. To those who participate, this is one of the great attractions of physics—the aliveness and vitality of the interplay of theoretical ideas and experiments which test and challenge them. Each new idea, each new experiment pushes back the frontier a bit further and gives us greater insight into the ways of Nature.

PHYSICS AND METAPHYSICS

We must always bear in mind that physical theories concern only the relationships between quantities that can be *measured.* It is for philosophers, not physicists, to ask such a question as "why does quantum theory work?" (Sometimes a physicist *does* ask a question of this kind but then he is wearing a different hat.) If a physicist can correlate a number of facts and deduce a general relationship from which he can calculate new quantities that are verified by experiment, that is all he can ask—this is success. It is quite another matter to ask for an all-encompassing, philosophically satisfying description of Nature.

1.8   *Some Elementary Mathematics*

THE LANGUAGE OF PHYSICS

Before beginning a meaningful discussion on any topic—but particularly a scientific topic—we must first establish a common ground of terminology; that is, we must first define the language we use. Because physics is a quantitative science in which the results of experimental observations and the predictions of theories can always be reduced to numbers, our language is mathematics. That it is possible to relate the results of mathematical manipulations to the real physical world is a truly remarkable fact. After all, mathematics is purely an invention of the mind, requiring no basis of physical reality, whereas physics is founded on experimental fact. Why, then, is mathematics so admirably suited to the description of the physical world? We do not know the answer, but the application of mathematical techniques to physical situations has been so successful that we seldom bother to ask *why* and continue to use the proven tool of mathematics.

THE "POWERS-OF-10" NOTATION

The physical world encompasses things very small and things very large—the size of an atom and the size of the Universe, the time required for light to travel from this page to your eye and the time that the Universe has existed. In dealing with these enormous ranges of distance and time (and other quantities), we encounter an annoying problem of notation. For example, if we wish to express the ratio of the diameter of a dime to the distance from the Earth to the Sun, we can write this as

$$\frac{\text{diameter of dime}}{\text{Earth-Sun distance}} = \frac{1}{15,000,000,000,000}$$

or, in decimal notation,

$$\frac{\text{diameter of dime}}{\text{Earth-Sun distance}} = 0.000\ 000\ 000\ 000\ 067$$

Obviously, neither of these methods of expressing the result is particularly convenient. To overcome this difficulty, we use the "powers-of-10" notation. Multiplying 10 by itself a number of times, we find

$$
\begin{aligned}
10 \times 10 &= 100 &&= 10^2 \\
10 \times 10 \times 10 &= 1000 &&= 10^3 \\
10 \times 10 \times 10 \times 10 \times 10 \times 10 &= 1,000,000 &&= 10^6
\end{aligned}
$$

That is, the number of times that 10 is multiplied by itself appears in the result as the superscript to the 10 (called the *exponent* of 10 or the *power* to which 10 is raised). Clearly, $10^1 = 10$, and, by convention, $10^0 = 1$.

Products of powers of 10 are expressed as

$$10^2 \times 10^3 = (10 \times 10) \times (10 \times 10 \times 10) = 10^5 = 10^{(2+3)}$$

That is, in general, the product of 10^n and 10^m is $10^{(n+m)}$:

$$10^n \times 10^m = 10^{(n+m)} \tag{1.1}$$

If the power of 10 appears in the denominator, the exponent is given a negative sign:

$$\frac{1}{10} = 0.1 = 10^{-1}$$

$$\frac{1}{1000} = 0.001 = \frac{1}{10^3} = 10^{-3}$$

$$\frac{1}{10^m} = 10^{-m} \tag{1.2}$$

Combining the rules gives the prescription for division:

$$\frac{10^n}{10^m} = 10^n \times \frac{1}{10^m} = 10^n \times 10^{-m} = 10^{(n-m)} \tag{1.3}$$

Using the powers-of-10 notation, we can now express the ratio of the diameter of a dime to the Earth–Sun distance as 6.7×10^{-14}. Expressed in this form, the information is much more easily assimilated than in the previously used form involving 13 zeroes.

A power of 10 is sometimes called an "order of magnitude." Thus, a dollar is an order of magnitude more valuable than a dime. Or, using the term in an approximate sense, we could say that the Earth is two orders of magnitude more massive than the moon. (Actually, the Earth is about 81 times as massive as the moon.) Learning to think in orders of magnitude (powers of 10) proves to be very convenient in gaining an appreciation for the physical world around us.

Example **1.1**

$$\frac{6,400,000}{400} = (6.4 \times 10^6) \times \frac{1}{4 \times 10^2}$$

$$= \frac{6.4}{4} \times \frac{10^6}{10^2} = 1.6 \times 10^4$$

Usually, we express numbers in this notation by writing the coefficient of the power of 10 as a number between 0.1 and 10. Thus, we write either 2.4×10^9 or 0.24×10^{10} instead of, for example, 24000×10^6 or 0.00024×10^{13}.

When discussing physical quantities, we sometimes use a prefix to the unit instead of the appropriate power of 10. For example, *centi-* means $1/100$ so that $1/100$ of a meter is called a *centimeter; mega-* means 10^6 so that $\$1,000,000 = 1$ megabuck. Table 1.1 lists some of the most frequently used prefixes.

An *integer* exponent represents the *power* to which a number is raised; for example, $10^2 = 100$ or $3^2 = 9$. A *fractional* exponent represents the *root* of a number; for example $(100)^{1/2} = \sqrt{100} = 10$ or $9^{1/2} = 3$. In general, the *m*th root of a number N is

$$\sqrt[m]{N} = N^{1/m} \tag{1.4}$$

which means that if $\sqrt[m]{N}$ is raised to the power *m*, the result is N itself:

$$(\sqrt[m]{N})^m = (N^{1/m})^m = N^{\left(\frac{1}{m} \times m\right)} = N^1 = N$$

Example **1.2**

$$\sqrt{400} = (400)^{1/2} = (4 \times 10^2)^{1/2} = (4)^{1/2} \times (10^2)^{1/2}$$
$$= 2 \times 10 = 20$$

Example **1.3**

$$\left[\frac{1600 \times 10,000}{2000}\right]^{1/3} = \left[\frac{(16 \times 10^2) \times 10^4}{2 \times 10^3}\right]^{1/3} = (8 \times 10^3)^{1/3}$$
$$= (8)^{1/3} \times (10^3)^{1/3} = 2 \times 10 = 20$$

EXPLANATION OF SYMBOLS

The ordinary form of an equation, in which the symbol $=$ is used to express the equality between two quantities, is familiar to everyone:

$$y = 16t^2$$

In certain situations, relations are written in different forms and convey

Table **1.1**   *Prefixes Equivalent to Powers of 10*

Prefix	Symbol	Power of 10
giga-	G	10^9 [a]
mega-	M	10^6 [a]
kilo-	k	10^3
centi-	c	10^{-2}
milli-	m	10^{-3}
micro-	μ	10^{-6}
nano-	n	10^{-9}
pico-	p	10^{-12}
femto-	f	10^{-15}

[a] $10^6 = 1$ *million*. In the US, $10^9 = 1$ *billion*, but the European convention is that $10^9 = 1000$ million and that 1 billion $= 10^{12}$; the prefix *giga-* is internationally agreed on to represent 10^9.

somewhat different information (but always *less* information than the complete expression of equality). In the case described by the equation above, for example, we might be interested in the fact that y depends on t^2 and we might not need to know the proportionality factor (that is, 16). We express this by saying that y *is proportional to* t^2, or, in symbols,

$$y \propto t^2$$

which is equivalent to the statement,

$$y = kt^2$$

where k stands for the proportionality constant (and may not even be known explicitly in some cases).

If we know a result only approximately, but not exactly, we use the symbol $\cong$, with the meaning *is approximately equal to*. Thus, if the *exact* expression relating y and t^2 were

$$y = 16.127t^2$$

we could write approximate relations as

$$y \cong 16.13t^2$$

or,

$$y \cong 16.1t^2$$

or,

$$y \cong 16t^2$$

The use of approximations of this type (for example, replacing 16.127 by 16) is a respectable procedure and may even be preferred to the "exact" calculation in many instances. It depends on how precise we wish our final result to be. There is no sense in carrying out a calculation to six decimal places if two will do. It is well to remember the maxim, "that which is good enough is best."

If we have only a vague idea of the magnitude of a quantity, then we use the symbol $\approx$ or $\sim$ to imply that the number is only *very* approximately known or *is of the order of magnitude of* some number. For example, we say that the galaxy Andromeda is about 3×10^{11} as massive as the Sun. But we do not know this number very precisely; it could be 2×10^{11}, or 6×10^{11}, or even farther from 3×10^{11}. Therefore, we say: *The mass of Andromeda is of the order of magnitude of* 3×10^{11} *times the mass of the Sun,* and we write

mass of Andromeda $\approx 3 \times 10^{11} \times$ mass of the Sun

or we say that the ratio of the mass of Andromeda to the mass of the Sun is $\sim 3 \times 10^{11}$.

The symbols $>$ and $<$ mean, respectively, *is greater than* and *is less than*:

area of New York $<$ area of Texas
mass of the Earth $>$ mass of the moon

If a certain quantity is *very much larger than* another quantity, we use the symbol $\gg$:

mass of Andromeda $\gg$ mass of Sun

Sometimes we know only that a certain quantity is larger (or smaller) than a poorly defined value. We then use the symbol $\gtrsim$ (or $\lesssim$) to mean *is larger (smaller) than about* some number. Thus, the gross national product is larger than about \$900,000,000,000, that is,

GNP $\gtrsim$ 900 gigabucks

The symbols that we will find useful in this book are listed in Table 1.2 along with their meanings.

APPROXIMATE CALCULATIONS

Some quantities in physics are known with great precision; uncertainties of less than 1 part in 10^6 are not at all rare. When dealing with such numbers it is frequently necessary (and certainly justifiable) to calculate a result that contains many figures. On the other hand, we often require only an approximate result in order to give us a close, but not precise, idea of the magnitude

Table **1.2**   *Mathematical Symbols and Their Meanings*

Symbol	Meaning
$=$	is equal to
$\propto$	is proportional to
$\cong$	is approximately equal to
$\approx$ or $\sim$	is *very* approximately equal to; is of the order of magnitude of
$>$ $(<)$	is greater (less) than
$\gg$ $(\ll)$	is much greater (less) than
$\geq$ $(\leq)$	is greater (less) than about

of a quantity. In such cases we can usually make approximations that greatly simplify the calculation.

Example **1.4**

Calculate

$$A = \frac{\pi \times \sqrt{2} \times 2.17}{(6.83)^2 + (1.07)^2}$$

We use the following approximate values:

$\pi = 3.1416 \cong 3$ $(6.83)^2 \cong 7^2 = 49 \cong 50$
$\sqrt{2} = 1.414 \cong 1.5$ $(1.07)^2 \cong 1$
$2.17 \cong 2$ $50 + 1 \cong 50$

Then,

$$A \cong \frac{3 \times 1.5 \times 2}{50} = \frac{9}{50} \cong \frac{10}{50} = 0.2$$

By using this procedure we can obtain a result that is sufficiently accurate for many purposes in a few seconds; a slide-rule calculation for this problem would require about a minute.

Even in the event that a more precise answer is required for a problem, it is valuable to make an approximate calculation first. By knowing beforehand the approximate answer, an error in the handling of more figures can often be spotted easily. Also, in a slide-rule calculation, the faulty placement of a decimal point is a common mistake; the approximate result is a useful guide in such cases.

CIRCULAR MEASURE

The most common unit of angular measure is the *degree,* defined to be 1/360 of a complete circle. For some purposes it is more convenient to use another unit, the *radian* (Fig. 1.7). We choose a unit of angular measure so that if $\theta = 1$ unit, the length of the portion s of the circumference of the circle (called the *arc*) that is intercepted by the pair of radii defining the angle is just equal to the radius r of the circle. This unit is called the *radian.* Then, we have, in general,

$$s = r\theta \tag{1.5}$$

where θ is measured in radians. The circumference c of a circle is 2π times the radius:

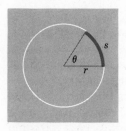

Fig. 1.7 *The central angle θ defines an arc of length s on the circumference of a circle of radius r.*

$$c = 2\pi r \tag{1.6}$$

Therefore, the 360° in a complete circle is equivalent to 2π radians. Thus,

$$1 \text{ radian} = \frac{360°}{2\pi} \cong 57°.3 \tag{1.7}$$

If we use a more accurate value of π (3.14159265 . . .), we find that 1 radian is 57°.2957795 . . . . Also, 1° = 0.0174533 . . . radian.

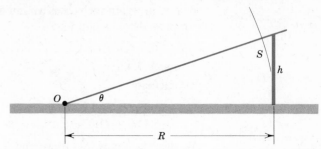

Fig. 1.8 *Estimating the height of an object by measuring the angle subtended at O;* $h \cong R\theta$.

The fact that $s = r\theta$ provides us with a method of closely estimating distances in certain circumstances. Suppose that a pole is placed vertically in the ground a distance R away from an observer located at a point O (Fig. 1.8). The observer measures the angle subtended by the pole and finds it to be θ. Mentally, the observer constructs a circle of radius R centered at O and passing through the bottom of the pole. The arc length S is then given by

$$S = R\theta \tag{1.8}$$

Since θ is a small angle, S is approximately equal to the height of the pole h, and we can write

$$h \cong R\theta \tag{1.9}$$

This method of obtaining an approximate value for h will be useful if θ is sufficiently small. Even for θ as large as $20°$, the error is only about 4 percent. If θ is a few degrees, the error is usually negligible for most purposes; for $\theta = 1°$, the error is 0.01 percent or 1 part in 10^4.

1.9 *Measurements and Graphs*

A PICTURE IS WORTH A THOUSAND TABLE ENTRIES

Just as a physicist constructs a model so that he may better visualize a complicated set of facts, he also uses graphical displays in order to present experimental data in a way easily assimilated. If an experiment is performed to measure, for example, the length of a certain rod, then the complete results of the experiment can be stated with a single number (and its associated uncertainty). But if the quantity measured can take on different values as we change some aspect of the environment, or if the quantity has a variation with time, then a single number no longer suffices to describe the situation. Suppose we suspend a long, thin rod from a rigid support and measure the length of the rod as we add weights to the lower end. We continue to add more and more weights until the point is reached at which the rod finally breaks. Such an experiment gives us two pieces of information—the way in which the rod stretches and the limiting value of the pull that we can exert without fracturing the rod. A list of all of the relevant numbers is, of course, useful to the structural engineer who must make calculations based on them, but a graph of the length of the rod *versus* the applied weight would give, at a glance, the overall picture for the particular rod used in the experiment. (Other rods of different materials and with different cross-sectional areas, would exhibit different "yield curves.") The results of such a hypothetical experiment are shown in the graph of Fig. 1.9. In this graph there are a number of experimental points; the solid dots are the measured values and the vertical lines represent the extent of the probable error in each measurement ("error bars"). From this graph it is easy to see that the length increases linearly (that is, the curve through the points is a straight line) for weights up to about 50 pounds (lb) but for greater weights the length increases more rapidly until, at approximately 95 lb, the yield point is reached and the rod breaks. The qualitative features of the results are much more easily seen in such a graphical display than if the results were only listed in a numerical table.

GRAPHICAL COMPARISON OF THEORY AND EXPERIMENT

In Fig. 1.9 a curve has been drawn arbitrarily through the measured points in order to assist in the visualization of the results. Of course, if we had a theory for the particular phenomenon we were investigating, then a vivid

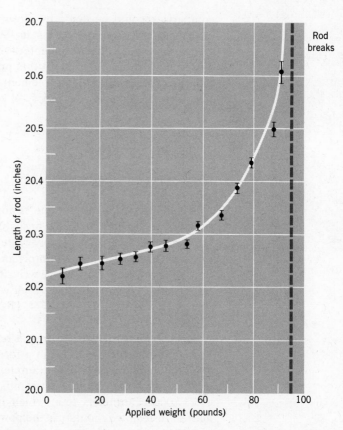

Fig. 1.9 *Results of a hypothetical experiment to measure the stretching and the breaking point of a rod.*

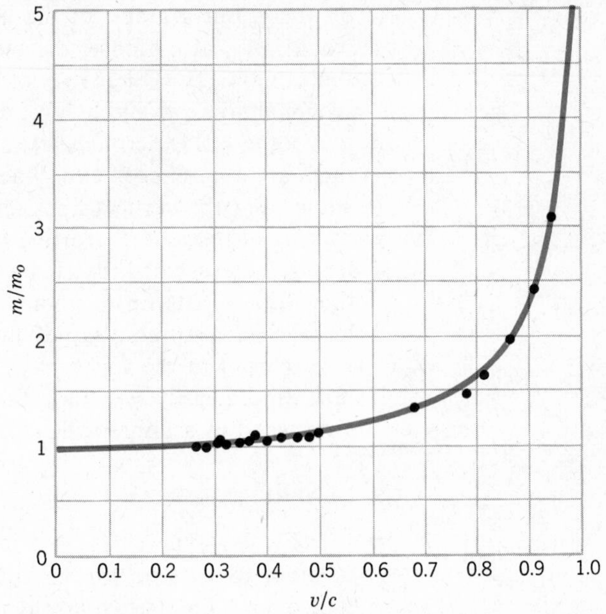

Fig. 1.10 *Variation of electron mass with velocity. The data represent measurements made during the period 1901–1909 and the values fall very close to the curve which is the prediction of relativity theory. Obviously, experiment has verified this particular theoretical prediction. (Error bars, which are as small as or smaller than the sizes of the points, are not exhibited in this graph.)* m/m_o *is the ratio of the electron mass to the mass at zero velocity.* v/c *is the ratio of the electron velocity to the velocity of light,* $c = 3 \times 10^{10}$ *cm/sec. The theory of this phenomenon is discussed in Section 11.3.*

comparison could be made by plotting the theoretical curve on the same graph with the experimental results. Such a comparison is made in Fig. 1.10 where the results of measurements of the electron mass are shown for various electron velocities. The curve is that predicted by relativity theory. It is clear that the experiments are in agreement with the theory and we may conclude that this prediction of the theory is verified.

1.10 *Significant Figures*

THE IMPLIED UNCERTAINTY

Frequently when we list the result of an experiment, for the sake of brevity, we do not give the probable error. That is, instead of writing 2.174 ± 0.002, we often merely write 2.174. By convention, when we use this shorthand notation, we imply that the uncertainty in the result is in the last digit given. Thus, if we write 2.174, we imply that the 4 is uncertain but not the 7. If the complete result were 2.174 ± 0.013, we would give the shortened result as 2.17, not 2.174. That is, we give only the *significant* figures of the result.

Because we can always express a result in many different units, we must be careful to write the result in a form that exhibits the significant figures in an unambiguous way. For example, if we have a result of 2.17 kilometers, implying that the 7 is the last significant figure, we should not write this as 217,000 centimeters (cm) which implies that there are six significant figures. Instead we should use the powers-of-10 notation and give the result as 2.17×10^5 cm. This procedure allows us automatically to terminate the number of digits with the last significant figure. If the first zero of 217,000 cm were significant, we would write 2.170×10^5 cm.[1]

When two numbers are combined, the result can only be as precise as the *least* accurate of the numbers. For example, if we consider the product of 2.17 ± 0.02 and 0.952 ± 0.001, we obtain for the result 2.1049. The last two digits are not significant, however, because there is an uncertainty of 1 percent in the factor 2.17 and this uncertainty is propagated to the product, which is then also uncertain by 1 percent. Therefore, the result must be stated as 2.10 ± 0.02. Similarly, if we *add* the same numbers, we must not express the result as 3.122 which implies that the final digit is significant. In this case, the uncertainty of ± 0.02 in the factor 2.17 (the dominant uncertainty) is propagated to the sum. Hence, the result must be stated as 3.12 ± 0.02.

Summary of Important Ideas

Physics is an *experimental* science in which progress toward a deeper understanding of Nature is made by the application of the *scientific method* to empirical facts and observations.

There is no set pattern to the *scientific method*—it is a basic philosophy, colored by personal taste, of the way observation and experiment, reason and logic are used to understand the world around us.

Physical models and theories are designed to correlate the facts that have been discovered about Nature. No theory is ever "true"—it only represents the *best present understanding* of the area to which it applies and it is always subject to modification (or to complete refutation) as new facts become available.

Physics attempts to deal with Nature in a precise and controlled way, always reducing findings to *numbers* that can be compared with the numerical predictions of theories.

Questions

1.1 Comment briefly on the following statement of Dr. James B. Conant: "Being well informed about science is not the same thing as understanding science."

1.2 According to Aristotle, if two objects were dropped from a certain height, the heavier object would fall faster. If two objects, one heavier than the other, are tied together with a certain length of string between them, what would Aristotle's "theory" predict for the motion? If the objects were tied together in contact, so that they form essentially a single object, what would Aristotle predict? Can you conclude that there is a logical fallacy in Aristotle's "theory"?

[1]The alert reader will find instances in this book where we have not rigidly followed this rule. (It is much easier to write 1370 than 1.37×10^3 even if the zero is not significant.) However, when the significant figures are really important to the point being made, the rule is followed.

1.3 The advancement of science, and physics in particular, has always resulted from the interplay of experiment and theory. Which has been more important in the past, experiment or theory? Which do you believe will play the dominant role in the future? Why?

1.4 Discuss the relationship (if any) between the *simplicity* of a physical theory and its *ease of comprehension*.

1.5 Einstein said that a great theory is one that possesses "inner perfection" and is verified by "external confirmation." Discuss the meaning of this statement. What is meant by each of the phrases?

1.6 List some discoveries in physics that seem to have had important applications in medicine and in communications. (Remember that such items as the invention of the telephone are not "physics discoveries.")

1.7 In driving through Washington, D.C., you notice that the cross streets are named Kenyon, Lamont, Morton, and Newton. After driving farther in the same direction you notice that the names are Jefferson, Kennedy, Longfellow, and Madison. What is your explanation of how the streets were named and how would you test your theory?

1.8 Discuss whether the following are tenable physical theories:
(a) 40,000 angels can stand on the head of a pin.
(b) The moon is tied to the Earth by a weightless and invisible rope through which any material object will pass without observable effect. (This is why the moon always turns the same face toward the Earth.)

1.9 Why do we continue to use a certain model when we do not know the fundamental physical basis for the model?

1.10 An experimenter measures a certain quantity and plots his results as in the accompanying graph. The curved line represents the theoretical prediction for his experiment. Comment on (a) the validity of the theory for the effect studied and (b) the method used by the experimenter for assigning probable errors to his measurements.

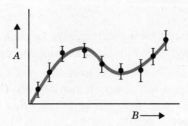

Problems

1.1 The result of a certain experiment gives the following progression of numbers: 1.00, 4.00, 9.00, 16.00, 25.00. Make a hypothesis about the phenomenon under study and predict the next few numbers that should be obtained. Suppose that an experiment is then carried out and the following results obtained: 35.9 ± 0.2, 49.2 ± 0.2, 63.6 ± 0.3, 81.3 ± 0.3. What conclusion would you make about your theory?

1.2 Write out the following numbers in full:
(a) 3.7×10^4 (c) 0.85×10^6
(b) 6.03×10^9 (d) 3×10^{12}

1.3 Write out the following in decimal notation:
(a) 6×10^{-4} (c) 0.39×10^{-5}
(b) 8.6×10^{-7} (d) 3×10^{-12}

1.4 Estimate the following quantities (one significant figure is sufficient) and express them as powers of 10:
(a) The ratio of your height to the diameter of your index finger.
(b) The ratio of your top running speed to the speed of a jet airliner.
(c) The ratio of the height of the Washington Monument to the diameter of a pin head.

1.5 Calculate the following using powers-of-10 notation:

(a) $240 \times 3{,}000{,}000 \times \dfrac{1}{600{,}000}$ (c) $(200)^3 \times (900)^{1/2}$

(b) $\sqrt{6400} \times 300 \times \dfrac{1}{\sqrt[3]{8000}}$ (d) $(30)^2 \times (2500)^{-1/2}$

1.6 Find *approximate* values (1 or 2 significant figures) for the following without calculating fully and without the use of a slide rule:

(a) $\dfrac{\sqrt{27} \times 4.82}{0.512}$ (c) $\dfrac{\sqrt{(3.41)^2 + (0.27)^2}}{\sqrt{3}}$

(b) $\dfrac{3\pi}{(9.6)^2}$ (d) $\dfrac{\pi^2 \times 9.68}{(3.1)^2 + (2.9)^2}$

1.7 A baseball player stands at home plate. Another player stands at first base, 90 feet (ft) away, and holds a 34-inch bat in a vertical position. The player at home observes that the bat subtends the same angle as does the height of the fence at the right field foul line. The fence marking states that it is 315 ft away from home plate. What is the height of the fence? (Hint: It's Milwaukee County Stadium.)

1.8 The Empire State Building is 1245 ft high. From a certain point on Long Island you observe that the top of the building is $3°$ above the horizon. How far are you from midtown Manhattan?

1.9 One directional radio receiver A is located 10 miles due north of a similar receiver B. A signal is received from a transmitter by A when A's antenna points $110°$ east of North. A signal from the same transmitter is received by B when B's antenna points $70°$ east of North. In what direction and how far away from the midpoint of AB is the transmitter?

1.10* A laser beam is projected from the Earth to the moon, a distance of 240,000 miles (mi). The beam cross section is a circle and the divergence is 2×10^{-5} radians (rad). (That is, the beam has a conical shape with an included angle of 2×10^{-5} rad.) What is the diameter of the light spot on the moon?

*The more difficult problems are indicated by asterisks.

1.11* A certain theory predicts that a physical quantity denoted by p will depend on the time according to the relation $p = 3.00t^2 + 1.00t + 0.65$. An experiment yields the following results:

t	p
0	0.70 ± 0.10
1	4.58 ± 0.15
2	14.83 ± 0.10
3	31.05 ± 0.50
4	52.00 ± 0.80

Graphically compare theory and experiment. What is your conclusion regarding the validity of the theory?

1.12 The table below lists the results of a certain experiment. Plot the results. What is your estimate of the way y depends on x? Show this in the graph.

x	y
1	16.2 ± 0.2
2	15.8 ± 0.3
3	15.7 ± 0.2
4	16.3 ± 0.3
5	16.1 ± 0.2
6	15.9 ± 0.2
7	16.3 ± 0.2
8	15.9 ± 0.2
9	15.9 ± 0.3
10	16.1 ± 0.3

1.13 An experiment gives the following table of results.

x	0	1	2	3	4	5	6	7
y	5.0	3.2	1.8	1.9	3.3	4.8	3.7	0.9

Make a plot of the data. Sketch a smooth curve that describes the experimental results in a satisfactory way. What is your estimate for the value of y that would be obtained for (a) $x = 2.5$, (b) $x = 5.5$, (c) $x = 8$?

1.14 If $x = 1.8 \pm 0.1$ and $y = 3.17 \pm 0.01$, express xy and $x + y$ with the proper number of significant figures.

Sundial (18th century)

2 *Length, Time, and Mass*

Man has long realized the necessity to measure things—the size of a farm, the amount of grain in a basket, the distance from one town to another, or the time required to make a certain journey. These ancient needs and the primitive methods devised to accomplish the task of specifying the *amount* of something have gradually evolved into a precise science of measurements. From crude water clocks and the gigantic stone calendar at Stonehenge, we have progressed to the point that an atomic clock now operates that is accurate to 1 second in 30,000 years. From distance measurements that were made by counting a man's strides, we now have length standards that are accurate to 1 in. in 10,000 miles.

It is basic to a system of precise physical measurements that we have available well-defined units and precise and reproducible standards. In this chapter we define the system of units we shall use for length, time, and mass. In the following chapter we discuss the measurements of these quantities over the ranges encountered in the physical Universe.

The ideas of length and time that are gained from everyday experience are immediately applicable in our physical theories. The concept of *length* is essentially geometrical in character, and, although *time* is a more difficult concept than that of distance, we can still appreciate time in terms of that which occurs between one event and another.

The concept of *mass,* on the other hand, differs from those of length and time in two fundamental respects. *First,* it is often stated that mass is a measure of the amount of matter in a body. Although we have some intuitive appreciation of what such a statement means, if we attempt to develop a theory of the way objects move and interact using only this definition of

Fig. 2.1 *This clock was built by the great English clockmaker John Harrison in about 1735. Harrison's clocks were the first with sufficient accuracy to permit truly reliable navigation on the open sea. On a five-month voyage in 1736, one of Harrison's clocks lost only 15 sec.*

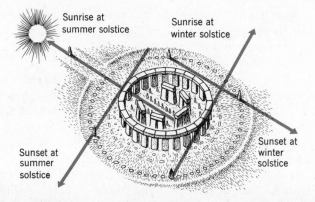

Fritz Henle—Monkmeyer

Fig. 2.2 *Stonehenge, in southern England, where 3,500 yr ago the high priest could stand in the center of the monument at the altar stone and see the rising Sun on the first day of summer through the pillars. The arrows drawn on the drawing above show the alignment of the various landmarks that pointed to the rising and setting Sun on the days of the summer and winter solstices. Only the sunrise at the summer solstice is viewed from inside the monument thus suggesting that this was a particularly significant day, perhaps the first day of the new year.*

mass, we rapidly come upon insurmountable difficulties. Indeed, on close examination we find that it is possible to give a proper definition of mass only within the framework of the theory of the dynamics of moving objects, as first developed by Newton and refined by Einstein in his relativity theory. Therefore, we shall postpone a discussion of a precise definition of mass until we discuss the subject of dynamics (Chapter 5). In the meantime, we continue using the vague "amount of matter" definition.

Second, it is not necessary that we associate length and time with matter and, therefore, these concepts do not depend on our knowledge of the properties of matter. *Mass,* however, is a fundamental property of matter and we cannot divorce one from the other. This fact has the important implication that mass cannot be considered to be infinitely divisible or *continuous* as can length and time.[1] That is, given any small unit of length or time, we know of no reason why we cannot conceive of an even smaller unit. Continuing this division process to the ultimate, we are led to the conclusion that space and time form a *continuum.* Matter, on the other hand, cannot be divided indefinitely; eventually we arrive at a point of final smallness, the constituents of the atom—electrons, protons, and neutrons. We cannot divide an electron in half (at least, not yet!) and so we have reached the limit. Matter, and therefore *mass,* comes in discrete bits; matter does not form a continuum. Of course, the sizes of atoms are sufficiently small that for essentially all everyday purposes we may consider matter to be continuous. But at the most fundamental level, the atomic nature of matter is of crucial importance. Although we do not discuss any details of the structure of matter until the next chapter, we shall make use of the atomic concept here in order to describe the way in which standards of measurements are established.

The *weight* of an object is not the same as its *mass.* The *weight,* as determined by a beam balance or a spring balance, is a measure of the gravitational attraction of some body (such as the Earth) on the object. The weight of a given body is different on Earth from the weight on the moon, for example, because the gravitational attraction is different in the two places. We will pursue this point further in Chapter 5.

2.1   *Length*

THE STANDARD METER

During the course of recorded history, man has devised an incredible number of units of length—cubits, spans, fathoms, chains, yards, furlongs, feet, statute miles, nautical miles, meters, and so forth. As far as widespread acceptance is concerned, only two systems survive today—the British system and the metric system. The British system (which is used in most English-speaking countries but which is certain to be replaced eventually by the metric system) is based on a standard *yard* (yd).[2] The derived units are the *foot* ($\frac{1}{3}$ yd), the *inch* ($\frac{1}{36}$ yd), and the *statute mile* (1760 yd). The metric system had its birth in France during the Napoleonic era (1801) and its use soon spread throughout continental Europe (usually by edict following a French conquest). The meter (m) was originally conceived as 10^{-7} of the distance from the equator to the north pole along a meridian passing through Paris (Fig. 2.3). The meter thus defined coincided reasonably closely with the British yard (1 m $\cong$ 39.37 in. $\cong$ 1.1 yd).

Soon after its definition in terms of the length of the Earth-quadrant, it

---

[1] According to some current speculations, even length and time may not be continuous at extremely small values which are far beyond our present ability to examine.

[2] The *yard* is now defined in terms of metric units (Table 2.1).

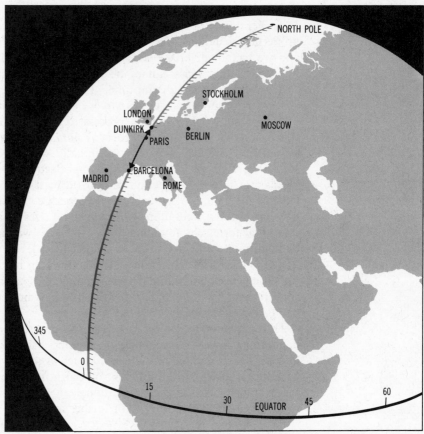

Fig. 2.3 *The original basis for the meter was 10^{-7} of the equator-to-pole quadrant of the Earth's meridian passing through Dunkirk and Barcelona (and, hence, also very close to the center of Paris). The Dunkirk-Barcelona distance was measured by a team of French surveyors and the extrapolation to the full quadrant was made by using astronomical determinations of latitude.*

was realized that a more practical standard meter was required. It was not until 1889, however, that the meter was officially defined as the distance between two parallel scribe marks on a certain bar of platinum-iridium. (In order to insure the reproducibility of measurements of this length, it was further specified that the bar be at the temperature of melting ice and be supported in a particular way.) The standard meter bar is housed in the International Bureau of Weights and Measures, near Paris, and copies (secondary standards) have been distributed to national standards agencies throughout the world. The United States owns Meter No. 27 (Fig. 2.4), which has been kept at the National Bureau of Standards since its delivery from France in 1890.

THE ATOMIC LENGTH STANDARD

The comparison of the length of an object to the standard meter can be made to a precision of 2–5 parts in 10^7 by using a high-power traveling microscope to view the scribe marks on the meter bar. The limitation is set by the roughness of the grooves that define the ends of the meter. The meter-bar definition of the meter suffers from two disadvantages: not only is the precision inadequate for many scientific purposes, but comparisons of lengths with a bar that is kept in a standards laboratory are quite inconvenient. These difficulties have been overcome with the definition, by international agreement in 1961, of a *natural* unit of length based on an atomic radiation. Because all atoms of a given species are identical, their radiations are likewise identical. Therefore, an atomic definition of length is reproducible everywhere. We now accept as the standard of length the wavelength (in vacuum) of a particular orange radiation emitted by krypton gas. (Actually, the definition is in terms of the radiation from the *isotope* of krypton with mass number 86; see Section 3.4.) The standard was arrived at by carefully measuring the length of the standard meter bar in terms of the wavelength of krypton-86 light. It was then decided that exactly 1,650,763.73 wavelengths would constitute 1 m. This definition is then consistent with the previous definition in terms of the distance between the scribe marks on the meter bar, but it has the advantage of being approximately 100 times as precise. Now the standard can be reproduced in many laboratories throughout the world.

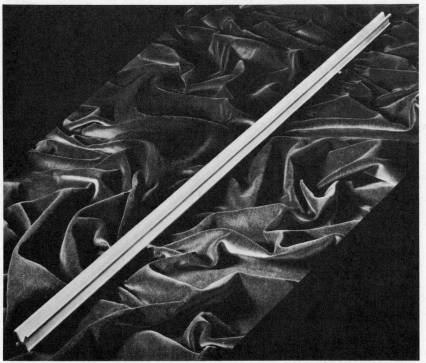

National Bureau of Standards

Fig. 2.4 *Meter No. 27, housed at the National Bureau of Standards, Gaithersburg, Maryland. Meter No. 27 was recompared with the primary standard in 1955.*

Table **2.1**   *Conversion Factors for Length*

1 cm = 0.3937 in.	1 in. = 2.54 cm ⎫
1 m = 3.281 ft	1 ft = 30.48 cm ⎬ exactly
1 km = 0.6214 mi	1 yd = 91.44 cm ⎭
= 3281 ft	1 mi = 5280 ft
	= 1.609 km

Table 2.1 gives the conversion factors connecting some of the units of length in the metric and British systems. Notice that the inch, the foot, and the yard are now defined *exactly* in terms of centimeters. Thus, the krypton-86 wavelength is the standard of length for both systems.

Table 2.2 shows in a schematic way the enormous range of distances we encounter in the Universe. Notice that there is a factor of 1000 between

Table **2.2**   *The Range of Distances in the Universe*

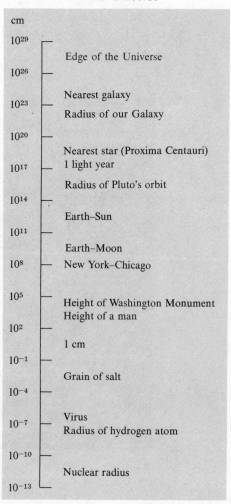

cm	
10^{29}	
	Edge of the Universe
10^{26}	
	Nearest galaxy
10^{23}	Radius of our Galaxy
10^{20}	
	Nearest star (Proxima Centauri)
10^{17}	1 light year
	Radius of Pluto's orbit
10^{14}	
	Earth–Sun
10^{11}	
	Earth–Moon
10^{8}	New York–Chicago
10^{5}	Height of Washington Monument
	Height of a man
10^{2}	
	1 cm
10^{-1}	
	Grain of salt
10^{-4}	
	Virus
10^{-7}	Radius of hydrogen atom
10^{-10}	
	Nuclear radius
10^{-13}	

Mount Wilson and Palomar Observatories

Fig. 2.5 *The great galaxy in the constellation Andromeda (M31) is at a distance of approximately 2.2 × 10⁶ L.Y. or 2.1 × 10²⁴ cm. It is slightly larger than our Galaxy and is about twice as massive. Two smaller galaxies (NGC 205 and NGC 221) are seen on opposite sides of M31. The Andromeda galaxy is actually one of our nearer neighbors in space.*

successive marks on the vertical scale. Between the smallest and the largest things about which we have any comprehension, the span is more than 40 orders of magnitude!

Table 2.3 gives the values of some of the distances that we shall find useful in our discussions.

Table **2.3**   *Some Useful Distances*

To Proxima Centauri (nearest star)	4.04×10^{18} cm $= 1.31$ pc
1 parsec (pc)[a]	3.086×10^{18} cm $= 3.26$ L.Y.
1 light year (L.Y.)[a]	9.460×10^{17} cm $= 0.306$ pc
1 astronomical unit (A.U.)[a] (Earth–Sun distance)	1.496×10^{13} cm
Radius of Sun	6.960×10^{10} cm
Earth–Moon	3.844×10^{10} cm
Radius of Earth	6.378×10^{8} cm
Wavelength of yellow sodium light	5.89×10^{-5} cm
1 Ångstrom (Å)	10^{-8} cm
Radius of hydrogen atom	5.292×10^{-9} cm
Radius of proton	1.2×10^{-13} cm

[a] These astronomical length units will be discussed in Chapter 3.

2.2   *Time*

STANDARDS BASED ON THE MOTION OF THE EARTH

Whereas length is essentially a geometrical concept, we have no such clear framework to support our notions of *time*. Aristotle spoke of time as "the number of motion," and Leibnitz wrote that "time is the abstract of all relations of sequence." In order to make measurements of lengths we can use a meter stick; we can move the meter stick from place to place—it endures. However, a standard interval of time can be used only once. The basic requirement for a time measurement is a repetitive process with a regular and countable pattern of recurrences. That is, we must have Aristotle's "number" and Leibnitz's "sequence." A possible repetitive pattern to use is the rising of the Sun. But one soon discovers that the interval between successive risings varies with the season of the year. We can improve the situation by choosing as a standard the *mean solar day,* the average (taken over a year) of the time required for the Earth to rotate once relative to the Sun. The *second* is then defined as 1/86,400 of the mean solar day. Time so defined is called *universal time* (U.T.). Despite the general success of this system, the determination of time by observation of the rotation of the Earth is inadequate for high precision work because of minor but quite perceptible anomalies in the speed of the Earth's rotation. These variations in speed amount to about 1 part in 10^8 during the course of a year. Con-

Fig. 2.6 *A cesium clock constructed at the National Bureau of Standards (Boulder Laboratories). This clock can measure time intervals to a precision equivalent to 1 sec in 30,000 yr.*

Table **2.4** *Relative Precision of Various Types of Clocks*

Type of Clock	Precision	
	1 sec in	1 part in
Hour glass	1.5 min	10^2
Pendulum clock	3 hr	10^4
Tuning fork	1 day	10^5
Quartz crystal oscillator	3 yr	10^8
Ammonia resonator	30 yr	10^9
Cesium resonator	3×10^4 yr	10^{12}
Hydrogen maser	3×10^6 yr	10^{14}

sequently, astronomical time is now computed on the basis of the length of the mean solar second at the beginning of the year 1900. This time is called *ephemeris time* (E.T.) and the *ephemeris second* is $1/31,556,925.97474$ of the tropical year 1900. The precision with which the ephemeris second is known is approximately 2 parts in 10^9. Universal and ephemeris time were in agreement in 1900, but because of cumulative effects they differ by half a minute in 1971.

THE ATOMIC TIME STANDARD

In order to improve the precision of time measurements, in 1967 a *natural* unit for time was adopted just as had been done previously for a length standard. Our present-day *atomic clocks* (Fig. 2.6) depend on the characteristic vibrations associated with cesium atoms (actually, cesium-133). The *physical second* is defined as the time required for 9,192,631,770 cycles of the particular vibrations in cesium-133. With this definition of the second, it is possible to compare time intervals to 1 part in 10^{12}, which corresponds to 1 sec in 30,000 years. Current research with other atomic vibrations (notably those associated with the hydrogen maser) indicates that we shall soon have a clock that will be precise to 1 part in 10^{14}, or to 1 sec in 3 million years!

Time standards for practical working purposes are provided by radio station WWV, located in Fort Collins, Colorado, and operated by the National Bureau of Standards. WWV operates on frequencies of 2.5, 5, 10, 15, 20, and 25×10^6 cycles per second (c/s) which are controlled to 1 part in 10^{10} by comparison with a cesium clock. A beat is given every second and 10 times per hour the time is given by voice.

Table 2.5 shows the range of time intervals that we encounter in the Universe. Notice that the span from the shortest to the longest time interval is greater than 40 orders of magnitude, about the same as the range of distances shown in Table 2.2.

2.3 *Mass*

OPERATIONAL STANDARDS

As was pointed out in the introductory section of this chapter, some study of dynamics is necessary before a proper definition of mass can be given. However, we can give an *operational* definition of mass in a straightforward way. Suppose we arbitrarily select a certain block of material and agree that this will be our standard of mass. We can then use a beam balance, such as that shown in Fig. 2.7, to compare the standard mass with another object or group of objects. This procedure can be followed to collect a set of secondary standards, all equivalent to the primary standard. By grouping the equivalent masses, we can produce secondary standards with masses 2, 5, 10, or any number times the mass of the primary standard. Similarly, we can work downward by making two equal masses (shown to be of equal mass by comparison on the beam balance) that together total to the mass of the primary standard. Therefore, once we settle on a primary standard, we can construct a wide range of secondary standards of greater and smaller masses. This is, in fact, the operational procedure we have adopted for establishing a practical set of mass standards, all of which has been accom-

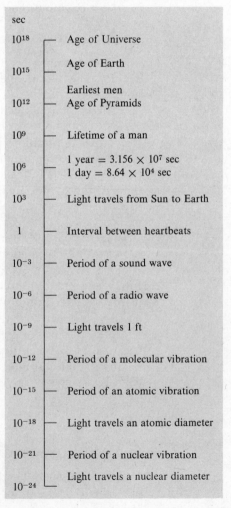

Table **2.5**   *The Range of Time Intervals in the Universe*

sec	
10^{18}	Age of Universe
10^{15}	Age of Earth
10^{12}	Earliest men Age of Pyramids
10^{9}	Lifetime of a man
10^{6}	1 year $= 3.156 \times 10^7$ sec 1 day $= 8.64 \times 10^4$ sec
10^{3}	Light travels from Sun to Earth
1	Interval between heartbeats
10^{-3}	Period of a sound wave
10^{-6}	Period of a radio wave
10^{-9}	Light travels 1 ft
10^{-12}	Period of a molecular vibration
10^{-15}	Period of an atomic vibration
10^{-18}	Light travels an atomic diameter
10^{-21}	Period of a nuclear vibration
	Light travels a nuclear diameter
10^{-24}	

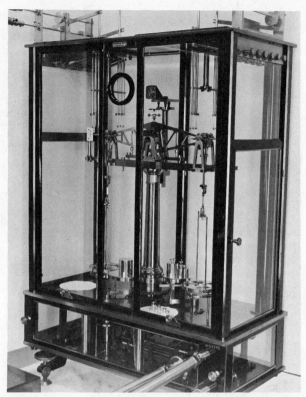

National Bureau of Standards

Fig. 2.7 *Beam balances have been used for mass comparison since the dawn of history.* (*Egyptian balances dating from c5000 B.C. have been found.*) *The photograph shows a sensitive balance used at the National Bureau of Standards with which mass comparisons can be made to 1 part in 10^8.*

plished without reference to the proper dynamical definition of mass.

The international standard of mass is a cylinder of platinum-iridium, which is defined as 1 kilogram (kg) = 10^3 grams (g). Secondary mass standards have been distributed, as have secondary meter bars, to standards laboratories throughout the world; the United States has Kilogram No. 20 (Fig. 2.8).

It would, of course, be highly desirable to have an atomic standard for mass just as we have for length and time. We do have such a standard, which is used in the comparison of the masses of atoms and molecules (see Section 3.3), but, unfortunately, at present we have no precision method of utilizing this standard above the level of individual atoms and molecules. When technology has developed to the point that we can determine precisely the mass of the standard kilogram in terms of the atomic mass standard, we shall certainly adopt the atomic unit of mass as our standard.

In the previous two sections we found that the ranges of distance and time encountered in the Universe were each approximately 40 orders of magnitude. The range of mass is enormously greater—from the mass of the electron to our estimate of the mass of the Universe, the span is about 80

Fig. 2.8 *The national standard of mass for the United States is Kilogram No. 20. This cylinder of platinum-iridium is 39 mm in diameter and 39 mm high. Recomparison with the international standard in 1948 showed that the U. S. standard is still equivalent to the international standard to within 1 part in 50 million.*

Table **2.6** *The Range of Masses in the Universe*

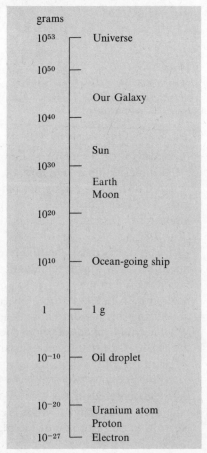

grams	
10^{53}	Universe
10^{50}	
	Our Galaxy
10^{40}	
	Sun
10^{30}	
	Earth Moon
10^{20}	
10^{10}	Ocean-going ship
1	1 g
10^{-10}	Oil droplet
10^{-20}	Uranium atom Proton
10^{-27}	Electron

Table **2.7** *Some Important Masses*

Object	Mass (g)
Sun	1.991×10^{33}
Earth	5.977×10^{27}
Moon	7.35×10^{25}
Proton	1.672×10^{-24}
Electron	9.108×10^{-28}

orders of magnitude! Table 2.6 shows some of the marker points on this gigantic range of masses. Values of some of the more important masses are given in Table 2.7.

DENSITY

How do we determine the mass of an object that is too large to place on a beam balance? It is a simple matter if we know the volume of the object. First, we take a small sample of the material of which the object is composed and measure its volume and mass. The ratio of these quantities determines the *density* of the material:

$$\text{density} = \frac{\text{mass}}{\text{volume}}$$

$$\rho = \frac{M}{V} \tag{2.1}$$

Once the density has been determined, the mass of the large object is given by the product of the density and the volume. If the object is not homogeneous (that is, if the density is not the same everywhere in the object), then an *average* density must be found before the mass can be determined.

Example **2.1**

What is the average density of the Earth?
According to Table 2.3, the radius of the Earth is

$$R_E = 6.378 \times 10^8 \text{ cm}$$

The volume is

$$V = \tfrac{4}{3}\pi R_E^3 = \tfrac{4}{3}\pi (6.378 \times 10^8 \text{ cm})^3$$
$$= 1.0814 \times 10^{27} \text{ cm}^3$$

According to Table 2.7, the mass of the Earth is

$$M_E = 5.977 \times 10^{27} \text{ g}$$

Therefore, the average density of the Earth is

$$\rho = \frac{M_E}{V}$$
$$= \frac{5.977 \times 10^{27} \text{ g}}{1.1 \times 10^{27} \text{ cm}^3}$$
$$= 5.527 \text{ g/cm}^3$$

The density of the rocky material in the Earth's crust averages about 3 g/cm³. We conclude, therefore, that the core of the Earth must be extremely dense in order to give the proper average value. (The density of the Earth's core is about 12 g/cm³, or only slightly less than the density of mercury.)

Some densities representative of those found in the Universe are given in Table 2.8.

CONSERVATION OF MASS

Mass is the first *property of matter* (as distinct from the basically geometrical concepts of length and time) that we have introduced. We have come to appreciate the permanence and the immutability of mass. A block of metal has a certain mass. This mass does not change with time. Nor does it change when its dimensions are changed; we can alter its shape by hammering or forging, but the mass does not change. We can even dissolve it in acid, but the combined mass of the acid and the metal (together with any gases that might be evolved in the process) remains unchanged.

These facts concerning mass constitute the first of a series of important fundamental statements called the *conservation laws* of physics. Instead of emphasizing the *differences* between various physical processes, in order to understand the fundamentals of Nature it is more profitable to seek those properties that are the *same* in any process. The conversion of water into ice and the conversion of water into its constituent gases, oxygen and hydrogen, are two different physical processes. But a common feature of these processes is that the *mass* of the material remains constant. A long series of experiments, all of which exhibit this feature, has led us to conclude that *mass is conserved* in all physical processes. This is the first of several *conservation laws* that we will discuss.

Stated in the simple terms "mass is conserved," the law is not entirely accurate. For all everyday processes that we encounter, mass is indeed a constant physical property. But, as we shall see, relativity theory predicts, and it is verified by experiment, that mass is interchangeable with another physical property, *energy*; the quantity that is actually conserved is the

Table **2.8**   *Some Representative Densities*

Type of Matter	Density (g/cm³)
Nuclear matter	10^{14}
Center of Sun	10^{2}
Lead	11.3
Aluminum	2.7
Water	1
Air	10^{-3}
Laboratory high vacuum	10^{-18}
Interstellar space	10^{-24}
Intergalactic space	10^{-30}

combination of mass and energy. We shall return to this point in Chapters 7 and 11 and shall use the concept of the interchangeability of mass and energy on frequent occasions in the discussion of nuclei and elementary particles.

2.4 *Systems of Units*

A QUESTION OF CONVENIENCE

We have already discussed the relationship between the metric and British units of length—1 inch is defined as *exactly* 2.54 cm. The unit of time in each system is the *second*. The unit of mass in the British system is called the pound[3] and is legally defined as *exactly* 453.59237 g. We therefore require only these two conversion factors (2.54 cm/in. and 453.59237 g/lb) in order to change between the British and metric systems of units. In the early part of this book we will use both systems. Thus, we will find speeds given in cm/sec as well as in mi/hr and masses stated in pounds and in grams. When we have completed the introductory material and begin the development of topics in modern physics, we completely forego the use of British units.

Within the metric system of units, two variations are in use. One is termed the *MKS system,* which is an abbreviation for the basic units of the system—meter-kilogram-second. The other is the *CGS system,* standing for centimeter-gram-second. There is little on which to make a choice between the two systems on the basis of the units for length and mass—simple powers of 10 relate the fundamental units. However, when electrical units are introduced, the definitions of the various quantities are sufficiently different in the two systems that the conversion factors are no longer simple. In discussions of electricity and magnetism, MKS units are customarily used (for example, volts and amperes). But in this book we are not concerned primarily with the electricity and magnetism of everyday things. We are instead interested in discussing topics of contemporary physics. Thus, we shall be dealing principally with atoms and nuclei and their interactions. It is customary in atomic and nuclear physics to use CGS units. The *physics* of a phenomenon does not depend in any way on the system of units used to describe it; therefore, one must make a choice on the basis of convenience. Since atomic and nuclear physics is usually described with the CGS system, we shall use that system.

THE USE OF UNITS

All physical quantities have *units* or *dimensions*. When we make numerical statements or write numerical equations concerning physical quantities, we must include the units of the quantities. If we make the statement, "The

---

[3] Or *pound-mass* to distinguish this unit from the traditional unit of *weight* in the British system which is still often called the "pound." To confuse matters even further, there is also the British monetary unit of the same name! In this book, "pound" will always mean a unit of *mass*.

distance traveled is equal to the speed multiplied by the time," we could express this more briefly as

distance = speed × time

or, by giving arbitrary symbols to the quantities, as

$d = s \times t$

This equation does not explicitly contain the units of the various quantities; the equation is valid for any system of units as long as they are used consistently. For example,

$$30 \text{ mi} = 15 \frac{\text{mi}}{\text{hr}} \times 2 \text{ hr}$$

Not only must the *numbers* balance in such an equation, but so must the *units.* On the right-hand side of the equation, the time unit "hour" occurs in both numerator and denominator and therefore cancels, leaving "miles" as the unit on both sides of the equation.

We can alter the unit of any quantity by using *conversion factors.* In order to convert 15 mi/hr to the corresponding number of feet/second, we use the conversions

$$1 \text{ hr} = 3600 \text{ sec or } \frac{1 \text{ hr}}{3600 \text{ sec}} = 1$$

$$1 \text{ mi} = 5280 \text{ ft or } \frac{1 \text{ mi}}{5280 \text{ ft}} = 1$$

Then, multiplying by factors that are *unity,* we have

$$15 \frac{\text{mi}}{\text{hr}} = 15 \frac{\text{mi}}{\text{hr}} \times \frac{1 \text{ hr}}{3600 \text{ sec}} \times \frac{5280 \text{ ft}}{1 \text{ mi}}$$

$$= \frac{15 \times 5280}{3600} \frac{\text{ft}}{\text{sec}}$$

$$= 22 \frac{\text{ft}}{\text{sec}}$$

Always give units when writing the numerical values of physical quantities. *Always* check equations to insure that the units on both sides are the same (or are equivalent in the sense that they are related by a conversion factor such as demonstrated above); if the units are *not* the same, something is wrong!

THE THREE BASIC UNITS

The units of all physical quantities can be expressed in terms of the basic units of length, mass, and time. When we introduce such quantities as *force* and *energy,* for convenience we shall give special names to the units (*dynes* and *ergs*), but these units are defined as certain combinations of length, mass, and time. These three units—centimeter, gram, and second—are all that we require; every physical quantity can be expressed in terms of these units.

Summary of Important Ideas

The basic unit of *length* is the *meter* or the *centimeter* (1 m = 100 cm). The *standard* of length is the wavelength of light from krypton-86.

The basic unit of *time* is the *second.* The *standard* of time is the vibration period of cesium-133.

The basic unit of *mass* is the *kilogram* or the *gram* (1 kg = 1000 g). The *standard* of mass cannot yet be stated in terms of atomic quantities with sufficient precision to be generally useful; therefore, the operational standard is a certain block of metal maintained in the international standards depository.

Density is mass per unit volume.

Mass is conserved; that is, mass can neither be created or destroyed—only rearranged. (We shall later see that mass and energy are intimately connected and that the conservation law properly refers to *mass-energy*.)

All physical quantities have *units,* and in equations relating various physical quantities, the *numbers* as well as the *units* on the two sides of the equation must be the same.

The fundamental physical units are those of *length, time,* and *mass.*

Questions

2.1 It has been argued by some persons that if we eventually give up the British system of measurements, we should adopt a duodecimal system (that is, a system based on *twelves*) instead of the metric system. Discuss the relative merits of decimal and duodecimal systems of measure.

2.2 Look up the definition of *time* in a dictionary. Ignoring those definitions that do not deal with the physical concept of time, comment on the definitions that pertain to time as we use the word in physics. Do these definitions give you a clear understanding of time? Try to devise a better definition of physical time.

2.3 You suspect that your stop-watch runs 0.01 sec slow every minute. Explain how you would use the WWV time signals to check this suspicion.

2.4 The *sidereal day* is the time interval between two successive passages of a star across the meridian. The *mean solar day* is the average time required for the Earth to rotate once relative to the Sun. Take account of the fact that the Earth rotates on its axis while revolving around the Sun and show that the sidereal day is shorter than the mean solar day.

2.5 Suppose you were faced with the task of selecting a material from which the standard kilogram was to be made. What properties would be desirable for the material and what properties would be undesirable? What procedures would you establish for the use and preservation of the standard?

2.6   The density of a long, thin piece of wire is to be determined. What measurements must be made? Which measurement must be made with the greatest care and why?

Problems

2.1   A certain athlete runs the 100-yd dash in 9.4 sec. What would be his expected time for the 100-m dash?

2.2   How many centimeters are there in 1 mi?

2.3   A jet airliner can make the trip from San Francisco to Washington, D.C. (2400 miles) in 4 hr. At this same speed, how long would be required for a round trip from the Earth to the moon?

2.4   What is the angle subtended at the Earth by the moon? What is the angle subtended by the Sun?

2.5   A certain electronic computer can perform 350,000 arithmetical operations per second. How many microseconds are required (on the average) for each operation?

2.6   A facetious unit of speed is furlongs/fortnight. Express this in mi/hr and in cm/sec.

2.7   A certain watch is claimed by the manufacturer to have an accuracy of 99.995 percent. By how many minutes might such a watch be in error after running a month?

2.8   The radioactive nucleus Ra^{226} (*radium* containing a total of 226 protons and neutrons) has an average lifetime of about 1620 yr. How many seconds is this?

2.9*   Compute your age in seconds to an accuracy of at least 3 parts in 10^4. (Remember leap years!)

2.10   A certain motor spins a shaft at a rate of 1000 revolutions per minute (rpm). If the motor runs for a week, how many revolutions will the shaft make?

2.11   The human heart is a marvelous machine. On the average, a man's heart beats at a rate of 72 per min. At the age of 70 yr, approximately how many times will his heart have beat?

2.12   Express 1 microcentury in more conventional units.

2.13*   How many hours are required for the Earth to rotate through 1 radian? In 1 second of time through how many seconds of arc does the Earth rotate? ($1° = 60' = 60$ arc min; $1' = 60'' = 60$ arc sec.)

2.14   What is the conversion factor between square inches and square centimeters? What is the conversion factor between cubic feet and cubic meters?

2.15   What is the volume of a hydrogen atom? Calculate and compare the densities of the hydrogen *atom* and the hydrogen *nucleus* (the *proton*).

2.16   A certain block is measured and found to have the following dimensions: length, 16.2 cm; width, 4.12 cm; height, 0.89 cm. The mass of the block is determined with an analytical balance and is found to be 206.35 g. What is the density of the block? (Remember the rules concerning significant figures.)

2.17 Use the fact that 1 qt = 946.3 cm^3, and comment on the accuracy of the old saying "A pint's a pound the world around."

2.18 The mass of the Andromeda galaxy (also known by its catalogue designation of M31) is approximately 8×10^{44} g. Estimate how many stars there are in M31. (Use the fact that the Sun is a rather typical star.)

2.19* Consider the Sun to consist entirely of hydrogen. (Actually, about 95 percent of the atoms in the Sun are hydrogen.) How many atoms of hydrogen are there in the Sun? What would be the size of the Sun if it had a density equal to that of the proton?

2.20 What is the density of the platinum-iridium alloy that was used to construct the standard kilogram? (See Fig. 2.8.)

Veil Nebula in Cygnus

3 *Galaxies and Atoms*

Man's curiosity about the natural happenings around him has never been limited only to what he could reach out and touch. Of course, ancient men wondered how plants grew or what was the nature of fire, but they also had a keen curiosity about the stars. In their efforts to understand the mysteries of the movements of the stars, the first science—astronomy—was born. Systematic astronomical observations were made by the Chinese at least 4000 years ago; and several centuries B.C. the Middle East astronomers were highly proficient in their art. By the middle of the first millenium A.D., the Mayans had developed a sophisticated calendar based on celestial movements. At about the same time, Hindu astronomers used their observations as a basis to calculate the Earth's diameter and the Earth–moon distance—calculations that were in error by only a few percent!

Ancient men wondered not only about the stars but also about the ultimate nature of *matter*. Although the concept of *atoms* was first developed by the Greek philosopher, Democritus (485–425 B.C.), no real progress was made in understanding the behavior of atoms until about 100 years ago because there existed no tools with which observations of atomic effects could be made. Thus, while ancient men were making astronomical observations and calculations, the nature of matter was the subject of only philosophical speculations.

3.1 *Planets and Stars*

DISTANCES IN THE SOLAR SYSTEM

How large is the Universe? How are we able to measure the enormous distances to the stars and to the farthest galaxies? Certainly, we cannot use the traditional methods developed for distance measurements on the Earth. Actually, a complicated chain of reasoning is necessary, leading from planetary distances to stellar distances to galactic distances, in order to provide information concerning the distances to the farthest reaches of the Universe. The first link in the chain is the determination of distances in the solar system.

Essentially all astronomical determinations of distance are based on a standard of length that is the distance from the Earth to the Sun.[1] This length standard is called the *astronomical unit* (A.U.). When Kepler analyzed the motions of the planets (particularly Mars), he was able to express planetary distances only in terms of the Earth–Sun distance. Figure 3.1 illustrates Kepler's method. From his data, Kepler knew that the time required for Mars to complete one orbit (the Martian *year*) was 687 days. Therefore, on two dates 687 days apart, Mars will be in the same position in its orbit but the Earth will occupy two different positions on these dates. In fact, in 687

[1] In its motion around the Sun the Earth does not describe a perfect circle (the motion is *elliptical*); therefore the Earth–Sun distance is not constant. The astronomical unit is defined in terms of the length of one axis of the elliptical orbit. Because the departure of the Earth's orbit from a circle is slight, the deviation of the actual Earth–Sun distance from 1 A.U. never exceeds 2 percent.

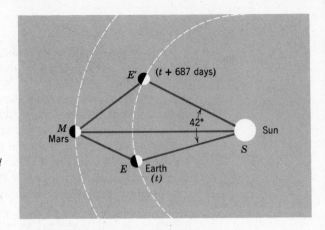

Fig. 3.1 *In 687 days Mars completes one orbit and the Earth is short of two complete orbits by 42°. By measuring the angles* $\angle \overline{MES}$ *and* $\angle \overline{ME'S}$, *Kepler was able to express the distance* $\overline{MS}$ *in terms of* $\overline{ES}$ (*i.e., the astronomical unit*).

days, the Earth will be short by 42° of completing two full orbits. Thus, in Fig. 3.1 the angle $\angle \overline{ESE'}$ is 42°. Furthermore, Kepler was able to determine from his data the angles $\angle \overline{MES}$ and $\angle \overline{ME'S}$ (that is, the angle between Mars and the Sun as viewed from the Earth on the two dates one Martian year apart). With this information, the Mars–Sun distance ($\overline{MS}$) can be expressed in terms of the Earth–Sun distance ($\overline{ES}$, the astronomical unit). Today, we would use trigonometric methods to make this calculation, but Kepler used geometrical constructions and found the ratio $\overline{MS}/\overline{ES} =$ 1.524. That is, the radius of the Martian orbit is 1.524 A.U. Similar measurements and calculations can be used to determine the distances to all of the planets in terms of the astronomical unit.

But how do we obtain a value (in cm) for the astronomical unit? In principle, we could use the *triangulation* or *range-finder* method illustrated in Fig. 1.8. However, a moment's reflection will reveal that triangulation measurements of the Earth–Sun distance are extremely difficult to make. Not the least of the problems is the fact that the Earth's diameter (our largest possible base line) subtends an angle of only 17.6 sec of arc at the distance of the Sun. (What are some other problems? See Question 3.2.)

The most precise method available at present for the determination of the astronomical unit involves a combination of triangulation and *radar-ranging* techniques. Figure 3.2 illustrates the method for the case of the planet

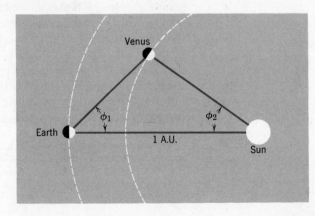

Fig. 3.2 *Triangulation method for the determination of the astronomical unit. Astronomical observations and a knowledge of the period of the Venus orbit are used to obtain the angles* ϕ_1 *and* ϕ_2. *Radar-ranging is used to measure the Earth-Venus distance. Then, knowing two angles and the length of one side of the triangle, the base (i.e., the astronomical unit) can be calculated.*

Venus. The angles ϕ_1 and ϕ_2 can be determined from astronomical observations and a knowledge of the period of the Venus orbit (see Problem 3.1). Then, the length of the base of the Earth–Venus–Sun triangle (that is, the astronomical unit) can be calculated by trigonometric methods if the length of either of the other two sides is known. A direct measurement of the Earth–Venus distance can be made by radar-ranging, in which a powerful transmitter sends out a burst of radio waves of extremely short duration. These waves travel through space and strike Venus; a small fraction of the reflected waves return to the Earth[2] where they are detected with a sensitive receiver (see Fig. 3.3). We know (also by direct measurements) that radio waves travel through space with the speed of light, 3×10^{10} cm/sec (or 186,000 mi/sec). Therefore, by measuring the time required for the round trip of the waves from the Earth to Venus and back, we have a determination of the Earth–Venus distance. Such experiments have yielded for the astronomical unit the value

$$1 \text{ A.U.} = 1.495\ 979 \times 10^{13} \text{ cm}$$

$$= 92{,}955{,}700 \text{ mi} \tag{3.1}$$

[2] Only about 10^{-23} of the initial pulse is detected by the receiver.

Fig. 3.3 *The radar receiving antenna at the Goldstone Tracking Station (Pioneer site) in California that was used in the first precision radar-ranging measurements of the distance from the Earth to Venus (1961).*

STELLAR DISTANCES

It would be a simple matter to find the distance to any star if all stars produced the same amount of light (in other words, if all stars had the same *intrinsic luminosity*[3]). To make such a distance determination, we would use the fact that the intensity of light that falls on a given surface from a point source of light decreases with the *square* of the distance from the source to the surface (see Fig. 3.4). Then, if the *apparent luminosity* of a certain star were 10^{-12} of the Sun's luminosity, we could immediately conclude that the star is 10^6 times as far from us as is the Sun; that is, the distance to the star is 10^6 A.U. or approximately 1.5×10^{19} cm.

Unfortunately, stellar luminosities are not all the same—there are, in fact, enormous variations in luminosity among stars, so that brightness measurements alone are not particularly useful in determining the distances to stars. However, as we shall see, when combined with observations of the *color* of the star's light, or, in certain cases, the way in which the light intensity varies with time, luminosity measurements can be extremely useful in estimating stellar distances.

We have actually only one unassailable method for measuring stellar distances—the triangulation or *parallax* method. If a star is sufficiently close to the solar system, then the apparent position of that star on the background of very distant stars will be slightly different depending on the position of the Earth in its orbit. The apparent excursion of the star will be a maximum if observations are made 6 months apart so that the Earth is on opposite sides of the Sun for the two measurements. As shown in Fig. 3.5, such measurements define the angle ϕ which is called the *parallax* for the star.[4]

---

[3] If two stars have the same *intrinsic luminosity,* they will appear to have the same brightness if they are at the same distance from the observer; if one star is at a greater distance, its *apparent luminosity* (i.e., its *brightness*) will be smaller.

[4] The *parallax* is defined for the base line equal to the *radius* of the orbit.

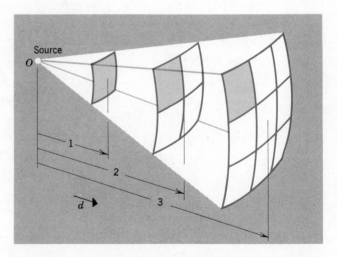

Fig. 3.4 *The intensity of light that is emitted uniformly in all directions from a source decreases as $1/d^2$ where d is the distance from the source. The intensity on one of the squares at a distance of three units from the source is 1/9 of the intensity on the square at a distance of one unit. Note that all of the squares have the same area.*

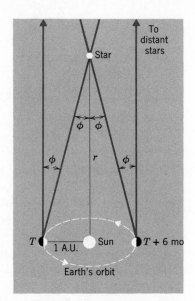

Fig. 3.5 *Parallax method for the determination of the distance of a star. The angle φ (one-half the included angle of the sighting lines) is the parallax and the distance to the star is equal to 1 A.U. divided by φ (in radians),* r = 1 A.U./φ.

The distance to the star is then equal to the length of the base line (that is, the orbit radius = 1 A.U.) divided by the parallax in radians; compare Fig. 1.8.

It is customary to give the astronomical parallax in *seconds of arc* instead of in radians. Therefore, using the conversion factor, $1° = 3600$ arc sec = 0.01745 rad, we can write for the distance r to a star that has a parallax ϕ,

$$r = \frac{3600 \text{ arc sec}}{0.01745 \text{ rad}} \times \frac{1 \text{ A.U.}}{\phi}$$

$$= 206{,}265 \times \frac{1}{\phi} \text{ A.U.} \tag{3.2a}$$

where ϕ is in seconds of arc. Thus, if a star has a parallax of 1 arc sec, it is at a distance of 206,265 A.U.; a parallax of $\frac{1}{2}$ arc sec means a distance of $2 \times 206{,}265$ A.U. Because no star has a parallax greater than 1 arc sec, all stars are at distances greater than 206,265 A.U. It therefore proves convenient to enlarge our standard of length when discussing stellar distances. For the new standard, we choose 206,265 A.U. and call this unit the *parsec* (which stands for *par*allax of 1 arc *sec*ond). The distance equation is now simply

$$r = \frac{1}{\phi} \tag{3.2b}$$

where r is in parsecs (pc) and ϕ is in seconds of arc. A star that has a parallax of 1 arc sec, for example, is at a distance of 1 pc; a star that has a parallax of 0.1 arc sec is at a distance of 10 pc. Equivalently, we can state that 1 pc is the distance at which a length of 1 A.U. would subtend an angle of 1 arc sec. Converting to centimeters, we find

$$1 \text{ pc} = 3.08 \times 10^{18} \text{ cm} \tag{3.3}$$

Another often used unit of stellar distance is the light year (L.Y.), which is the distance that light travels in one year. Since the speed of light is

3.00 × 10¹⁰ cm/sec and 1 yr = 3.16 × 10⁷ sec, the light year is

1 L.Y. = (3.00 × 10¹⁰ cm/sec) × (3.16 × 10⁷ sec)

$$= 9.46 \times 10^{17} \text{ cm} \tag{3.4}$$

Finally, the conversion between light years and parsecs is

$$1 \text{ pc} = 3.26 \text{ L.Y.} \tag{3.5}$$

THE NEAREST STARS

The number of stars whose distances from us can be measured by the parallax method is limited by the sensitivity with which parallax determination can be made. With modern equipment, a parallax as small as 0.005 arc sec (0.005″) can be detected, but if we require a measurement with a precision of 10 percent, then the smallest parallax is limited to an angle of approximately 0.05″. About 700 stars are sufficiently close so that parallax measurements to 10 percent or better have been made. The nearest of these stars are *Alpha Centauri* and *Proxima Centauri*, close neighbors in the sky, each of which has a parallax of 0.76″ and is therefore at a distance of 1/0.76 = 1.3 pc or 4.3 L.Y. Of those stars visible in the Northern Hemisphere, *Sirius* (the brightest star in the sky) is the closest—about 2.6 pc (8.5 L.Y.) The North Star or *Polaris* is at a distance of approximately 400 L.Y. (corresponding to a parallax of 0.008″.)

STELLAR MOTIONS

Stellar distances greater than ∼10²⁰ cm (corresponding to parallaxes less than about 0.03″) cannot be reliably determined by the parallax method; other techniques must be employed if we are to map the deeper regions of space. These techniques are necessarily indirect and are based on statistical analyses of the measurable properties of stars.

All stars are in motion, and if the speeds relative to the Earth are sufficiently great, these motions can be measured. Two effects can be detected. If the position of a star against the background of more distant stars appears to continually change with time, this motion (called the *proper* motion; see Fig. 3.6) can be measured and expressed in *seconds of arc per year.* (Proper motions of nearby stars can be as much as 10 arc sec/yr.) This information alone is not sufficient to determine the distance to the star because the *speed* of the star (that is, the *distance* perpendicular to the line-of-sight traveled per year) is not known.

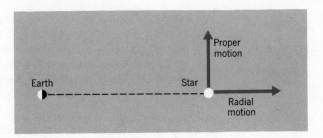

Fig. 3.6 *There are two aspects to the motion of a star relative to the Earth. The proper motion is that motion perpendicular to the Earth-star line and the radial motion is that motion along the Earth-star line (either toward or away from the Earth). In general, the motion of a star is a combination of these two types of motion.*

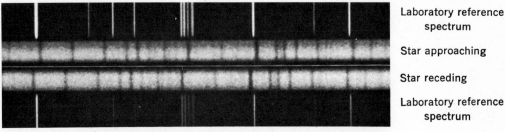

Laboratory reference
spectrum

Star approaching

Star receding

Laboratory reference
spectrum

Lick Observatory

Fig. 3.7 *The radial speed of a star's motion can be determined by measuring the shift in frequency of the characteristic features of its light. Shown here are two spectrograms (taken at different times) of the binary star* α^1 *Geminorum. Only one of the two stars in this binary emits enough light to be detected. Notice that the spectral lines from the star are shifted, with respect to the laboratory reference lines, in different directions corresponding to two phases of motion of the star. In one phase the star is moving toward the Earth and the frequency of the light is increased; in the other phase the star is moving away from the Earth and the frequency is decreased.*

There is, however, another effect that allows an absolute determination of a star's speed. If a star is moving away or toward the Earth (this motion is called the *radial* motion; see Fig. 3.6), the character of the light emitted by the star and observed on Earth is affected. Everyone is familiar with the fact that the tone (or *frequency*) of the sound of a train whistle is higher when the train is approaching than when receding. (As the train passes, the frequency changes and we hear a *whee-oo* sound.) This phenomenon of the dependence of frequency on the speed of the source toward or away from the observer is known as the *Doppler effect*. Light behaves in the same way as does sound in this regard. If a star is approaching us, the observed frequency of its light is increased and if the star is receding, frequency is lowered. By measuring the frequency shift of some of the characteristic features of the star's light, the speed of the radial motion can be determined (see Fig. 3.7). The importance of the Doppler-shift method lies in the fact that the determination of the radial speed by this method is *absolute* and does not depend on a knowledge of the distance to the star. In general, of course, the motion of a star is a combination of proper and radial motions, but these motions are independent and can be measured separately.

Suppose that we observe the motions of stars that are grouped in a *cluster* (that is, for a variety of reasons we believe that the stars in question all lie at approximately the same distance; see for example, the obvious grouping of stars in Fig. 6.2). For these stars we obtain measurements of proper motions (in *seconds of arc per year*) and radial speeds (in *kilometers per second*). In the case of "exploding" clusters, the stars are all moving outward from a common center and there are just as many stars moving in one direction as in any other direction. In this circumstance, the average speed in the radial direction is equal to the average speed in the direction perpendicular to the line-of-sight. A knowledge of the average perpendicular speed coupled with the average of the measured proper motions permits a calculation of the average distance to the star cluster.

Exploding star clusters represent a rather special case for the determination of stellar distances through studies of star motions, and for most clusters (which are not of the exploding type) different statistical techniques are required. Nevertheless, the use of various statistical methods has produced the most reliable information that we possess regarding distances to stars whose parallaxes are too small to measure.

COLOR-LUMINOSITY MEASUREMENTS

The *color* of light depends on the *frequency* of the light wave in a way similar to that in which the *tone* of a sound depends on the frequency of the sound wave. When light passes from one medium to another (as from air to glass or the reverse), if the light ray does not strike the interface perpendicular to its plane, the ray is bent, or *refracted,* and the amount of refraction depends on the frequency of the light wave. By passing a light ray through a triangular piece of glass, called a *prism,* refraction takes place both on entering and leaving, and the effect is enhanced.[5] Figure 3.8 shows a ray of white light that passes through a prism and is broken up into its component colors. Violet light (V), which has a high frequency, is more strongly refracted than the lower-frequency red light (R). A prism, therefore, produces a spatial separation of the frequency *spectrum* of the light ray. An instrument that produces a record (usually by photographic means) of the distribution of frequencies from a light source is called a *spectrograph* or a *spectrometer.*

By attaching a spectrograph to a telescope, it is possible to measure the amount of light of various frequencies that is emitted by a star. It is found that there is a wide range for the dominant colors of stars and the *color*

[5] If the two faces of the glass are parallel (as, for example, in window glass), the refraction effects cancel instead of adding and the breakup of the light into its various colors does not take place.

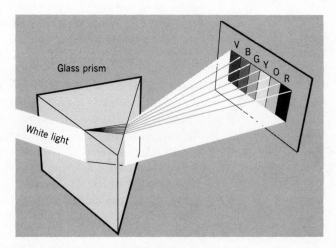

Fig. 3.8 *On passing through a glass prism, a ray of white light is broken up into its component colors. The higher frequency waves are refracted more than the lower frequency waves and so the spectrum shows red (R) at the right, followed by orange (O), yellow (Y), green (G), blue (B), and finally violet (V).*

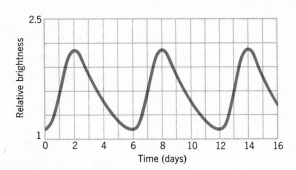

Fig. 3.9 *The light curve for δ Cephei, a typical member of the class of pulsating stars known as Cepheid variables. The period of the light intensity variations is approximately 6 days.*

of a star gives information regarding its temperature. There are blue-white (very hot) stars and red (relatively cool) stars, as well as a large number of intermediate classes. Cataloging the color classes and the luminosities of the stars for which we have distance measurements has shown that there exists a correlation (albeit approximate) between a star's color and its intrinsic luminosity. Once this relationship has been established for the nearby stars, it can be applied to the more distant stars. Thus, a star's approximate intrinsic luminosity can be deduced from its color. Then, a measurement of the apparent luminosity, together with the known inverse-square-law for the decrease of brightness with distance,[6] provides a method for determining the approximate distance to the star.

The color-luminosity method is capable of yielding only approximate values for stellar distances; nevertheless, it has provided valuable information in cases that do not permit the use of more precise techniques.

VARIABLE STARS

The Sun has been pouring out light at a constant rate for millions of years. Although we know that the rate was less when the Sun was a young star (the age of the Sun is about 4.5×10^9 yr) and that the rate will ultimately decrease as the Sun begins to die, for all practical purposes we can assume that the Sun is a constant source of light. Not all stars emit light at a constant rate for long periods of time. In fact, some stars (more than 10,000 are known) are observed to *pulsate;* that is, the light output varies in a regular way. Typical of this class is a star in the constellation *Cephus* that is labeled δ *Cephei*. In 1784 it was found that δ Cephei varies in brightness by about a factor of two, repeating its cycle every 6 days (see Fig. 3.9). The class of stars that have a behavior similar to that of δ Cephei are now known as *Cepheid variables*.

The importance of Cepheid variables and other variable stars lies in the fact that the *period* of the light curve for a given star (that is, the time between successive maxima or minima in light output—see Fig. 3.9) is related to the intrinsic luminosity of that star. Cepheid variables with periods between 1 and 150 days are known and the average intrinsic luminosities of these stars span a range of about a factor of 100. Again, it has only been through statistical analyses of large numbers of Cepheid variables that the

---

[6]This assumes that we can correct for the effect of light absorption by any clouds of gas or dust that may lie between the Earth and the star.

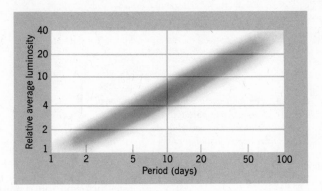

Fig. 3.10 *The period-luminosity curve (actually, a band) for Cepheid variables. There also exists a second class of these variable stars which has a slightly different period-luminosity curve.*

intrinsic luminosities have been determined,[7] but we now have a sufficient body of information to plot a fairly reliable graph of the average luminosity *versus* period for these stars. As can be seen in Fig. 3.10, the period-luminosity graph is actually a broad band instead of a line. To some extent, the existence of this band represents inadequacies in our measuring techniques; but the main reason is probably the natural variability of the period-luminosity relationship from star to star. With information such as that shown in Fig. 3.10, we can determine the intrinsic luminosity of a Cepheid variable to within about 50 percent by measuring the period of the light curve. Once the average *intrinsic* luminosity is known, a measurement of the average *apparent* luminosity can be used to deduce the distance to the star. Because it is relatively easy to make measurements of the period

[7]Of course, if we could make parallax measurements for Cepheid variables, there would be no necessity to use indirect methods to determine the intrinsic luminosities. Unfortunately, of the known Cepheid variables, only *Polaris* is sufficiently close to make parallax measurements.

Fig. 3.11 *Moonlight photograph of the 200-in. Hale telescope located on Mount Palomar in southern California. This is at present the largest optical telescope in the world, although a larger instrument is being constructed in the Soviet Union.*

Mount Wilson and Palomar Observatories

THE TWO HVNDRED INCH TELESCOPE

Mount Wilson and Palomar Observatories

Fig. 3.12 *Cut-away drawing of the 200-in. telescope on Mount Palomar. The construction of this telescope, completed in 1948, was first proposed by the astronomer George E. Hale and it is named in his honor.*

of the light curve and the apparent luminosity, Cepheid variables serve an extremely useful function in the determination of distances in deep space. The method is limited only by our ability to distinguish Cepheid variables with the largest telescopes, such as the 200-in. Hale telescope on Mount Palomar (shown in Figs. 3.11 and 3.12), and distances as large as 20 million L.Y. have been measured in this way.

3.2   *Galaxies*

ISLAND UNIVERSES

During the 18th century it was discovered that many of the objects in the sky, when viewed with the best telescopes of the day, did not appear as points of light; instead, these objects were *diffuse* and therefore could not be single stars. When the more powerful telescopes of the early 20th century were used to study these objects, it was found that some are glowing clouds of gas or clouds that reflect starlight (these we now refer to as *nebulae*), but that others are clearly composed of large numbers of individual stars. Because the distance determinations available at that time were quite poor, it was not at all clear whether these star groups were part of our local system of stars, the Milky Way, or whether they were separate "island universes." It was not until 1924 that Edwin Hubble of the Mount Wilson Observatory settled the question by analyzing the light curves of a number of Cepheid variables in three of the more prominent star groups. Hubble's distance determinations based on these measurements conclusively proved that these objects are not part of the Milky Way and are indeed "island universes." We now refer to these remote cousins of the Milky Way as *galaxies.* Hundreds of thousands of these galaxies have been observed and tabulated, but they are only a tiny fraction of the number that could be observed with our modern telescopes.

Mount Wilson and Palomar Observatories

Fig. 3.13   *A typical galaxy of the spiral type, designated by the catalogue number NGC (standing for New General Catalogue) 2841. Millions of stars are tightly grouped in the spiral arms and in the central region.*

Fig. 3.14 *The Whirlpool galaxy (NGC 5194) is a spiral galaxy very similar to our own. The small companion galaxy (NGC 5195), located about 14,000 L.Y. away from the larger galaxy, has no clear spiral structure and is classified as an irregular galaxy.*

OUR GALAXY

When it became clear that the Universe was populated with a large number of galaxies at great distances from the Earth, it was natural to inquire about the distribution of stars in our vicinity. What does our own Galaxy, the Milky Way, look like? Would our own Galaxy, if viewed from space, have an appearance similar to those galaxies that *we* can view? Of course, we are at a great disadvantage in making assessments about our own Galaxy because of our position *within* the Galaxy. Large regions of the Galaxy cannot be viewed because of obscuration by local clouds of dust. Therefore, much of the information concerning the overall aspects of our Galaxy has been obtained by indirect methods. The results of these studies have shown that our Galaxy is in no way unique—we live in an ordinary galaxy of ordinary size (diameter $\sim 10^{23}$ cm or 10^5 L.Y.). The Milky Way, if viewed from space, would appear very similar to the galaxy NGC 5194 (Fig. 3.14).

The Sun, moreover, does not occupy any special position within our Galaxy. Although it is difficult to pinpoint our location within the Galaxy, it is now believed that the Sun is located in the outer portion of one of the great spiral arms of the Milky Way (see Fig. 3.15). This information confirms our inconspicuousness in the Universe. Not only is the Earth just one of several planets that revolve around an ordinary star, but that star is just one of a hundred billion that comprise an ordinary galaxy.

GALACTIC DISTANCES

Although the use of Cepheid variables to determine the distances to galaxies was crucial in establishing that these star groups are not members

Fig. 3.15 *The Sun and the solar system are located in one of the great spiral arms of the local system of stars. Because of our location in the galaxy and because the galaxy is relatively flat, we see these stars as a concentrated band in the sky (the Milky Way). The central region of the galaxy is screened from our view by dust clouds.*

of the Milky Way, many other methods have also been employed in extra-galactic distance determinations. In fact, whenever individual stars can be isolated within a galactic structure, *some* method has been devised (although indirect and approximate) for obtaining distance estimates. By requiring that all of these methods (including some that we have not described) yield consistent distance values for a given galaxy, we have acquired a large amount of reasonably reliable extra-galactic distance information.

The study of the distribution of galaxies in space has shown that there are 16 galaxies, of various sizes and shapes, located within 2.5 million L.Y. of our own Galaxy. (There may be others that are obscured by dust clouds within our Galaxy.) Because there are relatively few galaxies at distances between 2.5 and 6 million L.Y., these 16 galaxies together with the Milky Way are referred to as the *local group* of galaxies (see Fig. 3.16). The clustering of galaxies is the rule, not the exception; almost all galaxies seem to be concentrated in groups. Small clusters may contain a dozen or so galaxies; giant clusters may contain a thousand or more galaxies. Figure 3.17 shows a cluster of galaxies in the constellation Hercules.

Within a distance of 50 million L.Y. from our Galaxy, thousands of other galaxies have been observed and *billions* are visible through the 200-in. Palomar telescope. Indeed, within the relatively small region of space defined by the "empty" bowl of the Big Dipper, the 200-in. telescope would reveal about a million galaxies!

THE COMPOSITION OF STARS AND GALAXIES

Of what materials are stars composed? Do galaxies consist of anything more than stars? Modern methods of analyzing the light from stars have shown that almost all stars consist primarily (60 percent or greater by mass) of the simplest element, hydrogen. The next most abundant element in stars is helium[8] and all of the heavier elements constitute no more than a few percent of the total mass.

In addition to stars, all galaxies contain huge clouds of hydrogen. Indeed, new stars are continually being formed by condensation from these clouds. Although a large amount of matter is contained in these galactic clouds, the *density* is extremely low, about 100 atoms/cm^3—far lower than in the best laboratory vacuum. In intergalactic space, the density of gas is even lower, about 1 atom/m^3. We live in a Universe that is mostly empty space; even the small amount of matter contained in the Universe is mostly hydrogen!

---

[8]The spectrum of helium was observed in light from the Sun before the element was discovered on Earth (hence, the name: *helios* = Sun).

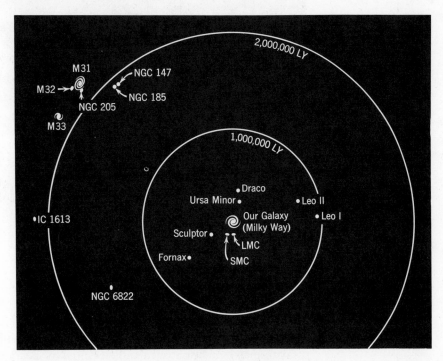

Fig. 3.16 *The local group of galaxies. The 16 near neighbors of our own Galaxy are shown projected onto a plane. The relative distances from the Milky Way are correctly represented, but because the other members of the local group are distributed in space, many of the distances between these galaxies are distorted. (SMC and LMC are abbreviations for Small Magellanic Cloud galaxy and Large Magellanic Cloud galaxy, respectively. Other abbreviations are for various catalogue names.) M31, which refers to object number 31 in Meisser's catalogue, is the great galaxy in Andromeda.*

Fig. 3.17   *A cluster of galaxies in the constellation Hercules.*

GALACTIC RED SHIFTS

The radial motions of individual stars (that is, motions either toward or away from us) can be detected and measured by means of the Doppler effect. If a star is moving *toward* us, its characteristic light features are shifted toward the blue, whereas if it is moving *away* from us, the light is shifted toward the red. When the light from distant galaxies is analyzed, a surprising result is obtained. All of these galaxies exhibit a *red shift* in their light, indicating that they are receding from us! Furthermore, the magnitude of the red shift—and, hence, the magnitude of the recessional speed—is greater for the more distant galaxies (see Fig. 3.18). In fact, there appears to be a direct proportionality between the radial speed S and the galactic distance r, namely,

$$S = Hr \tag{3.6}$$

where the proportionality constant H is called the *Hubble constant,* in honor of Edwin Hubble who established this relationship in the early 1930s. Some of the data that supports the direct proportionality between r and S is shown in Fig. 3.19. From such data we deduce that the value of H is approximately 90 km/sec per million pc (but the uncertainty is large—about 50 percent). That is, a cluster of galaxies that is 10^6 pc away recedes from us with a speed of about 90 km/sec and one that is 10^9 pc away recedes with a speed of about 10^5 km/sec. For the most distant galaxies thus far studied ($r \sim 2 \times 10^9$ pc), the red-shift measurements indicate that the recessional speed is more than half the speed of light!

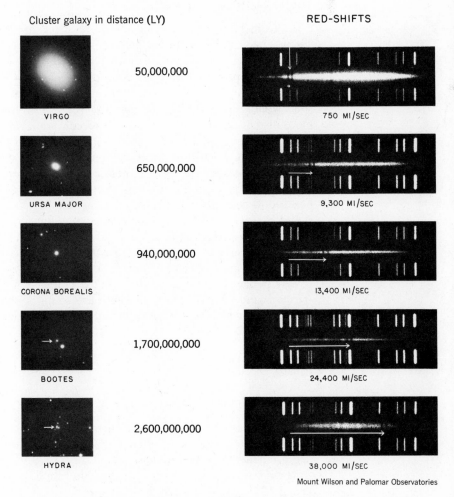

Cluster galaxy in distance (LY)　　　　　RED-SHIFTS

VIRGO — 50,000,000 — 750 MI/SEC

URSA MAJOR — 650,000,000 — 9,300 MI/SEC

CORONA BOREALIS — 940,000,000 — 13,400 MI/SEC

BOOTES — 1,700,000,000 — 24,400 MI/SEC

HYDRA — 2,600,000,000 — 38,000 MI/SEC

Mount Wilson and Palomar Observatories

Fig. 3.18 (a) *Photographs of individual galaxies in successively more distant clusters of galaxies.* (b) *The Doppler shift (red shift) in the light from these galaxies. The magnitude of the red shift increases with galactic distance.*

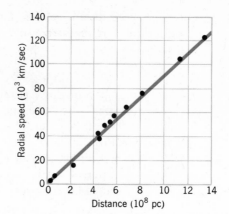

Fig. 3.19 *The distance-speed relationship for various clusters of galaxies. The distance values are only approximate.*

THE EXPANDING UNIVERSE

If all the galaxies, in every direction in space, are receding from us, do we not then occupy a special position at the center of the Universe? It is easy to show that this is not the case. Consider the following two-dimensional analogy. Suppose that we blow up a toy balloon that is covered uniformly with spots. As the balloon expands, an observer stationed on one of the spots would see all of the other spots receding from him. Furthermore, the more distant spots would be seen to recede more rapidly than the closer spots. The same results would be found by observers on each of the other spots. That is, in a uniform expansion, there is an increase of all of the inter-spot distances. In three dimensional space, the effects are identical. Consider a box of sand containing several small pebbles. If we imagine the box and its contents to expand in size, an observer stationed on any pebble would observe every other pebble receding from him.

The red-shift measurements are generally interpreted as clear evidence that the Universe is expanding.[9] Since the expansion is apparently uniform, there is no way to specify a "center" of the Universe.

Why is the Universe expanding? Is the Universe finite or infinite? How are stars and galaxies formed? These are some of the questions that we shall discuss when we return to the subject of astrophysics and cosmology in Chapter 17.

3.3 *Atoms*

THE BEGINNINGS OF ATOMIC THEORY

The sciences of astronomy and chemistry played crucial roles in establishing the foundations of physics. Early astronomical observations and measurements provided the basis for the theory of gravitation (Chapter 6) —the first of the great physical theories and the key to the understanding of large-scale phenomena in the Universe. Similarly, 19th century chemistry

---

[9] Other explanations have been proposed, but there is absolutely no evidence that the effect is due to any phenomenon other than expansion.

established the atomistic character of matter and opened the door for physicists to develop a *microscopic* description of Nature.

By the end of the 18th century, chemists had formulated a reasonably precise idea of what constitutes a pure, chemically identifiable substance. They were able to distinguish some of the basic materials that can be combined to form compounds but that cannot themselves be decomposed by chemical means—these are the chemical *elements*. These early chemists recognized that there is only a small number of elements (we now know just over a hundred), whereas the number of possible chemical compounds is enormous.

Modern atomic theory began with the enunciation by Antoine Lavoisier (1743–1794) of the *law of definite proportions*. This law states that when two elements combine to form a distinct chemical compound, they do not combine indiscriminately but with a definite ratio of masses that is characteristic of the combining elements and the compound that is formed. Thus, when water is formed from the elements hydrogen and oxygen, 1 g of hydrogen always combines with 8 g of oxygen to form 9 g of water. In addition, 1 g of hydrogen can combine with 16 g of oxygen to form 17 g of *hydrogen peroxide,* a compound with properties quite different from those of water. The same is true for every other chemical compound—there is always a definite proportion of each element present in the compound regardless of the origin of the particular sample.

In was John Dalton (1766–1844), an English chemist, who first clearly recognized that Lavoisier's law of definite proportions implied the existence of some fundamental unit for each chemical element. In other words, all matter must consist of basic particles or *atoms*. How else would it be possible to explain the fact that water always consists of 1 part (by mass) of hydrogen and 8 parts (by mass) of oxygen? The simplest explanation is that both hydrogen and oxygen consist of particles of definite, but different, masses and that they combine in some simple ratio to form water. That is, some number of hydrogen atoms combine with some number of oxygen atoms to form a water *molecule*. In the hydrogen peroxide molecule the ratio of oxygen to hydrogen must be exactly *twice* that in a water molecule. (As we shall see, an oxygen atom has a mass 16 times that of a hydrogen atom, and the water molecule contains *two* hydrogen atoms and *one* oxygen atom, thus accounting for the 8-to-1 mass ratio; the hydrogen peroxide molecule contains *two* hydrogen atoms and *two* oxygen atoms.)

The law of definite proportions alone provided sufficient evidence on which Dalton could base the *atomic concept*. However, there was no way to deduce, from the mass ratio of the elements in a compound, the relative masses of the constituent atoms without a knowledge of the *number* of atoms in the compound, and *vice versa*.

A few years after Dalton proposed his atomic theory, the French chemist Joseph Louis Gay-Lussac (1778–1850) discovered the *law of combining gas volumes*. Gay-Lussac found that when two gases combine to form a third gas, the volumes of the constituents and the product (all measured at the same temperature and the same pressure) are related to one another by ratios of small whole numbers. Thus, two volumes of hydrogen combine with one volume of oxygen to produce two volumes of water vapor (Fig.

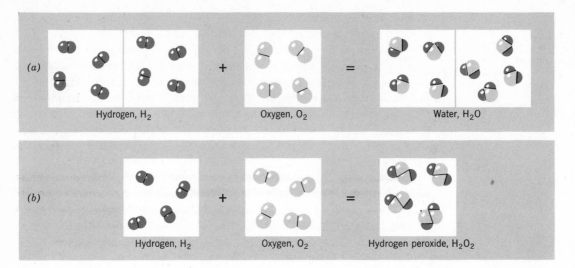

Fig. 3.20 (a) *Two volumes of hydrogen combine with one volume of oxygen to produce two volumes of water vapor.* (b) *One volume of hydrogen combines with one volume of oxygen to produce one volume of hydrogen peroxide vapor. Notice that each volume of gas contains the same number of molecules (Avogadro's hypothesis).*

3.20a). Furthermore, one volume of hydrogen combines with one volume of oxygen to produce one volume of hydrogen peroxide vapor (Fig. 3.20b).

By 1811, the Italian Count Amedeo Avogadro (1776–1856) was able to complete the foundations of modern atomic theory by offering a simple explanation of Gay-Lussac's law. At the same time he clarified the distinction between *atoms* and *molecules,* a point on which Dalton had been confused. Dalton believed that all elements existed as *atoms,* and that only compounds were in the form of *molecules.* To Avogadro, the term "molecule" meant, as it does today, the smallest unit of a substance that is capable of independent existence. An "atom" on the other hand, is the smallest unit of matter that can be identified as a particular chemical element. Thus, normal hydrogen gas consists of *molecules,* and each hydrogen molecule (as we now know) contains *two* hydrogen *atoms.*

Avogadro's important contribution that clarified the connection between the law of definite proportions and the law of combining gas volumes was his hypothesis that *equal volumes of all gases* (at the same temperature and same pressure) *contain equal numbers of molecules.* Avogadro did not know *how many* molecules there were in a given volume, but he appreciated that the number was extremely large.

Avogadro's inspired guess[10] explained the fact that 2 liters[11] of hydrogen can combine with 1 liter of oxygen to form 2 liters of water, whereas 1 liter of each gas can combine to form just 1 liter of hydrogen peroxide. Each molecule of water, according to Avogadro, must therefore contain the

[10] In spite of the brilliance of Avogadro's hypothesis, this important contribution went unappreciated for almost 50 years during which time confusion reigned among chemists trying to understand the relationship between elements and compounds.

[11] A *liter* (l) is defined as the volume of 1 kg of water (at maximum density) and is equal to 1000.028 cm^3. We shall use the approximate relation, $1\ l = 10^3\ cm^3$.

amount of hydrogen in one hydrogen molecule and one-half of the amount of oxygen in one oxygen molecule. But to determine the number of atoms in a molecule of hydrogen required additional information.

Quantitative chemical techniques can be used to analyze the composition of a gas and to determine the masses of the constituents that are contained in a given volume. If a number of hydrogen-containing gases are analyzed and the results stated in terms of the volume of gas (at standard temperature and pressure) necessary to contain 1 g of hydrogen, we can collect information such as that shown in Table 3.1. It requires, for example, 22.4 liters of hydrogen chloride to contain 1 g of hydrogen, but only $\frac{1}{6}$ of that volume of ether contains the same amount of hydrogen. The examination of a large number of compounds shows that it does not require more than 22.4 liters of any hydrogen-containing gas to produce 1 g of hydrogen. That is, in compounds such as hydrogen chloride, the hydrogen is spread as thinly as possible—one atom per molecule. It requires only $\frac{1}{2} \times 22.4$ liters of water vapor or hydrogen gas to contain 1 g of hydrogen. Since equal volumes of gas contain the same number of molecules, we conclude that there must be two atoms of hydrogen in every water molecule and in every molecule of hydrogen gas.

Table **3.1** *Amounts of Hydrogen in Various Gases*

Gas or Vapor	Mass of 1 liter [a] at Standard Temperature and Pressure (grams)	Volume of gas Required to Contain 1 g of Hydrogen (liters)
Hydrogen chloride	1.63	22.4
Hydrogen bromide	3.62	22.4
Hydrogen gas	0.09	$\frac{1}{2} \times 22.4$
Water	0.805	$\frac{1}{2} \times 22.4$
Formaldehyde	1.34	$\frac{1}{2} \times 22.4$
Ammonia	0.76	$\frac{1}{3} \times 22.4$
Methyl alcohol	1.43	$\frac{1}{4} \times 22.4$
Ethyl alcohol	2.05	$\frac{1}{6} \times 22.4$
Ether	2.05	$\frac{1}{6} \times 22.4$

[a] 1 liter $= 10^3$ cm^3

Having established that the lightest element, hydrogen, exists in the gaseous state as a *diatomic* (*two*-atom) molecule, we can now proceed to build up a system of relative atomic (or molecular) masses for all elements and compounds.

ATOMIC AND MOLECULAR MASSES[12]

The way in which relative atomic and molecular masses are determined is exemplified by the case of hydrogen and oxygen. From the discussions

[12] It is standard practice in chemistry to refer to atomic and molecular *weights* instead of *masses*. We shall use the more precise term but the distinction is not important for the purposes here.

above we know the following:

1. 1 g of hydrogen and 8 g of oxygen combine to form 9 g of water.
2. 2 liters of hydrogen and 1 liter of oxygen combine to form 2 liters of water vapor.
3. Equal volumes of gas contain the same number of molecules.
4. Hydrogen gas consists of diatomic molecules (symbol, H_2).

From this information we can draw the following conclusions (convince yourself that these statements actually follow from the above facts):

1. A molecule of water contains two atoms of hydrogen and one atom of oxygen (water = H_2O).
2. Oxygen gas consists of diatomic molecules (O_2).
3. The atomic mass of oxygen is 16 times that of hydrogen. (The *molecular* mass of oxygen is also 16 times that of hydrogen since both gases are in the form of diatomic molecules).
4. The molecular mass of water is 18 times the atomic mass of hydrogen.

Measurements of the types we have described for hydrogen, oxygen, and water can be made for other elements and compounds so that a system of relative atomic and molecular masses can be established. Modern (and more direct) techniques have been used to increase the precision with chemical measurements. Because oxygen combines readily with so many other ele-

Table **3.2** *Atomic Masses of Some Elements*[a]

Element	Symbol	AMU [b]
Hydrogen	H	1.00797
Helium	He	4.0026
Lithium	Li	6.939
Carbon	C	12.01115
Nitrogen	N	14.0067
Oxygen	O	15.9994
Neon	Ne	20.183
Sodium	Na	22.9898
Aluminum	Al	26.9815
Chlorine	Cl	35.453
Iron	Fe	55.847
Bromine	Br	79.909
Silver	Ag	107.870
Gold	Au	196.967
Lead	Pb	207.19
Uranium	U	238.03

[a] See inside front cover for a complete list.

[b] Based on carbon-12 (C^{12}) = 12 (exactly); see Section 3.4 for a discussion of *isotopes*. The values given are for the naturally occurring isotopic mixtures. Therefore, the mass of carbon is slightly greater than 12 due to the presence in Nature of 1.1 percent of the isotope C^{13}, and the mass of hydrogen is slightly greater than 1.007825 due to the presence of 0.015 percent of deuterium (H^2).

Table **3.3** *Approximate Molecular Masses of Some Compounds*

Compound	Symbol	Molecular Mass (AMU)
Water	H_2O	18
Hydrogen chloride	HCl	36.5
Hydrogen bromide	HBr	81
Ammonia	NH_3	17
Methyl alcohol	CH_3OH	32
Butane	C_4H_{10}	58
Proteins	(various)	$\sim 10^8$

ments and is therefore found in a large number of compounds, this element has been used as the basis for a scale of relative atomic and molecular masses. Accordingly, the mass of the oxygen atom was defined to be exactly 16 *atomic mass units* (AMU) and all other atomic and molecular masses were determined relative to oxygen.[13] Thus, the lightest element, hydrogen, has a mass of approximately 1 AMU (actually, 1.007825 AMU). Some atomic and molecular masses are given in Tables 3.2 and 3.3.

A variety of experimental techniques have been used to determine the connection between the AMU and the gram:

$$1 \text{ AMU} = 1.6605 \times 10^{-24} \text{ g} \tag{3.7}$$

and the mass of the hydrogen atom is then

$$1.007825 \times (1.6605 \times 10^{-24} \text{ g}) = 1.6735 \times 10^{-24} \text{ g}$$

The conversion factor between the AMU and the gram is not yet known with sufficient precision to permit the use of an atomic standard for all mass determinations. However, *relative* atomic masses can be measured with considerable precision. Therefore, we continue to use the arbitrary standard of the *gram* for all macroscopic (that is, large-scale) mass measurements, but we use the relative AMU scale for atomic and molecular measurements.

AVOGADRO'S NUMBER

A *mole* of a substance (element or compound) is defined to be an amount that has a mass in grams equal to the molecular mass of the substance expressed in AMU. Thus, a mole of hydrogen has a mass of 2 g, a mole of oxygen has a mass of 32 g, and a mole of water has a mass of 18 g. A mole of any substance therefore contains *exactly the same number of molecules* as a mole of any other substance. This number is called *Avogadro's number*:

[13] In 1961 it was internationally agreed that carbon-12 (C^{12}) would replace oxygen as the standard. The reason for this change was to bring into closer agreement the chemical scale (which had been based on the mass of the naturally-occurring isotopic mixture of oxygen) and the physical scale (which had been based on the mass of the O^{16} isotope). Atomic masses are now measured on a scale that has $C^{12} = 12$ (exactly). This change of standard caused only a small change in the way the mass of oxygen is listed (from 16 to 15.9994 AMU).

$$\boxed{\begin{aligned}\text{Avogadro's number} = N_0 &= \text{no. of molecules in 1 mole}\\ &= 6.022 \times 10^{23} \text{ molecules per mole}\end{aligned}}$$
(3.8)

The number of molecules N in a sample of mass M of a substance that has a molecular mass m is therefore given by

$$N = \frac{M}{m} N_0$$
(3.9)

where M/m is just the fraction of a mole present in the sample.

THE REALITY OF ATOMS

Although there are undeniable arguments based on chemical measurements in favor of the atomic theory of matter, how can we be *certain* that atoms exist? Since atoms are too small to be visible with even the most powerful microscopes,[14] is there any way in which we can obtain additional evidence for the existence of atoms?

In 1827 a Scottish physician, Robert Brown (1773–1858), used a microscope to examine some minute grains of pollen (similar to the pollen grain shown in Fig. 3.21) that were suspended in a drop of liquid. Brown observed that these pollen grains were not stationary in the liquid but were continually in a state of motion. Each grain followed a zigzag path, independent of the motion of neighboring grains. This type of continual, random motion of microscopic, inanimate objects is called *Brownian motion*.

Brown's discovery, similar to Avogadro's hypothesis, went unnoticed. Considerably later it was realized that the Brownian motion of tiny objects is the direct result of the impact of molecules in the suspension medium. The atoms and molecules of all materials are in a state of continual motion (thermal motion) and, although we cannot see these atoms and molecules, we can observe the effects of their collisions with objects of intermediate

[14] Some very large and complex molecules can be "seen" by using electron microscopes.

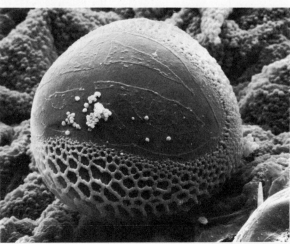

Fig. 3.21 *Micrograph of a pollen grain from Lilium Longiflorum (enlarged approximately 750 times) obtained with an electron microscope. Such pollen grains, if suspended in water and examined with a microscope, can be observed to move about due to the impacts of water molecules (Brownian motion). The "diameter" of a water molecule is about 10^{-6} of that of the pollen grain.*

Courtesy of Prof. J. Heslop-Harrison, University of Wisconsin

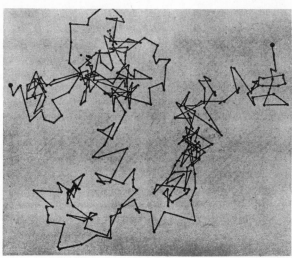

After J. Perrin

Fig. 3.22 *The successive positions of a smoke particle suspended in air, recorded at one-minute intervals. The zigzag motion is characteristic of a particle undergoing Brownian motion.*

size such as pollen grains or smoke particles. Brownian motion is therefore a clear indication of the existence of units of matter that are far too small to be observed directly.

The mathematical theory of Brownian motion based on molecular collisions was formulated by Albert Einstein in 1905. In 1908 the French physicist Jean Perrin (1870–1942) studied the motion of smoke particles (diameter $\sim 10^{-4}$ cm) suspended in air and plotted the positions of particular particles at regular intervals of time (see Fig. 3.22). Perrin's thorough investigation completely verified Einstein's theory and earned for Perrin the 1926 Nobel Prize in physics. It is interesting to note that Perrin's measurements, when interpreted with Einstein's theory, allowed an independent calculation of Avogadro's number. The value obtained from Perrin's experiment is in excellent agreement with that derived later from precise electrical measurements.

ATOMIC AND MOLECULAR DIMENSIONS

If atoms and molecules are so small that they cannot be observed, how can we learn about atomic and molecular sizes? A crude estimate of molecular size can be obtained in the following way. Suppose that we take a droplet of oil and, carefully depositing it on the surface of an expanse of still water, allow the oil to spread out. Oil and water do not mix; because the oil is less dense than the water, the oil will remain on the surface. We discover that a given amount of oil will always spread out over a certain area, and no more. If we attempt to force the oil to cover a greater area, we only tear holes in the oil film. The simplest explanation of this fact is that the oil spreads out until a film *one molecule thick* has been formed. If we know the volume of oil that was used and if we measure the area of the film on the water, we can readily calculate the thickness of the film because the volume of the film must equal the original volume of oil. Experiments of this type show that oil molecules are $\sim 10^{-7}$ cm in length; since the oil molecule contains many atoms, the dimensions of an individual atom are $\sim 10^{-8}$ cm (see Problem 3.23).

There is another way to estimate molecular size. We know that the mass of one mole of water is 18 g and since (liquid) water has a density of 1 g/cm^3, a mole of water will occupy a volume $V = 18$ cm^3. Therefore, in 18 cm^3 of water there are N_0 molecules and in 1 cm^3 the number of molecules is

$$N = \frac{N_0}{V} = \frac{6.02 \times 10^{23} \text{ molecules/mole}}{18 \text{ cm}^3/\text{mole}} \cong 3 \times 10^{22} \text{ molecules/cm}^3$$

The volume occupied by a single molecule is just the inverse of this number:

$$\frac{1}{N} = \frac{1}{3 \times 10^{22}} \cong 0.3 \times 10^{-22} \text{ cm}^3/\text{molecule}$$

If we consider each molecule to occupy a cubical volume, the length of the side of this cube will be the cube root of $1/N$:

$$d = \left(\frac{1}{N}\right)^{1/3} = (30 \times 10^{-24} \text{ cm}^3)^{1/3}$$

$$\cong 3 \times 10^{-8} \text{ cm}$$

Thus, the size of a water molecule is a few times 10^{-8} cm. A water molecule bears approximately the same relationship to 1 cm^3 as the solar system does to our entire Galaxy!

These methods give only estimates of atomic and molecular sizes. For example, $1/N$ is the volume *available* to each water molecule, but it is not necessary that each molecule *fill* such a volume—the molecule could actually be much smaller. By using modern techniques (for example, X-ray methods), it has been possible to verify that atoms have dimensions of approximately 10^{-8} cm. (But recall that, in accordance with the uncertainty principle, atoms have no "exact" size.) In fact, X-ray and spectroscopic techniques have allowed us to "map" the positions of atoms within molecules. For many molecules it has been possible to determine the spacings between atoms and the angles between the various inter-atomic directions, as in the case of the molecule of methyl alcohol, CH_3OH, illustrated in Fig. 3.23. Of course, the molecule is not a rigid structure; the distances between the atoms and their angular orientations are not fixed quantities. But by making many measure-

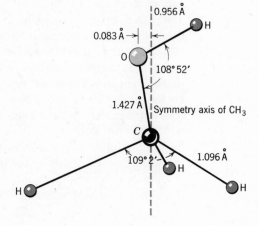

Fig. 3.23 *Structure of the molecule of methyl alcohol. Distances are measured in Ångstroms (1Å = 10⁻⁸ cm). The "sizes" of the atoms have been reduced in order to emphasize their positions in the molecule. (From P. Venketaswarlu and W. Gordy, J. Chem. Phys. 23, 1200, 1955.)*

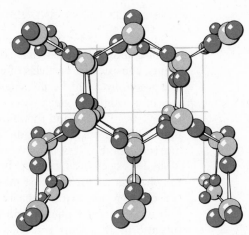

Fig. 3.24 *The crystal structure of ice. The black spheres represent oxygen atoms and the white spheres represent hydrogen atoms.*

From A. M. Buswell and W. H. Rodebush, Scientific American, April, 1956

ments, the *average values* of these quantities can be determined with considerable precision.

In many substances the molecules are arranged in regular arrays and form *crystals*. Crystal structures have also been analyzed by X-ray methods; for many crystals we know precisely the average locations of the atoms and the spacings between them. One of the simplest crystals is that of common salt (see Figs. 14.18 and 14.19); ice is also a relatively simple crystal (Fig. 3.24). The crystal structure of lead carbonate, as revealed by the technique of electron microscopy, is shown in Fig. 3.25. Molecular and crystal structures will be discussed further in Chapter 14.

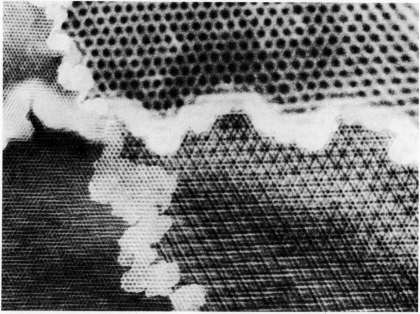

Courtesy of Siemens America, Inc.

Fig. 3.25 *The crystal structure of lead carbonate as revealed by an electron microscope. The magnification is 150,000 times.*

ATOMS, ELECTRONS, AND IONS

During the 1890s it became clear that atoms, although they were the smallest units of matter that could be identified as definite chemical elements, consisted of particles of even smaller size. In 1897 it was discovered that *electrons* could be detached from atoms and that these electrons are the basic units of negative electricity. Electrons are relatively easy to remove from atoms and therefore became the objects of a variety of experiments. Measurements made during the next few years showed that electrons are considerably less massive than atoms—even the mass of the lightest atom, hydrogen, was found to be 1837 times the mass of an electron. Today the electron is still the least massive of the material particles. As we shall see, electrons are extraordinarily active objects—they play crucial roles not only in all electrical phenomena (such as the conduction of electrical current) but also in all atomic and chemical processes.

Electrons are interesting and important in themselves, but a "free" electron exists only at the expense of leaving behind a positively-charged atom (called a *positive ion*). Some atoms can acquire an extra electron and become negatively-charged ions. These positive and negative ions have proved useful in many practical applications (as in batteries and transistors), and the study of ions has yielded much detailed information about atomic processes. In Chapters 12, 13, and 14 we shall describe some of the important results that have been obtained from investigations of microscopic phenomena involving atoms, electrons, and ions.

3.4   *Nuclei*

ATOMIC INTERIORS

By 1900 it had been established that all atoms in their natural states contain *electrons*. It was known that an electron has a small mass and carries a negative electrical charge. Scientists then wondered whether the mass of an atom was due exclusively to electrons (with the electrons' charge neutralized by a massless, positively-charged material) or whether there were other massive pieces of matter within the atoms. This question went unanswered until 1911, when Lord Rutherford completed the analysis of a series of experiments designed to study the inner structures of atoms. Rutherford found that he was able to interpret the experimental results only if the mass of an atom was considered to be concentrated almost entirely in a tiny central region of positive charge—the atomic *nucleus*.

Rutherford's discovery of atomic nuclei opened a new and exciting phase in the development of physics. Within a short time, Niels Bohr proposed an atomic model in which electrons revolve around the nucleus in much the same way that the planets revolve around the Sun. This was the first important step in the construction of a detailed model of atomic processes. Bohr's model of the hydrogen atom met with considerable success, but when the same basic model was applied to more complex atoms, formidable difficulties arose. The resolution of the difficulties with the "planetary" model of atoms led Schrödinger and Heisenberg to the formulation of the *quantum theory* of microscopic systems. This theory is the basis for all our modern descriptions of atomic and nuclear phenomena.[15]

PROTONS AND NEUTRONS

In all the atomic effects to which Bohr, Schrödinger, Heisenberg, and many others had turned their attention, the atomic nucleus appeared to play no role except as the positive charge that attracted the electrons—the nucleus was only the inert core of the active atom. The many chemical and optical effects that had been studied were solely the result of the atomic electrons. In the 1930s the nucleus itself came under detailed scrutiny. Nuclei were broken apart and their components studied. It was found that the nuclei of all atoms consist of only two fundamental types of matter. The first of these to be identified and studied was the *proton,* the nucleus of the hydrogen atom, which has a positive charge equal in magnitude to that of the electron and a mass that is 1836 times that of the electron. The second type of particle found in nuclei was the *neutron,* an object with a mass almost equal to that of the proton but without an electrical charge.

With James Chadwick's discovery of the neutron in 1932, many of the basic facts that had been learned about nuclei fell neatly into place. A chemical element is defined by the number of electrons in an electrically neutral atom of the element (a hydrogen atom has one electron, a helium atom has two, an oxygen atom has eight, and so forth). It was discovered that the nucleus of an atom contains the same number of positively-charged protons as there are electrons in the outer region of the atom. As far as we know today, the electrical charge carried by an electron is exactly equal to the charge on a proton,[16] but of opposite sign, so that an atom is electrically neutral. A hydrogen atom has one electron and one proton; a helium atom has two of each. But a helium atom is approximately *four* times as massive as a hydrogen atom. The reason is that a helium nucleus contains two *neutrons* in addition to its two protons. In fact, the nuclei of *all* atoms, except hydrogen, contain neutrons as well as protons.

The number of *protons* in an atom (or the number of *electrons* if the atom is electrically neutral) is called the *atomic number* of the particular element and is usually denoted by Z. Thus, the atomic number of hydrogen is 1, helium has $Z = 2$, oxygen has $Z = 8$, and so on. The total number of protons and neutrons in the nucleus of an atom is called the *mass number* and is denoted by A.

The proton and the neutron each have a mass of approximately 1 AMU:

$$\left. \begin{array}{l} \text{Proton mass} = m_{p} = 1.007825 \text{ AMU} \\ \text{Neutron mass} = m_{n} = 1.008665 \text{ AMU} \end{array} \right\} \tag{3.10}$$

Therefore, the mass of nucleus with mass number A is approximately equal to A AMU.

---

[15]The Rutherford experiment and the Bohr model are described in Chapter 13. Quantum theory is discussed in Chapters 12 and 13.

[16]Experimentally, we know that the two types of charge cannot differ in magnitude by more than 1 part in 10^{19}.

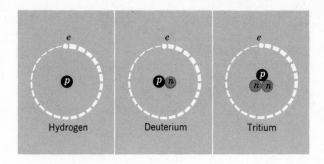

Fig. 3.26 *The three isotopes of hydrogen. Natural hydrogen consists predominantly of the* A = 1 *isotope. The* A = 3 *isotope (tritium) is unstable and undergoes radioactive decay. These simple schematic representations of atoms and nuclei are not realistic (see Sections 13.4 and 15.1). Atoms and nuclei are actually "fuzzy" objects (see Fig. 3.30).*

ISOTOPES

Nuclei of the same chemical element do not all have the same mass. Although most hydrogen atoms have nuclei that consist of a single proton, a small fraction of natural hydrogen atoms (about 0.015 percent) have one proton and one neutron in their nuclei. This "heavy hydrogen" is called *deuterium* (see Fig. 3.26). Another form of hydrogen atoms have nuclei with *two* neutrons; hydrogen with $A = 3$ is called *tritium*. The series of nuclei with a given value of Z but different values of A are called *isotopes* of the element.

Most elements have two or more stable isotopes; the average number is approximately 3, but tin has 10 isotopes.[17] Different isotopes of a given element are distinguished by using the mass number as a superscript. Thus, the stable isotopes of helium are He^3 and He^4 (Fig. 3.27) and the stable isotopes of oxygen are O^{16}, O^{17}, and O^{18}. A list of some of the isotopes of the lightest elements is given in Table 3.4.

NUCLEAR SIZES

The Rutherford experiments showed that atomic nuclei are very much smaller than atoms. How large *are* nuclei? Today we have a variety of methods to investigate nuclear sizes; one of the simplest involves the scattering of neutrons by nuclei. Let us consider a slab of material, for example, carbon, that is 1 cm thick. Because a carbon atom has a diameter of $\sim 10^{-8}$ cm, the slab will contain $\sim 10^8$ atomic layers. If the slab has a cross sectional area of 1 cm², the cube will contain approximately $(10^8)^3 = 10^{24}$

[17] There are 21 naturally occurring elements that have only a single stable isotope.

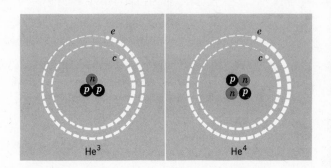

Fig. 3.27 *The stable isotopes of helium, He^3 and He^4. All other isotopes of helium are unstable. The abundance of He^3 in natural helium is only 1.5 parts per million.*

Table **3.4** *Properties of Some Light Elements*

Element	Z	A	Symbol	Remarks[a]
Hydrogen	1	1	H^1	Stable (99.985%)
	1	2	H^2 or D^2 (deuterium)	Stable (0.015%)
	1	3	H^3 or T^3 (tritium)	β Radioactive
Helium	2	3	He^3	Stable (0.00015%)
	2	4	He^4	Stable (99.99985%)
	2	6	He^6	β Radioactive
Lithium	3	6	Li^6	Stable (7.52%)
	3	7	Li^7	Stable (92.48%)
	3	8	Li^8	β Radioactive
Beryllium	4	7	Be^7	Radioactive (e capture)
	4	8	Be^8	α Radioactive
	4	9	Be^9	Stable (100%)
	4	10	Be^{10}	β Radioactive
Boron	5	10	B^{10}	Stable (18.7%)
	5	11	B^{11}	Stable (81.3%)
	5	12	B^{12}	β Radioactive

[a] The numbers in parentheses are the relative natural abundances of the isotopes.

atoms. If we were to look at the carbon cube with an imaginary super-microscope, we would "see" the atoms crowded together in the material (Fig. 3.28). Most of the atoms would be obscured by the first few atomic layers. Not so for the nuclei; by "looking through" the atoms we would be able to "see" essentially all of the 10^{24} nuclei—because of their small size, none would be hidden behind other nuclei.

The cross sectional area of each carbon nucleus is $\sigma = \pi R_C^2$, where R_C is the carbon nuclear radius. Therefore, the total cross sectional area of the carbon nuclei in the cube is approximately $10^{24}\sigma$. If we were to direct a beam of neutrons toward the carbon block, some of these neutrons would pass straight through the block because they are unaffected by the atomic electrons or by the electrical charge of the nuclei. Only those neutrons that actually strike the carbon nuclei would be deflected (Fig. 3.29). These scattered neutrons are therefore removed from the straight-through beam. The fraction of neutrons scattered out of the beam is $N_{\text{scattered}}/N_{\text{total}}$. This fraction must be the same as the fraction of the total cross sectional area of the block (1 cm^2) occupied by the carbon nuclei ($10^{24}\sigma$). That is,

$$\frac{N_{\text{scattered}}}{N_{\text{total}}} = \frac{10^{24}\sigma}{1 \text{ cm}^2}$$

or,

$$\sigma = \frac{N_{\text{scattered}}}{N_{\text{total}}} \times 10^{-24} \text{ cm}^2$$

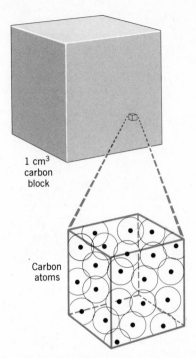

Fig. 3.28 *Even though most of the carbon atoms in the block are hidden by the first few atomic layers, the carbon nuclei are so small that we rarely find one nucleus directly behind another. There is no screening and all of the nuclei in the block can be "seen."*

When the experiment is carried out, one would find that approximately 30 percent of the incident neutrons are removed from the beam as a result of collisions with the carbon nuclei. Thus,

$$\frac{N_{\text{scattered}}}{N_{\text{total}}} \cong 0.3$$

so that

$$\sigma \cong 0.3 \times 10^{-24} \text{ cm}^2$$

Solving for the radius of the carbon nucleus, we find

$$R_C \cong 3 \times 10^{-13} \text{ cm}$$

Neutron scattering experiments of this general type have been performed with most of the naturally occurring elements as targets. It has been found that nuclear radii are given, to a good approximation, by the expression

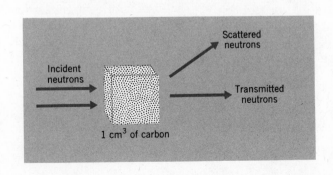

Fig. 3.29 *A beam of neutrons is directed toward the carbon block. Some neutrons proceed through the block without colliding with carbon nuclei. Those neutrons that strike carbon nuclei are deflected or scattered out of the beam.*

$$R \cong 1.4\, A^{1/3} \times 10^{-13}\ \text{cm}$$

(3.11)

where A is the mass number of the nucleus.

This expression for nuclear radii has the following meaning. Suppose that we calculate the *volume* of a nucleus by using Eq. 3.11; we find

$$V = \tfrac{4}{3}\pi R^3 \propto A$$

That is, the nuclear volume is simply proportional to the *total number* of protons and neutrons in the nucleus. The addition of protons and neutrons to a nucleus to form a new element does not squeeze the particles any tighter; each proton and neutron occupies essentially the same volume independent of the number of these particles in the nucleus.

The extreme smallness of nuclei is difficult to comprehend. If we were to magnify an atom until it was the size of the Houston Astrodome, the nucleus, located at mid-field, would be no larger than a pea! The atomic electrons would also be pea-sized objects whizzing around somewhere in the upper and lower decks. Atoms, similar to galaxies, are mostly empty space. In fact, if we could compact all of the matter in the known Universe into one huge super-nucleus (without changing the density of nuclear matter), it would fit well within the limits of the solar system.

RADIOACTIVITY

At about the time that Thomson was investigating the properties of electrons, another discovery of great importance was made by the French physicist, Henri Becquerel (1852–1908). Certain naturally occurring minerals were found by Becquerel to emit radiations of a type that had not previously been observed. Within a few years, the emanations from *radioactive* substances were classified into three groups:

1. *Alpha rays*—massive, positively-charged objects.
2. *Beta rays*—negatively-charged objects of small mass.
3. *Gamma rays*—neutral rays with no detectable mass.

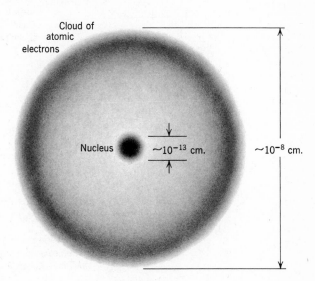

Fig. 3.30 *Schematic comparison of atomic and nuclear sizes. Atoms are about 100,000 times larger than nuclei.*

Detailed investigations of these radiations revealed that the beta rays are identical to electrons and that the alpha rays are nuclei of helium atoms. Gamma rays were found to have properties similar to light—the only difference being that the frequency of gamma rays is much higher than that of visible light.

Radioactivity is a *nuclear* phenomenon. Alpha and beta rays are emitted during the spontaneous disintegration of nuclei, and gamma rays result when the neutrons and protons within a nucleus spontaneously rearrange themselves (but without "disintegration").

The emission of an alpha ray (or α particle) by a nucleus necessarily changes both the atomic number and the mass number; that is, a new chemical element (a *daughter* element) is formed by the α decay of a *parent* nucleus. For example, when radium ($Z = 88, A = 226$) emits an α particle, radon ($Z = 86, A = 222$) is formed, as indicated schematically in Fig. 3.31. An α particle, therefore, has $Z = 2$ and $A = 4$—it is just the nucleus of a *helium* atom. Beta decay, on the other hand, does not involve the emission of a proton or neutron, so the mass number of the daughter nucleus is the same as that of the parent nucleus. But because the emitted particle carries

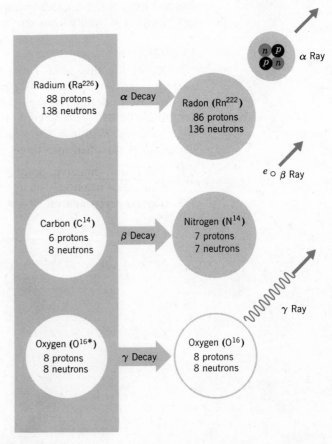

Fig. 3.31 *Examples of the three types of radioactive decay events. Alpha and beta decay involve nuclear disintegrations (i.e., changes in species) while gamma decay results from intranuclear rearrangements. The excited O^{16} nucleus that exists before γ decay takes place is indicated by $O^{16}*$.*

Table **3.5** *Some Radioactive Half-Lives*

Nucleus	Type of Decay	Half-Life
Thorium (Th^{232})	α	1.4×10^{10} y
Plutonium (Pu^{239})	α	100 y
Uranium (U^{229})	α	58 min
Carbon (C^{14})	β	5568 y
Cobalt (Co^{60})	β	5.3 y
Copper (Cu^{66})	β	5 min
Krypton (Kr^{94})	β	1.4 sec

a negative charge, the atomic number of the daughter nucleus is one unit *greater* than that of the parent. As shown in Fig. 3.31, when C^{14} ($Z = 6$, $A = 14$) decays by beta emission, N^{14} ($Z = 7, A = 14$) is formed. Beta decay is therefore equivalent to the transformation of one of the nuclear neutrons into a proton. The β particles emitted by nuclei are exactly the same as atomic *electrons*. Gamma-ray emission accompanies nuclear rearrangements that leave Z and A unaltered.

Since α and β emissions cause the parent nuclei to be transformed into different nuclear species, will not these disintegrations soon cause the complete depletion of the parent substance? Actually, this is not the case. Radioactive decay processes obey the following law: If we begin with an amount of a radioactive substance, then after a certain interval of time that is characteristic of the particular nucleus involved (called the *half-life* of the substance and denoted by $\tau_{1/2}$), *one-half* of the material will have disintegrated and one-half will remain. If we wait for another interval $\tau_{1/2}$, one-quarter of the original amount will remain. After each period of time $\tau_{1/2}$, there will remain one-half of the parent material that existed at the beginning of that time period. Radium-226, for example, has $\tau_{1/2} \cong 1600$ years; therefore, if we have 1 g of Ra^{226} to start with, after 1600 years we shall have $\frac{1}{2}$ g, after 3200 years we shall have $\frac{1}{4}$ g, after 4800 years we shall have $\frac{1}{8}$ g, and so forth (see Fig. 3.32). Thus, the amount of radium will become zero only after an infinitely long time.[18]

[18] This is mathematically true, but in radioactive decay we always deal with a countable number of nuclei. Eventually, any sample of material will be reduced to a single atom, and since a "half an atom" cannot exist, the last remaining nucleus will, at some time, decay and the original sample will be depleted to zero.

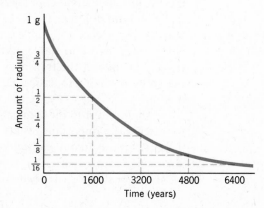

Fig. 3.32 *Radioactive decay curve of Ra^{226} ($\tau_{1/2} \cong$ 1600 y). In each interval of 1600 years, the amount of radium decreases by one-half.*

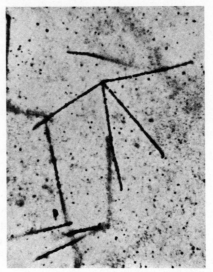

Fig. 3.33 *One method of rendering visible the path of a single nuclear particle is through the use of special photographic emulsions (called nuclear emulsions). This photomicrograph shows the tracks left by several α particles emitted in the radioactive decay of a single original thorium atom. First, thorium emits an α particle, leaving a radioactive atom; this atom emits an α particle leaving another radioactive atom; and so on. The length of the longest track is approximately 0.03 mm.*

Wills Physical Laboratory

The neutron—an essential ingredient in nuclei—is not a *stable* particle (as is the proton) when in the free state outside of nuclei. A free neutron actually undergoes radioactive β decay, forming a proton and an electron. The half-life of the neutron is approximately 12.8 min.

ELEMENTARY PARTICLES

Thus far we have discussed three fundamental units of matter—electrons, protons, and neutrons. All ordinary matter (and this includes all matter in the solar system and probably all matter in our Galaxy) consists of various combinations of just these three types of particles. But what *are* electrons, protons, and neutrons? What is the nature of these basic ingredients of matter?

In recent years, physicists have attempted to answer these questions by studying high-speed collision processes in which a fast proton (or electron or neutron) interacts with another proton or with a nucleus. As a result of these collisions, a number of types of particles of small mass have been found that are quite different from the ordinary particles found in atoms. These unusual particles are unstable and decay within very short periods of time (10^{-6} sec or less) into stable particles—electrons and protons (and gamma rays).[19] In spite of the transitory existence of these "elementary" particles, the study of their properties has yielded important clues to the nature of matter.

We shall return to the topic of elementary particles in Chapter 16 where we discuss the important results of elementary particle research and what it has taught us about the fundamental behavior of matter.

[19] Also, as we shall see, into *neutrinos*.

3.5 *The Domains of Physical Theories*

WHEN ARE OUR THEORIES VALID?

In physics we must deal with a variety of quantities, many of which have enormous ranges. *Lengths* vary from the size of elementary particles to the size of the Universe; *times* vary from the lifetimes of transitory particles to the age of the Universe; *masses* vary from the electron mass to galactic masses. How can we incorporate quantities on such vast scales into a theoretical description of the Universe? We have not been (and we may never be) successful in developing a single, all-encompassing theory that describes the variety of phenomena we encounter in microscopic events and in the macroscopic behavior of the matter in the Universe. Instead, we have constructed many different theories, each of which relates to a special area. Consequently, each of these theories is of limited validity.

Newton's famous laws of mechanics, for example, break down when an attempt is made to apply them to cases in which extremely high speeds are involved. Then, we use the *special theory of relativity*. But even this theory cannot be applied when we deal with extremely massive objects or when we attempt to interpret certain phenomena at the enormous distances of the galaxies; for such cases we invoke the *general theory of relativity*. When atomic or nuclear dimensions are involved, Newton's laws give way to *quantum theory* and to *relativistic quantum theory* when both small distances and high speeds are encountered.

How do we know when to use a particular theory? Unfortunately, there is no precise answer to this question. From experience we know that the relativity theory is certainly necessary when speeds approach the speed of light, and we know that the laws of Newtonian mechanics are adequate for the description of everyday objects if the speeds are very small compared to the speed of light. But at what point must we switch from Newtonian to relativistic mechanics? The answer, of course, depends on the specific case and on the accuracy desired for the particular problem. Nevertheless, we can give some crude guidelines to indicate the regions of validity of some of the more general theories that we currently use.

Figure 3.34 is a *distance-speed* diagram that indicates the areas in which we use five of the broader present-day theories. There is actually much overlapping in such a diagram and the separations between theory areas are meant to be schematic only. For example, although the general theory of relativity is shown as being applicable when astronomical distances are encountered, there is a crucial test of this theory that involves planetary motion and it is even possible to verify one of the predictions of the theory in the laboratory. Thus, the diagram means that *usually* we find it necessary to apply the considerations of general relativity only for astronomical distances.

The top of the diagram ends with speeds equal to the speed of light. Our present theories tell us that speeds for material particles greater than this limiting value have no physical significance. However, we cannot be absolutely certain that this is the case and, instead of being dogmatic about

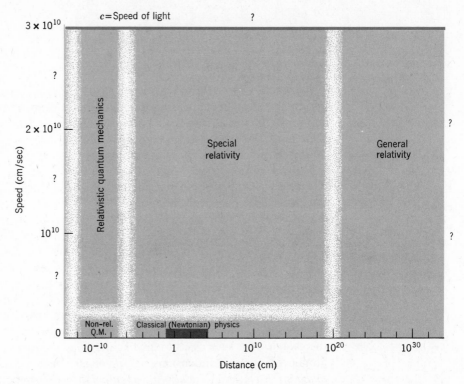

Fig. 3.34 *The distance-speed diagram shows schematically the regions of validity of our five broadest theories in physics. In the lower left-hand corner is the realm of atoms, molecules, and nuclei—nonrelativistic quantum mechanics. The region of our everyday experience is the small red line near the middle of the bottom scale.*

this point, we label the region $v > c$ with a question mark. Similarly, for distances smaller than the size of a proton and greater than the present estimate of the extent of the Universe we have no ideas of what physical theories apply or even whether it makes sense to inquire about theories for these areas.

The scales of distance and speed, which are shown in Fig. 3.34, encompass the total ranges of the quantities about which we have any knowledge in the physical Universe. It is a humbling thought to locate the region of our everyday experience in such a diagram—it is the small solid region near the middle of the bottom scale.

Summary of Important Ideas

The basic unit of length for astronomical distance measurements is the distance from the Earth to the Sun (the *astronomical unit*—A.U.).

The only *direct* method of determining the distances to stars is by the measurement of stellar *parallax*. *All other methods employ statistical techniques* based on the average properties of stars.

Stars are not all of the same *intrinsic luminosity* so measurements of relative brightness do not give reliable determinations of distance. However, stars of the same *color,* on the average, are approximately of equal luminosity and brightness measurements can be used to obtain estimates of distance.

The key to determining the distances to *galaxies* is based on the regular brightness variations of the *Cepheid variables.* The *period* of such a star is directly related to its average intrinsic luminosity.

The Universe appears to be *expanding.* All distant galaxies are receding from us with speeds that are proportional to their distances.

Distances to the farthest galaxies can be estimated by measuring the Doppler shifts (*red shifts*) in the frequencies of characteristic features of their emitted light.

Equal volumes of all gases (at the same temperature and pressure) contain equal numbers of molecules (*Avogadro's law*).

Atoms and molecules cannot be observed *directly* even with the most powerful microscopes, but certain *indirect* effects due to atoms and molecules (such as Brownian motion) are readily observable.

Atoms consist of tiny, massive cores (nuclei) surrounded by *electrons.* The average diameter of the electron portion of an atom is $\sim 10^5$ greater than that of the nuclear portion.

All nuclei (except that of hydrogen) contain *neutrons* as well as *protons.* The number of protons in the nucleus determines the atomic *species* and the number of neutrons determines the *isotope.*

Every radioactive material has a characteristic *half-life.* The amount of radioactive material in a given sample surviving at the end of a time interval equal to one half-life will always be *one-half* of the amount that existed at the beginning of that interval.

Questions

3.1 Is it possible to obtain a value for the astronomical unit by measurements made on the Earth-moon system? Explain.

3.2 Describe some of the difficulties that would be encountered in making a determination of the Earth-Sun distance by triangulation methods.

3.3 Argue that *all* of the stars that lie at the same distance from the Earth will have the same parallax, regardless of the positions of the stars relative to the plane of the Earth's orbit around the Sun.

3.4 Explain how an artificial Earth satellite could be used to make a precision measurement of the distance from New York to London.

3.5 A sidewalk "prophet" forecasts that all of the stars in the sky, except the Sun, will be extinguished tomorrow. Would you have any difficulty in checking his prediction?

3.6 Why are fewer galaxies observed along the belt of the Milky Way than in any other region of the sky?

3.7 Do you think that the Universe has always existed or that it had a *beginning*? Explain.

3.8 Was Avogadro's guess that equal volumes of gas always contain equal numbers of molecules a legitimate use of the scientific method?

3.9 Distinguish clearly, on the basis of atomic theory, the difference between *chemical* and *physical* changes in materials.

3.10 Naturally occurring chlorine has an atomic mass of 35.45 AMU. Do you expect chlorine to consist of a single isotope or to be a mixture of two or more isotopes? Explain.

3.11 List some of the uses of radioactivity that you have read about.

3.12 The age of the Earth is approximately 4.5 billion years. Do you expect any of the Th^{232} or Pu^{239} nuclei present at the time of the formation of the Earth to still be present? Explain. (See Table 3.5.)

Problems

3.1 Construct a diagram similar to Fig. 3.1 for the case of Venus. Describe how Kepler's method would be used to obtain the Venus-Sun distance in terms of the Earth-Sun distance. (See Table 6.1 for data on planetary orbits.)

3.2 Suppose that when the Earth lies directly between Mars and the Sun, a radar signal is beamed toward Mars. How long will it take before the return signal is received? (See Table 6.1 for data on planetary orbits.)

3.3 The great galaxy in Andromeda is approximately 2.2×10^6 L.Y. away. Express this distance in cm and in pc.

3.4* The star *Antares* (the brightest star in the constellation *Scorpio*) is at a distance of 52 pc. By indirect methods it has been determined that the angular diameter of Antares is 0.040 arc sec. Compare the diameter of Antares with that of the Sun. (The diameter of the Sun is approximately 1.4×10^{11} cm.)

3.5 Suppose that we wish to measure the distance from New York to San Francisco by triangulation (in one step), but that we have a base line only 10 cm long. What would be the angular difference in the sightings from the ends of the base line? Compare this value with that of the smallest measured stellar parallaxes.

3.6 If the Sun were suddenly to cease emitting light, how long would it be before we would observe the effect on Earth? What part of the Sun would appear to darken first? What would be the last portion to darken? How much time would there be between the first signs of darkening and complete darkening? (The diameter of the Sun is approximately 1.4×10^{11} cm.)

3.7 By a happy accident, *Polaris* (the North Star) is a relatively bright star and isolated from its neighbors in the sky; it serves navigators (and lost woodsmen) as an indicator of true north at any time of night and at any season of the year. Suppose that Polaris were at a distance of only 10 A.U.; would it then be useful as an indicator of true north?

3.8* A certain star has the same spectral classification as the Sun (i.e., the *color* is the same) and it is observed to have an apparent brightness that is 2.5×10^{-14} of that of the Sun. From other data it is estimated that $\frac{3}{4}$ of the light that could reach us from the star is actually absorbed by interstellar dust clouds. How far away (in pc) is the star?

3.9 Cepheid variable A has a period of 1 day and variable B has a period of 100 days. The apparent luminosity of both stars is the same. How much farther away is B than A? (Use Fig. 3.10.)

3.10 A certain cluster of galaxies is found to be receding from us with a speed of 80,000 km/sec. Express the distance to this cluster in parsecs and light years.

3.11 What would be the angular diameter of our Galaxy as measured by an observer in M31? (See Fig. 3.16.)

3.12 Assume that the galaxies have always had their present recessional speeds. At what time in the past did the Universe have "zero" size? That is, use the presently accepted value of the Hubble constant to estimate the age of the Universe.

3.13* The density of matter in interstellar space is about 1 hydrogen atom per cm^3. How many cubic light years of such space are required to contain the mass of the Sun (2×10^{33} g)?

3.14 What is the volume of 1 mole of hydrogen gas under standard conditions of temperature and pressure (STP)? (Refer to Table 3.1.)

3.15 How many molecules are there in a 120-g sample of sodium hydroxide (NaOH)?

3.16 Calculate the mass of 1 liter of carbon dioxide (CO_2) at STP.

3.17 How many molecules are there in 22.4 liters of any gas at STP? How many grams of any gas are there in 22.4 liters at STP?

3.18 How many grams of hydrogen are there in 1 kg of methyl alcohol? (Refer to Table 3.3.)

3.19 The density of water (at a temperature just below its boiling point) is 1 g/cm^3. If 1 cm^3 of water were vaporized (at a temperature just above its boiling point), what would be the volume of the vapor?

3.20 Hydrogen sulfide has the chemical formula H_2S. The mass of 22.4 liters of H_2S is approximately 34 g. What is the atomic mass of sulfur?

3.21 The density of octane (C_8H_{18}, a component of gasoline) is 0.703 g/cm^3. How many molecules of octane are there in 1 cm^3?

3.22* What is the average distance between molecules in a gas at standard temperature and pressure? (Assume that each molecule occupies an identical cubical volume and calculate the dimensions of the cube.)

3.23 In order to obtain an estimate of atomic and molecular sizes, Lord Rayleigh (1842–1919), an English physicist, performed the following experiment. He took an oil droplet (mass = 8×10^{-4} g, density = 0.9 g/cm^3) and allowed it to spread out over a water surface. The area of the film was found to be 0.55 m². Oil consists of chain-like molecules that have the property of only one end having an affinity for water. Therefore, when oil is spread on water, the oil molecules "stand on their heads." The film thickness therefore corresponds to the *length* of the oil molecule. Calculate this length from Lord Rayleigh's data. Each molecule of the type of oil used contains 16 atoms. Estimate the size of an individual atom.

3.24 The atomic masses of naturally occurring elements are sometimes quite different from an integer number of atomic mass units because the element is actually a mixture of isotopes. Copper, for example, consists of two isotopes, Cu^{63} (69.1%) and Cu^{65} (30.9%). What do you expect the atomic mass of natural copper to be (approximately)? Compare your result with that given in Table 13.24.

3.25 What is the approximate radius of the Al^{27} nucleus?

3.26 What approximate cross sectional area does a nucleus of copper (Cu^{65}) present to a neutron in a scattering experiment?

3.27 The distance between the two protons in a hydrogen molecule is 0.74Å ($1Å = 10^{-8}$ cm). If the scale of the molecule were increased until the protons had the size of a baseball, what would be the approximate separation of the protons?

3.28 What is the approximate *density* of nuclear matter? (Argue that all nuclei have approximately the same density.)

3.29 A sample of β-radioactive material is placed near a Geiger counter (a detector of β rays). The detector is found to count at a rate of 640 per sec. Eight hours later, the detector counts at a rate of 40 per sec. What is the half-life of the material?

3.30 An important method of determining the age of archeological items is by *radioactive carbon dating*. Radioactive C^{14} is produced at a uniform rate in the atmosphere by the action of cosmic rays. This C^{14} finds its way into living systems and reaches an equilibrium concentration of about 10^{-6} percent compared to normal, stable C^{12}. When the organism dies, C^{14} ceases to be taken up. Therefore, after the death of the organism, the C^{14} concentration decreases with time according to the radioactive decay law with $\tau_{1/2} = 5568$ years. An archeologist working a *dig* finds an ancient firepit containing some crude pots and bits of partially consumed firewood. In the laboratory he determines that the wood contains only 12.5 percent of the amount of C^{14} that a living sample would contain. What date does he place on the artifacts discovered in the dig?

3.31 When U^{238} (Z = 92) decays by α-particle emission, it forms another radioactive nucleus; when this nucleus decays it forms yet another radioactive nucleus, and so on. A long chain of radioactive decays is initiated by the original decay of U^{238}. In all, this chain includes 8 α decays and 6 β decays. What is the atomic number and the mass number of the end product? Identify the isotope. (Use Fig. 13.24.)

3.32 What is the speed of the Earth in its orbit around the Sun? Express the result in terms of the speed of light. Do you expect a nonrelativistic theory to be adequate to describe this motion?

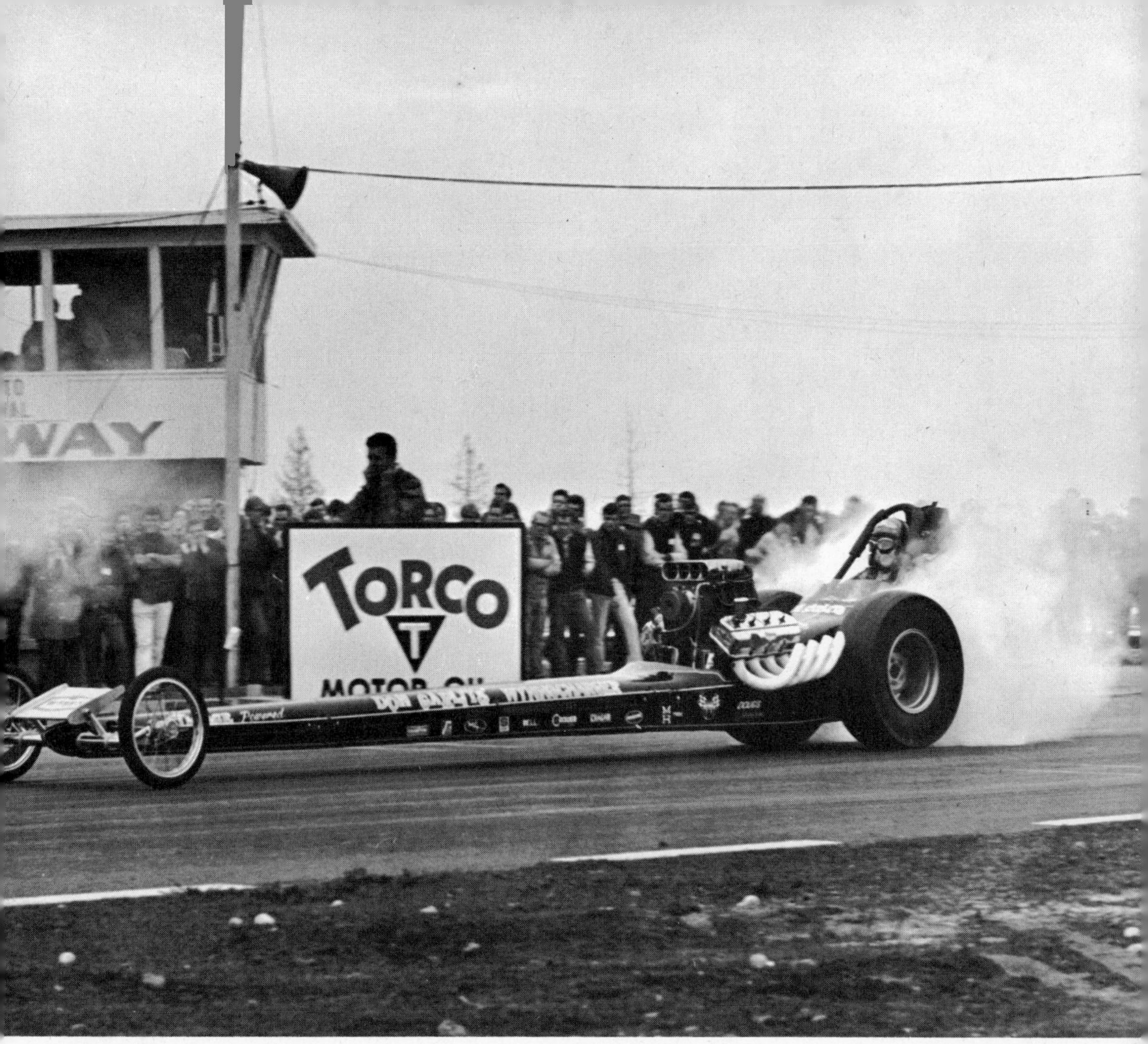

Drag Racer Courtesy Chrysler Corp.

4 *Motion*

We select the topic of *motion* for our first detailed study of the operation of basic physical principles. There are several reasons for this choice. Historically, the study of simple motions constituted the first extensive application of the scientific method to a problem of the real physical world. Through a series of ingenious experiments and well-constructed, logical arguments, Galileo Galilei (1564–1642) correctly formulated the laws governing the motion of falling bodies; he was the first to explain in detail the motion of projectiles. These advances, as trivial as they may seem today, mark the first systematic departure from the domination of "scientific" thought by the ideas of the ancient Greek philosophers, primarily Aristotle (384–322 B.C.). Although Aristotle was a precise logician and, on occasion, could even be a keen observer (as evidenced by his writings on biology), his approach to the explanation of natural happenings was basically *deductive,* without recourse to experiment. After pronouncing several elementary postulates that seemed to be in accord with everyday experience, Aristotle proceeded to draw various conclusions. His fundamental postulate regarding physical phenomena was that all matter was composed of the basic elements—Earth, Water, Air, and Fire. Each of these elements was believed to seek its "natural" position—Earth and Water, *down,* Air and Fire, *up.* Thus, a stone composed mainly of Earth, falls *down* and the larger and heavier the stone, the greater its rate of descent. Conclusions such as this were rarely, if ever, put to any quantitative experimental test. But even if

Fig. 4.1 (*left*)   Aristotle. *The teachings of this Greek philosopher dominated Western scientific thought until the Aristotelian system was overthrown by Galileo's formulation of the theory of mechanics based on experiment.*

Fig. 4.2 (*right*)   Galileo. *By basing conclusions regarding physical phenomena on experimental facts, Galileo demolished the Aristotelian system of natural philosophy and established the scientific method as the correct approach to the description of the physical world.*

an observation were at variance with an Aristotelian pronouncement, it was of no particular concern to the philosophers of that school of thought. It was clear to them that the Aristotelian doctrine was "philosophically true" because it was an integral part of a grand scheme that was intellectually and emotionally satisfying (being, in part, theological in design) and was not to be upset by the observation, here and there, of a few occurrences that violated the carefully constructed conclusions. Certainly, "observational truth" should not take precedence over "philosophical truth"! Galileo broke with this Aristotelian tradition and proceeded to base conclusions on experiment instead of on philosophy. In doing so, he became the first systematic practitioner of the modern scientific method.

The study of the motion of falling bodies and projectiles can be appreciated by appealing to only a few intuitive concepts and, as we shall see, to a single number determined from experiment. Furthermore, these basic concepts and their precise definitions will recur frequently as we proceed to study more complicated physical phenomena. Indeed, if there is one basic theme that extends throughout all areas of physics it is that of *motion*. Atoms in all forms of matter are continually in motion; the motion of electrons produces electrical current; the planets move around the Sun; and even the gigantic galaxies move through space. These facts dictate that a thorough understanding of *motion* be acquired at the outset of our survey of physics.

4.1 *Average Speed*

THE RATE OF MOVEMENT

The concept of speed is familiar to everyone; it is the rate at which something moves. The speedometer of an automobile registers the speed at which the vehicle is moving (although this instrument is frequently unreliable as any recipient of a speeding ticket will testify). If a certain automobile trip of 30 mi is to be completed in 1 hour, the speed required is, clearly, 30 mi/hr. Two points must be noted about this statement. First, no mention is made of the *direction* of travel; the trip could be made in a straight line, along a curving highway, or we could just go around the block a sufficient number of times to total 30 mi traveled. Later, we shall be concerned with the *direction* of motion, but now we consider only the simple case of *straight-line motion*. (We allow the possibility of moving in *either* direction along a given straight line.) Second, no mention is made of whether a constant speed of 30 mi/hr is maintained or whether the trip is made by stop-and-go driving. That is, the statement specifies only the *average* speed. If the first 15 mi were covered in 15 minutes (at a constant speed of 60 mi/hr) and, if because of heavy traffic, it required 45 minutes to negotiate the second 15 mi (so that the speed was 20 mi/hr for this part of the trip), the average speed for the entire 30-mi journey would still be

$$\text{average speed} = \frac{\text{total distance traveled}}{\text{total elapsed time}} \qquad (4.1)$$

or 30 m/hr. We denote speed by the symbol v and *average speed* by $\bar{v}$ (in anticipation of the later use of the term *velocity*). Also, x stands for the total distance traveled and t for the elapsed time. Therefore, in symbols, Eq. 4.1 becomes

$$\bar{v} = \frac{x}{t} \qquad\qquad (4.2)$$

In general, for any small interval of distance Δx that is traversed in the time interval Δt the average speed is given by[1]

$$\bar{v} = \frac{\Delta x}{\Delta t} \qquad\qquad (4.3)$$

Example **4.1**

In the two-part, 30-mi trip referred to above we had

15 mi at 60 mi/hr

15 mi at 20 mi/hr

Notice that the average speed is *not*

$$\bar{v} = \frac{60 \text{ mi/hr} + 20 \text{ mi/hr}}{2} = 40 \text{ mi/hr}$$

Instead, $\bar{v}$ must be calculated as follows. The time t_1 for the first part of the trip is

$$t_1 = \frac{x_1}{v_1} = \frac{15 \text{ mi}}{60 \text{ mi/hr}} = \frac{1}{4} \text{ hr}$$

and the time t_2 for the second part is

$$t_2 = \frac{x_2}{v_2} = \frac{15 \text{ mi}}{20 \text{ mi/hr}} = \frac{3}{4} \text{ hr}$$

Therefore, the actual average speed is

$$\bar{v} = \frac{x}{t} = \frac{x_1 + x_2}{t_1 + t_2}$$

$$= \frac{15 \text{ mi} + 15 \text{ mi}}{\frac{1}{4} \text{ hr} + \frac{3}{4} \text{ hr}} = \frac{30 \text{ mi}}{1 \text{ hr}} = 30 \text{ mi/hr}$$

Example **4.2**

Suppose that the first half of the distance between two points is covered at a speed $v_1 = 10$ mi/hr and that during the second half the speed is $v_2 = 40$ mi/hr. What is the average speed for the entire trip?

Again, the average speed is *not*

$$\bar{v} = \frac{10 \text{ mi/hr} + 40 \text{ mi/hr}}{2} = 25 \text{ mi/hr}$$

---

[1]The symbol Δx does *not* imply the product of Δ and x, but means "a small interval of x" or "an increment of x." In general, a Greek delta, Δ, in front of a quantity means an *increment* of that quantity (see Fig. 4.3).

Let $2x$ be the total distance traveled and let t_1 and t_2 denote the times necessary for the two parts of the trip. Then,

$$\bar{v} = \frac{2x}{t_1 + t_2}$$

$$t_1 = \frac{x}{v_1}; \quad t_2 = \frac{x}{v_2}$$

$$t_1 + t_2 = \frac{x}{v_1} + \frac{x}{v_2} = \frac{x(v_1 + v_2)}{v_1 v_2}$$

Therefore,

$$\bar{v} = \frac{2x}{\dfrac{x(v_1 + v_2)}{v_1 v_2}} = \frac{2v_1 v_2}{v_1 + v_2}$$

$$= \frac{2(10 \text{ mi/hr})(40 \text{ mi/hr})}{10 \text{ mi/hr} + 40 \text{ mi/hr}} = \frac{800}{50} \text{ mi/hr}$$

$$= 16 \text{ mi/hr}$$

4.2 Distance-Time Graphs

THE GEOMETRICAL REPRESENTATION OF VELOCITY[2]

If we know that an object that moves in a straight line was at a point labeled by x_1 at a time t_1 and at a point x_2 at a later time t_2, we can indicate these facts in a graph of position *versus* time, as in Fig. 4.3. The net distance traveled is $\Delta x = x_2 - x_1$ and the time interval during which the motion occurred is $\Delta t = t_2 - t_1$. Therefore, according to Eq. 4.3, the average velocity for the motion is the increment of distance moved divided by the corresponding increment of time:

$$\bar{v} = \frac{\Delta x}{\Delta t} = \frac{x_2 - x_1}{t_2 - t_1} \qquad (4.4)$$

[2] Here we use the term "velocity" interchangeably for "speed." In Section 4.7 we distinguish between the two terms.

Table **4.1**   *Some Typical Speeds*

Growth of hair (human head)	5×10^{-7} cm/sec
Rapidly moving glacier	3×10^{-4} cm/sec = 25 cm/day
Tip of sweep-second hand on wrist watch	10^{-1} cm/sec = 1 mm/sec
Running man	10^3 cm/sec = 10 m/sec
Racing car	7×10^3 cm/sec = 250 km/hr
Sound in air	3.3×10^4 cm/sec = 330 m/sec
X-15 rocket plane	2×10^5 cm/sec = 2 km/sec
Earth in orbit	3.0×10^6 cm/sec = 30 km/sec
Electron in hydrogen atom	2.2×10^8 cm/sec
Light in vacuum	3×10^{10} cm/sec

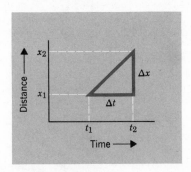

Fig. 4.3 *Distance-time graph.*

If the velocity with which an object moves is *constant*, then, clearly, the average velocity is always the same regardless of the particular time interval that is chosen for the computation. An object that moves with a constant velocity of 1 cm/sec will have traveled 1 cm after 1 sec, 5 cm after 5 sec, and 20 cm after 20 sec. The distance traveled is therefore a *linear function* of the time; that is, the graph of distance traveled *versus* time is a *straight line*. Figure 4.4 is a distance-time graph for an object moving with a constant velocity of 1 cm/sec. It is a simple matter to extract the average velocity for any given time interval from such a graph. Consider the case labeled (1) in Fig. 4.4. The time interval for this case is $\Delta t_1 = 4$ sec $- 2$ sec $= 2$ sec, and the distance interval is $\Delta x_1 = 4$ cm $- 2$ cm $= 2$ cm. The average velocity is therefore $\bar{v}_1 = \Delta x_1/\Delta t_1 = 2$ cm/2 sec $= 1$ cm/sec. Geometrically, the average velocity is just the length of the vertical dotted line (indicated by Δx_1 in Fig. 4.4) divided by the length of the horizontal dotted line (Δt_1). (Do not be confused by the fact that the lengths of these lines, as shown in Fig. 4.4, are not equal; time and distance are different physical quantities and we have chosen different numerical scales for the two axes of the graph.) Similarly, the average velocity for the case labeled (2) is $\bar{v}_2 = \Delta x_2/\Delta t_2 = 3$ cm/3 sec $= 1$ cm/sec. No matter what time interval is chosen, the average velocity will always be 1 cm/sec because the velocity is *constant*.

VELOCITY AND THE SLOPE OF THE DISTANCE-TIME GRAPH

Figure 4.5 shows three different distance-time graphs, all of which are straight lines; each corresponds to a different average velocity. As the

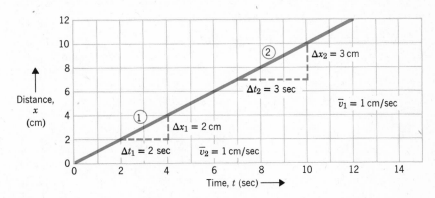

Fig. 4.4 *Graph of distance traveled* versus *time for an object moving at a constant velocity of 1 cm/sec.*

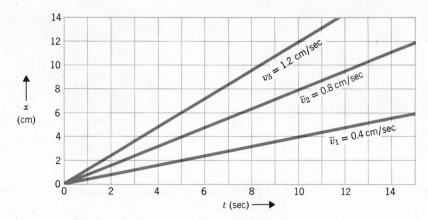

Fig. 4.5 *The greater the slope of the distance-time graph, the greater is the average velocity.*

steepness of the line increases, the average velocity becomes greater. This is a general result: *the slope of a distance-time graph determines the average velocity.*

In Fig. 4.6 we show a distance-time graph that is composed of several straight line segments. Here the average velocity will depend on the time interval chosen. But the procedure for the computation of the average velocity is exactly the same as before—the distance interval is divided by the corresponding time interval. For the first interval we obtain $\bar{v}_1 = 2$ cm/sec, whereas for the second interval we have $\bar{v}_2 = \frac{2}{3}$ cm/sec. (What is $\bar{v}$ for the interval between $t = 2$ sec and $t = 10$ sec?)

NEGATIVE VELOCITY

We have decided to limit the present discussion to the case of straight-line motion. But even in this restricted situation we always have the possibility of motion in either of the two directions along a straight line. If we elect to specify position on a straight line by marking off distances and labeling them with increasing positive numbers to the right, then motion toward the right takes place with *positive* velocity (that is, x *increases* with t). Similarly,

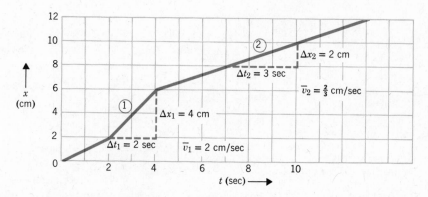

Fig. 4.6 *Distance-time graph for the motion of an object whose velocity is not constant.*

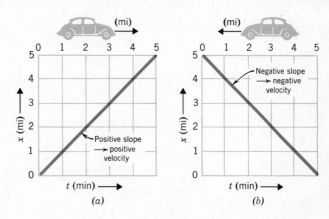

Fig. 4.7 *If distance increases to the right, motion to the right takes place with positive velocity and motion to the left takes place with negative velocity. An* upward *slope is positive and denotes a* positive *velocity; a* downward *slope is negative and denotes a negative velocity.*

motion to the left takes place with *negative* velocity (that is, *x decreases* with *t*). Notice that the identification of positive velocity with motion to the *right* is arbitrary. Once we establish a convention for labeling distance with increasing positive numbers, this direction (whether it be right or left) becomes the direction of positive velocity.

In Fig. 4.7a, an automobile moves with constant velocity from $x = 0$ to $x = 5$ mi during a period of 10 min. The velocity is

$$v_a = \frac{x_2 - x_1}{t_2 - t_1} = \frac{5 \text{ mi} - 0 \text{ mi}}{10 \text{ min} - 0 \text{ min}} = \tfrac{1}{2} \text{ mi/min} = 30 \text{ mi/hr}$$

so that $v_a > 0$. In Fig. 4.7b, the automobile starts at $x = 5$ mi and moves to $x = 0$ in 10 min. The velocity in this case is

$$v_b = \frac{x_2 - x_1}{t_2 - t_1} = \frac{0 \text{ mi} - 5 \text{ mi}}{10 \text{ min} - 0 \text{ min}} = -\tfrac{1}{2} \text{ mi/min} = -30 \text{ mi/hr}$$

so that $v_b < 0$.

In Fig. 4.7a, the distance-time graph has an *upward* slope, indicating *positive* velocity, whereas in Fig. 4.7b, the slope is *downward* indicating *negative* velocity.

Next, consider a certain round trip that is executed by automobile. Figure 4.8a shows a 20-mi round trip between points *A* and *B*. The first 2 mi were covered in 0.1 hr; the driver found himself at mile 6 after 0.2 hr; and he arrived at *B* after 0.3 hr. The driver remained at *B* for 0.1 hr before returning. He accomplished the return trip in 0.2 hr. The distance-time graph for the complete journey is shown in Fig. 4.8b. The average velocities for the 4 sections of the trip are:

$$\bar{v}_1 = \frac{2 - 0}{0.1 - 0} = \frac{2}{0.1} = 20 \text{ mi/hr}$$

$$\bar{v}_2 = \frac{10 - 2}{0.3 - 0.1} = \frac{8}{0.2} = 40 \text{ mi/hr}$$

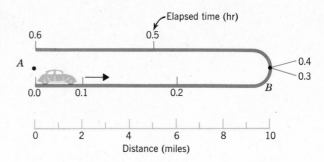

Fig. 4.8a *A 20-mile round trip between* A *and* B. *The time in hours is indicated at various points of the trip.*

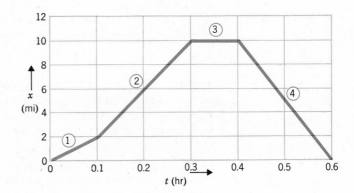

Fig. 4.8b *Distance-time graph for the trip of Fig. 4.8a.*

$$\bar{v}_3 = \frac{10 - 10}{0.4 - 0.3} = \frac{0}{0.1} = 0 \ \text{mi/hr}$$

$$\bar{v}_4 = \frac{0 - 10}{0.6 - 0.4} = \frac{-10}{0.2} = -50 \ \text{mi/hr}$$

Notice the following points:

1. The slope of the distance-time graph is steeper in interval (2) than in interval (1), and the average velocity is correspondingly greater.

2. The graph is flat (that is, it has *zero* slope) in interval (3) when the automobile was stopped. The formula for computing the average velocity automatically gives zero.

3. In interval (4) the end of the interval (at 0.6 hr) is at $x = 0$ mi, whereas the beginning (at 0.4 hr) is at $x = 10$ mi. Therefore, $\Delta x = 0 - 10 = -10$ mi, a *negative* quantity which gives a *negative* average velocity. The *downward* slope of the graph in interval (4) indicates negative velocity.

4.3   *Instantaneous Velocity*

THE LIMITING AVERAGE VELOCITY

If the motion of an object does not take place with constant velocity, then, in general, the average velocity depends on the particular time interval chosen for the calculation. For the trip shown schematically in Fig. 4.8a, if we choose the first 0.3 hr, we obtain $\bar{v} = 10$ mi/0.3 hr $= 33.33$ mi/hr, but if we choose the 0.3-hr interval from 0.1 hr to 0.4 hr, we find $\bar{v} = 8$ mi/0.3 hr $= 26.67$ mi/hr. It seems clear that it would be advantageous to have a method of specifying velocity that gives a unique answer without the necessity of always stating the time interval involved. For this purpose we need the concept of *instantaneous velocity*.

Figure 4.9 shows a distance-time graph that is *curved*. Start with point A, which corresponds to the displacement x_1 at the time t_1. If we take for the final position the displacement x_3, which occurs at time t_3 (point C), we have for the average velocity in this interval,

$$\bar{v}_{13} = \frac{x_3 - x_1}{t_3 - t_1}$$

If we next reduce the time interval to $t_2 - t_1$, we find an average velocity

$$\bar{v}_{12} = \frac{x_2 - x_1}{t_2 - t_1} > \bar{v}_{13}$$

so that $\bar{v}_{12}$ is *greater* than $\bar{v}_{13}$; that is, the slope of the line *AB* is greater than the slope of the line *AC*. If we continue to reduce the time interval (always starting at x_1), the average velocity will increase further.[3] We could, in fact, continue this process indefinitely; we could take smaller time intervals and obtain greater and greater average velocities. But if the average velocity for a very small time interval is calculated and this interval is decreased still further, very little change will be produced in the average velocity. We can therefore imagine a time interval so small that any further

[3]This *increase of* $\bar{v}$ as Δt is decreased is due to the fact that the distance-time graph has the particular curvature shown. What will happen to $\bar{v}$ as Δt is decreased if the curvature of the distance-time graph is *upward*?

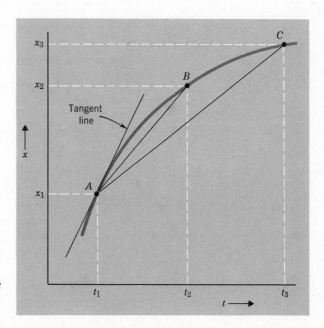

Fig. 4.9 *The average velocity between the initial point* A *and the final points* C *and* B *is greater for shorter time intervals. The* instantaneous *velocity at* A *results when the time interval is infinitesimally small and is equal to the slope of the tangent line at that point. For this case, note that* v $>$ v̄ *for* v̄ *calculated with any finite time interval starting at* t_1.

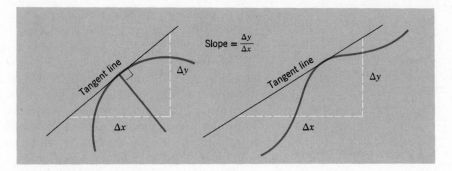

Fig. 4.10   *The line that is tangent to a curve at a certain point touches the curve only at that point. For a circle (a) the tangent line is perpendicular to the line connecting the point with the center of the circle.*

reduction will not alter the average velocity. This limiting average velocity we call the *instantaneous velocity, v.* Mathematically, we express this result as

$$v = \lim_{t_2 \to t_1} \frac{x_2 - x_1}{t_2 - t_1} = \lim_{\Delta t \to 0} \frac{\Delta x}{\Delta t} \qquad (4.5)$$

This equation states: "The instantaneous velocity v is given by the ratio of Δx to Δt in the limit that Δt approaches zero." Alternatively, "The instantaneous velocity is equal to the average velocity in the limit that the time interval becomes infinitesimally small."

Geometrically, the instantaneous velocity is equal to the slope of the line that is tangent to the distance-time graph at the point in question (see Figs. 4.9 and 4.10).

Example **4.3**

Suppose that the position of an object that moves along the x-axis is given by

$$x = 5t + 3$$

where x is in feet when t is in seconds. What is the instantaneous velocity of the object at any time?

At time t_1, the position is

$$x_1 = 5t_1 + 3$$

And at time $t_2 = t_1 + \Delta t$, the position is

$$x_2 = 5t_2 + 3$$
$$= 5(t_1 + \Delta t) + 3$$

Therefore,

$$v = \lim_{\Delta t \to 0} \frac{\Delta x}{\Delta t} = \frac{[5(t_1 + \Delta t) + 3] - [5t_1 + 3]}{\Delta t}$$

$$= \lim_{\Delta t \to 0} \frac{(5t_1 + 5\Delta t + 3) - (5t_1 + 3)}{\Delta t}$$

$$= \lim_{\Delta t \to 0} \frac{5\Delta t}{\Delta t}$$

$$= 5 \text{ ft/sec}$$

In this case, Δt cancels in the expression for v; that is, the instantaneous velocity is not a function of time and is constant (and is equal to the average velocity).

4.4 *Acceleration*

CHANGING VELOCITY

When you "step on the gas" in an automobile, you do so in order to increase the velocity, that is, you *accelerate*. When you apply the brakes, you do so in order to decrease the velocity, or you *decelerate*.[4] In either case, the essential feature of the motion is that there is a *change* of velocity.

Figure 4.11 shows a case in which the velocity increases linearly (that is, as a straight line) with the time, starting from $v = v_0$ at $t = t_0$. This is therefore a case of *accelerated* motion. Labeling the acceleration with the symbol a, we can express the average acceleration in a way analogous to that for the average velocity:

$$\boxed{\bar{a} = \frac{\Delta v}{\Delta t} = \frac{v_1 - v_0}{t_1 - t_0}} \tag{4.6}$$

The unit of acceleration must be the unit of velocity divided by the unit of time, in other words, (cm/sec)/sec, usually written as cm/sec^2.

Just as velocity is the rate of change of distance with time, $\Delta x/\Delta t$, acceleration is the rate of change of velocity with time, $\Delta v/\Delta t$. Compare Eq. 4.4 for $\bar{v}$ with Eq. 4.6 for $\bar{a}$.

[4] In physics we rarely use the term *deceleration*; instead we say that there has been a *negative acceleration*.

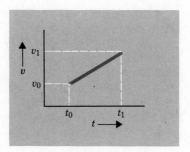

Fig. 4.11 *A case of constant acceleration; the velocity increases linearly with the time.*

UNIFORM ACCELERATION

In Fig. 4.11 the velocity-time graph is a straight line, so that $\bar{a} = a =$ constant. Although there are many physically interesting cases in which the acceleration changes with time, *we shall only consider cases in which the acceleration is constant (or uniform)*.[5]

If we displace the velocity-time graph (Fig. 4.11) to the left until the initial time occurs at $t_0 = 0$, the velocity-time graph has the form shown in Fig. 4.12. Then, the acceleration is given by[6]

$$a = \frac{\Delta v}{\Delta t} = \frac{v - v_0}{t} \tag{4.7}$$

from which

$$v - v_0 = at$$

or,

$$v = v_0 + at \tag{4.8}$$

That is, the velocity v at the time t is equal to the initial velocity v_0 plus the additional velocity acquired by virtue of the constant acceleration that acts during the time t.

Next, we seek an expression for the distance traveled in the event that there is a constant acceleration. If we consider the initial position to be at the origin of the distance scale so that $x_0 = 0$, the distance traveled is given, as always, by the product of the *average* velocity and the time:

$$x = \bar{v}t \tag{4.9}$$

Referring to Fig. 4.12, it is easy to see that the average velocity, for the case of uniform acceleration, is just the average of v_0 and v. Hence,

$$\bar{v} = \tfrac{1}{2}(v_0 + v)$$

[5] In Chapter 10 we will discuss a case in which the acceleration changes with time in a regular way.

[6] The velocity or position that one usually wishes to determine is the *final* value; we will suppress the subscripts on these values and use a subscript zero for the initial values. We will also usually set t = 0 at the initial point.

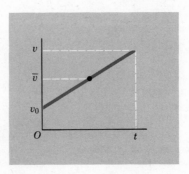

Fig. 4.12 *A velocity-time graph for motion with an initial velocity* v_0. *The average velocity during the time interval from 0 to* t *is* $\bar{v}$.

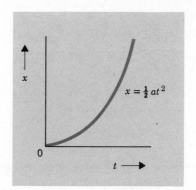

Fig. 4.13 *For the case of uniformly accelerated motion, the distance traveled is proportional to* t² *and, hence, the distance-time graph is a* parabola. (*Here,* $v_0 = 0$, $x_0 = 0$.)

Substitution of this expression for $\bar{v}$ into Eq. 4.9 gives

$$x = \tfrac{1}{2}(v_0 + v)t = \tfrac{1}{2}v_0 t + \tfrac{1}{2}vt$$

Using Eq. 4.8 for v, we have

$$x = \tfrac{1}{2}v_0 t + \tfrac{1}{2}(v_0 + at)t$$

or, finally,

$$x = v_0 t + \tfrac{1}{2}at^2 \tag{4.10}$$

That is, the distance traveled is equal to $v_0 t$ (the distance that would be traveled in the *absence* of acceleration, as we found previously) plus a term that depends on the acceleration and is proportional to the *square* of the elapsed time.

For the case of *uniform acceleration* and for the initial conditions, $x = 0$ and $v = v_0$ at $t = 0$, we can summarize the results as follows:

Acceleration:	$a = \text{const.}$
Velocity:	$v = v_0 + at$
Distance:	$x = v_0 t + \tfrac{1}{2}at^2$

(4.11)

Example **4.4**

A drag racer accelerates at a constant rate, starting from rest, and reaches a speed of 240 mi/hr in a distance of $\tfrac{1}{4}$ mi. What is the acceleration?

We have not derived a formula that allows us to calculate directly the acceleration from a knowledge of the final velocity and the distance traveled. We must therefore solve the problem in two steps, starting with a computation of the time required to go $\tfrac{1}{4}$ mi. We use

$$x = \bar{v}t$$

where $x = \tfrac{1}{4}$ mi and where the average velocity is

$$\bar{v} = \tfrac{1}{2}(v_0 + v) = \tfrac{1}{2}(0 + 240 \text{ mi/hr})$$
$$= 120 \text{ mi/hr}$$

The time required is, therefore,

$$t = \frac{x}{v} = \frac{1/4 \text{ mi}}{120 \text{ mi/hr}}$$

$$= \frac{1}{480} \text{ hr} = \frac{1}{480} \text{ hr} \times \frac{3600 \text{ sec}}{1 \text{ hr}}$$

$$= 7.5 \text{ sec}$$

Since the initial velocity is zero, the acceleration is given by

$$a = \frac{v}{t}$$

where v is the final velocity, 240 mi/hr. The result is then

$$a = \frac{240 \text{ mi/hr}}{7.5 \text{ sec}}$$

$$= 32 \text{ (mi/hr)/sec}$$

In this case we have given the result in *mixed* units, (mi/hr)/sec, instead of in mi/hr² or mi/sec² or ft/sec², because these mixed units seem to be the easiest to appreciate. There is no sense in establishing arbitrary rules regarding the use of units; we use those units that are most convenient or that convey the best impression of the result. The *physics* is never affected by the choice of units but the comprehension can suffer if a poor choice is made.

Example **4.5**

An automobile traveling at a speed of 30 mi/hr accelerates uniformly to a speed of 60 mi/hr in 10 sec. How far does the automobile travel during the time of acceleration?

$$30 \frac{\text{mi}}{\text{hr}} = 30 \frac{\text{mi}}{\text{hr}} \times \frac{5280 \text{ ft}}{1 \text{ mi}} \times \frac{1 \text{ hr}}{3600 \text{ sec}}$$

$$= 44 \text{ ft/sec (see footnote 7)}$$

$$a = \frac{\Delta v}{\Delta t} = \frac{88 \text{ ft/sec} - 44 \text{ ft/sec}}{10 \text{ sec}}$$

$$= 4.4 \text{ ft/sec}^2$$

$$x = v_0 t + \tfrac{1}{2}at^2$$
$$= (44 \text{ ft/sec}) \times (10 \text{ sec}) + \tfrac{1}{2} \times (4.4 \text{ ft/sec}^2) \times (10 \text{ sec})^2$$
$$= 440 \text{ ft} + 220 \text{ ft}$$
$$= 660 \text{ ft}$$

Suppose next that the automobile, traveling at 60 mi/hr, slows to 20 mi/hr in a period of 20 sec. What was the acceleration?

$$a = \frac{v_2 - v_1}{\Delta t} = \frac{20 \text{ mi/hr} - 60 \text{ mi/hr}}{20 \text{ sec}}$$

$$= -2 \text{ (mi/hr)/sec}$$

The automobile was *slowing down* during this period so the acceleration is *negative*.

---

[7] This is a useful relation to remember: 30 mi/hr = 44 ft/sec. Other speeds can often be converted by simple scaling; for example, 15 mi/hr = 22 ft/sec, and 120 mi/hr = 176 ft/sec.

4.5 *The Motion of Falling Bodies*

GALILEO'S EXPERIMENTS

When Galileo attacked the problem of the motion of falling bodies, he sought to find a simple relationship connecting quantities that he could measure. By dropping objects of different weights from high places (though probably not from the Tower of Pisa as legend would have it), Galileo quickly concluded that the weight of an object was not a factor in its falling motion. However, he did appreciate the fact that a larger object falls *slightly* faster than a small one. The smaller object has a larger ratio of surface area to mass and therefore experiences a greater retardation due to air friction (see Section 4.6). Galileo realized that the fact that the two objects fell at *almost* the same rate was much more significant from the standpoint of formulating the general rules of motion than is the fact that there is a *slight difference* in the motion. This approach to the study of Nature is imperative, for otherwise we would be immediately lost in a morass of detail—first, seek to explain the major fact, then consider the minor deviations. This is the scientific method at work.

Galileo began by rolling balls down inclined planes (see Fig. 4.14). In this way he was able to "dilute" the effect (gravity) that produced the motion of a freely falling body whose motion was too rapid for him to make accurate measurements. Because he lacked a clock to measure the short time intervals involved, he invented a *water clock* for the purpose. He used a large tank from which water was allowed to escape through a small pipe at the bottom. At the start of the motion to be studied, he began collecting the escaping water in a vessel and he removed the vessel from the stream at the end of the motion. By weighing the water collected he could compare the various short time intervals that were involved in his experiments to a precision of about 0.1 sec.

Galileo's hypothesis was that a falling object (or one rolling down an inclined plane) would acquire equal increments of velocity in equal intervals of time, that is, the motion would be one of *uniform acceleration*. But he could not test this hypothesis directly since to do so would necessitate measuring the velocity in several very short intervals during the motion.

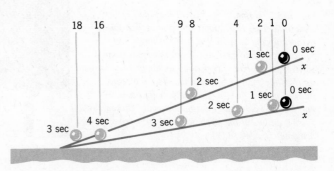

Fig. 4.14 *Galileo measured the times required for a ball to roll various distances down an inclined plane. He found that* x ∝ t² *for all angles of inclination and thereby verified that the balls were undergoing uniform acceleration.*

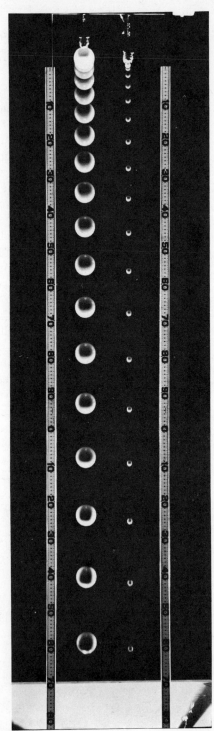

Fig. 4.15 *Two balls of unequal mass fall at the same rate. This is a stoboscopic photograph taken by opening the camera lens and flashing a light source at regular intervals. From the photograph, verify that the balls fall 4 times as far in 16 time units as they fall in 8 time units. The scales are marked in centimeters. Use the fact that* g = 980 *cm/sec² and calculate the time interval between successive flashes of the stroboscopic light. (Ans.: approximately 1/40 sec.)*

PSSC

(A moment's reflection will reveal that this is a rather difficult experiment.)

Galileo then reasoned that a body undergoing uniformly accelerated motion starting from rest would move, during any interval of time, a distance proportional to the square of that time. (In the previous section we made an algebraic calculation and found this result. This method was not available to Galileo; he used a geometric argument.) This conclusion can be tested by simple experiments because it involves measuring *distances* and times instead of *velocities* and times. Using his water clock, Galileo showed that $x \propto t^2$ for balls rolling down his inclined plane. Furthermore, he showed that this relation held for *all* the angles of inclination of the plane for which he could make measurements. By extrapolating his results (a procedure sometimes fraught with difficulties but correct in this case) to an angle of 90°, at which point the plane is no longer involved in the motion, Galileo concluded that a freely falling body obeys the same relation, namely, $x \propto t^2$, and therefore that the body undergoes uniform acceleration when falling freely.

THE ACCELERATION DUE TO GRAVITY

Present-day techniques permit the verification of Galileo's hypothesis regarding falling bodies to be made with high precision. An experiment using a stroboscopic flash is shown in Fig. 4.15. From measurements made with such techniques we find that the acceleration experienced by a body falling freely near the surface of the Earth is approximately 980 cm/sec², which is equivalent to 9.8 m/sec² or 32 ft/sec². We give this important number the symbol g:

$$
\begin{aligned}
g &= 980 \text{ cm/sec}^2 \\
&= 9.8 \text{ m/sec}^2 \\
&= 32 \text{ ft/sec}^2
\end{aligned}
\tag{4.12}
$$

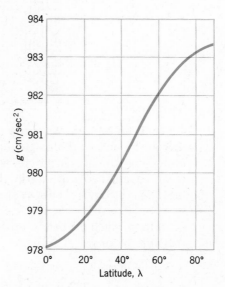

Fig. 4.16 *Variation of* g *over the surface of the Earth at sea level. The maximum variation is about 0.5%.*

Although we will always use the approximate values for g that are given above, it is of interest to note that because the Earth is not a homogeneous sphere (and also because it rotates), g varies from place to place on the surface of the Earth. At sea level, the variation as a function of latitude is shown in Fig. 4.16. The variation of g between $\lambda = 0°$ and $\lambda = 90°$ is approximately 0.5 percent.

Example **4.6**

A ball is released from rest at a certain height. What is its velocity after falling 256 ft?

Since the initial velocity is zero, we use

$y = \frac{1}{2}gt^2$

Solving for the time to fall 256 ft, we have

$$t = \sqrt{\frac{2y}{g}}$$

$$= \sqrt{\frac{2 \times 256 \text{ ft}}{(32 \text{ ft/sec}^2)}} = \sqrt{16 \text{ sec}^2} = 4 \text{ sec}$$

The velocity after 4 sec of fall is

$v = gt = (32 \text{ ft/sec}^2) \times (4 \text{ sec}) = 128 \text{ ft/sec}$

Example **4.7**

A ball is thrown upward with an initial velocity of 32 ft/sec from the top of a building. Calculate the velocity and the position as functions of the time.

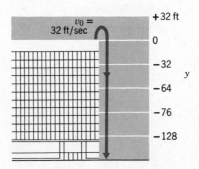

In this example we have *two* directions of motion to consider: first, the upward motion to the maximum height and then the downward motion toward the ground. Therefore, we must be careful to use the proper *signs* in our equations. We arbitrarily choose the positive direction to be *up*. Thus, the initial velocity will be $v_0 = +32$ ft/sec. But the acceleration is *downward*, so $a = -g = -32$ ft/sec². We choose the origin for distance ($y = 0$) at the point from which the ball is thrown. The equations for velocity and distance therefore become

$v = v_0 + at = (32 \text{ ft/sec}) - (32 \text{ ft/sec}^2) \times t$

$y = v_0 t + \frac{1}{2}at^2 = (32 \text{ ft/sec}) \times t - (16 \text{ ft/sec}^2) \times t^2$

From these equations we find

t(sec)	v(ft/sec)	y(ft)
0	32	0
1	0	16
2	−32	0
3	−64	−48
4	−96	−128

After 1 sec, the velocity of the ball has become zero; that is, the maximum height has been reached (16 ft) and the subsequent motion is downward. All velocities for $t > 1$ sec are therefore *negative*. At $t = 2$ sec, the ball has returned to its starting point ($y = 0$) and for all subsequent times, y is *negative*.

The diagrams below show the velocity and the distance as functions of the time.

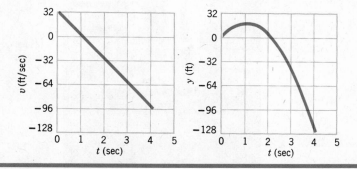

Example **4.8**

In example 4.4 the drag racer achieved an acceleration of 32 (mi/hr)/sec. Compare this value with g.

$$a = 32 \ \frac{\text{mi}}{\text{hr-sec}}$$

$$= 32 \ \frac{\text{mi}}{\text{hr}} \times \frac{1}{\text{sec}} \times \frac{5280 \ \text{ft}}{1 \ \text{mi}} \times \frac{1 \ \text{hr}}{3600 \ \text{sec}}$$

$$= 32 \times \frac{5280}{3600} \ \frac{\text{ft}}{\text{sec}^2}$$

$$= 1.46 \times 32 \ \text{ft/sec}^2$$

$$= 1.46 \ g$$

This acceleration is about the maximum that can be achieved by a vehicle that travels on wheels and depends on the friction between the wheels and the road for its thrust. Attempts to surpass this maximum value by using a more powerful engine will result merely in spinning tires. (Rocket-powered cars and sleds can, of course, achieve much greater accelerations.)

PRECISION MEASUREMENTS OF g

In Galileo's day, the value of g could be determined only very crudely. By the middle of the 19th century an accuracy of 1 part in 10^4 had been

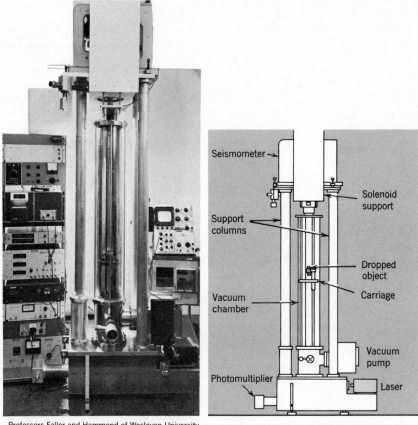

Professors Faller and Hammond of Wesleyan University

Fig. 4.17 *Photograph and schematic diagram of the apparatus used by Professors Faller and Hammond of Wesleyan University to measure g with high precision by using the technique of free fall in vacuum. A laser beam is used for distance measurements and electronic circuits provide fast timing. A seismometer is used to monitor and to correct for small earth tremors (micro-earthquakes).*

achieved and by the beginning of this century the precision had been increased by another order of magnitude. With current techniques, the motion of a freely falling body can be measured with considerable precision. Laser-beam techniques permit distance measurements to be made to better than 3×10^{-7} cm and fast electronic timing to 10^{-9} sec is possible. By applying these methods to a body falling freely in a vacuum, values of g that are precise to 1 part in 10^8 have been obtained and it appears possible to achieve an accuracy of 1 part in 10^9.

A significant increase in the precision with which g can be measured opens up the possibility of studying many interesting and important physical effects. The variations of g that occur over any small region of the Earth's surface provide an important clue to the structure of the interior of the Earth. The value of g averaged over the surface of the Earth is of considerable interest in astronomy because this value is involved in the determination of the distance to artificial satellites, to the moon, and to the planets. Also, if it were discovered that there is a long-term variation in the magnitude of *g* (distinct from the known variations associated with tidal effects), this would have a profound influence on our theories of the evolution of the Universe.

4.6 *Retarded Motion*

THE EFFECTS OF FRICTION

We have thus far been discussing motion from an *idealized* viewpoint; we have considered none of the ever-present effects that will produce deviations from the results we have derived. For example, in any real physical case, there are frictional effects that will retard the motion of a ball rolling down a plane; air resistance will retard falling objects. In addition, the value of *g* decreases with increasing height above the surface of the Earth (see Section 7.11). Thus, our results are really valid only for the case of an object falling in a vacuum through a small distance near the surface of the Earth (so that *g* is essentially constant for the motion).

The expression for the velocity of a falling object, Eq. 4.8, indicates that the velocity will continue to increase as long as the object is falling. However, in reality this is not the case; the resistance to the motion offered by the air increases with the velocity, so that for any falling object there is a limiting velocity for which there is no longer any net acceleration and the velocity ceases to increase. This effect is illustrated in Fig. 4.18.

The limiting velocity of a falling object is called its *terminal velocity* and

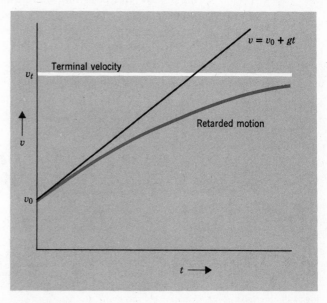

Fig. 4.18 *The motion of an object falling through the air is retarded by air resistance and its velocity does not increase linearly with the time. Instead, the velocity gradually approaches a limiting value, the* terminal velocity v_t. (*In this diagram, the velocity of fall is plotted as a* positive *quantity.*)

depends on the size and shape of the object as well as on its mass. An object that has a large surface area with which to contact the air will experience a large retardation effect. Therefore, a fluffy object, such as a loosely wadded paper ball, has a very low terminal velocity. A spherical object that has a smooth surface and a large mass-to-surface-area ratio, such as a cannon ball, will suffer only a small retardation effect and therefore will have a high terminal velocity. If a sky diver falls freely for several thousand feet before opening his parachute, he can reach a terminal velocity as high as 220 mi/hr.

The fact that a terminal velocity exists is indeed a fortunate circumstance. Otherwise a raindrop falling from a height of 10,000 ft would acquire a velocity of about 600 mi/hr by the time it reached ground level. A heavy rain storm would then be able to cause enormous damage.

4.7   Vectors

DIRECTION AND MAGNITUDE

All of the previous development has been limited to the case of motion along a straight line. We remove this restriction in further discussions and treat motion in two dimensions and eventually in three dimensions. These extensions require that we make use of the *vector property* of displacement and velocity.

If we wish to describe the motion of an automobile, we could say that the speed is 60 mi/hr. However, this is not a complete specification of the motion; more information is contained in the statement that the velocity is 60 mi/hr in the direction *northeast*. Velocity is a quantity that has both *magnitude* and *direction*. Such quantities are called *vectors*. Another such quantity is *displacement:* an object may move a certain distance but the vector description must include the *direction* of motion as well as the distance traveled. We shall encounter many other examples of vectors in later chapters: force, momentum, electric field, magnetic field, and so on.

Quantities that are completely specified by magnitude alone are called

Fig. 4.19 *Because of air friction, a sky diver will attain a maximum terminal velocity of free fall of approximately 220 mi/hr. When "spread-eagled," the terminal velocity is reduced to about 125 mi/hr.*

scalars. Mass, time, and temperature, for example, are scalar quantities. As we use the terms in physics, *speed* and *velocity* are not identical: speed is a scalar, velocity is a vector. We shall use the term "speed" when we are interested only in the rate at which an object moves and are not concerned with the direction of the motion. When we wish to convey the impression that direction as well as magnitude is important, the term "velocity" will be used.

In order to distinguish vectors from scalars, we will use boldface type for vectors. Thus, the velocity vector will be denoted by **v**. If we are interested only in the magnitude of the vector **v**, we will write this as v. (Of course, the magnitude of **v** is just the *speed*.)

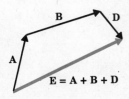

Fig. 4.20 *The addition of vectors* **A** *and* **B** *produces the vector* **C**.

SOME SIMPLE PROPERTIES OF VECTORS

We can write equations connecting vectors just as we can for scalars. Thus, the equation

$$\mathbf{A} = n\mathbf{B}$$

means that the magnitude of **A** is n times the magnitude of **B** and, furthermore, that the direction of **A** is the same as that of **B**. In some cases, n will be a scaler quantity with dimensions; then, the product $n\mathbf{B}$ will have units different from those of **B** and can represent a different physical quantity. In the next chapter we shall see that the *force* **F** acting on a body (a vector quantity) is equal to the product of the mass m of the body (a scalar) and the acceleration **a** of the body (also a vector); that is, $\mathbf{F} = m\mathbf{a}$, Newton's famous equation.

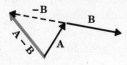

Fig. 4.21 *The addition of three vectors follows the same rule as for the addition of two vectors.*

In diagrams we represent vectors by arrows. The length of an arrow will be proportional to the magnitude of the vector and the direction of an arrow will be the direction of the vector. We will need only a few simple rules regarding the manipulation of vectors.

In Fig. 4.20 we indicate the rule for vector addition. Graphically, in order to add **A** to **B**, we place the origin of the vector **B** at the head of the vector **A**. Then, the line connecting the origin of **A** with the head of **B** is the sum vector: $\mathbf{C} = \mathbf{A} + \mathbf{B}$. More than two vectors can be added by simply continuing this procedure, as indicated in Fig. 4.21.

The negative of a certain vector **A** is another vector, $-\mathbf{A}$, which has the same magnitude as **A** but the opposite direction.

The subtraction of one vector from another is the same as adding the negative vector. Figure 4.22 shows the subtraction of **B** from **A** according to two equivalent diagrams. Notice that the vector **B** is placed in different positions in the two diagrams, but that it has the same magnitude and same direction in each case. Of course, if one vector is added to another that has the same magnitude but opposite direction, the result is zero: $\mathbf{A} + (-\mathbf{A}) = 0$.

Fig. 4.22 *The subtraction of* **B** *from* **A** *according to two equivalent diagrams. Notice that* $(\mathbf{A} - \mathbf{B}) + \mathbf{B} = \mathbf{A}.$

Just as we can add two vectors and find a single vector that represents the sum, we can also *decompose* a given vector into two vectors the sum of which yields the original vector. This process is particularly useful when the two vectors (called *component* vectors) lie at right angles to one another. We can represent this situation in an *x-y* plot, as in Fig. 4.23. Of course,

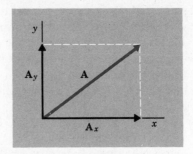

Fig. 4.23 *The vector* **A** *is decomposed into its component vectors,* $\mathbf{A}_x$ *and* $\mathbf{A}_y$: $\mathbf{A} = \mathbf{A}_x + \mathbf{A}_y$.

in such a situation we always have, according to the Pythagorean theorem of plane geometry,

$$A^2 = A_x^2 + A_y^2$$

Example **4.9**

An airplane, whose ground speed in still air is 200 mi/hr, is flying with its nose pointed due north. If there is a cross wind of 50 mi/hr in an easterly direction, what is the ground speed of the airplane?

$$v = \sqrt{v^2_{\text{airplane}} + v^2_{\text{wind}}}$$
$$v = \sqrt{(200)^2 + (50)^2}$$
$$= \sqrt{42,500}$$
$$= 206 \text{ mi/hr}$$

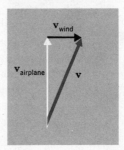

ACCELERATION IS A VECTOR

In addition to displacement and velocity, we also require for the complete description of motion, the vector property of *acceleration*. The *magnitude* of the acceleration of an object is the rate of change of the velocity, and the *direction* of the acceleration is the direction of the *change* in the velocity.

$\mathbf{v}_1 = 10 \text{ cm/sec}$
$t_1 = 0 \text{ sec}$

$\mathbf{v}_2 = 20 \text{ cm/sec}$
$t_2 = 1 \text{ sec}$

$a = 10 \text{ cm/sec}^2$

$\mathbf{v}_1 = 20 \text{ cm/sec}$
$t_1 = 0 \text{ sec}$

$\mathbf{v}_2 = 10 \text{ cm/sec}$
$t_2 = 1 \text{ sec}$

Fig. 4.24 *The* direction *of the acceleration vector is in the direction of the* change *in velocity.*

$a = -10 \text{ cm/sec}^2$

If an object is moving in a straight line in the $+x$ direction at the time $t_1 = 0$ with a velocity of 10 cm/sec and at a later time $t_2 = 1$ sec has a velocity of 20 cm/sec in the same direction, then the x-component of velocity has increased and the acceleration vector is in the $+x$ direction: $a = +10$ cm/sec². However, if the velocity at t_1 is 20 cm/sec and is 10 cm/sec at t_2, then the velocity has decreased in the $+x$ direction and therefore the acceleration vector is in the $-x$ direction: $a = -10$ cm/sec². Figure 4.24 illustrates this point.

Example **4.10**

At $t_1 = 0$ an automobile is moving eastward with a velocity of 30 mi/hr. At $t_2 = 1$ min the automobile is moving northward at the same velocity. What average acceleration has the automobile experienced?

The figure shows the velocity vectors, $\mathbf{v}_1$ and $\mathbf{v}_2$, and the vector $\Delta\mathbf{v}$ that represents the *change* in velocity. We have

initial velocity + change in velocity = final velocity

That is,

$$\mathbf{v}_1 + \Delta\mathbf{v} = \mathbf{v}_2$$

or,

$$\Delta\mathbf{v} = \mathbf{v}_2 - \mathbf{v}_1$$

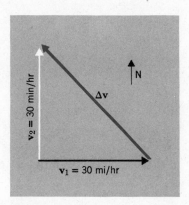

The *magnitude* of $\Delta\mathbf{v}$ is (refer to the figure and use the Pythagorean theorem)

$$\Delta v = \sqrt{(30 \text{ mi/hr})^2 + (30 \text{ mi/hr})^2}$$
$$= \sqrt{1800 \ (\text{mi/hr})^2}$$
$$= 42.4 \text{ mi/hr}$$

The magnitude of the average acceleration is

$$\bar{a} = \frac{\Delta v}{\Delta t}$$

$$= \frac{42.4 \text{ mi/hr}}{60 \text{ sec}}$$

$$= 0.71 \text{ (mi/hr)/sec}$$

The direction of $\Delta\mathbf{v}$, and hence the direction of **a** is, from the figure, in the direction *northwest*.

Before we can profitably proceed further with the discussion of vectors, we must introduce some of the essentials of trigonometry.

4.8   *Some Simple Aspects of Trigonometry*[8]

SINES AND COSINES

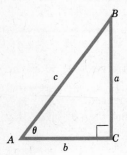

Fig. 4.25   *The angle* $\angle ACB$ *of the triangle is 90°.*

Consider the triangle (called a *right triangle*) shown in Fig. 4.25, in which the angle $\angle \overline{ACB}$ is a right angle, that is, 90°. The lengths of the sides opposite the vertices A, B, and C are, respectively, a, b, and c. From the Pythagorean theorem,

$$c^2 = a^2 + b^2 \tag{4.13}$$

We label by θ the angle $\angle \overline{BAC}$. The ratio of the length of the side opposite θ to the length of the hypotenuse is called the *sine* of the angle θ. We abbreviate this quantity as $\sin \theta$:

$$\sin \theta = \frac{a}{c} \tag{4.14}$$

Similarly, we define the ratio b/c to be the *cosine* of θ:

$$\cos \theta = \frac{b}{c} \tag{4.15}$$

Another quantity of interest is the *tangent* of θ:

$$\tan \theta = \frac{a}{b} \tag{4.16}$$

The tangent function is not independent of the sine and the cosine since

$$\tan \theta = \frac{\sin \theta}{\cos \theta} = \frac{a/c}{b/c} = \frac{a}{b}$$

The relationship between the sine and the cosine can be found by writing, from the defining equations,

$$c \sin \theta = a$$
$$c \cos \theta = b$$

If we *square* these equations and add them, we find[9]

---

[8] A convenient summary will be found inside the back cover.

[9] It is customary to write, for example, $(\sin \theta)^2$ as $\sin^2 \theta$. Remember that this is just a notational device and do not read into it any extra meaning.

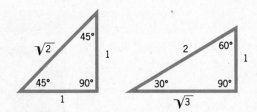

Fig. 4.26 *Two important triangles. The relative lengths of the various sides are indicated.*

$$c^2 \sin^2 \theta = a^2$$
$$c^2 \cos^2 \theta = b^2$$
$$\overline{c^2(\sin^2 \theta + \cos^2 \theta) = a^2 + b^2}$$

Comparing this result with Eq. 4.13, we see that

$$\boxed{\sin^2 \theta + \cos^2 \theta = 1} \tag{4.17}$$

If θ is near zero, then approximate expressions for $\sin \theta$, $\cos \theta$, and $\tan \theta$ are

$$\left. \begin{aligned} \sin \theta &\cong \theta \\ \cos \theta &\cong 1 \\ \tan \theta &\cong \theta \end{aligned} \right\} (\theta \text{ near zero}) \tag{4.18}$$

where θ on the right-hand sides of these equations is measured in *radians*.

Two triangles of particular importance, the 45°–45°–90° and 30°–60°–90° triangles, are shown in Fig. 4.26. Table 4.2 shows the values of the trigonometric functions for all of the angles involved plus the case $\theta = 0°$.

COMPONENTS OF A VECTOR

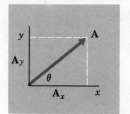

By using the basic rules of trigonometry, we can now express any vector in terms of components lying along the x- and y-axes:

$$\left. \begin{aligned} A_x &= A \cos \theta \\ A_y &= A \sin \theta \end{aligned} \right\} \tag{4.19}$$

Fig. 4.27 *Components of the vector* **A.**

Furthermore, if we know the components, we can always find the magnitude and direction of the composite vector. Referring to Fig. 4.27, if we

Table **4.2** *Values of Some Trigonometric Functions*

	$\theta = 0°$	$\theta = 30°$	$\theta = 45°$	$\theta = 60°$	$\theta = 90°$
$\sin \theta$	0	0.5	$\dfrac{1}{\sqrt{2}} = 0.707$	$\dfrac{\sqrt{3}}{2} = 0.866$	1
$\cos \theta$	1	$\dfrac{\sqrt{3}}{2} = 0.866$	$\dfrac{1}{\sqrt{2}} = 0.707$	0.5	0
$\tan \theta$	0	$\dfrac{1}{\sqrt{3}} = 0.577$	1	$\sqrt{3} = 1.732$	∞^a

[a] This symbol means "indefinitely large" or "infinite." Any non-zero number divided by zero is "infinite"; $\frac{1}{0} = \infty$.

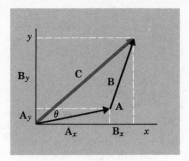

Fig. 4.28 $C = A + B$ *by components.*

know only A_x and A_y, then

$$A = \sqrt{A_x^2 + A_y^2} \tag{4.20}$$

We can also state: "The angle that **A** makes with the *x*-axis is that angle whose tangent is A_y/A_x." In symbols, this statement is written as

$$\theta = \tan^{-1}(A_y/A_x) \tag{4.21}$$

The exponent -1 on the tangent function does *not* refer to the reciprocal; this is merely a notational device that expresses the statement in quotation marks above. The equivalent statement is $\tan \theta = A_y/A_x$.

The sum of two vectors can be determined by adding the *x*-components and *y*-components; the combination of these sum components then defines the sum vector. In Fig. 4.28, the components of the vector **C** are

$$C_x = A_x + B_x$$
$$C_y = A_y + B_y$$

The magnitude and direction of **C** can be determined by using Eqs. 4.20 and 4.21 as follows:

$$C = \sqrt{(A_x + B_x)^2 + (A_y + B_y)^2}$$
$$\theta = \tan^{-1}\left(\frac{A_y + B_y}{A_x + B_x}\right)$$

Example **4.11**

A certain boat can move at a speed of 10 mi/hr in still water. The helmsman steers straight across a river in which the current is 4 mi/hr. What is his course and what is the velocity of the boat?

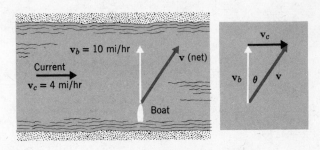

The speed of the boat is

$$v = \sqrt{v_b^2 + v_c^2}$$
$$= \sqrt{(10)^2 + (4)^2} = \sqrt{116}$$
$$= 10.8 \text{ mi/hr}$$

The angle θ, which determines the direction of the velocity, is

$$\theta = \tan^{-1}\left(\frac{v_c}{v_b}\right)$$
$$= \tan^{-1}\left(\frac{4}{10}\right)$$
$$= 22°$$

4.9 *Motion in Two Dimensions*

SEPARATION INTO VECTOR COMPONENTS

If we drop an object from a certain height, we know that the object will undergo accelerated motion straight downward. What will happen if, just as we drop the object, we also give it some initial velocity (v_{ox}) in the *horizontal* (x) direction? Clearly, the motion will no longer be straight downward but will be at some angle to the vertical. Now, we know that velocity is a vector quantity, so we can decompose the velocity vector in this case into a vertical component and a horizontal component. What equations describe the variation with time of these components? If we chose the upward direction as the direction of positive y, the acceleration due to gravity g acts only in the $-y$ direction. Therefore, the y-component of the velocity is

$$v_y = -gt$$

Since there is no horizontal component of the acceleration, the x-motion is simply

$$v_x = v_{ox}$$

These two equations are summarized by the important statement that *the resultant velocity vector consists of two components which act independently.* Only the vertical component of the motion undergoes acceleration, while the horizontal component proceeds at the constant initial velocity v_{ox}.

Figure 4.29 shows the way in which the horizontal and vertical velocity components combine to give the instantaneous velocity vector **v**. Figure 4.30 is a stroboscopic photograph of two balls that are dropped simultaneously, one with a horizontal velocity component. The picture reveals that the vertical motions are indeed identical.

If we also allow the vertical motion to have an initial velocity v_{oy}, then the equations that describe the motion are:

Acceleration:	$a_x = 0$	$a_y = -g$
Velocity:	$v_x = v_{ox}$	$v_y = v_{oy} - gt$
Displacement:	$x = v_{ox}t$	$y = v_{oy}t - \frac{1}{2}gt^2$

$$(4.22)$$

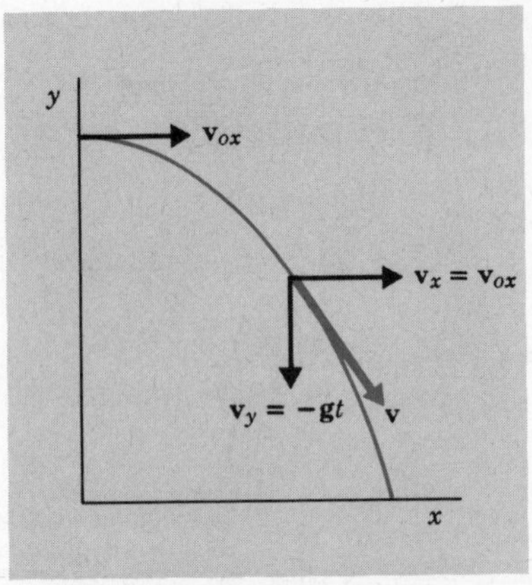

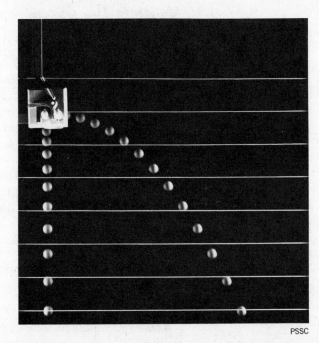

Fig. 4.29 *If an object is dropped and simultaneously given an initial horizontal velocity* $\mathbf{v}_{ox}$, *this horizontal velocity component remains constant while the vertical component increases linearly with the time. Thus, the motion follows a curved (actually,* parabolic) *path.*

Fig. 4.30 *The two balls were released simultaneously; the one on the left was merely dropped while the other was given an initial horizontal velocity. The vertical components of the motion of both balls are exactly the same. The stroboscopic photograph was taken with a flash interval of 1/30 sec. The distance between the horizontal bars is 15 cm. From the photograph verify that* $g = 980$ cm/sec^2. *What is the horizontal component of the velocity of the projected ball?*

PSSC

Example **4.12**

An object is released from a height of 64 ft and is given an initial horizontal velocity of 20 ft/sec. When the object strikes the ground, how far has it traveled in the horizontal direction?

Since $v_{oy} = 0$, we have

$$y = -\tfrac{1}{2}gt^2$$

The final value of y is -64 ft, so

$$-64 = -16t^2$$

Thus,

$$t = \sqrt{\frac{64}{16}} = \sqrt{4} = 2 \text{ sec}$$

Therefore,

$$x = v_{ox}t$$
$$= 20 \text{ ft/sec} \times 2 \text{ sec}$$
$$= 40 \text{ ft}$$

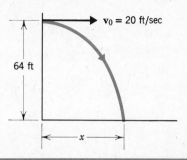

There is no conceptual difficulty in making the extension from motion in two dimensions to motion in three dimensions. However, because the geometry and trigonometry become more complicated we confine our attention to the case of two-dimensional motion. We sacrifice no new physics by this simplifying restriction.

4.10 *Projectile Motion*

A PRACTICAL APPLICATION

Because of the difficulty of following the motion of arrows or cannon balls in flight, the pre-Galilean conception of projectile motion was vague and inaccurate. One school of thought contended that the projectile rose to its maximum height and then fell straight downward (see Fig. 4.31). Galileo approached the problem as part of his study of mechanics and correctly analyzed the situation in detail. He did not have the advantage of such mathematical tools as calculus, or even algebra, but he made great use of

Fig. 4.31 *Medieval military engineers thought that a projectile rose in a straight line to its maximum height and then fell straight downward. Even with this faulty view of projectile motion there must have been an appreciation of the fact that maximum range is attained for an elevation angle of 45° since the cannon in the diagram is set in this position.*

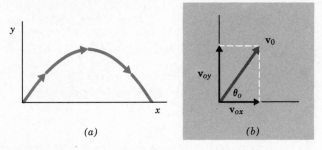

Fig. 4.32 (a) *The motion of a projectile.* (b) *The components of the initial velocity vector of the projectile.*

geometry and he performed extensive calculations to document his case. With the development we have made thus far, it is a simple matter to set down the important equations governing projectile motion.

In Fig. 4.32 we schematically show the path followed by a projectile fired at an angle θ_o with respect to the horizontal. The velocity vector of the initial motion v_o has x- and y- components, v_{ox} and v_{oy}, which are related to v_o by

$$\left. \begin{array}{l} v_{ox} = v_o \cos \theta_o \\ v_{oy} = v_o \sin \theta_o \end{array} \right\} \tag{4.23}$$

The subsequent motion is described by inserting these expressions into Eqs. 4.22:

$$\begin{aligned} v_x &= v_{ox} \\ &= v_o \cos \theta_o \end{aligned} \tag{4.24a}$$

$$\begin{aligned} v_y &= v_{oy} - gt \\ &= v_o \sin \theta_o - gt \end{aligned} \tag{4.24b}$$

Example **4.13**

A rifle is pointed at an angle of 45° above the horizontal and fires a bullet with a muzzle velocity of 1000 ft/sec. To what height does the bullet rise and what is its horizontal range?

First, we calculate the velocity components:

$$\begin{aligned} v_{ox} &= v_o \cos \theta_o \\ &= 1000 \cos 45° \\ &= 1000 \times 0.707 \\ &= 707 \text{ ft/sec} \end{aligned}$$

$$v_{oy} = v_o \sin \theta_o = 707 \text{ ft/sec}$$

In order to find the height to which the bullet rises, we calculate the value of y when the upward component of velocity has decreased to zero, that is, $v_y = 0$. First, we need the time required to reach the maximum height.

In the equation

$$v_y = v_{oy} - gt$$

we set $v_y = 0$ because this is the condition for maximum height. Then, solving for t,

$$t = \frac{v_{oy}}{g} = \frac{707}{32} = 22.1 \text{ sec}$$

Next, we use

$$y = v_{oy}t - \tfrac{1}{2}gt^2$$

Denoting by h the maximum value of y,

$$h = 707 \times 22.1 - 16(22.1)^2$$
$$= 15{,}600 - 7{,}800$$
$$= 7{,}800 \text{ ft}$$

In order to find the horizontal range R we make use of the fact that the total flight requires a time T that is *twice* the time that is necessary to reach maximum height; that is, $T = 2t = 44.2$ sec. Therefore, using

$$x = v_{ox}T$$

we find

$$R = 707 \times 44.2 = 31{,}200 \text{ ft.}$$

Thus, the range is about 6 mi. Of course, in a real physical case, the effect of air resistance will shorten the range; the degree of shortening depends on the size and shape of the projectile as well as on its mass-to-surface-area ratio. A rifle bullet will have a range of only about 1 mi but an artillery shell fired with the same muzzle velocity has a maximum range much closer to the theoretical limit we have derived.

Note that the range R is just equal to 4 times the height h. This is in fact a general result; when the elevation angle is 45° (which is the case for maximum range), we shall always have $R = 4h$ (see Problem 4.27).

THE RANGE OF A PROJECTILE

In the preceding example we found the range R by carrying through a specific numerical calculation. We can now derive a general algebraic expression for R.

The time required for the total flight of the projectile is $T = 2t$, where t is the time from firing until maximum height is reached and, according to one of the results in Example 4.13, is equal to v_{oy}/g. Thus,

$$\begin{aligned} T &= 2t \\ &= 2 \times \frac{v_{oy}}{g} \\ &= \frac{2v_o \sin \theta_o}{g} \end{aligned} \qquad (4.25)$$

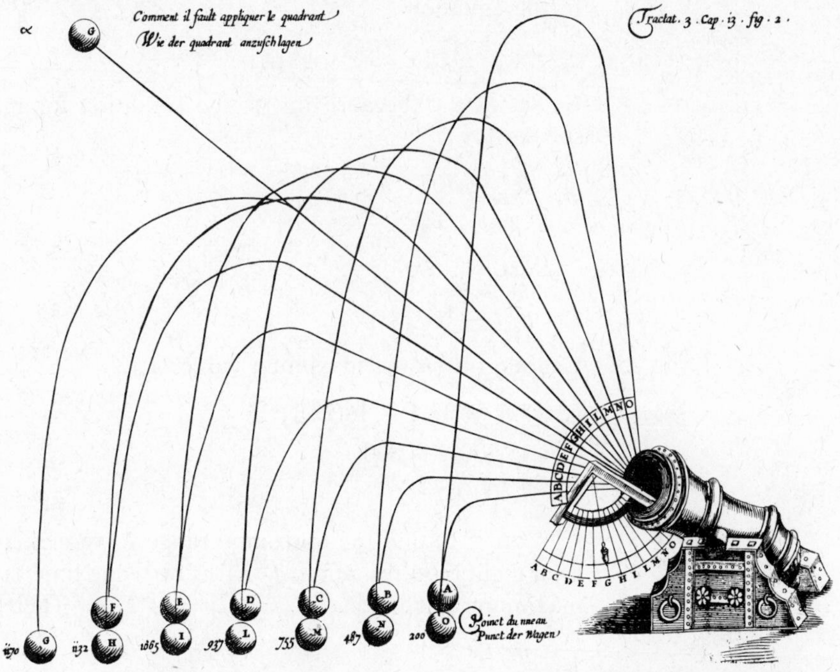

Rare Book Room, New York Public Library

Fig. 4.33 *This diagram from a gunner's handbook published in 1621 (during Galileo's lifetime) demonstrates a good understanding of the motion of real projectiles: the projectile paths are not true parabolas (because of air resistance effects); the same range can be obtained with two different elevation settings; and an elevation angle of 45° produces maximum range. Galileo's work was quick to influence military arts. (The quadrant used for setting elevations was an invention of Galileo.)*

The range is therefore

$$R = v_{ox}T$$
$$= (v_o \cos \theta_o) \times \left(\frac{2v_o \sin \theta_o}{g}\right)$$

or,

$$R = \frac{2v_o^2}{g} \sin \theta_o \cos \theta_o \tag{4.26}$$

By methods employing calculus (or just by calculating a variety of cases, as Galileo did), we can show that the maximum range occurs when the angle of elevation is $\theta_o = 45°$ (see Problem 4.26). Then,

$$R_{\text{max}} = \frac{2v_o^2}{g} \sin 45° \cos 45°$$
$$= \frac{2v_o^2}{g} \times \frac{1}{\sqrt{2}} \times \frac{1}{\sqrt{2}}$$

so that

$$R_{\max} = \frac{v_o^2}{g} \tag{4.27}$$

Therefore, an increase in the initial velocity by a factor of 2 will result in an increase of range by a factor of 4.

4.11 *Circular Motion*

ANGULAR VELOCITY

Another very important case of two-dimensional motion is that of motion in a circular path. For simplicity, we assume that the motion is *uniform,* that is, that the *speed* of the object is constant. If it takes 1 sec for the object to make a complete revolution, we say that the object moves at an *angular* rate of 1 rev/sec. But one complete revolution corresponds to 2π radians, so we can alternatively state that the object moves with an *angular velocity* of 2π rad/sec. It is customary to denote angular velocity (measured in radians per second) by the symbol ω.

If we do not have a complete revolution on which to base a calculation, we can still define the angular velocity in a manner entirely analogous to that used for ordinary (or *linear*) velocity. Thus, if an object moves from a point identified by the angle θ_1 to a point identified by the angle θ_2 in a time interval $t_2 - t_1$ (see Fig. 4.34), the *average angular velocity* is

$$\bar{\omega} = \frac{\theta_2 - \theta_1}{t_2 - t_1} = \frac{\Delta\theta}{\Delta t} \tag{4.28a}$$

Since it is only the angular *difference* that is important, the position labeled $\theta = 0$ is arbitrary.

Again, analogous to the case for instantaneous linear velocity, the *instantaneous* angular velocity is defined to be

$$\omega = \lim_{\Delta t \to 0} \frac{\Delta\theta}{\Delta t} \tag{4.28b}$$

For uniform circular motion, $\omega = \bar{\omega}$. We shall consider only such cases.

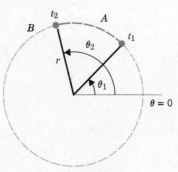

Fig. 4.34 *If the object moves from* A *to* B, *the average angular velocity is* $(\theta_2 - \theta_1)/(t_2 - t_1)$ *rad/sec.*

THE PERIOD

The *period* of circular motion is the time required for one complete revolution or cycle of the motion. Clearly, the period and the angular velocity are inversely related since the greater the angular velocity, the shorter the time required to make a revolution. The period is denoted by the symbol τ:

$$\tau = \frac{2\pi}{\omega} \tag{4.29}$$

If an object moves with uniform speed in a circular path with radius r, the distance traveled in 1 period is just the circumference of the circle, $2\pi r$. The time required for this motion is τ. Therefore, the magnitude of the linear velocity is

$$v = \frac{\text{distance}}{\text{time}}$$

$$= \frac{2\pi r}{2\pi/\omega}$$

Thus,

$$v = r\omega \tag{4.30}$$

For this case of circular motion, the velocity vector is continually changing in direction (see Fig. 4.35). However, if the motion is uniform, the *magnitude* of the velocity vector (the *speed*) is everywhere the same.

Example **4.14**

An automobile moves with a constant speed of 50 mi/hr around a track of 1 mi diameter. What is the angular velocity and the period of the motion?

$$\omega = \frac{v}{r} = \frac{50 \text{ mi/hr}}{0.5 \text{ mi}} = 100 \text{ rad/hr}$$

$$\tau = \frac{2\pi}{\omega} = \frac{2\pi}{100 \text{ rad/hr}} = 0.063 \text{ hr} = 3.8 \text{ min}$$

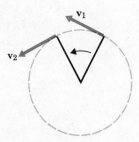

Fig. 4.35   *The velocity vector for circular motion changes direction continually.*

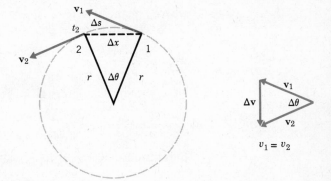

Fig. 4.36 *Velocity diagram of accelerated motion in a circle.*

4.12 *Centripetal Acceleration*

ACCELERATION TOWARD THE CENTER

If an object moves without acceleration, there is no change of velocity and the velocity vector is therefore constant in magnitude and direction. Similarly, if there is any change of velocity, then there must have been acceleration. This change need not be in magnitude; a change in the *direction* of the velocity vector, even if the magnitude remains constant, requires acceleration. Thus, an object moving uniformly in a circular path is continually accelerated.

We can derive an expression for the acceleration in circular motion by referring to Fig. 4.36. At the time t_1 the velocity vector of the moving object is $\mathbf{v}_1$. At the time t_2 the motion has progressed by an angle $\Delta\theta$ and the velocity vector is $\mathbf{v}_2$. If the motion is uniform, $v_1 = v_2$, even though the directions of $\mathbf{v}_1$ and $\mathbf{v}_2$ are different. In moving from point (1) to point (2), the object moves a distance Δs along the circumference of the circle. The length of the chord connecting the two points is Δx. In order to change the velocity vector from $\mathbf{v}_1$ to $\mathbf{v}_2$, the acceleration has produced a "change-in-velocity vector" which we label $\Delta\mathbf{v}$.

Because the velocity vectors, $\mathbf{v}_1$ and $\mathbf{v}_2$, are each perpendicular to radius lines at points (1) and (2), respectively, it follows from the geometry that the triangle formed by the two radius lines and Δx is *similar* to the triangle formed by $\mathbf{v}_1$, $\mathbf{v}_2$, and $\Delta\mathbf{v}$ (see Fig. 4.36). That is, both triangles are isosceles triangles with the same angle $\Delta\theta$ between the equal sides (Fig. 4.37). Therefore, the ratios of the lengths of the sides of the two triangles are equal:

$$\frac{\Delta v}{v} = \frac{\Delta x}{r}$$

Fig. 4.37 *The two triangles are similar. In the triangle at the right,* v *stands for* v_1 *and* v_2 *which are equal.*

where v stands for either v_1 or v_2 which are equal. Thus, the magnitude of the velocity change is

$$\Delta v = \frac{v \times \Delta x}{r}$$

Now, the acceleration is given by the standard expression

$$a_c = \lim_{\Delta t \to 0} \frac{\Delta v}{\Delta t}$$

Substituting for Δv,

$$a_c = \lim_{\Delta t \to 0} \frac{v \times \Delta x / r}{\Delta t}$$

$$= \lim_{\Delta t \to 0} \left(\frac{v}{r} \times \frac{\Delta x}{\Delta t} \right)$$

Because the quantities v and r are constants, they are not affected by the limiting process and we can rewrite the expression for a_c in the form

$$a_c = \frac{v}{r} \times \lim_{\Delta t \to 0} \frac{\Delta x}{\Delta t}$$

As Δt becomes infinitesimally small, the chord Δx becomes equal to the arc length Δs, and the limit of $\Delta s / \Delta t$ is just the linear velocity v. Therefore,

$$\boxed{a_c = \frac{v^2}{r}} \tag{4.31}$$

Referring to Fig. 4.36 we see that as Δt becomes small, $\Delta \theta$ also becomes small, and $\mathbf{v}_2$ almost coincides with $\mathbf{v}_1$. Then, $\Delta \mathbf{v}$ is a vector essentially perpendicular to both $\mathbf{v}_1$ and $\mathbf{v}_2$; that is, $\Delta \mathbf{v}$ points toward the center of the circle. Since the direction of the acceleration vector is the same as the direction of the change in velocity, the vector $\mathbf{a}_c$ is also directed toward the center of the circle. The acceleration is "center seeking" and is termed *centripetal* acceleration.

Using our previous result that $v = r\omega$ (Eq. 4.30), we can also express a_c as

$$a_c = \frac{v^2}{r} = \frac{(r\omega)^2}{r}$$

or,

$$\boxed{a_c = r\omega^2} \tag{4.32}$$

Example **4.15**

What are v, ω, and a_c for the motion of the Earth around the Sun?

$r = 1 \text{ A.U.} = 1.5 \times 10^{13} \text{ cm}$
$\tau = 1 \text{ yr} = 3.15 \times 10^7 \text{ sec}$

Therefore,

$$\omega = \frac{2\pi}{\tau}$$

$$= \frac{2\pi \text{ rad}}{3.15 \times 10^7 \text{ sec}} \cong 2 \times 10^{-7} \text{ rad/sec}$$

$$v = r\omega$$

$$\cong (1.5 \times 10^{13} \text{ cm}) \times (2 \times 10^{-7} \text{ rad/sec})$$
$$= 3 \times 10^6 \text{ cm/sec}$$
$$= 30 \text{ km/sec}$$

$$a_c = r\omega^2$$

$$\cong (1.5 \times 10^{13} \text{ cm}) \times (2 \times 10^{-7} \text{ rad/sec})^2$$
$$= (1.5 \times 10^{13}) \times (4 \times 10^{-14})$$
$$= 0.6 \text{ cm/sec}^2$$

Note that the acceleration due to the Earth's motion around the Sun is only about 0.06 percent of g.

Example **4.16**

We know that if we drop an object while giving it a horizontal velocity component, the object will fall toward the surface of the Earth with the horizontal velocity remaining constant. With what velocity must an object be projected so that the curvature of its path is just equal to the curvature of the Earth? In such a situation, the object would fall toward the Earth at the same rate that the surface of the Earth curves away from the instantaneous velocity vector; that is, the object would *fall* around the Earth. The height of the object above the surface of the Earth would therefore never decrease and the object would become a satellite of the Earth.

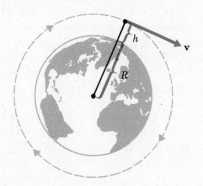

Suppose that we start with the object at a distance h above the surface of the Earth. The radius of the Earth is R so that the radius of the desired circular path of the object is $R + h$.

The centripetal acceleration required to maintain the circular motion is

$$a_c = \frac{v^2}{r} = \frac{v^2}{R + h}$$

This centripetal acceleration is furnished by gravity, so we can substitute g for a_c; thus,

$$g = \frac{v^2}{R + h}$$

As we shall see later when we study gravitation in more detail, the value of g depends on the distance from the center of the Earth. If h is small compared to R (let us take $h = 100$ mi $= 1.6 \times 10^7$ cm which is $\frac{1}{40}$ of the Earth's radius), the value of g will be essentially that appropriate for the surface of the Earth, *viz.*, $g = 980$ cm/sec². Therefore, solving for v, we find

$$
\begin{aligned}
v &= \sqrt{g(R + h)} \\
  &= \sqrt{(980 \text{ cm/sec}^2) \times (6.4 \times 10^8 \text{ cm} + 1.6 \times 10^7 \text{ cm})} \\
  &= \sqrt{(980 \text{ cm/sec}^2) \times (6.56 \times 10^8 \text{ cm})} \\
  &= \sqrt{62.7 \times 10^{10} \text{ cm}^2/\text{sec}^2} \\
  &= 7.9 \times 10^5 \text{ cm/sec} \\
  &= 7.9 \text{ km/sec}
\end{aligned}
$$

or approximately 5 mi/sec.

The period of the motion is

$$
\tau = \frac{2\pi}{\omega} = \frac{2\pi r}{v}
$$

$$
= \frac{2\pi \times (6.56 \times 10^8 \text{ cm})}{7.9 \times 10^5 \text{ cm/sec}}
$$

$$
= 5200 \text{ sec} = 87 \text{ min}
$$

Therefore, the satellite moving in a circular orbit at a height of 100 mi will require approximately an hour and a half to circle the Earth. Many of the artificial satellites that have been launched during the past few years have orbit characteristics similar to those in this example.

4.13   *The Ultimate Velocity*

THE VELOCITY OF LIGHT

According to the formulae we have developed, if an object starts from rest and is accelerated, the velocity after a time t will be $v = at$. Thus, for any non-zero value of a, if we make t sufficiently long, we can have an arbitrarily large velocity v. However, relativity theory tells us that this is not the case. One of the results of this theory (see Chapter 11) is that there exists a maximum velocity for any physically realizable motion. This ultimate velocity is the velocity of light in a vacuum, $c = 3 \times 10^{10}$ cm/sec.

According to relativity theory, the simple expression $v = at$ must be modified to[10]

$$
v = \frac{at}{\sqrt{1 + (at/c)^2}} \tag{4.33}
$$

---

[10] In this expression, a is the acceleration experienced by the moving object and t is the time as measured by an observer not undergoing acceleration.

For any value of *a,* after a sufficiently long time we will have $(at/c)^2 \gg 1$. Then, we can neglect 1 compared with $(at/c)^2$ in the radical, so that the velocity becomes, approximately,

$$v \cong \frac{at}{\sqrt{(at/c)^2}} = \frac{at}{at/c} = c$$

That is, for $t \to \infty$, the velocity approaches the velocity of light. Of course, v never equals c because the denominator is actually always slightly larger than at/c.

From this we must conclude that no matter how large an acceleration is impressed on a material object and no matter how long this acceleration is applied, the velocity can never exceed or equal the velocity of light. Thus, the *ultimate* physical velocity is the velocity of light.

Particles of small mass, such as electrons and protons, can be given extremely high velocities by devices of various sorts. Even the 20-kilovolt electron guns found in television sets produce electrons with velocities of approximately 8×10^9 cm/sec or 0.3 *c*. The new Soviet accelerator (Fig. 4.38) can produce protons with velocities exceeding 0.9999 *c*, and the Stan-

Fig. 4.38 *View of the beam lines of the huge Soviet accelerator at Serpukhov. The protons emerging from this machine are faster than those produced by any other accelerator in the world,* v = *0.999915c.*

Sovfoto

Fig. 4.39   *The Stanford Linear Accelerator (SLAC) near Palo Alto, California. This two-mile-long accelerator produces ultra-high velocity electrons for elementary particle research. The velocity of these electrons is only 1.5 cm/sec less than the velocity of light!*

ford Linear Accelerator (Fig. 4.39) can produce electrons with velocities exceeding 0.9999999999 c.

For the modest accelerations and relatively brief time intervals that we encounter in everyday experience, the term $(at/c)^2$ in Eq. 4.33 is, in fact, small compared with unity, so that the expression for v reverts to $v = at$. Only in the event that the velocity is comparable with the velocity of light is it necessary to employ the relativistic equations.

Summary of Important Ideas

Speed or *velocity* is the rate at which distance is traveled. *Velocity* is a *vector;* speed is a *scalar.*

On a distance-time graph, the *slope* of the curve is the *speed* or the *velocity.*

Acceleration is the rate of change of velocity. Acceleration is a *vector.*

For objects falling near the surface of the Earth, the horizontal and vertical motions are *independent.* The vertical motion undergoes an acceleration $g = 980$ cm/sec^2 = 32 ft/sec^2.

Projectile motion near the surface of the Earth is *parabolic* (in the absence of air friction).

An object moving in a circle has a *centripetal acceleration* that is directed toward the center of the circle.

No object can travel faster than the *speed of light.*

Questions

4.1 Is it possible for a moving body to have **v** and **a** *always* in *opposite* directions?

4.2 The acceleration applied to a certain body is constant in magnitude and direction. Describe situations in which (a) the velocity vector always has the same direction and (b) the body never moves in a straight line.

4.3 Why is it not possible to take account of air resistance effects in describing the motion of falling bodies by simply decreasing the value of *g?*

4.4 In the game of roulette, it is customary to set the ball into motion in the direction opposite to that of the revolving wheel. What is the difference between the motion of the ball relative to the wheel and the motion of the ball relative to the table? Can the ball ever have *zero* instantaneous velocity relative to the table even if the wheel is moving?

4.5 At what point or points in its path is the *speed* of a projectile a minimum? At what point or points is the speed a maximum?

4.6 If we wish to compute the initial velocity necessary for an ICBM to have a range of 5000 mi, we cannot use the simple expression in Eq. 4.27. What additional effects would we have to consider if we set out to derive an accurate expression for the range?

4.7 A particle moves uniformly in a circle. Describe the way in which the acceleration vector changes with time.

Problems

4.1 A driver on an Interstate highway notices the mileage markers at the following times:

Mile 120	11:30 A.M.
140	11:50 A.M.
150	12:40 P.M.
200	1:40 P.M.
208	1:46 P.M.
208	1:50 P.M.
218	2:05 P.M.

(a) Plot a graph of distance *versus* time for the trip.
(b) Indicate on the graph the average speeds for the various straight-line portions of the graph.
(c) What apparently happened near noon and at 1:46 P.M.?

4.2 During a certain automobile trip the speeds at different time intervals were as follows:

1:00 P.M.–2:00 P.M.	$v = 30$ mi/hr
2:00 P.M.–2:30 P.M.	$v = 40$ mi/hr
2:30 P.M.–3:00 P.M.	stopped
3:00 P.M.–4:30 P.M.	$v = 20$ mi/hr
4:30 P.M.–5:00 P.M.	$v = 60$ mi/hr

(a) What was the total length of the trip?

(b) What was the average speed?

(c) How long did it take to go the first 55 mi?

(d) What was the average speed for the first two hours? For the last two hours?

4.3 An automobile moves uniformly a distance of 60 ft in 2 sec; during the next 3 sec it moves only 30 ft. What was the average speed (a) during the first 2 sec, (b) during the next 3 sec, (c) during the 5-sec interval? (d) Compute the average speed for the first 3 sec.

4.4 An automobile starts from rest and after 3 sec is moving with a speed of 60 ft/sec. If the acceleration to this speed was uniform, how far did the automobile move in the first 2 sec of motion? How far did it move during the third second?

4.5 How far does an object released from rest fall during the 7th second of motion?

4.6 A ball is thrown upward from the top of a tower with an initial velocity of 128 ft/sec. What is the downward velocity of the ball 3 sec after it has passed the top of the tower on its way down?

4.7 A ball is thrown upward with an initial velocity of 96 ft/sec. How high will it rise? What will be the velocity of the ball when it returns to its original position?

4.8 At time $t = 0$ an object is thrown vertically into the air. It rises to a maximum height of 64 ft and then falls to the bottom of a pit which is 336 ft deep. At what time does the object strike the bottom?

4.9 Show that the velocity acquired by an object falling from a height h (starting from rest) is $v = \sqrt{2gh}$. (Note: This is an important result that we will use later.)

4.10 Plot a graph (v vs. t) of the instantaneous velocity of an object thrown upward with an initial velocity of 128 ft/sec. Determine *graphically* the approximate velocity after 2.75 sec and after 7.8 sec. At what time will the velocity be downward at 100 ft/sec?

4.11 On the moon the acceleration due to gravity is only $\frac{1}{6}$ as large as on the Earth. An object is given an initial upward velocity of 100 ft/sec at the surface of the moon. How long will it take for the object to reach maximum height? How high above the surface of the moon will the object rise?

4.12 A rocket is launched from the surface of the Earth with an upward acceleration of $4g$. After 10 sec, what is the velocity of the rocket and to what height has it risen?

4.13 Two automobiles, each traveling with a speed of 60 mi/hr, crash head-on. The impact velocity is the same as if the cars had been dropped from what height? If it requires $\frac{1}{50}$ sec for the cars to come to rest after the initial impact, what acceleration has each experienced?

4.14 The use of *Mach numbers* is one way to specify velocity. Mach 1 corresponds to the speed of sound in air and at sea level is approximately equal to 1100 ft/sec or 730 mi/hr; Mach 2 corresponds to twice the speed of sound, and so forth. A certain high-acceleration rocket can reach Mach 1.2 by the time it has traveled 1000 ft. What is the acceleration? Express the result as a multiple of *g*.

4.15   It requires approximately 2 sec for a parachute to deploy. During this time interval suppose that the speed of the parachutist is decreased uniformly from his free-fall velocity to essentially zero. If the parachutist is falling at the terminal velocity of 200 mi/hr when he opens his parachute, what acceleration does he experience during the deployment? Express the result in terms of *g*. Is parachuting safe for a person who "blacks out" at 5*g*?

4.16   Each vector in the diagram has a length of 2 units.

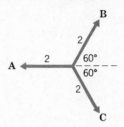

(a) What is the sum **B** + **C**?
(b) What is the sum **B** + **C** + ½**A**?
(c) What is the sum **A** + **B** + **C**?

4.17   The vector **A** has unit magnitude and the magnitude of **B** is $\sqrt{3}$. What is the direction and the magnitude of **A** + **B**?

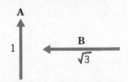

4.18   In still water a certain boat can travel at a speed of 10 mi/hr. The helmsman wishes to proceed *straight across* a river in which the current flows uniformly at a velocity of 5 mi/hr. In order to do this, the helmsman finds that he must steer the boat *upstream* at a certain angle. What is the angle?

4.19   At a certain instant an object has the following velocity components: $v_x = 300$ m/sec, $v_y = 400$ m/sec. What is the *speed* of the object at that instant?

4.20   An airplane is flying due north in still air at a velocity of 240 mi/hr. Suddenly the aircraft encounters a cross wind from the west whose velocity is 70 mi/hr. If the pilot does not alter his controls, what will be his new direction and new ground speed? Has the aircraft been accelerated?

4.21   Lay out the following vectors and *graphically* find the sum, first by adding **B** to **A** and then adding **C** to the sum, and next by adding **C** to **B** and then adding **A**.

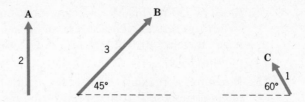

4.22 Find the sum of the three vectors in Problem 4.21 by adding *components*.

4.23 An object falls through a vertical distance of 256 ft while moving horizontally a distance of 100 ft. What is its horizontal velocity?

4.24 The acceleration of an object falling freely near the surface of the Earth is approximately 32 ft/sec². But this value differs slightly from one position on the Earth to another. The variation (near sea level) is about 0.5 percent. How much difference could such a variation make in the distance that an athlete can hurl the javelin? (Suppose the range is 300 ft.) Should track and field records be a function of where on the Earth's surface they were set?

4.25* During World War I, the Germans constructed an enormous railway gun (called Big Bertha by the Allies) that was used to bombard Paris from a distance of 75 mi. What was the *minimum* muzzle velocity necessary for the gun to have had such a range? (Recall that the effect of air resistance is always to reduce the range, so the actual velocity must have been somewhat greater than that computed with the simple formula.) What was the minimum time between firing and impact when the gun was adjusted for maximum range? What height did the shells reach?

4.26 By computing elaborate tables, Galileo showed that the maximum range of a projectile fired over level ground occurs for a muzzle elevation of 45°. He also showed that the range will be the same for muzzle angles of 45° + ϕ and 45° − ϕ, for any value of ϕ (between 0° and 45°). Show that this conclusion is correct by calculating the ranges for several values of ϕ. Sketch the two possible paths for some value of ϕ (see Fig. 4.33.)

4.27 A projectile is fired at an angle of 45° above the horizontal. Show that the range of the projectile is equal to four times the maximum height reached.

4.28 An object is projected vertically upward and returns to its original position in 2 sec. What was the initial vertical velocity? How high did the object rise? The object is next given the same initial velocity but at an angle of 45°. How high does it rise and what is its range?

4.29* The muzzle velocity of a bullet can be measured by firing the bullet through two rotating discs that are separated by a certain distance d and then measuring the angular displacement $\Delta\theta$ between the two holes. If the discs

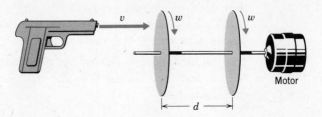

are rotating together with the same angular velocity ω, show that the muzzle velocity is given by $v = \omega d/\Delta\theta$. If the discs are 1 m apart and are driven by a 1000-rpm motor and if the angular displacement between the holes is found to be 29°, what is the velocity of the bullet?

4.30   What is the angular velocity and the period of a four-speed phonograph turntable when operating at each of its possible speeds ($16\frac{2}{3}$, $33\frac{1}{3}$, 45, and 78 rpm)? What is the centripetal acceleration on a particle at the outer edge of a moving $33\frac{1}{3}$ rpm 12-in. LP record?

4.31   What is the angular velocity of the Earth's rotation about its axis? What is the speed of a point on the Earth's surface at the equator? What is the speed at a point with latitude $\lambda = 45°$? (Express the latter two results in km/sec.)

4.32   An object is moving in a circular path of radius 9.4 ft and is experiencing a centripetal acceleration of 3 g. What is the speed of the object? What is the period of the motion?

4.33   An object moves in a circular path whose radius is 8 ft. If the period is 10 sec, express the centripetal acceleration experienced by the object in units of g. (Note: do not put in numerical values until you have an algebraic expression for a_c.)

4.34   Compute the centripetal acceleration of an object on the surface of the Earth at the equator and at the poles due to the Earth's rotation on its axis. If an object were dropped from a certain height above the surface of the Earth at the equator and at the North Pole, at which place would it strike the ground earlier?

4.35   An object, initially at rest, is subjected to an acceleration equal to g for a period of 1 yr. What will be the final velocity of the object? Is it permissible to use the nonrelativistic expression for the calculation?

International Weight-lifting Champion (USSR)

5 *Force and Momentum*

In the preceding chapter we discussed the topic of *kinematics,* the geometrical description of motion without reference to the *cause* of the motion. We now extend the discussion to include the *cause* of motion and treat the subject of *dynamics.* With regard to the introduction of new concepts, dynamics is fundamentally different from kinematics. The pattern of discussion of kinematics was simple and straightforward; each new quantity that was introduced was defined in terms of previously established quantities. But in order to make any progress in dynamics we find that it is necessary to introduce *two* new quantities simultaneously—that is, *force* and *mass.* It is not possible to give precise definitions for either force or mass that are based solely on kinematical concepts. Furthermore, the definitions cannot be made independent of one another. For these reasons the logical structure of dynamics is fraught with difficulties for which there appears to be no escape. No one has yet devised an unassailable, logical approach to dynamics—the force-mass dilemma is always with us. This fact is a great tribute to the insight of Isaac Newton who was able to construct a workable theory that was capable of explaining in detail many natural phenomena in spite of the basic weakness of its logical foundation.

Because of this logical difficulty, there is no "right way" to proceed with the discussion of dynamics. The approach we elect to use here is not unique, and several other avenues are possible. However, "the proof of the pudding is in the eating", and no matter how we arrive at the Newtonian laws of dynamics, the important point is that these laws correctly describe the motions of macroscopic bodies.

During this century it has been found that the Newtonian laws need modification when exceedingly small distances or extremely high velocities are encountered. The discrepancies between prediction and observation under these circumstances paved the way for the development of the theories of quantum mechanics and of relativity. These limitations of Newtonian theory do not in any way imply that the theory is obsolete or unimportant. Indeed, under almost all ordinary circumstances the Newtonian laws are a correct description of the dynamics of physical systems.

5.1 *The Intuitive Conception of Force*

FORCE AND INERTIA

Force is most commonly appreciated in terms of muscular action. We must exert a "great force" in order to push an automobile but a similar "great force" applied to a large truck produces no motion at all. We know that the truck is much larger than the automobile and contains a greater amount of matter; that is, the truck has a greater *mass* than the automobile. Evidently, the amount of motion that is produced by a given "force" depends on the mass of the body.

The push (or *force*) necessary to start the automobile into motion (even on a level pavement) is rather large, but once in motion it requires much less force to maintain a constant speed. To the ancients (who reported such

Fig. 5.1 *A large force is necessary in order to overcome the inertia of an automobile at rest and set it into motion. Once moving, it requires only a small force (to overcome friction) to maintain motion with a constant velocity.*

observations in terms of *carts* instead of *automobiles*), this was a most important fact from which they concluded (with some justification) that force is that which is necessary in order to maintain constant speed. The ancients, however, overlooked the crucial point that it requires a *greater* force to start the motion than to maintain it. We use the term *inertia* for that property of matter which causes it to resist a *change* in its state of motion. That is, the *inertia* of the automobile must be overcome in order to set it into motion. Similarly, once in motion at constant speed, its inertia prevents the automobile from being stopped unless a sizable force is applied against it.

FRICTION

Of course, we now understand why it is necessary to apply a force in order to maintain the constant speed of a moving object. In addition to any force that we may apply to an object, there will in general be opposing *frictional forces* in operation as well. Even a well-polished ball rolling over a smooth surface will eventually come to rest because of friction. If the ball is not well-polished and if the surface is not smooth, the motion will cease much more quickly. Friction comes about when the irregularities in the surface of an object tend to "snag" on similar irregularities on the surface of another object against which it is sliding or over which it is rolling. Obviously, the less regular the surfaces, the greater will be the friction. Since it is impossible to polish away *all* of the irregularity of any surface (all matter is composed ultimately of "irregular" atoms), friction can never be eliminated, although it can be reduced to exceedingly small values in certain circumstances. An electric motor, for example, in which the shaft rotates in precision, well-lubricated ball bearings, will require a considerable period of time to "run down" after the power is turned off. Friction also occurs when a body moves through a medium such as air or water. In order to move forward, the body must push aside the molecules of the medium; the forces (called *viscous* forces) that are exerted on the body by the molecules constitute a frictional drag. If an object were moving through space in a perfect vacuum, then, of course, it would experience no frictional force. But even in the best laboratory vacuum apparatus, each cm^3 contains $\sim 10^8$ gas atoms which will produce a small frictional force on any moving object.

Although we can never completely eliminate frictional effects, we shall proceed with the development of dynamics as if we could do so. That is, we shall ignore friction in our discussions, knowing that in real situations we must take these effects into account.

5.2 *Newton's Laws*

In 1687, after many years of work and many years of prodding by his friends to publish his findings, Isaac Newton finally produced his most important work, the famous *Principia*. Newton did not "invent" the subject of dynamics; on the contrary, he made the maximum use of previous work, especially the detailed experiments and analyses of Galileo. Newton's great contribution was to synthesize, from his own work and all that went before him, a complete description of the dynamics of bodies in motion. In this section we summarize Newton's results by stating his famous three laws in modern terms, and we briefly discuss the meaning of each law. (We assume here that the physical quantity *force* has both magnitude and direction, that is, that force is a *vector* quantity. In Section 5.7 we will pursue the matter of the vector property of force.)

NEWTON'S FIRST LAW (OR THE LAW OF INERTIA)

If the net force on an object is zero (i.e., if the vector sum of all forces acting on an object is zero), then the acceleration of the object is zero and the object moves with constant velocity. That is, if no force is applied to an object at rest, the object remains at rest; if the object is in motion, it maintains a constant velocity. In mathematical terms,

$$\mathbf{F}_{net} = 0 \implies \mathbf{a} = 0 \quad \text{or} \quad \mathbf{v} = \text{const.} \tag{I}$$

This law provides an explanation of the observation, mentioned in the preceding section, that once an automobile is moving, it requires only a small force to maintain constant velocity. If we did not have friction to overcome, the automobile would coast with constant velocity even though *no* force were applied. A hockey puck, sliding over smooth ice, will move a great distance at almost constant velocity even though there is no applied force. Eventually, of course, the puck will come to rest as a result of the non-zero friction that exists between the puck and the ice.

The first law, by itself, gives us only a crude notion regarding force. In fact, we have here only a definition of *zero* force. However, there is the implication that *force* is somehow intimately connected with *acceleration*. The second law states this connection.

NEWTON'S SECOND LAW

The accelerated motion of a body can only be produced by the application of a force to that body. The acceleration is proportional to the impressed force and the constant of proportionality is the *inertia* or *mass* of the body. In mathematical terms,

$$\boxed{\mathbf{F} = m\mathbf{a}} \tag{II}$$

Fig. 5.2 *Sir Isaac Newton (1642–1727), the great English mathematician and physicist who, as a young man, formulated a complete theory of dynamics, discovered the law of universal gravitation, and invented the calculus. In later years he became somewhat of a mystic but he served as President of the Royal Society from 1704 until his death.*

National Portrait Gallery

Because **a** is a well-defined quantity, this law expresses the relationship between force and mass but it defines neither. It remains for the third law to resolve the situation.

Notice that Eq. II is a *vector* equation. That is, the vectors **F** and **a** are related by the scalar quantity *m*. If the force **F** is in a certain direction, then the acceleration **a** is necessarily in that same direction.

It is essential to realize that whenever we wish to use Eq. II to describe the motion of an object, we must be certain that **F** in this equation is the *total* (or net) *force acting on the object in question*. Whether this object is exerting forces on other bodies is immaterial. We need only know the forces that are exerted *on* a body in order to calculate its motion.

NEWTON'S THIRD LAW

If object 1 exerts a force on object 2 then object 2 exerts an equal force, oppositely directed, on object 1. That is, a force is always paired with an equal reaction force. In mathematical terms,

$$\boxed{\mathbf{F}_{12} = -\mathbf{F}_{21}} \tag{III}$$

where the first subscript denotes the object receiving the force and the second subscript denotes the object exerting the force.

It is important to understand what the third law means and what it does *not* mean. First, consider a spring balance that is attached to a wall. If we pull on the spring balance it will register the force of the pull; call this force

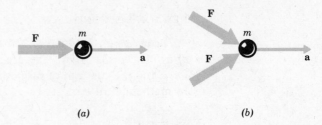

Fig. 5.3 *(a) Force and acceleration are in the same direction. (b) The acceleration is in the direction of the total (or net) force applied to the object.*

(a)   *(b)*

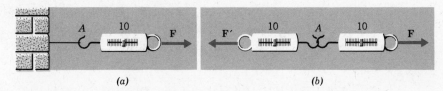

Fig. 5.4 (*a*) *A force* **F**, *of magnitude 10 units, is applied to a wall through a spring balance.* (*b*) *To measure the force that the wall exerts on the spring balance, we replace the wall with another balance.*

F and suppose that it has a magnitude of 10 units. The situation is that shown in Fig. 5.4a. According to Eq. III, the wall should be exerting a force −**F** on the spring. How do we know this? We can duplicate the effect of the wall by substituting for it another spring balance and by pulling on this second balance with the force necessary to make the first balance read 10 units while remaining stationary. This new force **F**′, can be read on the scale and is found to be 10 units (see Fig. 5.4b). The point at which the two springs are joined, or the point at which the spring is attached to the wall (these two equivalent points are labeled *A* in Fig. 5.4), is not in motion, so the acceleration is zero; thus, the net force must be zero, that is, **F** + **F**′ = 0, or **F** = −**F**′.

Next, consider two blocks that are resting on a *frictionless* surface, as in Fig. 5.5. A force **F** is applied to m_1. If m_1 and m_2 are in contact, is this force transmitted to m_2? The third law tells us that whatever force m_1 exerts on m_2, then m_2 must exert this same force on m_1. But the third law does *not* specify the magnitude of this force; only the *second* law can be used to find this quantity. The total mass of the two blocks is $m_1 + m_2$ and the only externally applied force is **F**; therefore, the acceleration of the two blocks together is

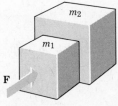

Fig. 5.5 *A force* **F** *is applied to* m_1. *Is this force transmitted to* m_2?

$$\mathbf{a} = \frac{\mathbf{F}}{m_1 + m_2}$$

The acceleration of m_2 is, of course, also **a**, so the force applied to m_2 that gives rise to this acceleration must be

$$\mathbf{F}_2 = m_2\mathbf{a} = \frac{\mathbf{F}m_2}{m_1 + m_2}$$

which is smaller than the external force **F** by the fraction $m_2/(m_1 + m_2)$. Similarly, the net force acting on m_1 is

$$\mathbf{F}_1 = m_1\mathbf{a} = \frac{\mathbf{F}m_1}{m_1 + m_2}$$

which is just the sum of the externally applied force **F** and the reaction force applied on m_1 by m_2, *viz.*, −$\mathbf{F}_2$:

$$\mathbf{F}_1 = \mathbf{F} - \mathbf{F}_2 = \mathbf{F} - \frac{\mathbf{F}m_2}{m_1 + m_2} = \frac{\mathbf{F}m_1}{m_1 + m_2}$$

Notice the important point, which we frequently find useful, that the motion of the two blocks considered together as a *system* can be determined without the necessity of calculating the internal forces (that is, the forces acting between the blocks). The acceleration of the system is found by dividing the externally applied force by the total mass of the system.

5.3   *The Third Law and the Definition of Mass*

MASS RATIOS AND ACCELERATION RATIOS

According to the view we have elected to take of Newtonian dynamics, the first and second laws are essentially only definitions, whereas the third law is indeed a physical *law*.[1] We consider Eq. III to be a real physical law because it allows us, at least in principle, to give a precise definition of *mass.* And, once mass is defined, *force* is defined through Eq. II.

Suppose that we have two objects that are isolated, that is, they interact only between themselves and with nothing else. An example of such a pair of isolated objects would be a star that has a single planet. Even the Sun and Earth are, to a good approximation, an "isolated" pair of objects because for many purposes we can neglect the presence of the moon and other planets. Or we could consider two spacemen in deep space; if Spaceman 1 pushes on Spaceman 2 with a force $\mathbf{F}_{21}$, then 2 pushes on 1 with an equal and opposite force. That is, the third law states

$$\mathbf{F}_{12} = -\mathbf{F}_{21}$$

where $\mathbf{F}_{12}$ means the force *on* 1 *due* to 2 and $\mathbf{F}_{21}$ means the force *on* 2 *due* to 1. Using Eq. II, we can write

$$m_1\mathbf{a}_1 = -m_2\mathbf{a}_2$$

Considering only the magnitudes of the accelerations,

$$m_2 = \left(-\frac{a_1}{a_2}\right) m_1 \tag{5.1}$$

Therefore, if we were to select m_1 as our standard mass (for example, the platinum-iridium cylinder shown in Fig. 2.8), we could determine m_2 by measuring the acceleration ratio, a_1/a_2. (The negative sign indicates only that the accelerations are in opposite directions.) Thus, the third law provides a method of uniquely defining *mass;* then, Eq. II gives a similarly precise definition of *force.*

THE VALIDITY OF THE THIRD LAW

If there is an interaction between two objects, A and B, Newton's third law requires that, due to the force of A on B, there exists an equal reaction force of B on A. However, consider the case in which A and B are separated by a distance d and exert electrical forces on one another. Electrical interactions are known to propagate with the velocity of light, $c = 3 \times 10^{10}$ cm/sec. Therefore, a change in the electrical force of A on B due to some

---

[1] As pointed out earlier, this is not the only possible interpretation; some authors prefer to view the second law as crucial and the third law as a definition.

change in the condition or position of A will not be felt by B until a time d/c after its occurrence. Similarly, the reaction to this force change will not be propagated back to A until a time $2d/c$ after the original action. This is clearly a violation of the third law which requires equal and opposite forces at all times. It follows that the third law cannot be valid for any case in which a finite time is required for the propagation of the force. Since *all* of the basic forces of Nature (see Chapter 6) have finite propagation times, the third law is *precisely* valid only in the event that the interacting objects are at rest so that there are no force changes to be propagated. However, this static case is of limited interest; after all, we wish to study the effects of *motion.* The third law is *approximately* valid (actually, to an extremely high degree) if the interacting objects are so close that the propagation time of the force is essentially zero. Thus, in all "contact" situations we can use the third law without fear of error.

The effect of the finite propagation time on the gravitational interaction of two objects is completely negligible except for the special case of astronomical bodies moving with extremely high velocities (Chapter 11).[2] For the motions of everyday macroscopic objects, deviations from the predictions of the third law are impossible to detect. We therefore consider the Newtonian laws to be completely accurate for all such motions.

5.4 *Frames of Reference*

INERTIAL REFERENCE FRAMES

Newton's laws involve the concept of *acceleration.* In order to measure acceleration we must specify a coordinate system or *frame of reference* with respect to which the measurement is made. Not all reference frames are equally useful. If we set out to study the dynamics of planetary motion, we would not choose a coordinate system that is fixed with respect to the Earth; in such a system the planets wander in a complicated manner across the sky and undergo apparent motions that are not indicative of the forces actually acting on them. That is, in an Earth-fixed coordinate system, Newton's laws cannot be used in any simple way to describe the motion of planets. However, if we choose a reference frame that is fixed with respect to the distant stars, the motion of the planets is found to conform to Newton's laws. *Any* reference frame in which Newton's laws are a correct description of the dynamics of moving bodies is called an *inertial reference frame.*

To Newton, the distant stars, which appeared to have fixed positions in space, satisfied the need for a basic inertial reference frame. We now know that these stars are not "fixed" but undergo continual and complicated motions. The specification of an inertial reference frame is therefore not a simple problem. However, for all but the most sophisticated analyses, we can consider the distant stars to be fixed and to constitute an acceptable inertial reference frame.

---

[2] There is, however, a gross violation of the third law for the case of the interaction of moving electrical charges and care must be exercised to calculate correctly the forces in such cases.

AN INFINITY OF INERTIAL REFERENCE FRAMES

The distant stars do not specify the only possible inertial reference frame. We can find many other reference frames in which Newton's laws are also valid. The Earth undergoes a complicated motion against the background of the distant stars. But if we confine our attention to small-scale phenomena that take place over relatively short periods of time, the motion of the Earth will not influence the phenomena to any appreciable extent. Therefore, in many practical situations, Newton's laws will be valid in a coordinate system fixed with respect to the Earth. Indeed, for most everyday applications of Newton's laws we find an Earth-fixed coordinate system to be quite adequate.

Suppose that we set up a laboratory in a large box that is at rest on the Earth. We equip ourselves with suitable meter sticks, clocks, and spring balances so that we can test Newton's laws. By making various measurements we can verify, for example, that $\mathbf{F} = m\mathbf{a}$ is a valid equation. The laboratory coordinate system therefore is an inertial reference frame. Next, someone removes our box laboratory and places it on a train that is moving with constant velocity. (Let us suppose that our box contains no windows so that we cannot measure the velocity of the train.) If we repeat the measurements that were made when the laboratory was at rest, what will we find? We will find exactly the same results! That is, Newton's laws are also valid in the moving reference frame. The reason is easy to see. A measurement of a certain acceleration requires the measurement of a time interval and the difference between two velocities:

$$\mathbf{a} = \frac{\mathbf{v}_2 - \mathbf{v}_1}{t_2 - t_1}$$

If we add a constant velocity $\mathbf{v}$ to the coordinate system (by transferring it to the moving train), we have for our new velocities, $\mathbf{v}'_2 = \mathbf{v}_2 + \mathbf{v}$ and $\mathbf{v}'_1 = \mathbf{v}_1 + \mathbf{v}$. Therefore, the new acceleration $\mathbf{a}'$ is

$$\mathbf{a}' = \frac{\mathbf{v}'_2 - \mathbf{v}'_1}{t_2 - t_1} = \frac{(\mathbf{v}_2 + \mathbf{v}) - (\mathbf{v}_1 + \mathbf{v})}{t_2 - t_1}$$

$$= \frac{\mathbf{v}_2 - \mathbf{v}_1}{t_2 - t_1} = \mathbf{a}$$

Thus, the acceleration measured in the moving reference frame is exactly equal to that measured in the reference frame at rest.[3]

We may therefore draw the following important conclusion: *If Newton's laws are valid in a certain reference frame, then they will also be valid in any other reference frame that moves with constant velocity with respect to the first frame* (see Fig. 5.6). Thus, there is no *one* reference frame that is preferred over all others and any statement regarding a frame at "absolute rest" is meaningless.

[3]The result of this nonrelativistic argument is still correct in relativity theory.

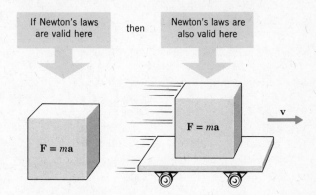

Fig. 5.6 *The addition of a constant velocity does not invalidate Newton's laws.*

5.5 *Mass and Weight*

WEIGHT IS A FORCE

It is important to distinguish between the concepts of mass and weight. *Mass* is an intrinsic property of matter. The mass of a body is a measure of the amount of material in the body. The body contains the same number of atoms regardless of its location, whether on the Earth, on the moon, or in space. Its mass is the same in all of these places, but the *weight* of the body is different. *Weight* is a measure of the gravitational force on a body. A body that has a certain weight at the surface of the Earth will have a different weight on the moon because the gravitational force exerted by the Earth on an object at the Earth's surface is greater (by about a factor of 6) than the gravitational force exerted by the moon on the same object at the moon's surface. In distant space, where the gravitational force due to any other object is negligible, the weight of a body is zero.

In general, the weight W of an object is related to its mass m by

$$W = mg \tag{5.2}$$

where g is the acceleration due to gravity *appropriate for the place at which the measurement is made.* Just as g varies slightly over the surface of the Earth and is different on the moon or other planets, so also does the weight vary.

By comparing Eq. 5.2 with Newton's second law, it is apparent that *weight is indeed a force* since it is equal to the product of a *mass* and an *acceleration*.

It is often said that an object falling freely is *weightless*. (For example, an astronaut in a space capsule orbiting the Earth; see Example 4.16.) Such a statement, according to our definition of weight, is misleading. At any given height above the surface of the Earth there will be a definite value of the gravitational force on a body; therefore the body has a definite weight. However, if the object is falling freely and we attempt to measure the weight of the object with a spring scale (which is also falling freely), then the scale will register no force, that is, a freely falling object is *apparently* "weightless"

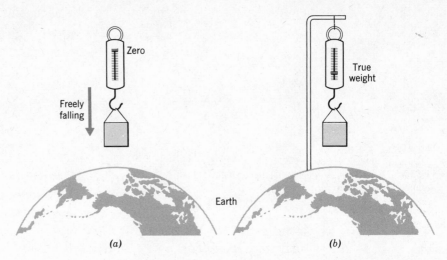

Fig. 5.7 *A freely falling body is apparently "weightless."*

(Fig. 5.7a). But the same spring scale, if attached to a rigid support, will measure the true weight of any object suspended from it at rest (Fig. 5.7b).

If a 1-kg block resting on a horizontal platform exerts a force, $F = mg = (1 \text{ kg}) \times (9.8 \text{ m/sec}^2) = 9.8$ newtons (N), on the block. Now, suppose that the platform is accelerated upward at a rate $a = 5$ m/sec^2. In this situation the platform is exerting an *additional* force of 5 newtons on the block. Therefore, if the block were resting on a scale, instead of directly on the platform, the scale would read $9.8 + 5 = 14.8$ N when the system is accelerating upward and we would say that the block "weighs" 14.8 N. But the force of gravity on the block is still 9.8 N. The apparent change in weight is due to the additional acceleration experienced by the block. (Similarly, if the platform were accelerated *downward* at the same rate, the "weight" of the block would be $9.8 - 5 = 4.8$ N.)

TWO TYPES OF MASS—ARE THEY THE SAME?

If we push an object across a horizontal frictionless surface and impart to it an acceleration a, we know, from Newton's second law, that a net force F was applied to the object. The resistance of the object to the push is due to its inertia or mass. Let us call this mass the *inertial mass* and denote it by m_I so that[4] $F = m_I a$. The presence or absence of gravity does not influence this result; the same equation would be valid if the experiment were carried out in a gravity-free region of space.

We know that the gravitational force between an object and the Earth (that is, the *weight*) is directly proportional to the mass of the object (see Eq. 5.2). We call this the *gravitational mass* of the object and we write $F = m_G g$. But is this *gravitational mass* m_G the same as the *inertial mass*

---

[4]Newton's equation, $\mathbf{F} = m\mathbf{a}$, is a *vector* equation. But for many purposes we require only the *magnitudes* of the force and the acceleration. Therefore, we will frequently write simply $F = ma$. One must always keep in mind the fact that the fundamental relationship is one involving *vectors*.

m_I to which Newton's second law applies? Nothing in our development of dynamics requires these two masses to be the same. This lack of knowledge is not a result of the fact that we have given only a summary of the important elements of dynamics; no one has yet been able to formulate a fundamental theory that prescribes the relationship between inertial mass and gravitational mass. However, it is possible to compare experimentally inertial and gravitational masses. Newton, in fact, addressed himself to this problem and made certain measurements, but his experiments were quite crude by today's standards. Sensitive experiments have recently been carried out which indicate that if there exists any difference between the two types of mass, this difference is less than about 1 part in 10^{11}. These experiments are of great importance in the general theory of relativity because it is a crucial assumption in this theory that inertial mass and gravitational mass are *exactly* equivalent. This important assertion is called the *principle of equivalence*. As our measurement techniques become more sophisticated, we can put this assertion to more stringent tests.

5.6 *Units of Force*

DYNES AND NEWTONS

Newton's second law specifies the dimensions for the quantity *force:*

$$F = ma$$

Therefore,[5]

$$[F] = [m] \times [a] = \text{g cm/sec}^2$$

We give the special name *dyne* to the unit of force in the CGS system:

$$1 \text{ dyne} = 1 \text{ g cm/sec}^2 \tag{5.3}$$

In the MKS system, we have

$$[F] = \text{kg m/sec}^2$$

and to this unit we give the name *newton* (N):

$$1 \text{ N} = 1 \text{ kg m/sec}^2 \tag{5.4}$$

Since $1 \text{ kg} = 10^3$ g and $1 \text{ m} = 10^2$ cm, the conversion factor between dynes and newtons is

$$1 \text{ N} = 10^5 \text{ dynes} \tag{5.5}$$

Example **5.1**

An object of mass 100 *g* is at rest. A net force of 2000 dynes is applied for 10 sec. What is the final velocity? How far will the object have moved in the 10-sec interval?

---

[5]We use the convention that square brackets around a symbol means "the dimensions (or units) of _____ are."

$$a = \frac{F}{m}$$

$$= \frac{2000 \text{ dynes}}{100 \text{ g}} = 20 \text{ cm/sec}^2$$

Since the acceleration is constant,

$$v = at$$

$$= (20 \text{ cm/sec}^2) \times (10 \text{ sec})$$

$$= 200 \text{ cm/sec}$$

$$s = \tfrac{1}{2} at^2$$

$$= \tfrac{1}{2} (20 \text{ cm/sec}^2) \times (10 \text{ sec})^2$$

$$= 1000 \text{ cm} = 10 \text{ m}$$

5.7   The Vector Property of Force

THE IMPORTANCE OF THE VECTOR SUM

In order to emphasize the vector nature of *force,* we now show some simple situations and vector diagrams. We continue to ignore friction in these situations.

Figure 5.8 illustrates three basic points: (a) a force is necessary to impart acceleration; (b) the forces applied to an object can cancel so that there is no acceleration; and (c) the net force is always the *vector* sum of the individual forces and may not be in the direction of any one force.

Figure 5.9 shows five additional situations:

(a) Here there are two forces acting on the object of mass m—the gravitational force $\mathbf{W}$ (that is, the weight of the object), and the upward pull of

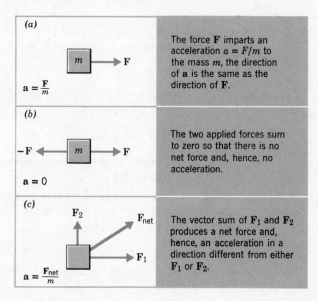

Fig. 5.8 *Three different force situations.*

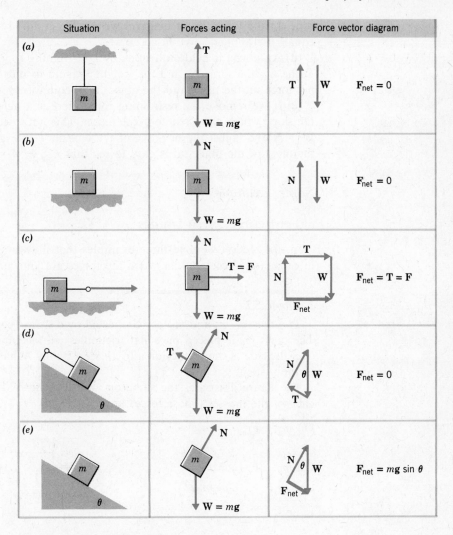

Fig. 5.9 *Analyses by force diagrams of five situations.*

the rope represented by the vector **T** (which stands for the *tension* in the rope). Of course, the rope also pulls *down* on the ceiling and the ceiling pulls *up* on the rope, but we are interested here only in the forces acting *on* the object because these and only these forces determine the acceleration of the object. The object is in equilibrium, so $\mathbf{W} = -\mathbf{T}$ and the net force is zero.

(b) In this example the mass is again in equilibrium so the gravitational force **W** is just balanced by the upward push of the floor represented by the vector **N** (which stands for the *normal* force, that is, that force perpendicular to the surface on which the object is resting). The net force acting on the object is zero.

(c) Next, we attach a rope of negligible mass to the object shown in Fig. 5.9b and pull with a force **F**. This force gives rise to a tension **T** in the rope. In the force vector diagram we see the addition $\mathbf{N} + \mathbf{T} + \mathbf{W}$ which equals the applied force $\mathbf{F}$.

(d) An object is held on a slope by a restraining rope. The rope tension $\mathbf{T}$, the normal force $\mathbf{N}$, and the weight $\mathbf{W}$, sum to zero so that there is no net force on the mass and the object is in equilibrium.

(e) If we remove the restraining rope, there is a net force acting down the slope which gives rise to acceleration. This net force is the vector sum $\mathbf{W} + \mathbf{N}$ and is just the negative of the tension $\mathbf{T}$ in Fig. 5.9d. In terms of the mass of the object, it is easy to see that $F_{\text{net}} = W \sin \theta = mg \sin \theta$.

5.8   Some Examples

CALCULATIONS WITH FORCES

In this section we give three examples that illustrate the methods used in calculating accelerations when the forces are known.

Example **5.2**

Two masses, m_1 and m_2, are connected by a string (considered massless). The mass m_2 is placed on a flat frictionless surface and m_1 is suspended over the side of the surface on a frictionless pulley. What is the acceleration of the system?

The figure illustrates the situation and also shows the forces acting on each of the masses. These forces are:

On m_1: $F_{\text{net},1} = m_1 g - T$
$$= m_1 a$$
On m_2: $F_{\text{net},2} = T$
$$= m_2 a$$

The net force on m_2 is just $\mathbf{T}$ since $\mathbf{N}$ and $\mathbf{W}_2$ cancel. In each case the acceleration is the same because the string connecting the masses remains taut.

Substituting for T in the expression for $F_{\text{net},1}$ gives

$$m_1 g - m_2 a = m_1 a$$

or,

$$m_1 a + m_2 a = (m_1 + m_2)a = m_1 g$$

from which we find

$$a = \frac{m_1}{m_1 + m_2} g$$

We could attack the problem in another way. First, we recognize that m_1 and m_2 are connected by a taut string and therefore can be considered to constitute a *system*. The mass of the system is $M = m_1 + m_2$ and Newton's second law is therefore written as

$$F_{net} = Ma$$

The net force on the system is just the gravitational force on m_1, in other words, $m_1 g$. That is, considering m_1 and m_2 as a *system,* we do not need to discuss the *internal* forces that hold it together (in this case, the tension in the string). The string remains taut and so there is always a fixed relationship between m_1 and m_2 which is determined by the length of the string; the pulley merely serves to change the direction of motion. Therefore, applying the second law, we have

$$Ma = m_1 g$$

or,

$$a = \frac{m_1}{M} g = \frac{m_1}{m_1 + m_2} g$$

which is the same result as before.

The solution to a problem is always simpler if all of the bodies involved act together as a system. Of course, before electing to use this approach in a problem, one must be careful to insure that there is indeed a *system* that acts together as a whole.

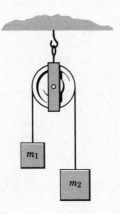

Example **5.3**

A pulley is suspended from a fixed support and over it are hung two unequal masses, m_1 and m_2 $(m_2 > m_1)$, which are connected by a massless string. What is the acceleration of the masses?

Here the two masses constitute a system and we do not need to consider the internal forces (the tension in the connecting string).

Since m_2 is the larger mass, the net force is the difference between the gravitational forces, $m_2g - m_1g$. Therefore,

$$F_{net} = Ma$$
$$m_2g - m_1g = (m_1 + m_2)a$$

from which we find

$$a = \frac{m_2 - m_1}{m_1 + m_2} g$$

If m_1 and m_2 are almost equal, then the acceleration will be very small; that is, the effect of gravity will be "diluted." George Atwood (1746–1807) used this fact to determine the value of g by measuring the accelerations of such systems. The accelerations were much smaller than g and therefore could be more easily measured than the acceleration of a freely falling body. This simple pulley system is sometimes called *Atwood's machine*.

5.9   *Momentum*

FORCE AND CHANGE OF MOMENTUM

Newton realized that quantities other than force, acceleration, and mass have physical importance. One of these quantities is *momentum* (which Newton called "quantity of motion"), defined to be the product of an object's *mass* and its *velocity*. We give to momentum (a *vector* quantity) the symbol **p**:

$$\text{Momentum} = \mathbf{p} = m\mathbf{v} \tag{5.6}$$

Frequently, we shall use the term *linear momentum* for **p** in order to distinguish it from *angular momentum,* which we introduce in Section 5.10.

Thus far, we have always considered a body's mass to remain constant. (But there is nothing in Newton's laws that *requires* mass to be constant with time.) In fact, there are important situations in which mass does change with time. For example, a rocket is propelled by the ejection of mass (usually in the form of gases at high velocity) and therefore the mass of the rocket-plus-fuel system decreases with time.[6] Newton allowed for the possibility that mass could change with time and actually originally stated his second law, not in terms of (mass × acceleration) as we have indicated, but in terms of the time rate of change of momentum. That is,

$$\overline{\mathbf{F}} = \frac{\Delta \mathbf{p}}{\Delta t} \quad \text{and} \quad \mathbf{F} = \lim_{\Delta t \to 0} \frac{\Delta \mathbf{p}}{\Delta t} \tag{5.7}$$

[6]This does not violate the law of mass conservation. Here, it is only the mass of the accelerating body that is changing; the sum of the masses of the body and the ejected gases is constant.

Using the definition of **p** and considering mass to remain constant in time, we have

$$\mathbf{F} = \lim_{\Delta t \to 0} \frac{\Delta(m\mathbf{v})}{\Delta t}$$

$$= m \lim_{\Delta t \to 0} \frac{\Delta \mathbf{v}}{\Delta t}$$

or, using the definition of acceleration,

$$\mathbf{F} = m\mathbf{a} \tag{5.8}$$

so that we recover the form of the second law that we used in Eq. II. *Equation 5.7 is the most general statement regarding force.* In the special case that the mass remains constant, Eq. 5.7 is equivalent to Eq. 5.8.

Example **5.4**

A 100-kg man jumps into a swimming pool from a height of 5 m. It takes 0.4 sec for the water to reduce his velocity to zero. What average force did the water exert on the man?

The man's velocity on striking the water was (see Problem 4.9)

$$v = \sqrt{2gh}$$

$$= \sqrt{2 \times (9.8 \text{ m/sec}^2) \times (5 \text{ m})}$$

$$= 10 \text{ m/sec}$$

Therefore, the man's momentum on striking the water was

$$p_1 = mv$$

$$= (100 \text{ kg}) \times (10 \text{ m/sec})$$

$$= 1000 \text{ kg-m/sec}$$

The final momentum was $p_2 = 0$, so that the average force was

$$\overline{F} = \frac{\Delta p}{\Delta t} = \frac{p_2 - p_1}{\Delta t}$$

$$= \frac{0 - 1000 \text{ kg-m/sec}^2}{0.4 \text{ sec}}$$

$$= -2500 \text{ N}$$

The negative sign means that the retarding force was directed opposite to the downward velocity of the man.

CONSERVATION OF LINEAR MOMENTUM

Next, we consider the importance of the concept of linear momentum. Suppose we have a system that is *isolated;* that is, the constituents of the system interact with one another but there is no outside agency that acts on them in any way. Truly isolated objects are not possible in the real physical world, but a group of objects whose mutual interaction is much greater than their interaction with other objects can frequently be treated as if they are isolated. For example, consider two hockey pucks that slide (almost) without friction over ice and collide with one another. The individual motions are influenced in a small way by the gravitational forces due to surrounding objects and to an even greater extent by friction, but the collisional interaction between the two pucks so overwhelms these other interactions that for most purposes the latter can be neglected.

If a group of objects constitutes an isolated system, there is no net force on this system. Since $\mathbf{F} = 0$, Eq. 5.7 requires that there be no change of linear momentum with time; in other words, $\mathbf{p} = $ const. This is an extremely important result:

If there is no external force applied to a system, then the total linear momentum of that system remains constant in time.

This is the statement of the *principle* (or *law*) *of the conservation of linear momentum.* The principle is valid not only in classical systems but in quantum mechanical situations as well.

The momentum to which the conservation principle applies is the *total linear momentum* of the system. If the system consists of two objects, m_1 and m_2, then

$$\mathbf{p} = \mathbf{p}_1 + \mathbf{p}_2$$
$$= m_1\mathbf{v}_1 + m_2\mathbf{v}_2 = \text{const.} \tag{5.9}$$

Suppose we have two masses at rest that are connected by a compressed spring, as in Fig. 5.10. (As in all such cases, we neglect the mass of the spring.) In this condition the total linear momentum of the system is zero and the conservation law tells us that it must always be zero. After the release of the spring, the objects move away from each other with velocities $\mathbf{v}_1$ and $\mathbf{v}_2$. Thus,

$$\underbrace{0}_{\substack{\text{before} \\ \text{release}}} = \underbrace{\mathbf{p}_1 + \mathbf{p}_2}_{\substack{\text{after} \\ \text{release}}}$$

so that

$$m_1\mathbf{v}_1 + m_2\mathbf{v}_2 = 0 \tag{5.10}$$

or, considering only the magnitudes of the velocities,

$$m_1v_1 = -m_2v_2 \tag{5.10a}$$

where the negative sign indicates that the velocities are oppositely directed.

Fig. 5.10 *The motion of the masses after release is determined by the conservation of linear momentum.*

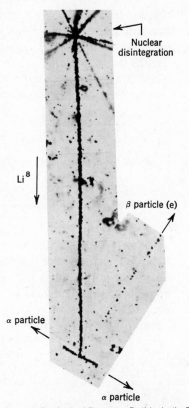

Nuclear
disintegration

Li⁸

β particle (e)

α particle

α particle

Fig. 5.11 *An example of the conservation of linear momentum in a nuclear process shown by the tracks of particles and nuclei recorded in a photographic emulsion. At the top of the photograph, a nuclear disintegration results when a cosmic-ray particle collides with a nucleus in the emulsion. From the nuclear disintegration there emerges a Li⁸ nucleus which travels a certain distance before coming to rest in the emulsion. Next, this nucleus, which is radioactive, decays to Be⁸ by the emission of a β particle (electron). Be⁸ is itself an unstable nucleus and it is the resulting decay that illustrates momentum conservation. Be⁸ breaks up into two identical α particles and these α particles leave the site of the original Be⁸ nucleus by moving in exactly opposite directions with equal momenta (as in the mechanical case shown in Fig. 5.10). The equal lengths of the two α-particle tracks in the emulsion is evidence for the equality of their momenta. (Because of their appearance in photographic emulsions, Li⁸ → Be⁸ → 2α decays are called "hammer tracks.")*

"The Study of Elementary Particles by the Photographic Method" by Powell

Example **5.5**

A high-powered rifle whose mass is 5 kg fires a 15-g bullet with a muzzle velocity of 3×10^4 cm/sec. What is the recoil velocity of the rifle?

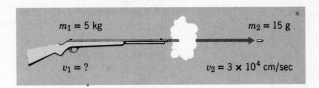

$m_1 = 5$ kg $m_2 = 15$ g

$v_1 = ?$ $v_2 = 3 \times 10^4$ cm/sec

From the conservation of momentum we have

$$m_1 v_1 = -m_2 v_2$$

$$v_1 = -\frac{m_2 v_2}{m_1}$$

$$= -\frac{(15\,\text{g}) \times (3 \times 10^4 \text{ cm/sec})}{5 \times 10^3 \text{ g}}$$

$$= -90 \text{ cm/sec}$$

This is a sizable recoil velocity and if the rifle is not held firmly against the shoulder, the shooter will receive a substantial "kick." However, if he *does* hold the rifle firmly against his shoulder, the shooter's body as a whole absorbs the momentum. That is, we must use for m_1 the mass of the rifle *plus* the mass of the shooter. If his mass is 100 kg, then the recoil velocity (now of the rifle plus shooter) is

$$v_1 = -\frac{(15g) \times (3 \times 10^4 \text{ cm/sec})}{5 \times 10^3 \text{ g} + 10^5 \text{ g}}$$

$$\cong -4.5 \text{ cm/sec}$$

This magnitude of recoil is quite tolerable.

Example **5.6**

In the preceding example, we needed the value of the muzzle velocity of the bullet in order to calculate the recoil velocity. Momentum conservation can also be used to measure such quantities. Suppose we fire the 15-g bullet into a 10-kg wooden block that is mounted on wheels and measure the time required for the block to travel a distance of 45 cm. This can easily be accomplished with a pair of photocells and an electronic clock. If the measured time is 1 sec, what is the muzzle velocity of the bullet?

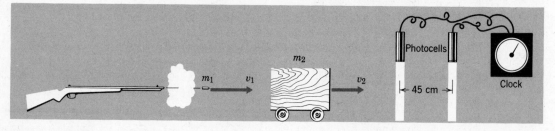

The recoil velocity of the block is 45 cm/sec, and from momentum conservation we have

$$m_1 v_1 = m_2 v_2$$

(Here, we do not have a negative sign because both velocities are in the same direction. Also, we take m_2 to be 10 kg, that is, we neglect the mass of the bullet embedded in the block.) Then,

$$v_1 = \frac{m_2 v_2}{m_1}$$

$$= \frac{(10^4 \text{ g}) \times (45 \text{ cm/sec})}{15 \text{ g}}$$

$$= 3 \times 10^4 \text{ cm/sec}$$

$$= 300 \text{ m/sec}$$

$$\cong 985 \text{ ft/sec}$$

In this example the bullet comes to rest in the block and imparts its momentum to the block. The process by which the bullet stops is a complicated one, but we need to know none of the details in order to calculate the velocity by using momentum conservation. This is indeed a powerful physical principle!

MOMENTUM CONSERVATION AND THE THIRD LAW

Previously, we used Newton's third law to express, in Eq. 5.1, the ratio of two isolated and interacting masses in terms of the ratio of their accelerations. Thus, the third law permits a definition of mass in terms of kinematical quantities. Equation 5.10a, written in the form

$$\frac{m_1}{m_2} = -\frac{v_2}{v_1}$$

states that we can use momentum conservation to determine the ratio of two masses by measuring the ratio of their velocities. But we obtained the momentum conservation law from Eq. 5.7 which is a statement of Newton's *second* law. Can we therefore avoid the necessity of introducing the third law, since it apparently serves no function other than to define mass, and obtain all of the machinery of dynamics from the second law?

The answer to this question is an emphatic "no." The reason is that we have really used the third law in obtaining the momentum conservation law. Consider the statement that led directly to momentum conservation: *If a group of objects constitutes an isolated system, there is no net force on this system.* This statement can be true only if the third law is obeyed, for if object 1 exerts a force on object 2 there will be a net force on the system unless object 2 exerts an equal and opposite force on object 1. An isolated system cannot accelerate itself; this is guaranteed by the third law, in the guise of momentum conservation.

CONSERVATION OF THE MOMENTUM VECTOR

Momentum is a *vector* quantity, so the principle of momentum conservation can be applied component by component. That is, if **p** is conserved, then p_x and p_y (and also p_z if the problem is three-dimensional) are *individually* conserved.

Example 5.7

A railway gun whose mass is 70,000 kg fires a 500-kg artillery shell at an angle of 45° and with a muzzle velocity of 200 m/sec. Calculate the recoil velocity of the gun.

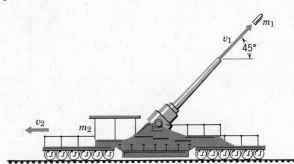

First, consider the horizontal or x-component of the shell's momentum:

$$p_{1x} = m_1 v_1 \cos 45°$$
$$= (500 \text{ kg}) \times (200 \text{ m/sec}) \times 0.707$$
$$= 7.07 \times 10^4 \text{ kg-m/sec}$$

This must equal (except for the sign) the recoil momentum of the gun which moves only horizontally:

$$p_2 = m_2 v_2 = -7.07 \times 10^4 \text{ kg-m/sec}$$

Therefore,

$$v_2 = -\frac{7.07 \times (10^4 \text{ kg-m/sec})}{7 \times 10^4 \text{ kg}}$$
$$\cong -1 \text{ m/sec}$$

or, approximately 2 mi/hr.

What has happened to the *vertical* component of the recoil momentum, $p_{1y} = 7.07 \times 10^4$ kg-m/sec? Since the railway platform is in contact with the Earth, the Earth absorbs the vertical momentum. The Earth *does* recoil, but because of the extremely large value of the Earth's mass compared to that of the railway gun, the recoil velocity cannot be measured.

5.10   *Torque and Angular Momentum*

A TORQUE CAN PRODUCE ROTATION

If a force is applied to the handle of a door, the door can be opened or closed; that is, a rotation around the hinge line can take place (Fig. 5.12a). But if the same force is applied in such a way that the force vector passes *through* the hinge line, there can be no rotation (Fig. 5.12b). A force that is applied "off center," and can therefore produce rotation constitutes a *torque.*

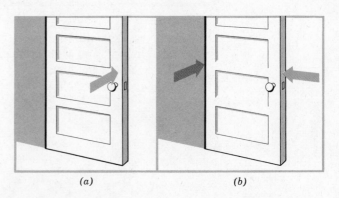

(a)   (b)

Fig. 5.12   (a) *The applied force causes rotation around the hinge line.* (b) *The applied forces each pass through the hinge line and no rotation results.*

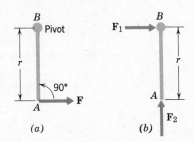

Fig. 5.13 (*a*) *The torque is* $T = r \times F$. (*b*) *The force* $\mathbf{F}_1$ *is directed at the pivot point so that* $r = 0$ *and there is no torque. The force* $\mathbf{F}_2$ *has no component perpendicular to the line AB so there is no torque.*

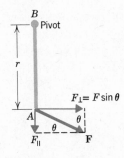

Fig. 5.14 *The torque is* $T = rF \sin \theta$. *Note that* θ *is the angle between* **F** *and the line that connects the pivot point to the point at which the force is applied.*

The quantitative measure of a torque is the distance from the point of application of the force to the center of rotation multiplied by the component of the force perpendicular to this distance. In Fig. 5.13a, the force **F** acts at the point A which is a distance r from the center of rotation. The force vector is perpendicular to the line AB and therefore the entire force is effective in producing the torque:

$$\text{torque} = T = r \times F \tag{5.11}$$

In Fig. 13b, neither F_1 nor F_2 can produce a torque around the point B.

If the force is applied at an angle to the line AB, as in Fig. 5.14, then only the *perpendicular* component of the force $F_\perp$ can contribute to the torque:

$$T = r \times F_\perp$$

or, since $F_\perp = F \sin \theta$,

$$T = rF \sin \theta \tag{5.12}$$

(Clearly, the other component of the force, $F_\parallel$, can produce no rotation, since its direction is in line with the pivot point.)

It is not necessary to have a pivoted rod in order to define torque. The force F applied to the free particle of mass m in Fig. 5.15 constitutes a torque, $T = r \times F$, referred to the arbitrary point O. If the constant force F acts for a certain time, the particle will accelerate, but since it will move in the direction of the force, the *torque* around O will remain the same. (Check this by noting that the product of $F_\perp$ and the distance from m to O remains constant.)

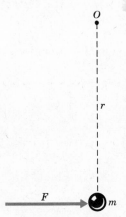

Fig. 5.15 *The torque on* m *around the arbitrary point* O *is* $T = r \times F$.

THE CONSERVATION OF ANGULAR MOMENTUM

From Newton's laws we have obtained the important result that if the net force on a body is zero, the linear momentum will remain constant. It is also true that if the net *torque* on a body is zero, the *state of rotation* will remain constant. This fact constitutes another important conservation law, on a level equal to the conservation of linear momentum. To state this new conservation principle, we define a quantity called the *angular momentum*, which is the product of the distance from the object to the axis of

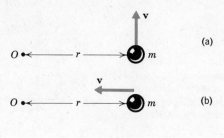

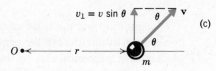

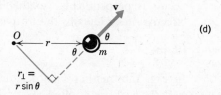

Fig. 5.16 (a) *The angular momentum about* O *is* L = rmv. (b) *The angular momentum about* O *is zero.* (c) *The angular momentum about* O *is* L = rmv$_\perp$ = rmv *sin* θ. (d) *The angular momentum about* O *is* L = r$_\perp$mv = rmv *sin* θ.

rotation and the perpendicular component of the linear momentum, as in Fig. 5.16. Angular momentum is denoted by *L*:

$$
\begin{aligned}
L &= r \times p_\perp \\
&= rmv_\perp \\
&= rmv \sin \theta
\end{aligned}
\tag{5.13}
$$

where θ is the angle between **p** (or **v**) and the line connecting *m* and the axis point O; then, $v_\perp = v \sin \theta$, as in Fig. 5.16c.

Notice, in Fig. 5.16d, that we can write equivalent expressions for *L* as

$$L = rmv_\perp$$

or,

$$L = r_\perp mv \tag{5.13a}$$

where $r_\perp = r \sin \theta$ is the perpendicular distance from the axis point O to the line defined by the vector **v**. Therefore,

$$L = (r \sin \theta) \times mv$$
$$= rmv \sin \theta$$

as before.

Because the *linear* velocity *v* is equal to the product of *r* and the *angular* velocity ω (Eq. 4.30), we can use $v = r\omega$ to express the angular momentum as

$$L = mr^2 \omega \sin \theta \tag{5.14}$$

In terms of the angular momentum thus defined, we can state: *If no external torque is applied to a body or system of bodies, the angular momentum remains constant.* This is the statement of the principle (or law) of *conservation of angular momentum.* This is a well-established law of physics; no exceptions or contradictions are known. As is the case for linear momentum, the conservation of angular momentum is valid in quantum mechanical as well as classical systems.

Just as the *force* on a body is equal to the time rate of change of *linear momentum* (Eq. 5.7), the *torque* is equal to the time rate of change of *angular momentum:*

$$\bar{T} = \frac{\Delta L}{\Delta t} \text{ or } T = \lim_{\Delta t \to 0} \frac{\Delta L}{\Delta t} \tag{5.15}$$

The angular momentum conservation principle follows directly from this statement since, if $T = 0$, then L does not change with time.

It should be noted that angular momentum is a well-defined quantity even if the object is not moving in a curved path. For example, in Fig. 5.16a, if the mass m moves in a straight line with the velocity vector as shown, the perpendicular distance from its line of motion to the point O is always r, therefore, the angular momentum about this point is always $L = mvr$ (in the absence of any external torque). In Fig. 5.15, on the other hand, the torque on m around O is constant and therefore the angular momentum around O increases uniformly with time (that is, at a constant rate).

Example **5.8**

A ball of mass 100 g is attached to the end of a string and is swung in a circle of radius 100 cm with a constant linear velocity of 200 cm/sec. While the ball is in motion, the string is shortened to 50 cm. What is the change in the velocity and in the period of the motion?

The initial angular momentum is

$L = mvr$

$= (100 \text{ g}) \times (200 \text{ cm/sec}) \times (100 \text{ cm})$

$= 2 \times 10^6 \text{ g-cm}^2/\text{sec}$

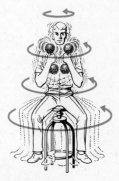

Fig. 5.17 *A student is set to rotating on a stool while holding two massive dumbbells at arm's length. When he draws the dumbbells to his sides,* r *decreases so that the angular velocity* ω *must increase in order to conserve angular momentum (see Eq. 5.14).*

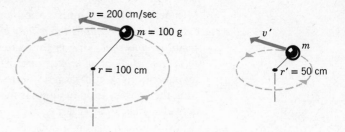

The initial period is

$$\tau = \frac{2\pi}{\omega} = \frac{2\pi}{(v/r)} = \frac{2\pi r}{v}$$

$$= \frac{2\pi \times (100 \text{ cm})}{200 \text{ cm/sec}}$$

$$= \pi \text{ sec}$$

Shortening the string does not apply any torque to the ball because the applied force lies along the line connecting the ball with the center of rotation. Therefore, the final angular momentum is equal to the initial angular momentum:

$$L' = mv'r' = L$$

Thus, the final velocity is

$$v' = \frac{L}{mr'}$$

$$= \frac{2 \times 10^6 \text{ g-cm}^2/\text{sec}}{(100 \text{ g}) \times (50 \text{ cm})}$$

$$= 400 \text{ cm/sec}$$

The new period is

$$\tau' = \frac{2\pi r'}{v'}$$

$$= \frac{2\pi \times (50 \text{ cm})}{400 \text{ cm/sec}}$$

$$= \frac{\pi}{4} \text{ sec}$$

Therefore, decreasing the radius by a factor of 2 has increased the linear velocity by the same factor but has decreased the period by a factor of 4.

Example **5.9**

A satellite of mass m moves around the Earth as shown (actually, the path is an *ellipse*). Which instantaneous velocity is greater, v_1 (at point P) or v_2 (at point A)?

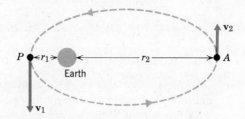

Considering the Earth as a fixed object and neglecting the influence of the Sun and other planets, the angular momentum of the satellite around the Earth is constant. Therefore,

$$mv_1r_1 = mv_2r_2$$

Since $r_1 < r_2$, we must then have

$$v_1 > v_2$$

The velocity is greatest when the satellite is nearest the Earth; this point is called the *perigee* (labeled P in the diagram). The velocity is least at the farthest point from the Earth—the *apogee* (A) of the orbit.

THE DIRECTION OF ANGULAR MOMENTUM

Angular momentum is actually a *vector* quantity. The *magnitude* of the angular momentum vector has already been defined, but because *angular* motion is involved, the *direction* of the vector must be specified separately. If a particle is moving in a circular path around a certain point, the angular momentum vector is defined to have the direction in which a right-hand screw would advance if moved in the same sense (see Fig. 5.18). Thus, for motion in a plane, the angular momentum vector is *perpendicular* to that plane. In an x-, y-, z-coordinate system, a particle that moves in the x-y plane around the origin has an angular momentum vector that lies along the z axis. Alternatively, if the fingers of the right hand are curled in the direction of the motion of the particle, the thumb will point in the direction of **L**.

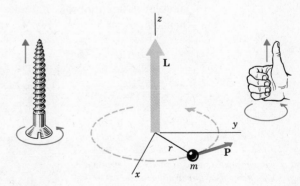

Fig. 5.18 *The direction of the angular momentum vector is the same as the direction of advance of a right-hand screw moving in the same sense; or, if the fingers of the right hand are curled in the direction of motion, the thumb points in the direction of* **L.**

5.11   *Conservation of Momentum and the Discovery of the Neutrino*

THE "MISSING" MOMENTUM IN β DECAY

In physics we always seek to interpret the observational facts regarding natural phenomena in the simplest possible terms. The various conservation laws provide us with the means toward this end. We know that in every physical process certain quantities do not change—mass (actually, mass-energy), linear momentum, and angular momentum. In the following chapters we shall find that other quantities are also conserved. These conservation laws have been enormously useful in providing the essential clues in solving many fundamental problems in physics.

One of the successes of the conservation principles in recent years was the discovery of an elusive elementary particle, called the *neutrino*. In radioactive β decay, the atomic nucleus emits an electron and in the process is transformed into a new atomic species that differs in atomic number by one unit. For example, radioactive C^{14} (carbon-14) emits an electron and becomes N^{14} (nitrogen-14) (see Fig. 3.30). Let us consider the momentum balance in such a process. In Fig. 5.19 we have initially a nucleus A that is at rest. Thus, the initial linear momentum is zero: $\mathbf{p}_1 = 0$. On decay, an electron is emitted with a certain momentum $\mathbf{p}_e$. Conservation of linear momentum therefore requires that the transformed nucleus B recoil with a momentum $\mathbf{p}_B$ that is equal in magnitude but opposite in direction. That is,

Before   After
  decay: $\mathbf{p}_1 = 0$ decay: $\mathbf{p}_B + \mathbf{p}_e = 0$

The two momentum vectors are said to be *co-linear*.

However, in actual β decay it was discovered that the electron momentum and the recoil momentum were, in general, *not* co-linear (Fig. 5.20). It appeared that linear momentum was not conserved since no object was observed that could carry off the "missing" linear momentum.

An additional problem arose concerning *angular* momentum. Every nucleus and every elementary particle has associated with it a certain amount of angular momentum. We can picture this angular momentum arising from a spinning motion of the particle (Fig. 5.21), but in reality the angular momentum is an *intrinsic* property of the particle and does not depend on mechanical motion. Nevertheless, this picture is a useful one; this type of angular momentum of elementary particles and nuclei is called *spin angular momentum,* or merely *spin*. In radioactive β decay it was found that the

Fig. 5.19 *If radioactive β decay consisted solely of the emission of an electron accompanied by the recoil of the transmuted nucleus, the two momentum vectors would be equal in magnitude and opposite in direction.*

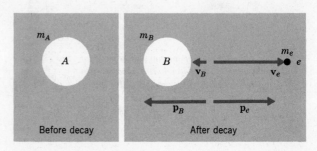

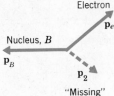

Electron

$\mathbf{p}_e$

Nucleus, *B*

$\mathbf{p}_B$

$\mathbf{p}_2$

"Missing"
momentum
(neutrino, *ν*)

Fig. 5.20   *In actual β de-cay it is found that the momenta of the electron and the recoil nucleus are not colinear. There is some "missing" linear momentum which is attributed to the emission of an additional "particle," the neutrino.*

spin of nucleus *A* did not equal the (vector) sum of the spins of nucleus *B* and the electron. It therefore appeared that angular momentum was not conserved.[7]

THE POSTULATE OF THE NEUTRINO

There seemed to be no acceptable escape from this dilemma; apparently linear momentum and angular momentum (and energy) were not conserved in the elementary process of β decay. However, the conservation principles were (and are) so much a part of the thinking concerning physical processes that there was a great reluctance to forego these enormously successful and useful laws. In order to "solve" the problem, Wolfgang Pauli, in 1930, postulated the existence of a new elementary particle that we now call the *neutrino*. If Pauli's neutrino was emitted along with the electron in β decay, the neutrino could carry off the "missing" linear momentum, angular momentum, and energy. But to do all of this and to account for the fact that it had thus far escaped detection, the neutrino had to be endowed with unlikely properties—it must carry no charge and have no mass, it must have almost no interaction with ordinary matter, but it must carry off linear momentum, angular momentum, and energy while traveling with the velocity of light! The neutrino was therefore a "ghost" particle and many regarded it merely as a crutch on which the conservation principles were propped. However, in 1953, Clyde Cowan and Fredrick Reines conducted a remarkable series of experiments in which the neutrino was actually detected. Thus, not only was the faith in the conservation laws vindicated, but the way was opened for investigation of this elusive new elementary particle—one that is of great current interest in high energy physics research and in astrophysics.

5.12   *Center of Mass*

THE SIGNIFICANT POINT FOR SYSTEMS

We have stated the principles of conservation of linear momentum and angular momentum as applied to *particles*. But what is the significance of these principles for aggregates of particles (*systems*) or macroscopic bodies? In order to apply the conservation principles with complete generality, we must find the single point within a body or system of bodies that always moves in accordance with the conservation laws. This point is called the *center of mass* of the system. When considering the implications of the conservation laws on the motion of a system of bodies, we must always refer to the motion of the center of mass of that system.

CALCULATION OF THE POSITION OF THE CENTER OF MASS

For a simple system consisting of two particles we can calculate the position of the center of mass as follows. Imagine that the two masses are

Spin
angular
momentum

Proton

Fig. 5.21   *The intrinsic angular momentum of a proton, for example, can be pictured as arising from a spinning motion of the particle.*

[7]It was also found that energy conservation appeared to be violated. *Energy* is discussed in Chapter 7. The argument is compelling even without the introduction of energy considerations.

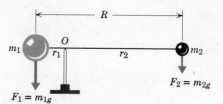

Fig. 5.22 *Two unequal masses are balanced on a pivot.*

connected by a rigid, massless rod, as shown in Fig. 5.22. This system can be balanced on a pivot at a certain point O that corresponds to the center of mass. Because there is no rotation of the system around O, we know that the torque around O produced by the gravitational force on m_1 equals the torque around O produced by the gravitational force on m_2. That is,

$$m_1 g r_1 = m_2 g r_2$$

or,

$$m_1 r_1 = m_2 r_2 = m_2 (R - r_1) \qquad (5.16)$$

where $R = r_1 + r_2$ is the distance between m_1 and m_2.

We can always calculate the position of the center of mass of two objects by using Eq. 5.16. For systems consisting of three or more particles or for rigid bodies, the expressions for the coordinates of the center of mass are easy to derive but they are more complicated and we have no need for them here.

CENTER OF MASS AND THE LAWS OF DYNAMICS

We can summarize the importance of the center-of-mass concept with the following statements:

1. In the absence of external forces, the center of mass of a system moves with constant velocity.

2. If a force is applied to a system as a whole, the center of mass undergoes an acceleration $\mathbf{a} = \mathbf{F}/M$, where M is the total mass of the system.

3. In the absence of external torques, the total angular momentum of the system around its center of mass remains constant.

Fig. 5.23 *Time-lapse (stroboscopic) photograph of a wrench sliding across a smooth surface while rotating about its center of mass. The position of the C.M. is marked with black tape as a cross. Use a ruler and verify that the C.M. moves in a straight line, as required by Newton's laws, even though the wrench as a whole undergoes a complicated motion.*

Summary of Important Ideas

Except for objects moving with velocities approaching the velocity of light and for effects that take place on an atomic scale, *Newton's laws* are essentially a correct description of the dynamics of physical systems.

If Newton's laws are valid in a certain frame of reference, they are also valid in *any* frame that moves with *constant linear velocity* relative to the first frame.

A body can be accelerated only if a *force* is applied to it. Force is the time rate of change of momentum.

The motion of a body is determined by the net force acting *on* the body and does not depend on the forces that the body exerts on other bodies.

Mass is an intrinsic property of matter; *weight* is the *gravitational force* on an object.

According to the *principle of equivalence,* gravitational mass and inertial mass are identical.

Force, linear momentum, and angular momentum are all *vectors.*

If no external force is applied to a system, the linear momentum of the center of mass of the system remains *constant* in time.

If no external torque is applied to a system, the angular momentum of the system around its center of mass remains *constant* in time.

Questions

5.1 Explain how it is possible for an experienced fisherman to land a 100-lb fish with a line rated at 10-lb breaking strength. Does the fisherman ever use only the line to pull the fish out of water and into the boat?

5.2 A heavy weight is suspended from the ceiling by a length of string. Hanging downward from the weight is another length of the same string. Which string will break when (a) a steady pull is exerted on the lower string and (b) a sudden pull is exerted on the lower string? Explain.

5.3 List the controls in an automobile which, when activated, tend to make a passenger alter his position in his seat. (There are at least four.) What is the nature of the acceleration produced by each?

5.4 A rocket is coasting at constant velocity in free space. When the rocket engine is turned on, the rocket begins to accelerate. Explain how this is possible since there is no matter in the vicinity that can exert a force on the rocket.

5.5 What is the principle of operation of a *recoilless* rifle?

5.6 What is the difference between the form of the recoil of a rifle on firing a bullet depending on whether the bore is smooth or rifled? (The rifling imparts a spin to the bullet.)

5.7   A wheel can turn freely in a horizontal plane on a vertical axle that is stuck in the ground. A cat is sitting on the rim of the wheel.

(a) The wheel is not in motion. Suddenly the cat begins to walk around the rim. What happens to the wheel and why?

(b) The wheel is spinning at a constant rate with the cat stationary at one point on the rim. Suddenly the cat begins to move toward the center of the wheel. What happens to the wheel and why?

5.8   Consider a stool top that can move freely about a pivot at the center. A boy sits on the stool and holds a spinning top with its axis vertical; the stool is at rest. Suddenly, the boy grasps the top and stops the spinning. What is the result?

5.9   Would it be possible to design a practical helicopter with only one set of rotating blades? Explain.

5.10 What is the advantage of throwing a football so that it spirals about its axis?

5.11 An artillery piece fires a shell. While still in flight the shell explodes and breaks into many pieces. What statement can you make regarding the motion of the shell fragments? (Consider the trajectory of the center of mass of the shell fragments.)

Problems

5.1   A ball is swung in a vertical circle with constant speed at the end of a 1-m rope. What must be the period of the motion for the ball to be "weightless" at the top of the swing? (Here, "weightless" means that there would be zero tension in the rope and yet it would remain straight.)

5.2   A 200-lb man stands on a set of scales in an elevator. The elevator begins to descend with an acceleration $a = 5$ ft/sec^2. What is the reading on the scales? What is the reading when the elevator is ascending with $a = 3$ ft/sec^2?

5.3   A 10-kg object is pulled over a rough surface by a 20-newton force. The object accelerates at a rate of 1.5 m/sec. What is the frictional force between the object and the surface? [The *net* force on the object is equal to (*applied force*) − (*frictional force*).]

5.4   Two forces, one twice as large as the other, both pull on an object in the same direction ($m = 100$ g) and impart to it an acceleration $a = 3$ m/sec. What are the magnitudes of the forces? If the smaller force is removed, what is the new acceleration?

5.5   A 100-g object is at rest at the origin of an x-y coordinate system. At a certain instant two forces are applied to the object—250 dynes in the $+y$ direction and 500 dynes in the $+x$ direction. Describe the subsequent motion of the object.

5.6   A 1-kg block is pulled across a horizontal surface by a force of 2 N that is directed at an angle of 45° above the horizontal. If there is a frictional retarding force of 0.5 N, what is the acceleration of the block?

5.7 A 40-kg block is sliding over a rough surface. At a certain instant the velocity of the block is 2 m/sec. If the block is slowed to rest by friction in a uniform manner after the block has traveled 40 m, what was the average frictional force?

5.8 A pendulum consists of a 100-g ball suspended from a 1-m string. The ball is drawn to one side until the string makes an angle of 30° with the vertical and is then released. Sketch the force diagram at the instant of release. What is the initial acceleration of the ball? (Find the magnitude and the direction.)

5.9 When a certain type of bullet ($m = 10$ g) is fired from a rifle, a force of 3×10^8 dynes is exerted for a millisecond. What is the muzzle velocity of the bullet?

5.10 A fire hose ejects 50 kg of water per second at a speed of 40 m/sec. What force must be exerted by the firemen holding the hose in order to keep it stationary?

5.11 A ball ($m = 100$ g) is thrown against a wall with a velocity of 10 m/sec and it rebounds with a velocity of the same magnitude. If the ball was in contact with the wall for 1 msec (10^{-3} sec), what was the average force exerted on the ball by the wall? (Find the magnitude and the direction of the force.)

5.12 In Example 5.6 it is found that the bullet penetrates 10 cm of wood. Assume that there is a uniform decrease of the velocity to zero. How long did it take for the bullet to come to rest? What average force did the wood exert on the bullet?

5.13 In "shoot-'em-up" movies, the villain is often knocked down by a single bullet from the hero's gun. A .357 magnum bullet has a mass of approximately 10 g and a muzzle velocity of 4×10^4 cm/sec. What would be the recoil velocity of an average-size man struck by such a bullet? Is it likely that he would be knocked down?

5.14 An object of mass 100 g is moving in the x-direction with a velocity of 45 cm/sec and another object of mass 200 g is moving in the y-direction with a velocity of 15 cm/sec. The two objects collide and stick together. Describe the subsequent motion.

5.15 A cannon of mass 1000 kg is mounted on wheels and rests on a flat surface. If a shell of mass 10 kg is fired horizontally with a muzzle velocity of 200 m/sec, what will be the recoil velocity of the cannon? If the cannon were attached rigidly to the deck of a ship of mass 20×10^6 kg which is moving with a velocity of 10 km/hr, what will be the velocity of the ship after the cannon is fired *forward?*

5.16* Two children ($m = 30$ kg each) are sliding on ice in "saucers" ($m = 1$ kg each). They are moving side by side in parallel paths with the same velocity, $v = 3$ m/sec. Each child has with him a 2-kg block. If each child throws his block with a velocity of 10 m/sec to the other who catches it, describe the subsequent motion of each child. (Assume the saucer–ice friction is negligible.)

5.17 A 10-kg cart is moving over a horizontal surface at a constant velocity of 2 m/sec. A 2-kg lump of clay is dropped into the cart from a height of

4 m and it sticks to the bed. Describe the subsequent motion. What happened to the vertical momentum?

5.18   An object ($m = 350$ g) is at rest at the origin of a certain coordinate system. An internal explosion breaks the object into three pieces. One piece ($m_1 = 100$ g) goes off in the $-x$ direction with a velocity of 5 m/sec. A second piece ($m_2 = 200$ g) goes off in the $-y$ direction with a velocity of 2.5 m/sec. What is the velocity and direction of the third piece?

Space Walk

6 *The Basic Forces of Nature*

In our everyday experience we encounter a variety of forces—the muscular force exerted to open a door, the frictional force in the door hinges, and the elastic force in the door spring; the force that the atmosphere exerts on a barometer and the force that the Earth exerts on the moon; the electrical force that starts an automobile engine, the hydraulic force that operates the brakes, or the mechanical force that stops the car if we are unfortunate enough to collide with a lamp post. In spite of the large number of names that we have given to forces that we use or must overcome, there are only *two* basic forces that govern the behavior of all everyday objects—the *gravitational* force and the *electrical* force. All of the various forces mentioned above are actually only different manifestations of these two fundamental forces.

In this chapter we examine in detail the gravitational force and that part of the electrical force that acts between charges at rest—the *electrostatic* force. However, these forces are not sufficient to describe *nuclear* phenomena. The study of processes involving nuclear and elementary particles has shown that there are two additional forces in Nature, the so-called *strong* nuclear force and the *weak* force. Because the gravitational and electrostatic forces have *long ranges* (that is, they are effective over large distances), these forces are exclusively responsible for all large-scale phenomena—those that we encounter in everyday experience and those that we can see taking place in distant stars and galaxies. The nuclear force and the weak force, on the other hand, are of extremely short range and their effects are evident only on the scale of nuclear sizes. Nevertheless, these forces play a crucial role in our existence. All life on Earth is sustained by the light that we receive from the Sun and this light is the end result of nuclear processes that take place deep in the Sun's interior.

Thus, none of the four basic forces is superfluous; all are required for the orderly operation of the Universe. Of course, there is always the possibility that Nature is more complicated than we are aware of, but at the present time we know of only four basic forces—the gravitational force, the electrical force, the strong nuclear force, and the weak force. There seems to be no need to invoke any additional type of force to account for any observed process.[1] It is interesting that Nature, in her simplicity, has contrived to run her entire Universe by employing only *four* basic forces.

6.1 *The Gravitational Force*

NEWTON'S CALCULATION

The acceleration of an object can only be produced by the application of a force. *Gravitational force* produces an acceleration of 32 ft/sec^2 for

[1]There are proponents of a certain brand of general relativity who propose that there are two distinct types of gravitational forces, but this suggestion has not as yet received any firm experimental confirmation.

objects falling freely near the surface of the Earth. Although this force is clearly in operation on the Earth, we can ask the question, "Is this force unique to the Earth or does it operate throughout the Universe?" Newton addressed himself to this question but he knew of no way to investigate the "universality" of gravitation except within the solar system. He therefore chose the Earth-moon system for his study and hypothesized that the force that holds the moon in its orbit around the Earth is the same as the force that attracts objects near the surface of the Earth. Newton knew, from previous triangulation measurements, that the Earth-moon separation r_m is approximately 60 times the radius of the Earth R_E, which is about 4000 mi (see Fig. 6.1); therefore, $r_m \cong 240,000$ mi. He also knew the period of the moon's rotation around the Earth — $\tau = 27.3$ days. In order to maintain this motion, the moon must experience a centripetal acceleration equal to the square of its linear velocity divided by the radius of the orbit — $a_c = v^2/r_m$. The linear velocity of the moon is

$$v = \frac{\text{circumference of orbit}}{\text{period of rotation}}$$

$$= \frac{2\pi r_m}{\tau} = \frac{2\pi \times (240,000 \text{ mi})}{(27.3 \text{ days}) \times (86,400 \text{ sec/day})}$$

$$= 0.64 \text{ mi/sec}$$

Therefore, the centripetal acceleration of the moon is

$$a_c = \frac{v^2}{r_m}$$

$$= \frac{(0.64 \text{ mi/sec})^2}{240,000 \text{ mi}} \times (5280 \text{ ft/mi})$$

$$= 0.009 \text{ ft/sec}^2$$

This acceleration is smaller than the value of g at the surface of the Earth by the factor

$$\frac{a_c}{g} = \frac{0.009 \text{ ft/sec}^2}{32 \text{ ft/sec}^2} \cong \frac{1}{3600}$$

The rule that governs the diminution of a quantity as it spreads out uniformly from the source without being absorbed (for example, *light;* see

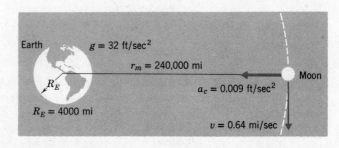

Fig. 6.1 *The centripetal acceleration of the moon in its orbit is due to the gravitational attraction of the Earth.*

Fig. 3.4) is that the intensity decreases with the *square* of the distance:

Intensity $\propto \dfrac{1}{r^2}$

Newton therefore made the assumption that gravity, just as light, follows the inverse square law of intensity. Thus, if the moon is 60 times farther away from the center of the Earth than is an object at the surface of the Earth, the gravitational force, and hence the acceleration, should be smaller for the moon by a factor $1/(60)^2$ or $1/3600$. This is just the value of the ratio a_c/g.

Although the calculated and observed accelerations agree, thus confirming Newton's hypothesis, it is important to realize two points regarding the derivation:

1. Newton *assumed* that the force of gravity was inversely proportional to the square of the separation of the interacting bodies. This assumption gave the correct result for the Earth-moon system, but there was no guarantee that it would be valid for other systems with different separations. Newton went on to apply the $1/r^2$ law to planetary motion and again found it valid. In more recent times, binary star systems (that is, two stars rotating around one another) have been found to obey the Newtonian form of the gravitation law. It therefore appears that the gravitational force varies as $1/r^2$ throughout the Universe. No exceptions to this rule have ever been found. (However, see the remarks at the end of Section 6.2.)

2. In comparing the centripetal acceleration of the moon with the value of *g*, both distances were measured from the *center of the Earth*. When he originally made his calculation (in 1666), Newton could not justify this choice and therefore refrained from publishing his results. In fact, it was not until 1687, by which time Newton had developed the necessary mathematical tools (the *calculus*) to justify his earlier calculation, that his law of gravitation was formally announced. Newton had succeeded in showing, with his calculus, that if two uniform spherical objects attract each other with a force that varies as $1/r^2$, the force will always be correctly calculated if the mass of each body is considered to be *concentrated entirely at its center*. This important result was the key to the universal gravitation law—it showed that all gravitational calculations could be made by considering the entire mass of a spherical body, such as the Earth or the moon, to be located at its center.

THE UNIVERSAL GRAVITATIONAL FORCE LAW

Newton's results are summarized by the statement that *the gravitational force which two particles or spherical objects mutually exert on one another is inversely proportional to the square of the distance between their centers and directly proportional to the product of their masses.*[2] That is,

$$F_G = G\frac{m_1 m_2}{r^2}$$

(6.1)

---

[2]If the objects are so large that they cannot be considered to be point particles or are not spherical, a more complicated statement is necessary which amounts to considering each object to be composed of a large number of point particles.

Mount Wilson and Palomar Observatories

Fig. 6.2 *A globular cluster of stars in the constellation Hercules. Hundreds of thousands of stars are held together in this cluster by gravitation. Richard Feynman, winner of a share of the 1965 Nobel Prize in physics, has said that "if one cannot see gravitation acting here, he has no soul."*

where the constant of proportionality G is called the *universal gravitational constant*. The value of G is (see Section 6.3)

$$\left.\begin{aligned} G &= 6.673 \times 10^{-8} \text{ dyne-cm}^2/\text{g}^2 \quad \text{(CGS)}\\ &= 6.673 \times 10^{-11} \text{ N-m}^2/\text{kg}^2 \quad \text{(MKS)} \end{aligned}\right\} \tag{6.2}$$

Example **6.1**

With what force does the Earth attract the moon?

From Tables 2.3 and 2.6 (or from the table of astronomical data inside the front cover) we have the following values:

$r_m = 3.84 \times 10^{10}$ cm

$m_m = 7.35 \times 10^{25}$ g

$m_E = 5.98 \times 10^{27}$ g

Therefore, using Eq. 6.1, we have

$$F_G = G \frac{m_m m_E}{r_m{}^2}$$

$$= (6.67 \times 10^{-8} \text{ dyne-cm}^2/\text{g}^2) \times \frac{(7.35 \times 10^{25} \text{ g}) \times (5.98 \times 10^{27} \text{ g})}{(3.84 \times 10^{10} \text{ cm})^2}$$

$$= 2.0 \times 10^{25} \text{ dynes}$$

We can obtain this result in another way by using the value of the moon's centripetal acceleration calculated above. The force on the moon is just its acceleration times its mass:

$$F_G = m_m a_c$$

$$= (7.35 \times 10^{25} \text{ g}) \times (0.009 \text{ ft/sec}^2) \times (30.48 \text{ cm/ft})$$

$$\cong 2 \times 10^{25} \text{ dyne}$$

SYNCHRONOUS SATELLITES

One of the most significant recent advances in the field of communications has been the establishment of a system of artificial satellites that remain in fixed positions at a certain height over the Earth's equator. These *synchronous satellites* are equipped to provide relay facilities for many channels in the intercontinental communications network (Fig. 6.3).

The height at which these satellites must be placed into orbit is determined by the fact that the period of rotation must exactly equal the period of rotation of the Earth. If the periods are equal, the satellite and the Earth will rotate together (*synchronously*) and the satellite will remain in a fixed position with respect to the Earth.

The linear velocity of the satellite in terms of its distance R measured from the center of the Earth is

$$v = \frac{2\pi R}{\tau}$$

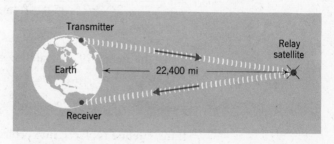

Fig. 6.3 *Synchronous satellites relay radio communications between points on the Earth that would not otherwise be able to communicate via radio because of the straight-line nature of the propagation of high-frequency radio waves.*

where τ is the rotation period, equal to 1 day. The centripetal acceleration is

$$a_c = \frac{v^2}{R} = \frac{(2\pi R/\tau)^2}{R}$$
$$= \frac{4\pi^2 R}{\tau^2} \tag{6.3}$$

Now, the force (and, hence, the acceleration) on a satellite of mass m due to the gravitational attraction of the Earth at a distance R is

$$a = \frac{F_G}{m} = \frac{GM_E}{R^2} \tag{6.4a}$$

We know that at the surface of the Earth (where $R = R_E = 4000$ mi), the acceleration of the satellite would just be g:

$$g = \frac{F_G}{m} = \frac{GM_E}{R_E^2} \tag{6.4b}$$

Dividing Eq. 6.4a by Eq. 6.4b, we have

$$\frac{a}{g} = \frac{R_E^2}{R^2} \tag{6.5}$$

The only acceleration experienced by the satellite is just the centripetal acceleration. Therefore, substituting a_c from Eq. 6.3 for a in Eq. 6.5, gives

$$\frac{4\pi^2 R}{\tau^2 g} = \frac{R_E^2}{R^2}$$

from which

$$R^3 = \frac{\tau^2 g R_E^2}{4\pi^2} \tag{6.6}$$

Using $\tau = 1$ day $= 86{,}400$ sec and the values for g and R_E, we have, by taking the cube root,

$$R = \left[\frac{(86{,}400 \text{ sec})^2 \times (32 \text{ ft/sec}^2) \times (4000 \text{ mi})^2}{4\pi^2 \times (5280 \text{ ft/mi})} \right]^{1/3}$$
$$= 26{,}400 \text{ mi}$$

Thus, the height of the satellite above the surface of the Earth is

$$h = R - R_E$$
$$= 26{,}400 - 4{,}000 = 22{,}400 \text{ mi}$$

If the satellite is placed into orbit at this height above the Earth's equator, it will be a *synchronous* satellite. (The behavior of the satellite if it is at this height, but not above the equator, is the subject of Question 5.1.)

Fig. 6.4 *The Syncom communications satellite, first placed into orbit in 1963. This photograph was taken in a simulated space environment and shows the firing of the rocket motor that places the satellite into its synchronous orbit from its launch orbit.*

6.2 *Planetary Motion*

KEPLER'S LAWS

Between 1609 and 1611, Johannes Kepler (Fig. 1.3) enunciated his famous three laws of planetary motion. Kepler's conclusions were based on his analysis of the extensive data relating to planetary positions (particularly pertaining to the planet Mars) that had been acquired by Tycho Brahe (Fig. 1.2) during many years of observation. (Refer to Section 3.1 for a discussion of Kepler's distance determinations.)

The statements of Kepler's laws are:

I. The motion of a planet is an ellipse with the Sun at one focus.

II. The line connecting the planet with the Sun sweeps out equal areas in equal times.

III. The period of a planet's motion and its distance from the Sun are related by $R^3/\tau^2 = $ constant, where the constant is the same for all planets.

We shall discuss laws I and III, but the second law is of no particular importance in the subsequent developments.

KEPLER'S FIRST LAW

By geometrical construction from his position data, Kepler showed that planetary orbits are elliptical (but nearly circular). Newton used more sophisticated mathematics to prove the more general result that all orbits of objects interacting via a $1/r^2$ force are *conic sections*. The four possible types of curves that are in this category are obtained from the intersections of a plane with a cone (see Fig. 6.5). If the plane sections the cone at right angles to the axis of the cone, the result is a *circle*. By sectioning the cone at an angle, an *ellipse* is produced. If the sectioning angle is increased until it coincides with the cone angle (that is, until the plane is parallel to one of the straight lines that runs the length of the cone on its surface), a *parabola* results. A further increase in sectioning angle yields an *hyperbola*. Since specific sectioning angles are required to produce the circle and the parabola, these curves are actually only special cases of the ellipse and the hyperbola, respectively. Thus, in Nature we do not find orbits that are *exactly* circular or parabolic, although in certain cases these shapes are approached closely. The orbits of Venus and Neptune, for example, are the most nearly circular of all planetary orbits.

There is a simple method for constructing an ellipse that follows from an alternate definition. First, select the two points, F_1 and F_2, that are the

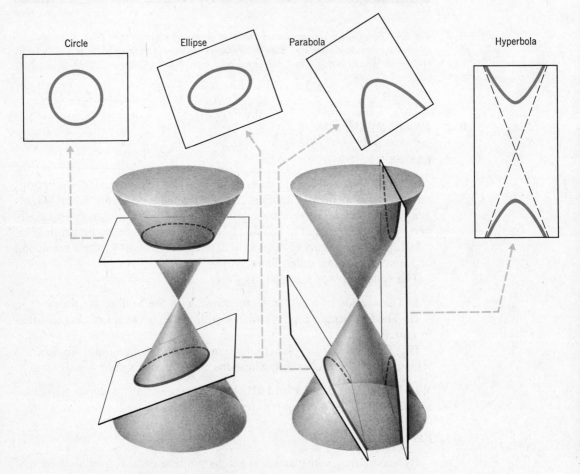

Fig. 6.5 *The four possible conic sections.*

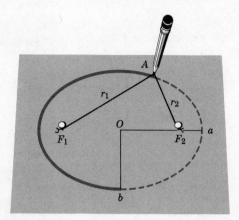

Fig. 6.6 *An ellipse is formed by the line connecting all points (such as A) for which* $r_1 + r_2 = constant.$ *$\overline{Oa}$ is the semi-major axis and $\overline{Ob}$ is the semi-minor axis.*

foci of the ellipse (Fig. 6.6). Next, attach the ends of a length of string (which is longer than the distance from F_1 to F_2) to pins at F_1 and F_2. Then, with the tip of a sharp pencil, extend the string until it is taut. The pencil tip will be at a distance r_1 from F_1 and a distance r_2 from F_2 (Fig. 6.6). Finally, the ellipse is constructed by drawing the curve which the pencil follows as it is moved in such a way that maintains the tautness of the string. The corresponding definition is an ellipse is the curve that is formed by connecting all points of equal values of the sum $r_1 + r_2$, where r_1 and r_2 are the distances from a given point on the curve to the foci.

Kepler described planetary orbits as ellipses with the Sun at one focus. (The other focal point has no physical significance for the case of planetary motion.) A more detailed analysis shows that it is not the center of the Sun that lies at one focus, but rather the *center of mass* of the Sun-planet system. Because the Sun is so much more massive than any planet, the center of mass of any Sun-planet combination lies very near the Sun's center.

The ellipse is the general form of the orbit for *bound motion,* that is, the two objects are bound together by their mutual gravitational attraction and cannot escape from each other. The hyperbola is the general form of the orbit for *unbound* motion. For example, if two isolated stars move toward each other, as in Fig. 6.7, they will execute hyperbolic orbits relative to their

Table **6.1** *Planetary Data*

Planet	Semi-major axis (A.U.)	Period (years)	Mass (Earth masses)	Diameter (Earth diameters)
Mercury	0.387	0.241	0.054	0.37
Venus	0.723	0.615	0.814	0.96
Earth	1.000	1.000	1.000	1.00
Mars	1.524	1.880	0.107	0.52
Jupiter	5.204	11.865	317.4	10.95
Saturn	9.580	29.650	95.0	9.13
Uranus	19.141	83.744	14.5	3.73
Neptune	30.198	165.95	17.6	3.52
Pluto	39.439	247.69	0.18	0.47

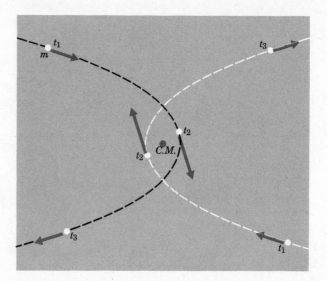

Fig. 6.7 *The hyperbolic orbits of two "colliding" stars that interact via a gravitational force. The positions and the velocity vectors of the two stars are shown for the times* t_1, t_2, *and* t_3. *The center of mass of the system remains fixed.*

common center of mass. The position of the center of mass will, of course, move with constant velocity. If, at any instant, the center of mass is at rest with respect to some coordinate system, it will remain at rest as long as the two interacting stars are not influenced by any outside force.

KEPLER'S THIRD LAW

We have already seen in Eq. 6.6 that for the particular case of a satellite moving around the Earth in a circular orbit of radius R,

$$R^3 = \frac{\tau^2 g R_E^2}{4\pi^2}$$

The same result would be valid for a planet moving around the Sun if we substituted for g and R_E the values of the acceleration due to gravity at the surface of the Sun, g_S, and the radius of the Sun, R_S, respectively. The quantities g_S and R_S are the same for each case of a planet circling the Sun, so that we have

$$\frac{R^3}{\tau^2} = \text{constant} \qquad (6.7)$$

with the same constant applying for each case. This is the statement of Kepler's third law.

Although we obtained the result for the special case of circular motion, the law is also valid for elliptical motion where R refers to the length of the semi-major axis (see Fig. 6.6).

DOES THE GRAVITATIONAL FORCE DEPEND EXACTLY ON $1/r^2$?

We have been using Newton's universal gravitational force law with a dependence on distance of the form $1/r^2$. But how do we know that the exponent is not 2.000 001 or 1.999 999? Is the exponent *exactly* 2? A sensitive test of a possible deviation of the exponent from 2 can actually be made by observations of planetary orbits.

Newton showed that if the gravitational force varies exactly as $1/r^2$, then the elliptical orbits described by the planets must remain in *fixed* positions. In particular, the point of the ellipse that is closest to the Sun (called the *perihelion*) must remain fixed in its relation to the distant "fixed" stars. Of course, there are small deviations from exact elliptical orbits (perturbations) due to the influence of the other planets, but these deviations are small because of the dominant gravitational force of the Sun; furthermore, mathematical methods exist for the precise calculation of these perturbations. Therefore, if any motion of the perihelion (apart from that expected due to other planets) is observed, this would indicate that the exponent in the force law expression is not *exactly* 2.

About a hundred years ago, a small unexplained motion in the perihelion of Mercury was observed. The perihelion moves forward (*precesses*) at a very slow rate so that the orbit has the appearance of a slowly rotating ellipse (Fig. 6.8). After subtraction of all of the effects due to other planets, there is a net precession that amounts to 43 sec of arc per century. That is, it would require more than 800 yr (or ~3300 revolutions) for the perihelion to move by 1°! To measure such a small quantity with precision (the uncertainty in the result is only 1 percent) is a truly remarkable achievement.

At first, the precession of Mercury's perihelion was thought to indicate the presence of an unobserved planet near the Sun that would influence the motion of Mercury and whose effects would not have been included in the previous perturbation calculations. This planet (prematurely named *Vulcan*) was sought for many years without success, and so it was believed that Newton's gravitational force law was slightly in error. Early in this century, Einstein showed that there was indeed a small correction needed in the gravitational force law due to relativistic effects. This correction depends on the planet's velocity and therefore is important only for Mercury, which has the highest velocity of any planet. With this correction, the precession of Mercury's perihelion is entirely accounted for. (But see the comments in Section 11.5.)

Therefore, there is an "error" in the $1/r^2$ form of the gravitational force law, just as there is in Newton's laws of dynamics, which manifests itself

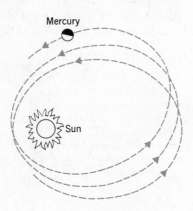

Fig. 6.8 *The perihelion of Mercury's orbit is observed to precess slowly. (The elongation of the orbit and the amount of precession have been greatly exaggerated.)*

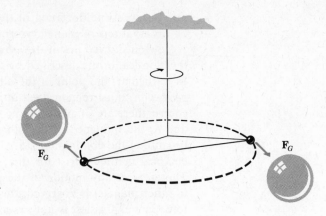

Fig. 6.9 *Schematic diagram of Cavendish's experiment to determine* G.

for objects in motion with high velocities. But Newton's form of the force law is valid for essentially all practical purposes and it appears to be entirely correct for *static* gravitational forces.

6.3   *The Measurement of G*

THE CAVENDISH EXPERIMENT

The value of the constant G that appears in the expression for the gravitational force must be determined by experiment. The smallness of G demands that any laboratory measurement of its value be performed with extreme care. Such a measurement was first made in 1798 by the English chemist, Henry Cavendish (1731–1810), using an instrument now known as a *Cavendish torsion balance.* The operation of the Cavendish balance is shown schematically in Fig. 6.9 and a sketch closely following that in Cavendish's original paper is shown in Fig. 6.10.

The idea of the Cavendish experiment was actually to measure the force of attraction between objects of known masses. As shown in Fig. 6.10, Cavendish mounted two small lead balls (2 in. in diameter, $m = 0.775$ kg) on opposite ends of a 2-m rod. The rod was supported by a fine wire. Two larger lead balls (8 in. in diameter, $m = 49.5$ kg) could be brought close to the small balls. The attractive force between the two sets of balls caused the smaller balls to move toward the larger balls, thus twisting the suspension wire by a small amount. The force constant for twisting the wire was

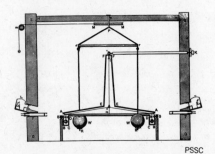

Fig. 6.10 *Sketch of Cavendish's apparatus as it appeared in his original paper. Notice that all of the manipulations, including the movement of the large balls* W, *were performed from outside the enclosure* G. *The measurements of the deflection angles were made with the telescopes* T. *Candles provided the illumination.*

PSSC

obtained in a separate, calibration experiment using known forces. The degree of twist was then a measure of the force between the balls.

The result of the Cavendish experiment was a value of G only 1 percent different from that now accepted, namely, 6.673×10^{-8} dyne-cm^2/g^2.

6.4   *The Electrostatic Force*

POSITIVE AND NEGATIVE ELECTRICITY

It has been known for about 200 years that in Nature there are two basic types of electricity and, following the scheme of Benjamin Franklin, we designate these *positive* and *negative* electricity. The basic carriers of negative electricity are, of course, the *electrons,* which constitute the outer portions of all atoms. The atomic cores, the nuclei, are the seats of positive electricity, the carriers of which are *protons.* All macroscopic matter is basically electrically *neutral,* because the magnitude of the negative electrical charge carried by an electron is equal to that of the positive electrical charge carried by a proton and all atoms in their natural states contain equal numbers of protons and electrons.

The distribution of electrical charge is almost always accomplished by the movement of *electrons;* the more massive positively-charged nuclei remain essentially stationary in almost all electrical processes. That is, a material is given a negative charge by the addition of excess electrons or a positive charge by the removal of electrons; the number of *atoms* is not changed in either case.

Certain types of materials (such as metals) have an interesting property—a fraction of the atomic electrons are not bound to any particular atom but are free to move about in the material. Such materials are termed *conductors.* If an electrical charge is placed on such a material, it will quickly distribute itself. On the other hand, if an electrical charge is placed on an *insulating* material, the charge will remain localized. Insulators (for example, glass, plastics) do not have free electrons and therefore electricity does not readily flow in such materials.

A basic property of electricity is that *like charges repel* and *opposite charges attract.* These facts can easily be demonstrated by some simple experiments. When a glass rod is rubbed with a silk cloth, the friction between the materials causes electrons to be transferred from the glass to the silk. Thus, the glass rod becomes *positively* charged. When a rubber rod is rubbed with a piece of fur, electrons are transferred to the rubber rod and it becomes *negatively* charged. If we suspend a pair of light-weight balls (such as pith balls) on strings, as in Fig. 6.11, and touch each of the balls with a glass rod that has been rubbed with a silk cloth, the balls will both acquire positive charges and will repel each other. On the other hand, if one of the balls has been touched instead with a rubber rod that has been rubbed with fur, it would acquire a negative charge and would therefore be attracted to the other ball.

These experiments clearly establish the rules for attraction and repulsion. They also demonstrate that the glass rod carried a charge opposite to that carried by the rubber rod, but other experiments are necessary to show which

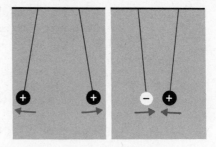

Fig. 6.11 *Like charges repel; opposite charges attract.*

was positively charged and which was negatively charged (according to the convention that electrons carry negative charge).

THE ELECTROSCOPE

The *electroscope* is a simple instrument for the detection of the presence of electrical charge (but it cannot distinguish between positive and negative). An electroscope (Fig. 6.12) is simply a more sophisticated version of the charge "detector" that consists of a pair of suspended pith balls. Instead of pith balls, two thin metallic sheets or leaves are used. If the electroscope is to be a sensitive detector of charge, the leaves must have very little mass so that, when charged, their displacement is accentuated. A common material for the leaves is gold, which can be pounded or rolled into extremely thin sheets. In order to eliminate the effects of air movements on the lightweight leaves, it is customary to enclose the leaves in a transparent container made of an insulating material (such as glass or clear plastic). The connection to the leaves is made by a thin conducting rod at the top (see Fig. 6.12). At the top of the rod there is usually mounted a large metal ball or plate. If an electrical charge (of either sign) is placed on the leaves via the outside ball, the electrostatic repulsion will cause the leaves to separate; the greater the amount of charge, the greater will be the separation.

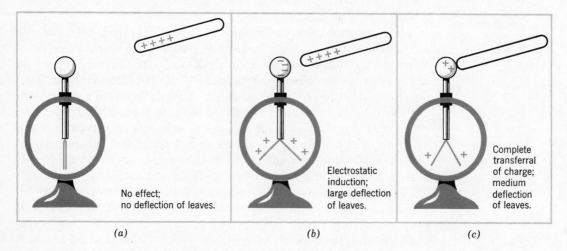

(a)   (b)   (c)

Fig. 6.12 *Three stages in the charging of an electroscope.*

The stages in the charging of an electroscope are as follows. Refer to Fig. 6.12. In (a) a positively-charged rod is being brought up to the electroscope which is initially in the neutral condition with its leaves collapsed. In (b) the charged rod is sufficiently close to the large ball to attract some of the electrons from the leaves to the large ball. This is the process of *electrostatic induction;* that is, a charge is *induced* in the ball due to the presence of the charged rod. The leaves are then left with a large positive charge and therefore experience a large deflection away from one another. If the rod is brought into contact with the large ball, as in (c), the charge on the rod is transferred to the electroscope. Since the charge is shared between the ball and the leaves (even if all of the charge on the rod is transferred), the leaves have a smaller positive charge than in (b) and therefore experience a smaller deflection. Notice that in all three stages of charging, the *total* amount of charge in the system is the same.

THE CONSERVATION OF CHARGE

An object can be given an electrical charge by the transferral of electrons to or away from the object. Charge may be *lost* by the object but it is then *gained* by some other object. It is one of the fundamental conservation laws of Nature that *the total amount of electrical charge in an isolated system remains constant.*

We shall see later that it is possible, under certain conditions and subject to certain restrictions, to *create* a pair of electrical charges—a *negative* electron (that is, an ordinary electron) and a *positive* electron (that is, a *positron,* a particle identical in every respect to an electron except that it carries a *positive* charge). Although processes such as these are possible, they do not violate the law of charge conservation because the total charge of the created pair is *zero* so that the net charge of the system remains constant.

No electron or proton has ever been observed to be created or destroyed *by itself.* Creation and destruction processes *always* occur with positive-negative pairs of particles. No deviation from the law of charge conservation has ever been discovered.

COULOMB'S LAW OF ELECTROSTATIC FORCE

In 1785, the French physicist Charles Augustin de Coulomb (1736–1806), extensively studied electrostatic forces using a sensitive torsion balance (Fig. 6.13) similar to that used by Cavendish in his investigation of the gravitational force constant (Fig. 6.10). From his measurements, Coulomb concluded that the electrostatic force between two charged objects varies as the inverse square of the distance between them.[3] That is, the electrostatic force has the same dependence on distance as does the gravitational force. The electrostatic force is proportional to the product of the charges involved (again, of the same general form as the gravitational case with, of course, *charge* substituted for *mass*). Therefore, the expression for the electrostatic

---

[3]Coulomb concluded that the power of the distance was 2 ± 0.02. Measurements made recently at Princeton University have shown that the exponent is 2 to within 1 part in 10^{12}.

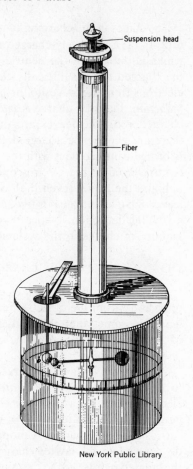

Suspension head

Fiber

Fig. 6.13 *Coulomb's torsion balance for the study of electrostatic forces as it appeared in his 1785 paper. The force is measured between the charged balls* a *and* b.

New York Public Library

force law (also called *Coulomb's law*) is

$$F_E = -\frac{q_1 q_2}{r^2} \quad \text{(CGS)} \tag{6.8}$$

where q_1 and q_2 are the charges on the two objects and r is their separation. Notice that the expression for F_E contains a *negative* sign, whereas Eq. 6.1 for F_G does not.[4] The reason for the introduction of the sign difference is to maintain the convention that a *positive* force is *attractive*, while a *negative* force is *repulsive*. The expressions for F_E and F_G must therefore differ in sign because mass is always positive and the gravitational force is always attractive. Like charges repel and therefore, when q_1 and q_2 have the same sign, F_E must be negative to indicate a repulsive force. Similarly, when the

---

[4]In some books you will find that Coulomb's law for F_E is written without the negative sign. We have elected to include the sign explicitly so that for *any* type of force calculation the sign of the result automatically and always indicates whether the force is attractive or repulsive.

charges have opposite signs, their product is negative which cancels the overall negative sign and produces a positive (attractive) force.

Notice also in Eq. 6.8 that we do not have an "electrostatic force constant" analogous to G for the gravitational case. This fact results from our choice of CGS units. In this system we define the unit of charge to be consistent with the units of force and distance when the proportionality constant is *unity*. Since *force* has been defined (relative to mass) by Newton's equation, $F = ma$, this choice of the proportionality constant leaves us no freedom in the unit for charge:

$$\text{CGS:} \quad [q] = \text{dyne}^{1/2}\text{-cm} = \text{statcoulomb (statC)} \tag{6.9}$$

where we give the special name *statcoulomb* to the unit of charge.

In the MKS system (which we will not use for the calculation of electrostatic forces), there *is* a proportionality constant with units in the expression for Coulomb's law. In the MKS system,

$$\text{MKS:} \quad [q] = \text{coulomb (C)} \tag{6.10}$$

where

$$1 \text{ C} = 3 \times 10^9 \text{ statC} \tag{6.11}$$

The elementary charge in Nature is the charge on the electron or the proton. Although the signs are, of course, opposite, the magnitudes of these charges are identical.[5] We denote the elementary charge by the symbol e:

$$\left. \begin{array}{l} e = 4.803 \times 10^{-10} \text{ statC} \\ = 1.602 \times 10^{-19} \text{ C} \end{array} \right\} \tag{6.12}$$

Example **6.2**

Calculate the electrostatic force *on q_1 due to q_2 (F_{12})* and the force *on q_2 due to q_1 (F_{21})* for the case illustrated in the figure.

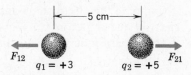

Using Coulomb's law,

$$F_{12} = -\frac{(+3)(+5)}{(5)^2} = -0.6 \text{ dyne (i.e., to the } left)$$

$$F_{21} = -\frac{(+5)(+3)}{(5)^2} = -0.6 \text{ dyne (i.e., to the } right)$$

Both forces have the same magnitude (Newton's third law for electrostatic forces) and the negative sign for each indicates that the force on each charge is *repulsive*.

---

[5] See the footnote on p. 81.

Example **6.3**

Suppose that all of the electrons in a gram of copper could be moved to a position 30 cm away from the copper nuclei. What would be the force of attraction between these two groups of particles?

The atomic mass of copper is 63.5. Therefore, 1 g of copper contains a number of atoms given by Avogadro's number divided by the mass of 1 mole (that is, 63.5 g):

$$\text{No. atoms} = \frac{6.02 \times 10^{23} \text{ atoms/mole}}{63.5 \text{ g/mole}} = 0.92 \times 10^{22} \text{ atoms/g}$$

The atomic number of copper is 29; in other words, each neutral copper atom contains 29 electrons. Therefore, the number of electrons in 1 g of copper is

$$\text{No. electrons in 1 g} = 29 \times 0.92 \times 10^{22} = 2.7 \times 10^{23} \text{ electrons}$$

Thus, the total charge on the group of electrons is

$$q_e = 2.7 \times 10^{23} \times (-e)$$
$$= 2.7 \times 10^{23} \times (-4.8 \times 10^{-10} \text{ statC})$$
$$= -1.3 \times 10^{14} \text{ statC}$$

A similar positive charge resides on the group of copper nuclei. Hence, the attractive electrostatic force is

$$F_E = \frac{q_e^2}{r^2} = \frac{(1.3 \times 10^{14} \text{ statC})^2}{(30 \text{ cm})^2}$$
$$= 1.9 \times 10^{25} \text{ dyne}$$

This force is as great as the gravitational force between the Earth and the moon! (See Example 6.1.)

Example **6.4**

Compare the electrostatic and gravitational forces that exist between an electron and a proton.

The electrostatic force law and the gravitational force law both depend on $1/r^2$:

$$F_E = -\frac{q_1 q_2}{r^2}$$

$$F_G = G \frac{m_1 m_2}{r^2}$$

Therefore, the ratio F_E/F_G is independent of the distance of separation:

$$\frac{F_E}{F_G} = -\frac{q_1 q_2}{G m_1 m_2}$$

For the case of an electron and a proton this becomes

$$\frac{F_E}{F_G} = -\frac{(-e)(+e)}{G m_e m_p} = \frac{e^2}{G m_e m_p}$$

Substituting the values of the quantities, we find

$$\frac{F_E}{F_G} = \frac{(4.80 \times 10^{-10} \text{ statC})^2}{(6.67 \times 10^{-8} \text{ dyne-cm}^2/\text{g}^2) \times (9.11 \times 10^{-28} \text{ g}) \times (1.67 \times 10^{-24} \text{ g})}$$

$$= 2.3 \times 10^{39}$$

Thus, the electrostatic force between elementary particles is enormously greater than the gravitational force. Therefore, only the electrostatic force is of importance in atomic systems. In nuclei, the strong nuclear force overpowers even the electrostatic force but not to the extent that electrostatic forces are completely negligible. Many important nuclear effects are the result of electrostatic forces.

Example **6.5**

In the Bohr model of the hydrogen atom, the electron is considered to move in a circular orbit around the nuclear proton. The radius of the orbit is 0.53×10^{-8} cm. What is the velocity of the electron in this orbit?

If we assume that the only force acting is the electrostatic force,

$$F_E = -\frac{q_1 q_2}{r^2} = \frac{e^2}{r^2}$$

$$= \frac{(4.80 \times 10^{-10} \text{ statC})^2}{(0.53 \times 10^{-8} \text{ cm})^2}$$

$$= 8.2 \times 10^{-3} \text{ dyne}$$

In order to maintain a circular orbit, the electron must be subject to a centripetal acceleration:

$$a_c = \frac{v^2}{r}$$

The mass of the electron times this acceleration must equal the force on the electron:

$$m_e a_c = \frac{m_e v^2}{r} = F_E$$

or, solving for v,

$$v = \sqrt{\frac{r F_E}{m_e}}$$

$$= \sqrt{\frac{(0.53 \times 10^{-8} \text{ cm}) \times (8.2 \times 10^{-3} \text{ dyne})}{(9.11 \times 10^{-28} \text{ g})}}$$

$$= 2.18 \times 10^8 \text{ cm/sec}$$

Hence, the electron velocity is about 1 percent of the velocity of light.

ELECTRICAL FORCES IN GENERAL

In this section we have discussed only the force that acts between electrical charges at rest—the electrostatic force. When charges are in motion relative to one another we shall find (in Chapter 9) that an additional force—the *magnetic* force—comes into play. The term *electromagnetic force* is frequently used to indicate that both electrostatic and magnetic effects are present. However, magnetic forces have no existence independent of electrical charges; these forces arise *exclusively* from charges in motion. Therefore, we will use the term *electrical force* in a general sense to indicate an electrostatic force (if the charges are at rest) and to include the possibility of a magnetic force (if the charges can be in motion).

6.5   *The Nuclear and Weak Forces*

WHAT HOLDS NUCLEI TOGETHER?

The two forces that we have described thus far—the gravitational and electrical forces—are the only forces needed to account for the motions of all everyday objects and even the behavior of atomic systems. However, when we look deeper into the atom and probe into the nature of the forces that operate within nuclei, we find that gravitational and electrical forces are no longer adequate to describe the effects that are observed.

We know that nuclei are extremely small—typical radii are only a few times 10^{-13} cm. These nuclei contain positive charges (up to ~ 100 e) and we know that electrostatic forces, especially at small distances, can be quite large. Obviously, there must be some extremely strong attractive force that acts within nuclei and overcomes the repulsive Coulomb force that tends to repel the protons from one another.

Example **6.6**

Calculate approximately the electrostatic force that exists between two protons in a typical nucleus.

We know, from Eq. 3.12, that the radius of a nucleus with mass number A is given approximately by

$$R = 1.4A^{1/3} \times 10^{-13} \text{ cm}$$

From this expression we find that the radius of an iron nucleus ($A = 56$) is about 5.4×10^{-13} cm. Let us take 2×10^{-13} cm as a typical separation for two protons in a nucleus. Then, the electrostatic force between the protons is

$$F_E = -\frac{e^2}{r^2}$$

$$= -\frac{(4.80 \times 10^{-10} \text{ statC})^2}{(2 \times 10^{-13} \text{ cm})^2}$$

$$\cong -6 \times 10^6 \text{ dynes}$$

This is an enormous repulsive force; it is approximately equal to the gravitational force on a 6-kg mass near the surface of the Earth! This result emphasizes the extraordinary strength of nuclear forces since these forces must overcome the electrostatic repulsion and hold the nucleus together.

THE STRONG NUCLEAR FORCE

The force that acts at the small distances within nuclei and maintains the stability of nuclei in spite of their tendency to fly apart because of Coulomb repulsion is called the *strong nuclear force*. (The designation "strong" is to distinguish this force from the *weak* force that acts between nuclear and elementary particles, which we shall describe below.)

The strong nuclear force acts attractively between protons and protons (*p-p*), between protons and neutrons (*p-n*), and between neutrons and neutrons (*n-n*). We know that the *p-n* and *n-n* forces are essentially identical, and, apart from the Coulomb portion, the *p-p* force is the same as the *p-n* and *n-n* forces. Since protons and neutrons have so many similar properties (except primarily for the lack of charge on the neutron), these particles are collectively called *nucleons*. Therefore, when we discuss the *nucleon-nucleon* (*N-N*) force, we include the three possible combinations listed above.

The *N-N* force has a dependence on distance that is distinctly different from that of the gravitational and electrostatic forces. These latter forces vary as $1/r^2$ and therefore are termed *long-range* forces; that is, gravitational and electrical effects are manifest over large distances. The *N-N* force, on the other hand, is *short-range* and is effective only over distances up to $\sim 10^{-13}$ cm (nuclear dimensions). When a proton is removed from a nucleus, by the time a separation distance of a few times 10^{-13} cm has been reached, there is no longer any nuclear force attracting the proton toward the nucleus—only the repulsion of the Coulomb force remains.

The *N-N* force is not entirely attractive. At a distance of $\sim 10^{-14}$ cm, the *N-N* force becomes *repulsive*. That is, the *N-N* force has a *hard core* that resists pushing two nucleons too close together. The dependence of the *N-N* force on distance is shown schematically in Fig. 6.14. It is possible to show

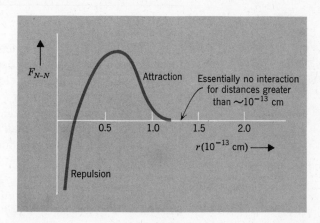

Fig. 6.14 *The nucleon-nucleon force is strongly attractive, but only in the range from $\sim 10^{-14}$ to $\sim 10^{-13}$ cm.*

this force only *schematically* because it is not a simple central force[6] as are the gravitational and electrostatic forces. There are important quantum mechanical effects that are crucial in the *N-N* case and make the detailed description of the *N-N* force extremely complicated. We do not yet know, in fact, all of the details of the *N-N* force; this is one of the central problems facing nuclear physics today.

THE WEAK FORCE

The process of nuclear β decay involves the emission from the parent nucleus of an electron and a neutrino (see Section 5.11). Just as a proton and a neutron interact primarily by the *strong* nuclear force, the electron (e) and the neutrino (v) interact (*exclusively*) by the *weak* force. This weak force is responsible for the process of nuclear β decay. Our understanding of the weak force is at present very meager; we have no detailed knowledge, for example, of the dependence of the force on distance, except that it is of short range, certainly no greater than that of the strong nuclear force and possibly of *zero* range.

A weak force exists between *all* pairs of elementary particles. It is the exclusive force between electrons and neutrinos, but the same force (albeit much weaker than the electrical or strong nuclear force) exists even between two protons. The relative sizes of the various forces that act between pairs of elementary particles are listed (in order of magnitude) in Table 6.2. The nucleon-nucleon force is arbitrarily assigned unit magnitude.

Table **6.2** *Comparison of the Forces that Act between Elementary Particles*

Force	Ratio of Strengths at Small Distances ($\sim 10^{-13}$ cm)			
	e-v	e-p	p-p	p-n n-n
Strong nuclear	0	0	1	1
Electrostatic	0	10^{-2}	10^{-2}	0
Weak	10^{-13}	10^{-13}	10^{-13}	10^{-13}
Gravitational	0	10^{-41}	10^{-38}	10^{-38}

Because the neutrino interacts with all other elementary particles only through the weak force and because neutrinos can now be produced in copious amounts at large accelerator installations, the study of neutrinos and the weak interaction is at present being extensively pursued.

The physics of nuclei and elementary particles will be discussed further in Chapters 15 and 16. We shall find that there are other elementary particles, in addition to those mentioned here, that participate in strong and weak interactions.

---

[6]A *central force* between two particles is one that acts only along the line that connects the particles. The *N-N* force has parts that are central and parts that are non-central.

Summary of Important Ideas

There are only four basic forces in Nature—gravitational, electrical, strong nuclear, and weak.

The gravitational and electrical forces depend on $1/r^2$ and are *long-range* forces. The nuclear force and the weak force have *short ranges*.

All motions of two isolated objects that interact via a $1/r^2$ force are *conic sections*. Planets move in *elliptical* orbits around the Sun.

For purposes of gravitational calculations, the entire mass of a uniform spherical object can be considered to be concentrated at its *center*. For purposes of electrical calculations, the entire charge of a uniform spherical distribution of charge can be considered to be concentrated at its *center*.

The net electrical charge of an isolated system remains *constant*.

The *strong* nuclear force is responsible for the stability of nuclei. The *weak* force is responsible for nuclear β decay.

The order of *strengths* of the basic forces is (1) strong nuclear, (2) electrical, (3) weak, and (4) gravitational.

Questions

6.1 A synchronous satellite is placed in orbit, not at the equator, but at a certain latitude λ. What is the motion of the satellite relative to the surface of the Earth?

6.2 The mass of the moon is approximately $\frac{1}{81}$ of the mass of the Earth but the gravitational attraction on an object at its surface is only about $\frac{1}{6}$ of the gravitational attraction on the same object at the surface of the Earth. Why?

6.3 An artificial satellite is in circular orbit around the Earth. Friction between the satellite and the air in the upper atmosphere has the effect that the satellite's speed *increases*. Explain why. (Consider angular momentum.)

6.4 Comets are members of the solar system. Explain why some comets (such as Halley's comet) appear remarkably bright for a brief period and then become unobservable, even with powerful telescopes, for a number of years.

6.5 Astronomers have evidence that the Earth's rate of rotation is slowing down due to effects of the moon's gravitational attraction. How will this affect the length of a day? Will there be a similar effect on the length of a year?

6.6 Two protons approach each other in space. Describe their relative orbits if they approach (a) head-on and (b) with initially parallel but displaced paths.

6.7   In the Cavendish experiment, can the force on one of the small balls due to the farther large ball be neglected in comparison to the force due to the nearer large ball?

6.8   Two objects carry small electrical charges. Explain how an electroscope could be used to determine whether the charges are of the same or opposite signs.

6.9   A conducting rod *A* is supported on an insulating stand. Another conducting rod *B* is suspended near the end of rod *A* by strings. What will happen to *B* when a positively-charged object *D* is brought close to the end of rod *A* that is farthest from rod *B*? What would happen if *C* were *negatively* charged?

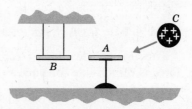

6.10   Two pith balls, suspended on threads as in Fig. 6.11, carry equal and opposite charges so they are attracted toward each other and the threads are deflected. If an *uncharged* copper sphere is placed midway between the pith balls, in what way will the deflection of the threads be changed? Explain in detail.

6.11   Someone proposes that instead of there being two kinds of electric charges, + and −, there are three kinds, α, β, and γ, with the property that like charges repel and unlike charges attract. Devise an experiment that would decide between the two theories.

6.12   Someone proposes that the attraction between the Earth and the Sun is not due to the gravitational force but is the result of the Sun and the Earth having an imbalance of electrical charge. Describe experiments (either practical experiments or "thought" experiments) that could test this hypothesis.

6.13   A proton is moving directly toward a certain atom. Describe the forces that the proton experiences as a function of its distance from the nucleus of the atom.

6.14   What forces act between the following pairs of elementary particles: *v-p*, *v-n, e-n?*

Problems

6.1 Near the surface of the Earth, we know that the acceleration due to gravity is $g = 32$ ft/sec^2. What values would be found at the following distances above the Earth's surface: 100 mi, 1000 mi, 4000 mi?

6.2 The mass of the moon is about $\frac{1}{81}$ of the mass of the Earth. At what point on a line between the Earth and the moon will an object experience zero net gravitational force?

6.3 Use the data in Table 6.1 and compute the value of g on Mars. What is the weight of a 100-gram mass on Mars?

6.4 What is the value of g at the surface of the Sun?

6.5* The moon's orbital velocity is 0.64 mi/sec. In traveling at this velocity for 3 sec, how far has the moon "fallen" toward the Earth? Compare this with the distance that an object will fall near the surface of the Earth in 3 sec. Calculate the ratio of the two distances and account for the value.

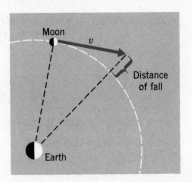

6.6 Two identical spherical objects are separated by a distance of 1 m. If the gravitational force between them is 6.7×10^{-4} dyne, what is the mass of each object?

6.7* If the mass of the Earth were 4 times its actual value, what would be the length of the lunar month? (Assume the Earth-moon distance is the actual distance.)

6.8 What is the gravitational force on a 100-kg synchronous satellite?

6.9 If the moon were suddenly placed in a circular orbit around the Earth at a distance twice that at present, what would be the new period? What would be the new orbital velocity?

6.10 Use the data in Table 6.1 and verify Kepler's third law by comparing R^3/τ^2 for 4 or 5 of the planets.

6.11 An astronomer claims to have found a new planet whose orbit is midway between the Earth and the Sun and whose period is 240 days. What is your opinion of his claim and why?

6.12 Two spherical objects, interacting only via their mutual gravitational forces, execute circular orbits in space around their common center of mass. If one object is three times as massive as the other, sketch the orbits of each. Compare the periods of the orbits of the two objects.

6.13 Assume that the Earth is the only planet in the solar system. What is the radius of the orbit that the Sun executes around the center of mass of the Earth-Sun system? What is the period of this motion of the Sun?

6.14 What was the attractive force between one of the large balls and one of the small balls in the Cavendish experiment when the centers of the balls were separated by 20 cm?

6.15 Two identical spherical copper balls carry charges of $+28$ statC and -4 statC and their centers are separated by 2 cm. If the balls are brought together until they *touch* and then returned to their original positions, what will be the electrostatic force between them?

6.16 A certain object is given a positive electrical charge by removing 1 picogram of electrons. (1 pg $= 10^{-12}$ g.) How many electrons were removed and what is the charge on the object?

6.17* A droplet of water has a radius of 0.1 mm. If there is 1 extra electron per 10^9 water molecules, what is the charge on the droplet?

6.18 A 1-kg mass, when allowed to fall freely near the surface of the Earth, experiences an acceleration of 980 cm/sec^2. If this mass carries a charge of $+1$ statC, what would the charge on the Earth have to be in order to double the acceleration? (Consider the Earth to be uniformly charged.)

6.19 A particle of mass $m = 0.1$ g carries a charge q. At a distance of 10 cm

above m there is a spherical object that carries a charge $Q = +10^3$ statC. What must be the value of q if there is no net force on m? (The mass m is located at only a small height above the surface of the Earth.)

6.20 Three charges are located at the following positions along a straight line: $q_1 = +10$ statC at $x = 0$; $q_2 = -40$ statC at $x = 20$ cm; $q_3 = +45$ statC at $x = 50$ cm. What is the net force (magnitude and direction) on q_2?

6.21 A proton and an electron are at rest a distance 10^{-8} cm apart. What is the electrostatic force exerted by one of the particles on the other? When they begin to move under the influence of this force, what is the initial acceleration of each particle?

6.22 Two α particles (helium nuclei) are separated by a distance of 1 Å. What electrostatic force does each α particle exert on the other?

6.23 Two completely ionized lithium ions ($Q = +3e$) exert a repulsive force of 210 dynes on each other. How far apart are they?

Underwater H-bomb explosion at Bikini Atoll

Joint Task Force One

7 *Energy*

Without a doubt, *energy* is the single most important physical concept in all of science. A clear understanding of energy and an appreciation of its importance was not fully realized until 1847 when the German physicist Hermann von Helmholtz (1821–1894) enunciated the general law regarding energy. Since that time the consideration of energy in physical (and biological) processes has been a crucial ingredient in the efforts to understand natural phenomena.

Closely associated with energy is the concept of *work*. Historically, there grew up intuitive conceptions regarding *energy* and *work* just as there did for length, time, and mass. Some of these ideas of the layman are, in fact, quite closely related to the precisely formulated ideas of the physicist. It could be said, for example, "Eat a good meal and you will have a lot of energy," and "A person who has a great deal of energy can do a large amount of work." These statements correspond closely to those that the physicist would make: "The stored chemical energy in foodstuffs can be transferred to biological systems," and "Energy is the capacity to do work."

In this chapter we first define *work* and then discuss the connection between work and *energy*. There are three basic forms of energy: (1) *kinetic* energy due to the state of motion of a body; (2) *potential* energy due to the effects of forces exerted on a body by other bodies; and (3) *mass* energy associated with the inertial mass of a body according to the famous Einstein equation, $\mathcal{E} = mc^2$.

The importance of the concept of energy lies in the fact that various forms of energy within an isolated system can be transformed into one another *without loss* of energy. That is, in any physical process, *energy is conserved*. The law of energy conservation, which is obeyed in every known process, is the most fruitful law in physics for the analysis of phenomena of every sort.

7.1 *Work*

MOTION AGAINST A FORCE

If we exert a force on an object and move it a certain distance, we say that we have done *work*. Lifting a weight, for example, requires exerting a force sufficient to overcome the downward gravitational force. If we apply such a force sufficiently long to raise the weight to a height *h*, we have done a certain amount of work. Or, if we push an object across a rough surface with a force sufficiently large to overcome friction and move it a distance *s*, we have again done a certain amount of work. If we push the object a distance 2*s*, we will have done twice the work. The amount of work required also increases with the force necessary to displace the object. That is, the amount of work done is proportional to both the applied force and the distance through which the force acts.

Suppose that we apply a constant force to an object and thereby cause it to move with *constant velocity* over a rough surface. Since the acceleration of the object is zero, we know that there is no *net* force being applied. The

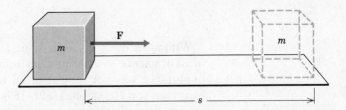

Fig. 7.1 *When the displacement is in the same direction as the applied force, the work done is* W = Fs.

externally applied force in this case just balances the retarding frictional force so that $F_{net} = 0$. Nevertheless, work *is* being done by the external force because the external force is working against the frictional force. The act of displacing an object against a retarding force (such as friction or gravity) by the application of an external force constitutes work, regardless of whether the object is accelerated or moves with constant velocity.

It is possible that work is done only during a portion of a displacement. If an object rests on a frictionless surface, the application of a horizontal force will accelerate the object. If the force is then removed, the object will "coast" at constant velocity. Work is done during the first phase of the motion (while the force is applied), but not during the second phase (when there is no applied force).

WORK = FORCE × DISTANCE

Work is the product of the force applied to an object by an outside agent and the distance through which the force acts on the object. In Fig. 7.1, the work done by the force F in displacing the block a distance s is $W = Fs$. In this case the direction of the displacement is the same as the direction of the applied force. In Fig. 7.2, however, the force is applied at an angle θ relative to the direction of motion. This force can be considered to be the vector sum of two independent forces, the x- and y-components:

$$F_x = F \cos \theta$$
$$F_y = F \sin \theta$$

The component F_x has the direction of the displacement; therefore the work done by this component of the force is $W = F_x s = Fs \cos \theta$. There is

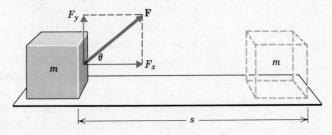

Fig. 7.2 *Only the component of the force in the direction of the displacement does any work,* W = F$_x$s = Fs *cos* θ.

no displacement in the *y*-direction; therefore, the component F_y *does no work*. In general, then, our definition of work must be modified:

Work is the product of the component of force in the direction of the displacement and the magnitude of the displacement produced by the force, or

$$W = Fs \cos \theta \qquad (7.1)$$

where θ is the angle between the force vector **F** and the direction of the displacement. Although work depends on the product of two quantities that are vectors, work is a *scalar* quantity.

ERGS AND JOULES

In the CGS system, the unit of work is $[W] = [F] \times [S]$ or $[W] = $ dyne-cm. Since 1 dyne = 1 g-cm/sec^2, the basic unit of work is 1 g-cm^2/sec^2. To this unit we give the special name *erg:*

CGS: 1 dyne-cm = 1 g-cm^2/sec^2

$$= 1 \text{ erg}$$

In the MKS system, we have

MKS: 1 N-m = 1 kg-m^2/sec^2

$$= 1 \text{ joule (J)}$$

$$= 10^7 \text{ ergs}$$

The designation *joule* for the unit of work in the MKS system is in honor of James Prescott Joule (1818–1889), the English physicist whose efforts greatly clarified the concepts of work and energy. The term *erg* is from the Greek word (*ergon*) for work.

Example **7.1**

A horizontal force of 5 N is required to maintain a velocity of 2 m/sec for a box of mass 10 kg sliding over a certain rough surface. How much work is done by the force in 1 min?

First, we must calculate the distance traveled:

$s = vt$

$\quad = (2 \text{ m/sec}) \times (60 \text{ sec})$

$\quad = 120 \text{ m}$

Then,

$W = Fs$

$\quad = (5 \text{ N}) \times (120 \text{ m})$

$\quad = 600 \text{ N-m} = 600 \text{ J}$

Example **7.2**

To a good approximation, the force required to stretch a spring is proportional to the distance the spring is extended. That is,

$$F = kx$$

where k is the so-called *spring constant* or *force constant* and depends, of course, on the dimensions and material of the spring. Many elastic materials, if not stretched too far, obey this simple relationship—called *Hooke's law*[1] after Robert Hooke (1635–1703), a contemporary of Newton.

Suppose that it requires 100 dynes to extend a certain spring 5 cm. What force is required to stretch the spring from its natural length to a length 20 cm greater?

The force constant is

$$k = \frac{F_1}{x_1}$$

$$= \frac{100 \text{ dynes}}{5 \text{ cm}}$$

$$= 20 \text{ dynes/cm}$$

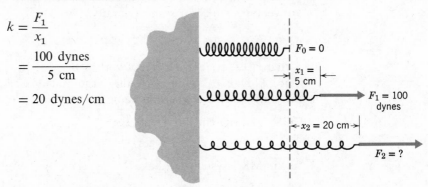

Therefore, the force required to extend the spring to 20 cm is

$$F_2 = kx_2$$

$$= (20 \text{ dynes/cm}) \times (20 \text{ cm})$$

$$= 400 \text{ dynes}$$

How much work is expended in stretching the spring to 20 cm?
The *average* force required to stretch the spring to 20 cm is

$$\bar{F} = \frac{F_2 + F_0}{2} = \frac{F_2 + 0}{2} = \frac{kx_2}{2}$$

and the work expended is the average force multiplied by the distance:

$$W = \bar{F}x_2 = \tfrac{1}{2} kx_2^2$$

$$= \tfrac{1}{2} \times (20 \text{ dynes/cm}) \times (20 \text{ cm})^2$$

$$= 4000 \text{ ergs}$$

7.2 Power

WORK PER UNIT TIME

Power is the rate at which work is done. If an amount of work W is done during an interval of time t, the average power expended is

$$\bar{P} = \frac{W}{t} \tag{7.2}$$

---

[1]This is not a *law* in the same sense that Newton's statements are laws. Hooke's relation is merely a convenient approximate expression that is useful but does not constitute a fundamental rule of Nature. Unfortunately, historically many such expressions have been labeled "laws."

Just as we extended the definition of average velocity and established the concept of instantaneous velocity, we may define *instantaneous power* as

$$P = \lim_{\Delta t \to 0} \frac{\Delta W}{\Delta t} \tag{7.3}$$

The units of power are

CGS: 1 erg/sec
MKS: 1 J/sec = 10^7 ergs/sec = 1 watt (W)

The special name for the unit of power in the MKS system is in honor of James Watt (1736–1819), the Scottish engineer who was instrumental in making the steam engine commercially practical. Watt introduced the British engineering unit of power, approximately equal to the power generated by a working horse, which he called *horsepower* (hp):

1 hp = 746 W

$\cong \frac{3}{4}$ kilowatt (kW)

The amount of energy used during a certain operation or process is frequently expressed as the product (power) × (time). For example, the amount of energy generated by a 1-kW power plant operating for 1 hr is 1 kW-hr. The use of 1 kW-hr of electrical power in a household at the present time costs about 2–3 cents; the rates vary with the amount of power consumed and with the location and type of power plant that supplies the power. Near large hydroelectric plants the rates tend to be somewhat lower.

In Eq. 7.2 we can write the work W as the product (force) × (distance):

$$\bar{P} = \frac{W}{t} = \frac{Fs}{t} = F\left(\frac{s}{t}\right) = F\bar{v} \tag{7.4}$$

where we have used the fact that s/t is the average velocity $\bar{v}$. The *instantaneous* power P can obviously be expressed in terms of the instantaneous velocity v:

$$P = Fv \tag{7.5}$$

Example **7.3**

A constant horizontal force of 10 N is required to drag an object across a rough surface at a constant speed of 5 m/sec. What power is being expended? How much work would be done in 30 min?

$P = Fv$

$\quad = (10 \text{ N}) \times (5 \text{ m/sec})$

$\quad = 50 \text{ J/sec}$

$\quad = 50 \text{ W}$

$W = Pt$

$\quad = (50 \text{ W}) \times (\frac{1}{2} \text{ hr})$

$\quad = 25 \text{ W-hr}$

The work, of course, is done against the force of sliding friction.

7.3   *Kinetic Energy*

THE ENERGY ASSOCIATED WITH MOTION

Suppose we have a *free* object—for example, an object in space initially at rest (with respect to some coordinate system) and subject to no forces. Then, if we apply a certain constant force F, the object will be accelerated. After it has moved a distance s, the object will have acquired a velocity $v = at$. The distance traveled is $s = \frac{1}{2} at^2$ and the product $F \times s$ is

$$F \times s = (ma) \times (\tfrac{1}{2} at^2)$$
$$= \tfrac{1}{2} m \times (at)^2$$
$$= \tfrac{1}{2} m \times v^2$$

or,

$$Fs = \tfrac{1}{2} mv^2 \tag{7.6}$$

Thus, an amount of work, $W = Fs$, has been performed on the object and we say that the object has acquired an amount of *energy* equal to $\frac{1}{2} mv^2$. This energy, which the object possesses *by virtue of its motion*, is called *kinetic energy:*

Kinetic energy: $KE = \tfrac{1}{2} mv^2$ $\tag{7.7}$

Clearly, the unit of kinetic energy is the same as that for work.

Example 7.4

A free particle, which has a mass of 20 grams is initially at rest. If a force of 100 dynes is applied for a period of 10 sec, what kinetic energy is acquired by the particle?

In order to calculate the kinetic energy we must first compute the final velocity acquired by the particle:

$$v = at = \left(\frac{F}{m}\right)t$$

$$= \left(\frac{100 \text{ dynes}}{20 \text{ g}}\right) \times (10 \text{ sec}) = 50 \text{ cm/sec}$$

Then,

$$KE = \tfrac{1}{2} mv^2$$
$$= \tfrac{1}{2} \times (20 \text{ g}) \times (50 \text{ cm/sec})^2 = 25{,}000 \text{ ergs}$$

How much work was done by the applied force? The distance moved is

$$s = \tfrac{1}{2} at^2 = \tfrac{1}{2}\left(\frac{F}{m}\right)t^2$$

$$= \tfrac{1}{2} \times \left(\frac{100 \text{ dynes}}{20 \text{ g}}\right) \times (10 \text{ sec})^2 = 250 \text{ cm}$$

so that the work done is

$$W = F \times s$$

$$= (100 \text{ dynes}) \times (250 \text{ cm}) = 25{,}000 \text{ ergs}$$

Thus, the work done is transformed entirely into the kinetic energy of the particle.

7.4 *Potential Energy*

THE ENERGY ASSOCIATED WITH POSITION

Suppose we lift an object of mass m, originally at rest, to a position that is a height h above the initial position and leave the object again at rest. Clearly, we have done work against the gravitational force, but there is no net change in velocity and therefore we have imparted no kinetic energy. However, the object *does* possess energy by virtue of its *position*. We can easily see this by allowing the object to fall toward its original position. After falling through a distance h, it will have acquired a velocity $v = \sqrt{2gh}$ (see Problem 4.9) and its kinetic energy will be

$$KE = \tfrac{1}{2} mv^2 = \tfrac{1}{2} m(2gh) = mgh$$

This amount of energy, mgh, may be accounted for as follows. In order to balance the gravitational force on an object of mass m, the application of a force mg is required. If a force infinitesimally greater[2] than mg acts through a distance h (the height to which the object is raised), an amount of work $W = Fs = mgh$ is done. The object then possesses an energy mgh that has the potential of being transformed into kinetic energy (by falling through the height h). We call this energy, which the body possesses *by virtue of its position*, the *potential energy* of the body. That is, the work done against

[2] By choosing a force infinitesimally greater than mg, we are assured that we can actually *move* the object (not just balance the gravitational force); but, at the same time, we make no appreciable error by equating the work done to mgh.

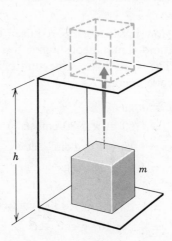

Fig. 7.3 *An object acquires a potential energy mgh by being raised to a height* h.

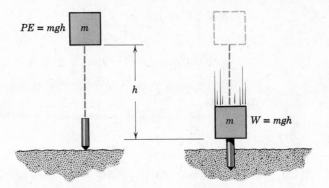

Fig. 7.4 *The potential energy of the block is transformed in kinetic energy by falling and does work by driving the stake into the ground.*

the gravitational force is stored by the object raised and is retained as potential energy:

> Potential energy: $PE = mgh$ $\qquad\qquad$ (7.8)

ENERGY AND WORK

As we have mentioned earlier, *energy is the capacity to do work.* The potential energy acquired by an object in being raised to a certain height can be converted into work in a number of ways. One way of accomplishing this is by means of a *pile-driver* as illustrated schematically in Fig. 7.4. By falling through a height h a block of mass m transforms its potential energy mgh into kinetic energy $\frac{1}{2} mv^2$, where $v = \sqrt{2gh}$. The moving block, in being stopped, can do an amount of work $mgh = \frac{1}{2} mv^2$. In Fig. 7.4, this amount of work is expended in driving the stake into the ground and is equal to $\overline{F} \times \Delta s$, where $\overline{F}$ is a very large average force and Δs is the small distance through which the stake moves.

Example 7.5

How much work is required to raise a 100-g block to a height of 200 cm and simultaneously give it a velocity of 300 cm/sec?

The work done is the sum of the potential energy, $PE = mgh$, and the kinetic energy, $KE = \frac{1}{2} mv^2$:

$PE = mgh$

$\quad = (100 \text{ g}) \times (980 \text{ cm/sec}^2) \times (200 \text{ cm})$

$\quad = 1.96 \times 10^7 \text{ g-cm}^2/\text{sec}^2$

$\quad = 1.96 \times 10^7 \text{ ergs}$

$KE = \frac{1}{2} mv^2$

$\quad = \frac{1}{2} \times (100 \text{ g}) \times (300 \text{ cm/sec})^2$

$\quad = 4.5 \times 10^6 \text{ g-cm}^2/\text{sec}^2$

$$W = PE + KE$$
$$= 1.96 \times 10^7 \text{ ergs} + 0.45 \times 10^7 \text{ ergs}$$
$$= 2.41 \times 10^7 \text{ ergs}$$
$$= 2.41 \text{ J}$$

7.5 *Conservative Forces*

WORK IS INDEPENDENT OF THE PATH

It is important to realize that because the gravitational force is always directed vertically downward, no force, and hence no work, is required (in the absence of friction) to move an object horizontally at constant velocity. Therefore, if we choose two different paths to raise an object to the height h, as in Figs. 7.5a and 7.5b, the amount of work is the same, namely, *mgh*. Displacement is actually a *vector,* since both a magnitude and a direction are specified. Therefore, any arbitrary displacement can always be resolved into horizontal and vertical components. Only the vertical motion requires work to be done; the horizontal motion requires no expenditure of work (i.e., this motion is obtained "free"). Thus, the movement of an object from one position to another against a constant gravitational force requires the same amount of work *regardless of the path taken* (as in Fig. 7.5c).

A force that has the property that the amount of work done against it depends only on the initial and final positions of the object moved (and not on the path followed) is called a *conservative force.* The (approximately) constant gravitational force that exists in the vicinity of the surface of the Earth is clearly a conservative force. In fact, the gravitational force that acts over large distances and varies as $1/r^2$ is also a conservative force. In this case, any motion along a circular arc (with constant r) is "free," but motion along a radius requires work. The electrostatic force is another conservative force. In general, any force whose direction is always along the line connecting the two bodies that are interacting is a conservative force. That is, all *central* forces are conservative. The gravitational and electrostatic forces are conservative central forces.

Frictional forces are *nonconservative* forces because the amount of work

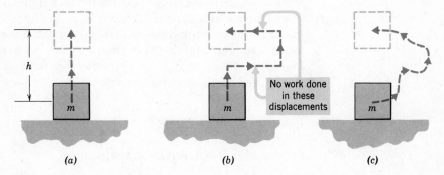

(a)   (b)   No work done in these displacements   (c)

Fig. 7.5 *An amount of work mgh is required to raise an object of mass* m *to a height* h *regardless of the path chosen.*

expended against friction generally depends on the total path length, not merely on the positions of the end points. It *does* require work, for example, to slide a block across a table and back to the starting point.

7.6   *Energy is Only Relative*

ONLY ENERGY DIFFERENCES ARE IMPORTANT

Unlike length and mass, *energy* has no absolute value. Suppose we ask the question "How much potential energy does a certain body have?" The answer is not simply *"mgh,"* because then one can ask *"h* above what?" and *h* is not just the height above the Earth's surface, because one could always drop the object into a hole and release some extra potential energy as kinetic energy. Therefore, the potential energy of the object at the surface of the Earth surely is not *zero.* The absolute value of the potential energy at any particular point has, in fact, no physical significance; it is only the *difference* in potential energy between two points that is important. In moving an object between two points, only the difference in potential energy can be converted into kinetic energy. Since we can always add a constant amount to the value of the potential energy at each of the two positions without altering the *difference* in potential energy between the positions, the absolute value of the potential energy is arbitrary. Although, as we shall see later, it is often convenient to specify the position of zero potential energy, this choice is arbitrary and one should distinguish clearly between *convenience* and *physical requirement.*

Kinetic energy is also a relative concept. The kinetic energy of a moving automobile, for example, appears to have different values for an observer standing on the road and for an observer in a train traveling on a track alongside the road. It is the *relative* velocity that determines the kinetic energy through the relation $\frac{1}{2} mv^2$. The kinetic energy of an object has different values with respect to different moving coordinate systems. Again, it is only the *change* in kinetic energy that is important because it is just this change in energy that appears as work in any frame of reference. James Clerk Maxwell, the famous English mathematical physicist who formulated the theory of electromagnetism, expressed the situation as follows:

We must, therefore, regard the energy of a material system as a quantity of which we may ascertain the increase or diminution as the system passes from one definite condition to another. The absolute value of the energy in the standard condition is unknown to us, and it would be of no value to us if we did know it, as all phenomena depend on the variations of the energy, and not on its absolute value.

7.7   *Conservation of Energy*

OUR MOST USEFUL PHYSICAL PRINCIPLE

In the simple examples given in the preceding sections it was implicit that potential energy could be transformed into kinetic energy, and *vice versa,* without any loss of energy. That is, a mass m in falling through a height h acquires a kinetic energy $\frac{1}{2}mv^2$, where $mgh = \frac{1}{2}mv^2$, or $(PE)_{\text{initial}} = (KE)_{\text{final}}$. Thus, energy (in one form or another) has been *conserved* during the process. This is, in fact, a general result embodied in the principle (or *law*) of *the conservation of energy*.

Energy conservation is a far-reaching principle, and, just as for linear momentum and angular momentum, there is no known exception to the rule that energy is conserved in every physical process. Indeed, we have the attitude that if we do not find a balance of energy in a certain process, we invent a form of energy that exactly makes up the deficit! This is not really a trick or a dishonest attempt to cover up our ignorance about Nature, for once we invent a new form of energy we must thereafter use the same definition and always incorporate this new form into our calculations in the same way. If we have made a poor choice, we will rapidly come upon a contradiction. Following this attitude, we have invented thermal energy, electromagnetic energy, nuclear energy and many other forms. Using these ideas consistently, we find no contradictions. Henri Poincaré (1854–1912), the great French mathematical physicist, expressed this outlook in the following way:

As we cannot give a general definition of energy, the principle of the conservation of energy simply signifies that there is *something* which remains constant [in every physical process]. Well, whatever new notions of the world future experiments may give us, we know beforehand that there will be something which remains constant and which we shall be able to call *energy*.

A principle is useful only if it allows us to gain some insight into the way Nature works. From this standpoint, energy conservation is surely the most useful single principle in science. Together with momentum conservation (see Section 5.11), the forcing of an energy balance in radioactive β decay led to the postulate of the neutrino, one of the most interesting of the fundamental particles. By utilizing the principle of energy conservation we have gained considerable insight into the complicated processes that occur in biological systems. In spite of the great difficulty of making precise physical measurements on living organisms, the conservation of energy has been verified to an accuracy of 0.2 percent in the metabolic processes in small animals.

Many problems in which complicated forces are involved and therefore the solutions to which are extremely difficult to construct by using Newton's laws can, nevertheless, be solved in a simple way by using the conservation laws, particularly energy conservation. Nothing, it is said, succeeds like success and energy conservation is certainly a successful principle.

Example **7.6**

A 1-kg block slides down a rough inclined plane whose height is 1 m. At the bottom, the block has a velocity of 4 m/sec. Is energy conserved?

At top: $PE = mgh$

$$= (1 \text{ kg}) \times (9.8 \text{ m/sec}^2) \times (1 \text{ m})$$

$$= 9.8 \text{ J}$$

At bottom: $KE = \frac{1}{2}\,mv^2$

$\qquad\qquad = \frac{1}{2} \times (1\ \text{kg}) \times (4\ \text{m/sec})^2$

$\qquad\qquad = 8\ \text{J}$

Apparently, energy is not conserved. But we know that friction is present between the block and the rough plane. A certain amount of energy (1.8 J) has evidently been expended in overcoming this friction. This amount of energy appears as *thermal energy* and could be detected by measuring the temperature rise in the block and the plane after the slide is completed. (Thermal energy will be discussed in Sections 7.15 through 7.17.)

Table 7.1 shows the range of energies that are encountered in various physical processes.

*Table **7.1**   Range of Energies*

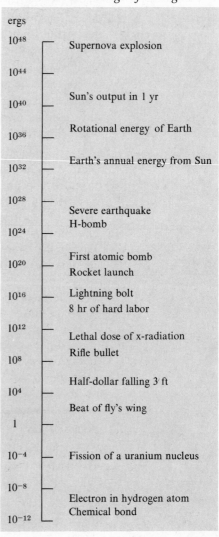

ergs	
10^{48}	Supernova explosion
10^{44}	
10^{40}	Sun's output in 1 yr
10^{36}	Rotational energy of Earth
10^{32}	Earth's annual energy from Sun
10^{28}	
10^{24}	Severe earthquake H-bomb
10^{20}	First atomic bomb Rocket launch
10^{16}	Lightning bolt 8 hr of hard labor
10^{12}	Lethal dose of x-radiation
10^{8}	Rifle bullet
10^{4}	Half-dollar falling 3 ft
	Beat of fly's wing
1	
10^{-4}	Fission of a uranium nucleus
10^{-8}	
10^{-12}	Electron in hydrogen atom Chemical bond

7.8 *Forms of Energy*

KINETIC AND POTENTIAL ENERGY

When the center of mass of an object is moving, the object possesses kinetic energy; in particular, it possesses a *translational* kinetic energy (denoted by KE_{trans}). Suppose, on the other hand, that the center of mass of the object is stationary but that the object is spinning. There is no net translational motion of the object and therefore $KE_{trans} = 0$. But the constituent parts of the object *are* in motion and therefore possess kinetic energy. This form of kinetic energy, which arises from spinning motion, is called *rotational* kinetic energy (KE_{rot}).

The potential energy that a body possesses can be the result of any of the possible forces that act on the body. For everyday objects, the only basic forces that act are the gravitational and electrical forces. Consequently, there are only two basic types of potential energy—gravitational potential energy (PE_G) and electrical potential energy (PE_E).

SPECIAL FORMS OF ENERGY AND THEIR ORIGINS

Table 7.2 lists some of the special forms of energy to which we frequently refer. *Mechanical* energy takes many forms but in each case it is relatively easy to identify the basic type or types of energy from which the particular form is derived. For example, the energy of a baseball thrown into the air with a spinning motion consists of KE_{trans}, KE_{rot}, and PE_G. For some of the other types of energy, the origin is not always so obvious. The *elastic* energy in a compressed spring, for example, arises from the potential energy stored as a result of work done in displacing the atoms of the material

Table **7.2** *Some Forms of Energy and Their Origins*

Type of Energy	Examples	Basic Form
Mechanical	Moving sled	KE_{trans}
	Rotating fly-wheel	KE_{rot}
	Water in a storage tower	PE_G
Elastic	Compressed spring	PE_E
	Rubber band	PE_E
Chemical	Fuel (gas, coal, etc.)	PE_E
	Food	PE_E
	Explosives	PE_E
Sound	Sonic boom	KE_{trans}
	Seismic waves	KE_{trans}
Thermal	Molecular agitation	KE_{trans}
Electromagnetic	Charges at rest	PE_E
	Electric current	$KE_{trans} + PE_E$
	Light	PE_E
	Radio Waves	PE_E

relative to one another. Since electrical forces exist among these atoms, the energy is PE_E. Also, when a molecule of a chemical fuel such as butane or a food substance such as glucose is formed, work is done against the electrical forces between the constituent atoms and the energy is stored as PE_E. Sound and thermal energy are forms of KE_{trans}; in sound waves the molecules move in an orderly fashion, first in one direction and then in another, but thermal energy involves the random motion of molecules. The energy contained in electromagnetic radiations, such as light or radio waves, is basically PE_E which is released as KE_{trans} when the wave interacts with an electrical charge.

Energy can take many forms and can be changed from one form to another. But in any physical process, if a strict accounting is made, it is always found that the total amount of energy is *constant*—energy is conserved.

7.9 *Elastic and Inelastic Collisions*

THE IMPORTANCE OF INTERNAL ENERGY

In addition to the energy that a body possesses by virtue of its motion or position, there is also the so-called *internal energy* of a body. An increase in the internal energy can take place, for example, if some of the kinetic energy of the body *as a whole* is converted into increased relative motion of the atoms or molecules that make up the body. This motional energy of the atoms or molecules does not manifest itself on a scale that can be detected by observing the motion or position of the body as a whole; instead, the result is an increase in the *temperature* of the body. Later in this chapter we discuss in detail a theory of heat based on the motion of atoms and molecules. For the present purpose it suffices to realize that there exists the possibility of altering the internal energy of a body.

When two objects collide, we know from our previous discussion that momentum and energy are conserved. We do not require, however, that kinetic energy be conserved; it is necessary only that the *total* energy be conserved. If some of the kinetic energy is converted into thermal energy, we say that the collision is *inelastic*. In general, an inelastic collision is one in which there is a change of the internal energy of one or both of the colliding objects. If there is no change of the internal energy, the collision is *elastic*. If the amount of kinetic energy converted into internal energy is the maximum allowed by the conservation of momentum, the collision is said to be *perfectly inelastic*.

Figure 7.6 shows two elastic and two inelastic collisions. If a ball dropped onto a fixed surface rebounds to its original height, an *elastic* collision has occurred. Of course, this is an idealized process and does not occur in Nature. During the collision there is always some loss of kinetic energy to the motion of the molecules in the ball. Therefore, all such *real,* macroscopic collisions are inelastic and the ball will never rebound to its original height. Collisions between macroscopic objects are always inelastic to some degree although in favorable cases the fraction of the kinetic energy that is converted into internal energy may be quite small.

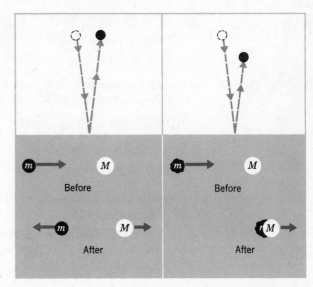

Fig. 7.6a *Elastic collisions.*

Fig. 7.6b *Inelastic collisions.*

If a body of mass m collides with an object of mass M (where $m < M$), as in Fig. 7.6, the smaller mass will rebound, thereby imparting some energy and momentum to the larger mass, which will recoil along the original direction of motion. Elastic collisions of this type can actually occur between elementary particles (such as an electron incident on a proton) because there is no way to change the "internal energy" of such particles.

As a final example (see Fig. 7.6 b), suppose that a blob of putty is projected toward an object at rest. On collision the putty sticks to the object instead of rebounding as in the previous case. In this situation conservation of momentum alone determines the final velocity of the combination. A maximum amount of kinetic energy has been converted into internal energy and the collision is said to be *perfectly inelastic.*

Example 7.7

A ball of mass $m_1 = 100$ g traveling with a velocity $v_1 = 50$ cm/sec collides "head on" with a ball of mass $m_2 = 200$ g which is initially at rest. Calculate the final velocities, v_1' and v_2', in the event that the collision is *elastic.*

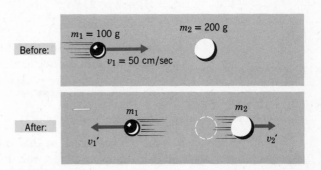

First, we use momentum conservation to write

p (before)   = p (after)

$m_1 v_1 + m_2 v_2 = m_1 v_1' + m_2 v_2'$

In order to prevent the equations from becoming too clumsy, we suppress the units (which are CGS throughout); then we have

$100 \times 50 + 0 = 100 v_1' + 200 v_2'$

Dividing through by 100 gives

$$50 = v_1' + 2 v_2' \tag{1}$$

From energy conservation, we have (since there is no *PE* involved and since the collision is elastic)

$$KE \text{ (before)} = KE \text{ (after)}$$

$$\tfrac{1}{2} m_1 v_1^2 + \tfrac{1}{2} m_2 v_2^2 = \tfrac{1}{2} m_1 v_1'^2 + \tfrac{1}{2} m_2 v_2'^2$$

$$\tfrac{1}{2} \times 100 \times (50)^2 + 0 = \tfrac{1}{2} \times 100 \ v_1'^2 + \tfrac{1}{2} \times 200 \ v_2'^2$$

Dividing through by $100/2 = 50$ gives

$$2500 = v_1'^2 + 2 v_2'^2 \tag{2}$$

We now have two equations, (1) and (2), each of which contains both of the unknowns, v_1' and v_2'. We can obtain a solution by solving Eq. 1 for v_1',

$$v_1' = 50 - 2 v_2' \tag{3}$$

and substituting this expression into Eq. 2:

$$2500 = (50 - 2 v_2')^2 + 2 v_2'^2$$

or,

$$2500 = 2500 - 200 v_2' + 4 v_2'^2 + 2 v_2'^2$$

From this equation we find

$$6 v_2'^2 = 200 v_2'$$

so that

$$v_2' = \tfrac{200}{6} = 33\tfrac{1}{3} \text{ cm/sec}$$

Substituting this value into Eq. 3 we find

$$v_1' = 50 - 2 \times 33\tfrac{1}{3}$$

$$= -16\tfrac{2}{3} \text{ cm/sec}$$

The negative sign means that after the collision, m_1 moves in the direction *opposite* to its initial direction.

Example **7.8**

A cue ball is struck so that it moves with a velocity of 100 cm/sec. The cue ball strikes an object ball and is deflected through an angle of 60°. In what direction does the object ball recoil and what is its velocity?

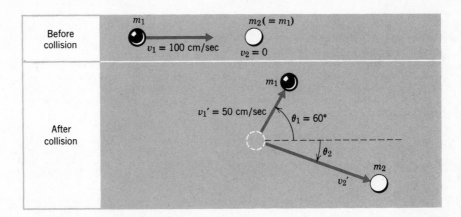

We proceed as in Example 7.7 by using the equations that represent momentum and energy conservation. But now we have a *two*-dimensional problem with an additional momentum equation.

Before collision the x- and y-components of the momentum are

$$p_x = m_1 v_1$$
$$p_y = 0$$

These quantities must be individually conserved. *After* collision we have, using the trigonometric expressions for the x- and y-components of the velocities,

$$m_1 v_1' \cos\theta_1 + m_2 v_2' \cos\theta_2 = m_1 v_1 \quad (x\text{-components})$$
$$m_1 v_1' \sin\theta_1 - m_2 v_2' \sin\theta_2 = 0 \qquad (y\text{-components})$$

where a negative sign is required in the equation for the y-components since these momenta are oppositely directed.

Since the collision is *elastic,* we also have an energy equation:

$$\tfrac{1}{2} m_1 v_1^2 = \tfrac{1}{2} m_1 v_1'^2 + \tfrac{1}{2} m_2 v_2'^2$$

Because $m_1 = m_2$, the momentum and energy equations become

$$v_1' \cos\theta_1 + v_2' \cos\theta_2 = v_1 \tag{1}$$
$$v_1' \sin\theta_1 - v_2' \sin\theta_2 = 0 \tag{2}$$
$$v_1^2 = v_1'^2 + v_2'^2 \tag{3}$$

These equations are all that we require to solve the problem because v_1 and θ_1 are known; that is, we have three equations for the three unknowns, v_1', v_2', and θ_2. We proceed as follows. First, we solve Eq. 2 for v_1'. Then, we substitute this result into Eq. 1 and solve for v_1. Thus, we have equations for both v_1 and v_2' that can be substituted into Eq. 3. This equation then contains the single unknown, θ_2. The value of θ_2 is substituted back into the previous equations in order to solve for v_1' and v_2'. The actual calculation requires considerable algebraic manipulation and is left as an exercise for the interested reader. The results are:

$$v_1' = 50 \text{ cm/sec}$$
$$v_2' = 86.6 \text{ cm/sec}$$
$$\theta_2 = 30°$$

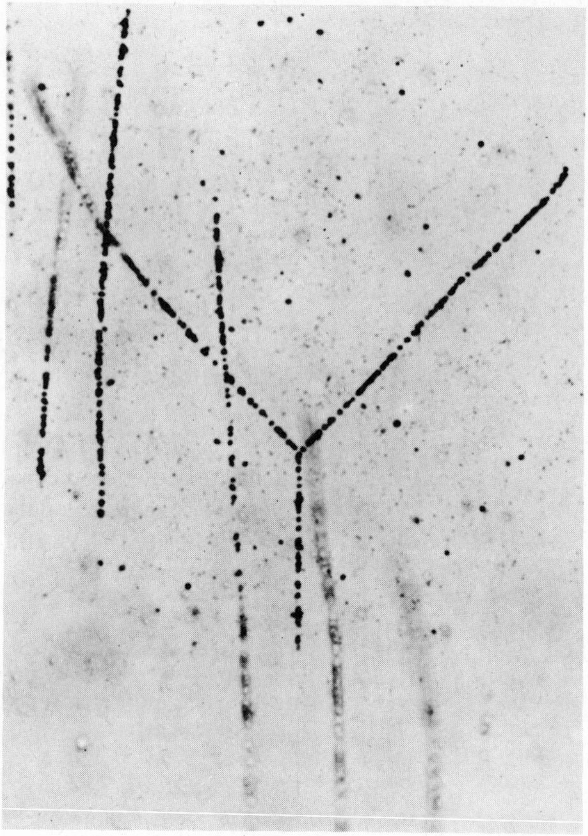

"Nuclear Physics in Photographs" by Powell and Occhaialini

Fig. 7.7 *The tracks are the record in a photographic emulsion of the collision of a moving proton with a proton at rest in the emulsion. The angle between the tracks after collision is 90°.*

The important point to notice is that the angle between the deflected cue ball and the recoiling object ball is $60° + 30° = 90°$; that is, the two velocity vectors, v'_1 and v'_2, are at *right angles*. This is, in fact, a general result for the collision of a particle or ball in motion with another particle or ball of *equal mass* which is at rest; the two objects always leave the point of collision at right angles. Thus, when a moving proton collides with another proton at rest the result is always[3] that the two protons take perpendicular paths away from the point of collision. An example of such an event is shown in Fig. 7.7.

Example **7.9**

A pendulum with a bob of mass M is raised to height H and released. At the bottom of its swing, it picks up a piece of putty whose mass is m. To what height h will the combination $(M + m)$ rise?

There are three phases to the problem—the fall of M, the collision of M and m, and the rise of $M + m$. The first and last phases involve energy conservation and the second phase involves momentum conservation.

---

[3]"Always" if the velocities involved are sufficiently small so that relativistic effects can be neglected. (See also Section 11.3.)

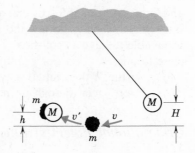

(1) Fall: $(PE)_{\text{initial}} = (KE)_{\text{final}}$

$$MgH = \tfrac{1}{2}Mv^2$$

from which

$$v = \sqrt{2gH} \tag{1}$$

(2) Collision: $p_{\text{initial}} = p_{\text{final}}$

$$Mv = (M + m)v' \tag{2}$$

(3) Rise: $(KE)_{\text{initial}} = (PE)_{\text{final}}$

$$\tfrac{1}{2}(M + m)v'^2 = (M + m)gh$$

from which

$$v' = \sqrt{2gh} \tag{3}$$

Substituting Eqs. 1 and 3 into Eq. 2 gives

$$M\sqrt{2gH} = (M + m)\sqrt{2gh}$$

Cancelling $\sqrt{2g}$ and squaring, we have

$$M^2H = (m + M)^2h$$

so that the final height is

$$h = \left(\frac{M}{m + M}\right)^2 H$$

As an exercise, show that the amount of kinetic energy converted to internal energy is $mMgH/(m + M)$.

COLLISIONS WITHOUT CONTACT

When one billiard ball strikes another billiard ball, we say that a *contact* collision has taken place. That is, the molecules on the surface of one ball have come into close proximity with those on the surface of the other ball. The term "contact" as used here is somewhat vague, however, since it is not clear what is meant by saying that one molecule "touches" another molecule. What actually happens in a so-called "contact" collision is that the molecules interact by means of electrical forces to the extent that there occurs a mutual repulsion. By extending this reasoning, we can state that, in general, the effect of a collision between two objects is due to the electrical

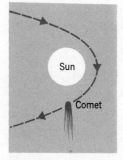

Fig. 7.8 *A comet "collides" with the Sun without any "contact."*

and/or gravitational forces that act between them. (Of course, for collisions involving nuclear particles, the nuclear force must be included.) Thus, it is possible to have a "collision" in which the objects never really come very close to one another. For example, a comet is swept around the Sun (Fig. 7.8) by virtue of the mutual gravitational attraction. This is, in fact, a form of "collision" even though no physical "contact" is involved.

7.10   *Gravitational Potential Energy*

WORK REQUIRED TO LIFT AN OBJECT

Near the surface of the Earth, where we can consider the gravitational force to be constant, we know that to lift an object of mass m through a height h requires an amount of work $W = mgh$. After lifting, we say that the object has a gravitational potential energy $PE_G = mgh$. If we consider distances that are no longer negligible compared to the radius of the Earth, then we must take account of the fact that the gravitational force varies as the square of the distance from the center of the Earth.

We now calculate the amount of work required to raise an object of mass m from an initial position that is a distance r_1 from the center of the Earth, to a final position that is a distance r_2 from the center of the Earth, without any change in the kinetic energy of the object (see Fig. 7.9). The force on m at r_1 is

$$F_1 = G \frac{Mm}{r_1^2} \qquad \text{at } r_1 \tag{7.9a}$$

and at r_2 the force is

$$F_2 = G \frac{Mm}{r_2^2} \qquad \text{at } r_2 \tag{7.9b}$$

We know that the amount of work done on an object is equal to the average force exerted multiplied by the distance through which the object is moved:

$$W = \bar{F} \times s \tag{7.10}$$

For the situation in Fig. 7.9, the distance moved is clearly $s = r_2 - r_1$, but what is the average gravitational force on m in the interval from r_1 to r_2? In Example 7.2 we calculated the average force by taking the *arithmetic*

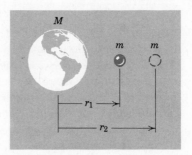

Fig. 7.9 *Work done by an outside agency is required to move* m *from* r$_1$ *to* r$_2$.

average of the forces at the initial and final positions. But in that case we were dealing with a force that varies *linearly* with distance. In the present case, we have a $1/r^2$ force and the arithmetic average is no longer appropriate. We need a procedure that yields an average force that is closer to F_1 than to F_2 because the force falls off rapidly in going from r_1 to r_2. Such an average (and the one appropriate for a $1/r^2$ force) is the *geometric* average, namely,

$$\bar{F} = G \frac{Mm}{r_1 r_2} \qquad (7.11)$$

(We have not proved that this average is correct; we have only argued that it is reasonable. In spite of this arbitrariness, the result is nevertheless *exact*.) Substituting this expression for $\bar{F}$ and $s = r_2 - r_1$ into Eq. 7.10, we find

$$W_{12} = G \frac{Mm}{r_1 r_2} \times (r_2 - r_1)$$

$$= GMm \left(\frac{1}{r_1} - \frac{1}{r_2} \right) \qquad (7.12)$$

The increase in gravitational potential energy of m in moving from r_1 to r_2 is just the work required to effect this change of position; that is,

$$PE_G = GMm \left(\frac{1}{r_1} - \frac{1}{r_2} \right) \qquad (7.13)$$

Notice that the gravitational potential energy depends only on the initial and final positions of the mass, r_1 and r_2. Although we considered here a particularly simple path by which the mass was moved from r_1 to r_2, in fact, PE_G depends only on r_1 and r_2 for *any* path that connects the initial and final positions. That is, the gravitational force is a *conservative force* (see Section 7.5).

POTENTIAL ENERGY NEAR THE SURFACE OF THE EARTH

Let us use the general (and exact) result for PE_G that we have just derived and compute the potential energy gained by an object in raising it to a height h above the surface of the Earth. If h is small compared with R_E, the radius of the Earth, then we expect that we will obtain our previous result, namely, $PE_G = mgh$. From Eq. 7.13, the gain in potential energy in raising m from R_E to $R_E + h$ is

$$PE_G = GMm \left(\frac{1}{R_E} - \frac{1}{R_E + h} \right)$$

$$= GMm \frac{h}{R_E(R_E + h)} \qquad (7.14)$$

If h is small compared to $R_E(h \ll R_E)$, we can neglect h in the denominator and write

$$PE_G \cong \frac{GMmh}{R_e^2} \qquad (7.15)$$

Now, the force on m at the surface of the Earth is

$$F_G = G \frac{Mm}{R_E^2} \tag{7.16}$$

But this force is just mg, the weight of the object, so

$$mg = G \frac{Mm}{R_E^2}$$

Therefore,

$$GMm = mgR_E^2 \tag{7.17}$$

Substituting this value for GMm into Eq. 7.17, the potential energy becomes

$$PE_G \cong \frac{(mgR_E^2)h}{R_E^2}$$

or,

$$PE_G \cong mgh \qquad \text{(for } h \ll R_E \text{)} \tag{7.18}$$

which is our previous result.

ESTABLISHING THE "ZERO" OF POTENTIAL ENERGY

According to our general expression (Eq. 7.13), the increase in gravitational potential energy in raising an object of mass m from the surface of the Earth (radius R_E) to a position that is a distance r from the Earth's center is

$$PE_G = GMm \left(\frac{1}{R_E} - \frac{1}{r} \right) \tag{7.19}$$

Let us arbitrarily establish a reference level for gravitational potential energy by taking the potential energy to be zero at infinite distance. If we allow r to become indefinitely large, $r \to \infty$, then Eq. 7.19 becomes

$$PE_G = GMm \left(\frac{1}{R_E} - \frac{1}{\infty} \right)$$

$$= \frac{GMm}{R_E} \qquad \text{(as } r \to \infty \text{)}$$

In order to make $PE_G = 0$ as $r \to \infty$, clearly we must add a constant, $-GMm/R_E$, to our expression for PE_G. We are able to do this because potential energy is only a *relative* concept (see Section 7.6); we take advantage of this fact and require $PE_G = 0$ as $r \to \infty$ by adding a constant amount to the potential energy at every position. Thus, for a general point at a distance r from the Earth's center, we have, by adding $-GMm/R_E$ to Eq. 7.19,

$$\boxed{PE_G = - \frac{GMm}{r}} \tag{7.20}$$

This expression is, of course, a general result and is not limited to the case of the Earth's gravitational attraction. If, for example, $M = $ mass of

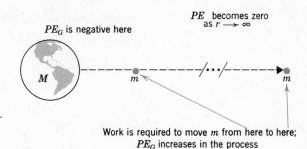

Fig. 7.10 *Work is required to* increase *the gravitational potential energy from a negative value to a less negative value or to zero (as* r → ∞).

the Sun, m = mass of the Earth, and r = radius of the Earth's orbit, then PE_G becomes the gravitational potential energy of the Earth due to the Sun's attraction.

According to our choice of position for zero gravitational potential energy, the value of PE_G is always *negative* and increases (that is, becomes less negative) as r is increased, becoming zero as $r →$ ∞. Work is always required to increase the potential energy (that is, to separate the bodies); the work required is a maximum when the bodies are infinitely far from each other in the final position (see Fig. 7.10).

POTENTIAL ENERGY DIFFERENCES

It must be remembered that only *changes* in potential energy are physically meaningful. Therefore, we are usually concerned with calculating $\Delta(PE_G)$ in moving an object of mass m from r_1 to r_2 as measured from M. According to Eq. 7.12,

$$\Delta(PE_G) = W_{12} = GMm\left(\frac{1}{r_1} - \frac{1}{r_2}\right) \tag{7.21}$$

If $W_{12} > 0$, then work was done by some outside agent *against* the attractive gravitational force. If $W_{12} < 0$, then work was done *by* the gravitational force and this amount of energy can be used, for example, to increase the kinetic energy of the bodies. If two masses with a certain separation are released, the attractive gravitational force will cause the masses to accelerate toward one another and the separation will decrease. Consequently, the gravitational potential energy will decrease, and there will be an equivalent gain in kinetic energy.

Example **7.10**

The *escape velocity* v_e is the minimum velocity with which an object must be propelled from the surface of the Earth if it is to move an infinite distance away, thereby "escaping" from the gravitational attraction of the Earth. (We neglect here the fact that a real object launched from the surface of the Earth would encounter air resistance before leaving the atmosphere.) In order to calculate v_e, we note that the initial kinetic energy at the Earth's surface, $\frac{1}{2}mv_e^2$, will be completely expended in raising the initial gravitational potential energy, $-GMm/R_E$, to zero as $r →$ ∞. That is, the total energy for the object, K.E. + P.E.$_G$, becomes exactly zero as $r →$ ∞.

Using energy conservation, we can write

$$(KE + PE_G)_{\text{surface of Earth}} = (KE + PE_G)_{r \to \infty}$$

$$\tfrac{1}{2}mv_e^2 - \frac{GMm}{R_E} = 0$$

from which

$$v_e = \sqrt{2GM/R_E}$$

Substituting $G = 6.67 \times 10^{-8}$ dyne-cm^2/g^2, $M = 5.98 \times 10^{27}$ g, and $R_E = 6.37 \times 10^8$ cm, we find

$$v_E = 1.13 \times 10^6 \text{ cm/sec}$$
$$= 11.3 \text{ km/sec}$$
$$= 25{,}280 \text{ mi/hr} \qquad \text{(for escape from the Earth)}$$

Actually, such a velocity is not sufficient to allow an object to escape from the solar system because the gravitational influence of the Sun is so much greater than that of the Earth. In order to calculate the velocity of escape *from the Sun,* starting from the position of the Earth, we use the same expression for v_e but substitute the mass of the Sun for M (1.99×10^{33} g) and the Earth-Sun distance for R_E (1.50×10^{13} cm). Then, we find

$$v_e = 42.1 \text{ km/sec} \qquad \text{(for escape from the Sun)}$$

The result we have just obtained allows us to make a good estimate for the impact velocity of meteorites when they strike the Earth. A particle falling toward the Sun from infinitely far away[4] (starting from rest) will acquire, by the time it has reached the position of the Earth's orbit around the Sun, a velocity just equal to that calculated above, namely, 42.1 km/sec. The velocity of a meteorite relative to the Earth is the vector sum of this velocity and the orbital velocity of the Earth around the Sun (29.9 km/sec). If the meteorite collides "head on" with the moving Earth, the impact velocity will be approximately $42.1 + 29.9 = 72.0$ km/sec, whereas an "overtaking" meteorite will have an impact velocity of approximately $42.1 - 29.9 = 12.2$ km/sec. Most meteorites have impact velocities in the range 10–70 km/sec.

7.11   *Electrostatic Potential Energy*

COMPARISON OF GRAVITATIONAL AND ELECTROSTATIC POTENTIAL ENERGIES

We consider now only that portion of electromagnetic energy that is associated with charges at rest—*electrostatic* energy.

Except for the important fact that electric charge can be either positive

---

[4] Actually, meteorites arise from material in the solar system and so do not originate infinitely far away. But the velocity acquired by an object falling from the outer portion of the solar system is not too different from that of an object originating infinitely far away. Therefore, the numerical result here is not seriously in error.

or negative, whereas mass is always positive, the case of electrostatic potential energy is the same as that of gravitational potential energy. We have already seen that the gravitational force between two objects with masses m_1 and m_2, separated by a distance r, is

$$F_G = G \frac{m_1 m_2}{r^2} \qquad (7.22)$$

From this fact, we found that the gravitational potential energy in such a case is

$$PE_G = -G \frac{m_1 m_2}{r} \qquad (7.23)$$

where we have again used the convention that $PE_G = 0$ when $r \to \infty$.

For the electrostatic case, the force between the charges q_1 and q_2, separated by a distance r, is

$$F_E = -\frac{q_1 q_1}{r^2} \qquad (7.24)$$

where we understand that a *positive* force is *attractive* (that is, when q_1 and q_2 have opposite signs) and a *negative* force is *repulsive* (that is, when the charges have the same sign). Because the expression for F_E is of the same form as that for F_G, it follows that the expression for the electrostatic potential energy PE_E must be of the same form as that for PE_G, except that we must carry along the sign difference between F_G and F_E. Thus, we have

$$PE_E = \frac{q_1 q_2}{r} \qquad (7.25)$$

WORK AND POTENTIAL ENERGY CHANGES

The work done, or change in PE_E, in moving a charge q_1 from r_1 to r_2 as measured from q_2 is

$$W_{12} = \Delta(PE_E) = q_1 q_2 \left(\frac{1}{r_2} - \frac{1}{r_1} \right) \qquad (7.26)$$

Compare this expression with Eq. 7.21 for $\Delta(PE_G)$.

The difference of sign between PE_G and PE_E or between $\Delta(PE_G)$ and $\Delta(PE_E)$ is easy to understand by referring to the concept of *work*. This is illustrated schematically in Fig. 7.11 for the case of electrostatic potential energy. Whenever work is done *against* a conservative force (such as F_E or F_G), the potential energy *increases;* $W > 0$ and $\Delta(PE) > 0$. If work is done *by* such a force, the potential energy *decreases;* $W < 0$ and $\Delta(PE) < 0$. The first two cases[5] in Fig. 7.11 apply also for the gravitational case.

[5] Only the first two cases since F_G is always *attractive*.

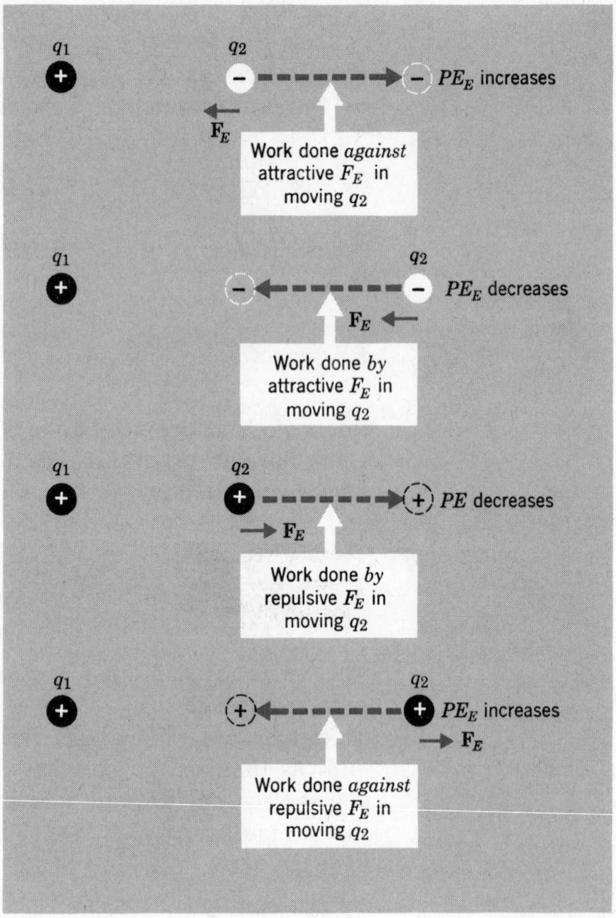

Fig. 7.11 *The changes in* PE$_E$ *are shown for the four possible cases of a charge* (q$_2$) *moved in the presence of a positive charge* (q$_1$).

Example 7.11

Suppose we have two charges, $q_1 = +4$ statC and $q_2 = -6$ statC, with an initial separation of $r_1 = 3$ cm. What is the change in potential energy if we increase the separation to 8 cm?

Using Eq. 7.26,

$$\Delta(PE_E) = q_1 q_2 \left(\frac{1}{r_2} - \frac{1}{r_1} \right)$$

$$= (+4)(-6) \times \left(\frac{1}{8} - \frac{1}{3} \right)$$

$$= (-24)\left(-\frac{5}{24} \right)$$

$$= +5 \text{ ergs}$$

In this case there is a net *increase* in the electrostatic potential energy (that is, $\Delta(PE_E) > 0$) because work was done by an outside agent against the attractive electrostatic force.

7.12 *Potential Difference and the Electron Volt*

STATVOLTS AND VOLTS

Equation 7.26 states that the work done in moving a charge q from a distance r_1 to a distance r_2 measured from a fixed charge q' is

$$W = q\,q'\left(\frac{1}{r_2} - \frac{1}{r_1}\right) \tag{7.27}$$

If we know q, q', and the distances, we can always calculate W. If we change q to a new value, then W also changes. However, the ratio W/q does not depend on q; W/q is a property only of the charge q'. (We call q' the *source* charge.) The quantity W/q is the work *per unit charge* required to move from r_1 to r_2 in the presence of the electrostatic force due to the source charge q'. This is an extremely useful concept and we therefore give W/q a special name—*potential difference*,[6] V:

$$\text{Potential difference: } V = \frac{W}{q} \tag{7.28}$$

Note that W is the difference in potential energy (for a particular value of q), but W/q does not have the dimensions of energy, but of energy (or work) *per unit charge*. If 1 *erg* of work is required to move 1 *statC* of charge from r_1 to r_2 in the presence of q', we say that the potential difference between r_1 and r_2 is 1 *statvolt* (statV). That is, in CGS units,

CGS: 1 statV = 1 erg/statC (7.29a)

In the MKS system, the unit of work is the *joule,* the unit of charge is the *coulomb,* and the unit of potential difference is the *volt* (V):

MKS: 1 V = 1 J/C (7.29b)

Since 1 J = 10^7 ergs and 1 C = 3×10^9 statC,

1 statV = 300 V (7.30)

The MKS unit of potential difference, the *volt,* is just that unit which is familiar in terms of household electricity. The potential difference between the two wires of common household electrical circuits is 110 V; the potential difference between the terminals of a flashlight battery is 1.5 V.

THE ELECTRON VOLT

A unit of energy that is quite useful in dealing with problems in atomic and nuclear physics is obtained in the following way. Suppose that a charge

[6] Sometimes, V is called simply the *potential.*

e, equal to the charge on an electron (which is the same as the charge on a proton, disregarding the sign), is moved from one position to another between which exists a potential difference of 1 V. How much work has been done *on* the charge? Using Eq. 7.28, we have

$$W = qV = e \times V$$
$$= (1.602 \times 10^{-19} \text{ C}) \times (1 \text{ V})$$
$$= 1.602 \times 10^{-19} \text{ J} \tag{7.31}$$

Or, in CGS units,

$$W = (4.80 \times 10^{-10} \text{ statC}) \times (\tfrac{1}{300} \text{ statV})$$
$$= 1.60 \times 10^{-12} \text{ erg}$$

This unit of *work* or *energy* is given the special name *electronvolt,* and is denoted by the symbol *eV.*

$$\left. \begin{array}{l} 1 \text{ eV} = 1.602 \times 10^{-19} \text{ J} \\ \phantom{1 \text{ eV}} = 1.602 \times 10^{-12} \text{ erg} \end{array} \right\} \tag{7.32}$$

Larger units of energy are convenient for many problems, especially in nuclear physics; those most frequently used are

1 kiloelectronvolt (keV) $= 10^3$ eV $= 1.602 \times 10^{-9}$ erg

1 megaelectronvolt (MeV) $= 10^6$ eV $= 1.602 \times 10^{-6}$ erg

1 gigaelectronvolt (GeV) $= 10^9$ eV $= 1.602 \times 10^{-3}$ erg

---

Example **7.12**

A proton, starting from rest, falls[7] through a potential difference of 10^6 V. What is its final kinetic energy and final velocity?

For the kinetic energy we have, simply,

$$KE = 10^6 \text{ eV} = 1 \text{ MeV} = 1.602 \times 10^{-6} \text{ erg}$$

In order to compute the final velocity, we use

$$KE = \tfrac{1}{2} m_p v^2$$

or,

$$v = \sqrt{\frac{2KE}{m_p}}$$
$$= \sqrt{\frac{2 \times (1.60 \times 10^{-6} \text{ erg})}{1.67 \times 10^{-24} \text{ g}}}$$

so that

$$v = 1.38 \times 10^9 \text{ cm/sec}$$

---

[7] When an object falls from a certain height above the surface of the Earth, it moves from a position of high PE_G to a position of lower PE_G. When an electrical charge moves from a position of high PE_E to one of lower PE_E, we use the gravitational terminology and say that the charge "falls" through a certain potential difference although the actual motion may be horizontal or even upward.

If we had considered an electron, instead of a proton, falling through a potential difference of 10^6 V, we could not have computed the final velocity in such a simple way. Relativity theory provides the correct method of calculation. The relativistic effect, which is manifest for a 1-MeV electron, is negligible for a proton of the same energy owing to the much larger mass of the proton. However, for protons with energies of about 100 MeV or more we must also use the relativistic expression for computing the velocity.

THE ELECTRONVOLT AS A GENERAL ENERGY UNIT

The *electronvolt* is commonly used as a unit of energy even in the event that the particle has not fallen through any potential difference. For example, a neutron (which has approximately the same mass as a proton but no electric charge and, therefore, cannot experience an electrostatic force) which is moving with a velocity of 1.38×10^9 cm/sec is said to have an energy of 1 MeV. That is, the unit "1 MeV" means that the particle has a definite amount of kinetic energy and it does not matter how the particle acquired this energy.

The eV notation is rarely used for objects larger than atomic size because the unit is too small to be convenient. The energy of a 0.1-g meteorite which strikes the Earth with a velocity of 50 km/sec, for example, has an energy of approximately 8×10^{14} GeV.

THE VAN DE GRAAFF ACCELERATOR

The *electrostatic generator* or *Van de Graaff accelerator* gives high velocities to charged particles, such as protons, by allowing them to fall through a large potential difference. The operation of these machines, which are much used in current research problems in nuclear physics, is illustrated schematically in Fig. 7.12.

An endless belt made of an insulating material (usually rubberized cotton) is driven by a drive motor located at ground (that is, *zero*) potential. Electrons are removed from the belt as it passes a screen that is maintained at a high positive potential difference relative to ground by a high-voltage power supply. Since electrons have been *removed*, the portion of the belt that moves upward from the screen carries a *positive* charge. At the upper end of the Van de Graaff column, the positive charge is neutralized by the flow of electrons from the hemispherical dome that caps the machine. Thus, the dome acquires a large positive charge and has a high positive potential difference relative to ground. It is possible to charge the dome to about 10 million volts! In practice, the entire assembly is contained within a tank that holds a gas at high pressure to suppress the sparking of the dome to neighboring objects that are at ground potential. If a source of positive ions (such as protons) is placed in the dome, the high positive potential will repel these ions; they will be accelerated to the ground level. When the dome has a potential of 5 million volts relative to ground, the protons will be accelerated to an energy of 5 MeV. Similarly, if doubly-ionized helium ions (charge $= 2e$) are accelerated through the same potential difference, they

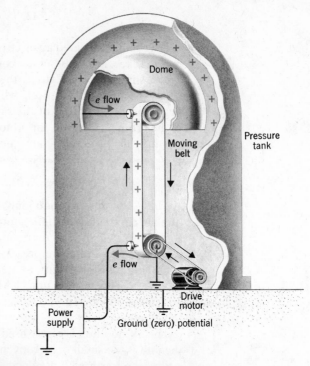

Dome

e flow

Moving belt

Pressure tank

e flow

Power supply

Drive motor

Ground (zero) potential

Fig. 7.12 *Schematic of a Van de Graaff accelerator.*

Fay Photo Service, Inc.

Fay Photo Service, Inc.

Fig. 7.13 *The internal structure of a 5.5-million volt Van de Graaff accelerator. The rings serve to distribute the voltage drop from dome to ground in a uniform manner. The picture at the right shows the pressure tank that surrounds the accelerator during operation. (Courtesy of High Voltage Engineering Corporation.)*

242

will acquire an energy of 10 MeV. These energies are sufficiently high so that the particles can produce nuclear reactions by disintegrating other nuclei that they strike.

7.13 *Potential Energy Diagrams*

TOTAL ENERGY = PE + KE

Figure 7.14a shows a roller-coaster car and the hill-and-valley track over which the car runs. Suppose the car starts from rest at a height h_1 above ground level. From experience we know that the speed of the car will be greatest in the valleys of the track and will be least on the hills. This fact is due to the interchange of potential energy and kinetic energy. Since the potential energy at any point is proportional to the height of that point above the reference (or ground) level, we can convert the track diagram directly into a *potential energy diagram,* as in Fig. 7.14b. From this curve we can read directly the *PE* at any position. The position $s = s_1 = 0$ corresponds to the starting point where $(PE)_1 = mgh_1$ and $(KE)_1 = 0$. Thus, the total energy $\mathcal{E}$ at $s = s_1$ is $\mathcal{E} = (PE)_1 + (KE)_1 = mgh_1$. If frictional losses are neglected, energy conservation requires that the total energy at any other position is also mgh_1. At $s = s_2$, where the car is at a height h_2, the potential energy is $(PE)_2 = mgh_2$; the kinetic energy at that position must be the difference between $\mathcal{E}$ and $(PE)_2$, or $(KE)_2 = \mathcal{E} - (PE)_2 = mg(h_1 - h_2)$. That is, the kinetic energy at any position is given graphically by the difference between the total energy line and the potential energy curve.

THE ELECTROSTATIC CASE

Figure 7.15 shows the potential energy diagram for an electron in the presence of a proton. The curve represents the general expression for the

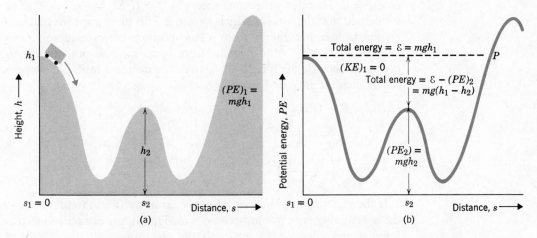

Fig. 7.14 (a) *A roller coaster with a car starting from a height* h_1. (b) *The potential energy diagram corresponding to the roller-coaster of* (a). *The height of the curve at any position is equal to the* PE *and the difference between the total energy line and the curve is equal to the* KE *at that position. Notice that the car cannot pass over the hill on the right since to move above the height of the point* P *requires an energy greater than the total energy* $\mathcal{E}$ *of the car.*

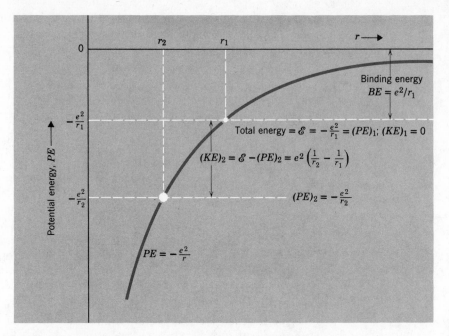

Fig. 7.15 *Potential energy diagram for an electron in the presence of a proton. The energy lines are drawn for the case in which the electron is at rest at the position* $r = r_1$.

potential energy:

$$PE_E = \frac{q_1 q_2}{r} = \frac{(e)(-e)}{r} = -\frac{e^2}{r}$$

If the electron is at rest at the distance $r = r_1$, the kinetic energy is zero: $(KE)_1 = 0$. Therefore, the total energy is $\mathcal{E} = (PE)_1 = -e^2/r_1$. (Do not be concerned that the total energy is *negative*. This merely means that we have chosen to measure energy from a zero position corresponding to $r \to \infty$.) In moving closer to the proton (from r_1 to r_2), the potential energy is *decreased* because the force is attractive. The decrease in *PE* is compensated by an increase in *KE*. Thus,

$$(KE)_2 = \mathcal{E} - (PE)_2$$
$$= -\frac{e^2}{r_1} - \left(-\frac{e^2}{r_2}\right)$$
$$= e^2\left(\frac{1}{r_2} - \frac{1}{r_1}\right)$$

If the electron were moved infinitely far away from the proton ($r \to \infty$), the potential energy would become zero. That is, the attractive electrostatic force would decrease to zero and the electron would be *free* from the influence of the proton. In the *free* case, the total energy is $\mathcal{E} = 0$ (or > 0). In going from the position $r = r_1$ to the *free* position ($r \to \infty$) requires that the total energy be raised from $-e^2/r_1$ to zero. In other words, at $r = r_1$ the electron is *bound* by an amount of energy e^2/r_1; this is called the binding energy (*BE*) of the system.[8] In the hydrogen atom, $BE = 13.6$ eV. Thus, an amount of energy equal to (or greater than) 13.6 eV is required to ionize a hydrogen atom.

7.14 *Heat as a Form of Energy*

THE FIRST LAW OF THERMODYNAMICS

All material things consist of atoms and molecules, and these atoms and molecules are continually in motion, whether the object is gas, liquid, or solid. Therefore, even if an object is motionless in a position of zero potential energy (relative to some base position), energy is associated with the internal motion of the constituent atoms and molecules. That is, there is always a certain *internal energy* for any collection of atoms or molecules. If we alter the system by causing the atoms to move more violently, we say that we have added *thermal energy* to the system (by doing work *on* the system or by adding *heat* to the system), thereby increasing the internal energy. *Thermodynamics* is the branch of physics concerned with the interplay of heat, work, and energy.

The conservation of energy is a general physical law and it must apply when we consider (as we have not done in any detail until now) the internal energy of an object in addition to its "external" forms of energy. Let us denote by U the internal energy of a body and by Q the amount of thermal energy (or *heat*) supplied to the body. If a body takes in an amount of heat Q, the internal energy is increased by exactly that amount: $\Delta U = Q$. Alternatively, the body could do a certain amount of work when supplied with the heat Q; for example, if the body is a gas, it could expand against a restraining force as in Fig. 7.16. The increase in internal energy is diminished if the body does work when supplied with heat. The principle of energy conservation states that the change in internal energy is equal to the heat supplied *to* the body *minus* the work done *by* the body:

$$\Delta U = Q - W \qquad (7.33)$$

This equation is called the *first law of thermodynamics*, but it is simply a statement of energy conservation when thermal energy is included.

[8] In the convention that we have elected to use, the total energy of a *bound* system is always *negative;* the total energy of an *unbound* system is *positive* (or zero).

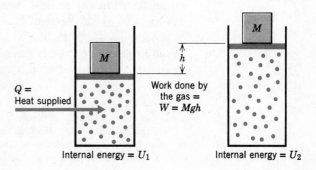

Fig. 7.16 *An amount of heat Q is supplied to a gas and the expanding gas does an amount of work* W. *The increase in the internal energy of the gas is* $\Delta U = Q - W$. *Notice that a portion of the thermal energy is transformed into gravitational potential energy; i.e., the work done by the expanding gas is* W = mgh.

$Q =$ Heat supplied

Work done by the gas = $W = Mgh$

Internal energy = U_1

Internal energy = U_2

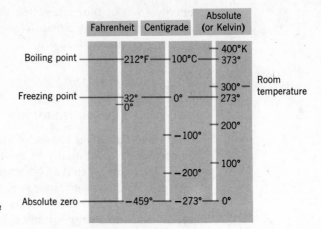

Fig. 7.17 *Comparison of the Fahrenheit, centri-grade, and the absolute (or Kelvin) temperature scales. (The use of the absolute scale is discussed in Section 7.15.)*

TEMPERATURE (AN OPERATIONAL DEFINITION)

The quantitative specification of internal energy or thermal energy requires the concept of *temperature*. Temperature is a familiar idea, indicating the degree of "hotness" or "coldness" of an object. We shall frame a precise definition of temperature in the next section when we discuss the microscopic theory of heat; an operational definition will suffice for the moment. The difference in the "hotness" of boiling water and the "coldness" of ice is divided into 100 equal parts, called *1 Centigrade degree* (°C). The temperature of melting ice is arbitrarily designated 0°C and that of boiling water 100°C when the measurements are carried out at sea level. It has been found that the volume of certain substances, such as mercury, increases uniformly as the temperature is increased.[9] Therefore, if an amount of mercury is confined within a narrow column in a tube of glass, the level will rise as the temperature is increased and will fall when the temperature is decreased. By marking the points corresponding to the temperatures of ice (0°C) and boiling water (100°C) and dividing the interval into 100 units,[10] we have a conventional mercury *thermometer*. We can now imagine the use of such an instrument in the investigation of various heat phenomena.

HEAT UNITS

Having defined a method for measuring temperature T, we can use this method to arrive at a specification of the amount of heat energy Q that is transferred in any process. Again we use water as the standard substance and say that the amount of heat required to raise the temperature of 1 g of water by 1°C (from 14.5°C to 15.5°C) is equal to *1 calorie* (cal). (The amount of heat required to raise 1 kg of water by 1°C is 1 kilocalorie, often written as 1 Calorie,[11] with a capital rather than a lower case "c".) The

---

[9] The point of how we know this to be true will become clear with the more complete discussion in the following section.

[10] On the *Fahrenheit* scale the freezing and boiling points of water are 32°F and 212°F, respectively. The Fahrenheit scale is rarely used in scientific work.

[11] The Calorie is the unit usually used in specifying the energy content of foods.

calorie is a unit of energy (or work) and a definite relationship exists between the calorie and the erg or joule. This conversion factor can be determined by doing a known amount of work on a known mass of water and measuring the resultant increase in temperature. Joule accomplished this by measuring the work necessary to force water to flow through narrow pipes and simultaneously measuring the resulting increase in temperature brought about by friction. The value he obtained in 1845 (4.14×10^7 ergs/cal) is quite close to the presently accepted value:

$$
\left. \begin{array}{l}
1 \text{ cal} \quad = 4.186 \times 10^7 \text{ ergs} \\
\qquad\quad = 4.186 \text{ J} \\
1 \text{ Calorie} = 4186 \text{ J}
\end{array} \right\} \tag{7.34}
$$

Since the *erg* and the *joule* are precisely defined units, the *calorie* is superfluous. However, because of historical tradition the unit is still in use in thermodynamics and we are therefore forced to add another conversion factor to our growing list.

SPECIFIC HEAT

The quantity of heat required to raise the temperature of an object by an amount ΔT is proportional both to ΔT and to the mass m of the object. Calling the proportionality constant c, we can write

$$
Q = cm \, \Delta T \tag{7.35}
$$

The quantity c is called the *specific heat* of the substance. The dimensions of c are cal/g-°C. Usually c is a function of the temperature, but over a small range of temperatures (for example, near room temperature), c is essentially constant. The values of the specific heats of a few substances are given in Table 7.3.

All the values of c listed in Table 7.3 are less than unity, some substantially less. That is, water (for which $c = 1$, by definition) has an abnormally high specific heat. Water, one of our most plentiful resources, is therefore an efficient conveyor of thermal energy (as, for example, in the common hot-water heating system).

Table **7.3** Specific Heats of Some Materials near Room Temperature	
Substance	*c (cal/g-°C)*
air	0.17
aluminum	0.219
copper	0.0932
ethyl alcohol	0.535
gold	0.0316
iron	0.119
lead	0.0310
mercury	0.0333

Example **7.13**

A 10-g lead bullet is traveling with a velocity of 10^4 cm/sec and strikes a heavy wood block. If, in coming to rest in the block, half of the initial kinetic energy of the bullet is transformed into thermal energy in the block and half into thermal energy in the bullet, calculate the rise of temperature of the bullet. (The block remains stationary during the collision.)

The initial kinetic energy is

$$\tfrac{1}{2}mv^2 = \tfrac{1}{2} \times (10 \text{ g}) \times (10^4 \text{ cm/sec})^2$$
$$= 5 \times 10^8 \text{ ergs}$$

The temperature rise is

$$\Delta T = \frac{Q}{cm}$$

where $Q = \tfrac{1}{2} \times (5 \times 10^8 \text{ ergs})$. The specific heat of lead is 0.0310 cal/g-°C (see Table 7.3). Therefore,

$$\Delta T = \frac{\tfrac{1}{2} \times 5 \times 10^8 \text{ ergs}}{(4.186 \times 10^7 \text{ ergs/cal})(0.0310 \text{ cal/g-°C}) \times 10 \text{ g}}$$
$$= 19.3°C$$

7.15 *The Microscopic Theory of Heat*

PRESSURE

If we apply a force F to a flat surface whose area is A, this force is transmitted to the substance, for example, a gas, on the other side of the surface (Fig. 7.18). From the standpoint of the effect on the gas, it is not the *force* that is of primary importance, but the *force per unit area* on the surface which distributes the force to the gas. This quantity we call the *pressure:*

$$\text{Pressure} = \frac{\text{Force}}{\text{Area}}$$

$$\boxed{P = \frac{F}{A}}$$

(7.36)

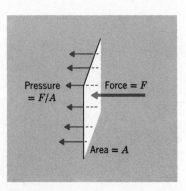

Fig. 7.18 *Pressure is force per unit area.*

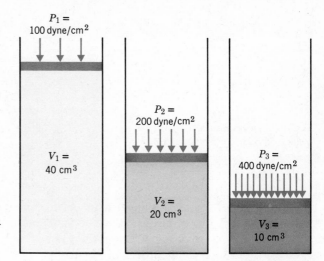

Fig. 7.19 *Robert Boyle discovered the inverse relationship between the volume of a gas and the applied pressure for constant temperature.*

The pressure that the air in the atmosphere exerts on all objects at sea level is 1.0×10^6 dynes/cm^2. This pressure is termed 1 *atmosphere* (1 atm):[12]

$$1 \text{ atm} = 1.0 \times 10^6 \text{ dynes/cm}^2 \qquad (7.37)$$

An alternate method of specifying pressure is to give the height of a column of mercury (symbol: Hg) that a vacuum will support at that pressure. One atm corresponds to 760 mm (or 30 in. Hg). Atmospheric pressures on weather maps (see Fig. 8.1) are usually expressed in units of in. Hg.

For the purposes of specifying laboratory conditions, the standard pressure is considered to be 1 atm. The conditions *standard temperature and pressure* (STP) or *normal temperature and pressure* (NTP) refer to $P = 1$ atm and $T = 0°C$.

THE LAWS OF BOYLE AND CHARLES–GAY-LUSSAC

In 1662, Robert Boyle (1627–1691), an English chemist, discovered that the volume of a gas confined to a cylinder is inversely proportional to the pressure exerted on the gas through a piston, as long as the temperature remains constant (see Fig. 7.19). Furthermore, this rule is obeyed regardless of the shape of the container, or the area of the piston, or the nature of the gas. The term *Boyle's law* is given to this fact, which can be expressed as $P \propto 1/V$, or,

$$PV = \text{constant} \qquad (T \text{ constant}) \qquad (7.38a)$$

The French physicists Jacques Charles (1746–1823) and Joseph Louis Gay-Lussac (1778–1850) independently discovered the relationship between the temperature of a gas and its volume (at constant pressure) and so extended Boyle's law. This relation states that the volume of a gas is proportional to its temperature, $V \propto T$, or,

$$\frac{V}{T} = \text{constant} \qquad (P \text{ constant}) \qquad (7.38b)$$

---

[12] This is the average value; depending on local weather conditions, excursions of a few percent from this value occur.

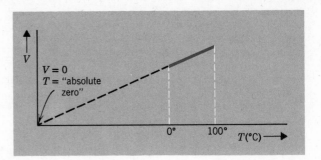

Fig. 7.20 *The extrapolation of the* V-T *curve defines the "absolute zero" of temperature,* $T_0 = -273°C.$

This linear relationship can be extrapolated to the point at which $V = 0$, as in Fig. 7.20. The temperature thus defined is called the "absolute zero" of temperature, the value of which is $T_0 = -273°C$.

It proves convenient in most thermodynamic calculations to measure temperature from the point of "absolute zero." On this scale, we refer to the *absolute temperature,* which we denote by T^*. The unit of absolute temperature is the same as that on the Centigrade scale, but in honor of Lord Kelvin (1824–1907), the Scottish mathematical physicist who contributed greatly to the advancement of the subject of thermodynamics, we call the unit of absolute temperature the *Kelvin degree* and denote it by °K.

THE IDEAL GAS

A real gas will liquify or even solidify before the temperature $T^* = 0°K$ is reached. That is, Eqs. 7.38a and 7.38b are only approximate descriptions of real gases. However, we can invent a substance, called an *ideal gas,* which *perfectly* obeys the laws of Boyle and Charles–Gay-Lussac. Combining these laws we can write the *equation of state* for an *ideal gas:*

$$\frac{PV}{T^*} = \alpha \quad \text{(ideal gas)} \tag{7.39}$$

where α is a constant and depends on the molecular properties of the gas. Fortunately, *real* gases, especially at very low gas densities, are closely described by this equation of state.

If we test the ideal gas law (Eq. 7.39) with different masses M of the same gas, or if we use different gases, each with the same total mass M but with different molecular masses m, we find that α is directly proportional to M and inversely proportional to m: $\alpha \propto M/m$, or,

$$\alpha = R\frac{M}{m} \tag{7.40}$$

where R is a constant, called the *universal gas constant,* which does not depend on the specific properties of the gas used. The ratio of the mass of a gas M to its molecular mass m is the *number of moles n* of the gas (see Eq. 3.9):

$$\text{Number of moles} = n = \frac{M}{m} \tag{7.41}$$

Thus, the ideal gas law becomes

$$PV = nRT^*$$ (7.42)

Notice that in the ideal gas law equation the temperature must be given in degrees *absolute*.

The quantities P, V, T^*, and n can be measured, and R is therefore an experimentally determined quantity:

$$R = 8.314 \times 10^7 \text{ ergs/mole-}°K$$
$$= 1.986 \text{ cal/mole-}°K$$ (7.43)

The ideal gas law is one of the important relations of thermodynamics—it unites the *mechanical* concepts of pressure and volume with the *thermodynamic* concept of temperature. In fact, we have in this law our fundamental physical definition of temperature:

$$T^* = \frac{PV}{nR}$$ (7.44)

We can use this relation to determine experimentally the temperature of a medium by constructing a *gas thermometer*. Such an instrument consists of a bulb that contains a definite volume of gas and a method for measuring

Table **7.4**　*Range of Temperatures in the Universe*

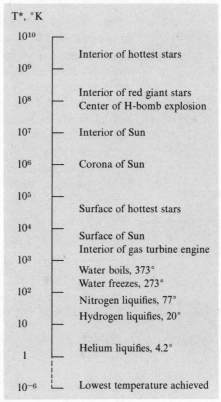

T^*, °K	
10^{10}	
10^9	Interior of hottest stars
10^8	Interior of red giant stars Center of H-bomb explosion
10^7	Interior of Sun
10^6	Corona of Sun
10^5	
	Surface of hottest stars
10^4	Surface of Sun Interior of gas turbine engine
10^3	
	Water boils, 373°
10^2	Water freezes, 273°
	Nitrogen liquifies, 77°
10	Hydrogen liquifies, 20°
1	Helium liquifies, 4.2°
10^{-6}	Lowest temperature achieved

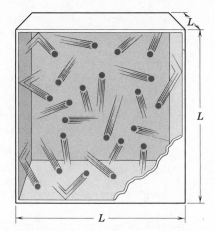

Fig. 7.21 *Molecules move at random in a gas.*

the pressure.[13] (We should use an *ideal* gas, but real gases of simple molecular structure actually work well.) Both the volume and the pressure can be measured and the temperature can therefore be determined. A gas thermometer is useful over a wider temperature range than is a mercury thermometer because mercury freezes at $-38.9°C$ and boils at $356.6°C$. In principle at least, a helium thermometer could be used down to the temperature at which it liquifies, $4.2°K$.

KINETIC THEORY

We now proceed to develop a *microscopic* theory, based on our knowledge of the behavior of molecules, that will lead to a connection with the *macroscopic* properties of gases as described by the ideal gas law equation. We begin by considering a certain large number N of ideal gas molecules that are confined within a cubical box whose sides have a length L. Within this box the molecules move at *random* with velocities v. These velocities are not all the same but this need not concern us at the moment because we will eventually require only the *average* velocity.

Consider first a single molecule of mass m which has an x component of velocity equal to v_x, as in Fig. 7.22. The initial momentum is $p_x = mv_x$. If the molecule collides elastically with the wall, the momentum after collision will be $p'_x = -mv_x$. Thus, the *change* in momentum during the collision is

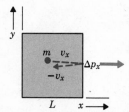

Fig. 7.22 *A molecule collides elastically with the wall of the container and alters the sign (but not the magnitude) of its velocity.*

$$\Delta p_x = p_x - p'_x = 2\,mv_x$$

and this amount of momentum is delivered to the right-hand wall of the cube (Fig. 7.22).

A collision with the right-hand wall occurs once every round trip of the molecule; hence, the time interval between successive collisions is

$$\Delta t = \frac{2L}{v_x}$$

[13] The pressure can be measured, for example, in terms of the height of a column of mercury or oil that the pressure will support.

The *average* force (averaged over the time Δt) that is exerted on the right-hand wall by the collision of this single molecule is

$$\bar{F}_x = \frac{\Delta p_x}{\Delta t} = \frac{2mv_x}{2L/v_x} = \frac{mv_x^2}{L} \tag{7.45}$$

We must now take into account the fact that N molecules contribute to the *total* force on the right-hand wall. We can compute the total force by using the *average* of the x component of the velocity for the N molecules. However, this is not quite correct because in Eq. 7.45 it is the *square* of v_x that is related to the force. Therefore, in computing the total force we must use the average of v_x^2 $(\overline{v_x^2})$ instead of the square of the average of v_x $(\overline{v}_x^2)$.

Example **7.14**

Show that $\overline{v}^2$ does *not* equal $\overline{v^2}$.

Consider four particles with the following velocities: 1, 2, 3, and 4 cm/sec. The square of the average of v is

$$\overline{v}^2 = \left(\frac{1+2+3+4}{4}\right)^2 = \left(\frac{10}{4}\right)^2 = (2.5)^2 = 6.25 \ (\text{cm/sec})^2$$

whereas the average of v^2 is

$$\overline{v^2} = \frac{(1)^2 + (2)^2 + (3)^2 + (4)^2}{4}$$

$$= \frac{1 + 4 + 9 + 16}{4} = \frac{30}{4} = 7.5 \ (\text{cm/sec})^2$$

so that there is a substantial difference between the two methods of averaging.

If the individual velocities are $+1$, -2, -3 and $+4$ cm/sec, there will, of course, be no change in the value of $\overline{v^2}$, but the average velocity $\overline{v}$ will be *zero*.

If we multiply the right-hand side of Eq. 7.45 by N and take the average value of v_x^2, we obtain the total force exerted on the wall of the cube:

$$F = \frac{Nm\overline{v_x^2}}{L}$$

The average of v_x^2 is very simply related to the average of the square of v; by using the Pythagorean theorem for three dimensions we have

$$\overline{v^2} = \overline{v_x^2} + \overline{v_y^2} + \overline{v_x^2} = 3\,\overline{v_x^2}$$

where we have used the fact that

$$\overline{v_x^2} = \overline{v_y^2} = \overline{v_z^2}$$

which is a consequence of the fact that a large number of molecules are moving at random. Therefore,

$$\overline{v_x^2} = \tfrac{1}{3}\,\overline{v^2}$$

so that

$$F = \frac{Nm\overline{v^2}}{3L}$$

The area of each wall is L^2, so the pressure is

$$P = \frac{F}{A} = \frac{F}{L^2} = \frac{Nm\overline{v^2}}{3L^3}$$

Since the volume of the box is $V = L^3$, we can rewrite this expression as

$$PV = \tfrac{1}{3} Nm\overline{v^2} \tag{7.46}$$

Now, the average kinetic energy of one of the molecules is

$$\overline{KE} = \tfrac{1}{2} m\overline{v^2}$$

so that Eq. 7.46 becomes

$$PV = \tfrac{2}{3} N \overline{KE} \tag{7.47}$$

Equation 7.42, the ideal gas law, is a description of the *macroscopic* properties of a gas. Equation 7.47, on the other hand, is based on a *microscopic* view of the interaction of gas molecules with the walls of a container. The left-hand sides of these two equations (PV) are identical. Therefore, we can equate the right-hand sides of these equations and obtain the vital link between the macroscopic concept of temperature and the microscopic description of molecular motion. That is,

$$\tfrac{2}{3} N \overline{KE} = nRT^*$$

or,

$$\overline{KE} = \tfrac{3}{2} \frac{nR}{N} T^* \tag{7.48}$$

Now, n is just the number of moles of the gas and N is the number of molecules; the ratio N/n is the number of molecules per mole, which is Avogadro's number (see Eq. 3.9):

$$\frac{N}{n} = N_0 = 6.02 \times 10^{23} \text{ molecules/mole} \tag{7.49}$$

The combination of quantities nR/N now becomes R/N_0. Because this is the ratio of two constants, we can represent the ratio by a simple constant:

$$k = \frac{R}{N_0} = \frac{8.314 \times \text{ergs (mole-}^\circ\text{K)}^{-1}}{6.02 \times 10^{23} \text{ (mole)}^{-1}}$$

$$= 1.380 \times 10^{-16} \text{ erg/}^\circ\text{K}$$

$$= 8.62 \times 10^{-5} \text{ eV/}^\circ\text{K} \tag{7.50}$$

This physical constant k is called the *Boltzmann constant* in honor of Ludwig Boltzmann (1844–1906), an Austrian physicist who was one of the leading contributors to the microscopic (or *kinetic*) theory of heat.

Equation 7.48 can now be expressed as

$$\overline{KE} = \tfrac{3}{2} kT^* \tag{7.51}$$

Having obtained this connection between the macroscopic temperature and the microscopic kinetic energy by equating our experimental and theoretical expressions for PV, we must now ask how we can verify our connecting equation. One of the most direct ways to establish Eq. 7.51 is by the measurement of molecular velocities.

MOLECULAR VELOCITIES

Equation 7.51 can be expressed as

$$\overline{KE} = \tfrac{1}{2} m\overline{v^2} = \tfrac{3}{2} kT^*$$

from which

$$\overline{v^2} = \frac{3kT^*}{m}$$

If we take the square root of $\overline{v^2}$, we do not obtain $\bar{v}$ (compare Example 7.14), but a different quantity, which we call the *root-mean-square* (or *rms*) velocity:

$$v_{rms} = \sqrt{\frac{3kT^*}{m}} \tag{7.52}$$

The velocities of the molecules in a gas are actually distributed according to a function that can be obtained from kinetic theory by using more powerful mathematical techniques than we have introduced. This distribution of velocities is called the *Maxwellian distribution*, in honor of James Clerk Maxwell (1831–1879), the Scottish theoretical physicist who contributed greatly to kinetic theory and developed the theory of electromagnetism. Figure 7.23 shows the Maxwellian distribution functions $f(v)$ for three different temperatures. As the temperature is increased, the peak of the distribution moves toward higher velocities, in accordance with Eq. 7.52 which states that v_{rms} increases as $\sqrt{T^*}$.

The accuracy of the Maxwellian formula for the distribution of molecular velocities has been directly tested and verified in numerous experiments, thus establishing that Eq. 7.52 for v_{rms} is correct. (It is not difficult to verify, for example, that $v_{rms} \propto \sqrt{T^*}$ and $v_{rms} \propto 1/\sqrt{m}$.)

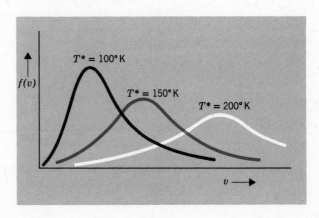

Fig. 7.23 *Maxwellian distribution functions for the molecules of a gas at three different temperatures. f(v) is the number of molecules having velocities in a small range around the velocity v. The peaks of the distributions (corresponding to the most proba-ble velocities) occur at velocities* $v_p \cong 0.82\ v_{rms}$; *that is, the root-mean-square velocity is slightly greater than the most probable velocity. Notice that as the temperature increases, the molecular velocities are spread over a wider range and fewer molecules have the most probable velocity.*

Table **7.5**   *Approximate Values for Various Quantities in the Atmosphere at Sea Level*

No. of molecules	$2.7 \times 10^{19}/cm^3$
Pressure	1.0×10^6 dynes/cm^2
Density	1.3×10^{-3} g/cm^3
Molecular velocity	$\begin{cases} 4.8 \times 10^4 \text{ cm/sec (oxygen, } O_2) \\ 5.1 \times 10^4 \text{ cm/sec (nitrogen, } N_2) \end{cases}$
Mean distance between molecular collisions	8×10^{-6} cm
Collision frequency	6×10^9 sec^{-1}
Typical distance between molecules	3.5×10^{-7} cm
Mean molecular mass	4.8×10^{-23} g

Example **7.15**

Compute the *rms* velocity of an oxygen molecule at room temperature.

Since room temperature is approximately 20°C, for T^* we use 273° + 20° = 293°K. The mass of an oxygen molecule is approximately 32 AMU; thus, $m = 32 \times 1.66 \times 10^{-24}$ g $= 5.3 \times 10^{-23}$ g. Therefore,

$$
\begin{aligned}
v_{rms} &= \sqrt{\frac{3kT^*}{m}} \\
&= \sqrt{\frac{3 \times (1.38 \times 10^{-16} \text{ ergs/}°K) \times (293°K)}{5.3 \times 10^{-23} \text{ g}}} \\
&= 4.78 \times 10^4 \text{ cm/sec.}
\end{aligned}
$$

which is approximately 1000 mi/hr! Of course, molecules do not travel very far at these speeds because of frequent collisions that change the direction of motion. Therefore, it takes a certain amount of time for the scent of perfume from a bottle uncorked in one part of a room to penetrate (or *diffuse*) to the other side of the room.

Example **7.16**

Compute the average kinetic energy in electronvolts of a gas molecule at room temperature.

According to the result expressed in Eq. 7.51, absolute kinetic energy does not depend on the mass of the molecule; therefore, the molecules of *all* gases have the same absolute kinetic energy (but *different* values of v_{rms}) at a given temperature.

$$
\begin{aligned}
\overline{KE} &= \tfrac{3}{2} kT^* \\
&= \tfrac{3}{2} \times (8.62 \times 10^{-5} \text{ eV/}°K) \times (293°K) \\
&= 0.038 \text{ eV}
\end{aligned}
$$

Example **7.17**

How many molecules are there in 1 cm^3 of air at STP?

Combining Eqs. 7.47 and 7.51, we can express the ideal gas law equation as

$PV = NkT^*$

or,

$$N = \frac{PV}{kT^*}$$

$$= \frac{(1.0 \times 10^6 \text{ dynes/cm}^2) \times (1 \text{ cm}^3)}{(1.38 \times 10^{-16} \text{ erg/}^\circ\text{K}) \times (273\,^\circ\text{K})}$$

$$= 2.7 \times 10^{19} \text{ molecules (in 1 cm}^3)$$

THE ASSUMPTIONS IN KINETIC THEORY

In deriving the microscopic version of the ideal gas law equation, we have made explicit or implicit use of several assumptions:

1. A gas consists of a large number of molecules that have no appreciable size compared to the average distance between molecules.
2. No forces act on the molecules except during collision with the walls; that is, there is no interaction between molecules.
3. The molecules are in random motion $(\overline{v_x^2} = \overline{v_y^2} = \overline{v_z^2})$.
4. The molecules obey Newton's laws.
5. Collisions of the molecules with the walls are elastic.

It is indeed remarkable that such a simple, classical model of an ideal gas is capable of yielding a gas law equation that is so closely obeyed by real gases. This is especially true when it is realized that the assumptions in the theory are only approximations to the real physical case. The molecules in a gas *do* interact with one another and the molecules *do* have finite sizes. It has been possible to take account of these facts in the more complete theory. And, indeed, the theory has provided us with an independent determination of molecular sizes ($\sim 10^{-8}$ cm) and a description of the way in which molecules interact. Modifications having to do with quantum effects have been made in the theory to improve its accuracy and now the *kinetic theory of gases* is one of the most useful (and most highly developed) theories in physics.

7.16 *The Second Law of Thermodynamics*

THE DIRECTION OF HEAT FLOW

Having established the fact that every macroscopic substance has a certain amount of internal energy associated with the motion of its constituent molecules, we can ask the question "Is it possible to extract this internal energy and use it to do work on another substance?" Suppose we have a certain mass m_1 of a material that is at a temperature T_1^*. If we place this object in contact with another mass m_2 of a material that is at a *lower* temperature T_2^*, as in Fig. 7.24, then we know from experience that the pair of objects will eventually come to a common temperature between T_1^* and T_2^*. In other words, some of the internal energy of m_1 has been used to do work on m_2 and has increased the internal energy and temperature of m_2.

Heat energy can flow from a hotter body to a colder body. But the reverse is not true. We cannot use the internal energy of m_2 to increase the temperature of m_1 while the temperature of m_2 is *decreased*. Thus, unless work is done by an outside agent, heat energy always flows from objects at higher temperatures to objects at lower temperatures. This is the substance of the *second law of thermodynamics.*

ORDER AND DISORDER

The second law of thermodynamics can be stated in terms of the probability that a system will be in one of the possible configurations that are available to it. For example, if we have a large number of gas molecules in a box, it is most probable that the molecules will be distributed uniformly throughout the volume of the box; it is extremely unlikely that the molecules will be found, at some instant, all in one corner of the box. The latter (unlikely) configuration is one of a high degree of *order,* whereas the uniform distribution of randomly moving molecules constitutes *disorder.* Any ordered system in Nature if left to itself, always tends to proceed spontaneously to a configuration with a lesser degree of order; that is, the trend in natural occurrences is always towards *disorder.*[14] The situation shown in Fig. 7.24 is an example of such a process: the initial configuration is one in which the molecules in m_1 have a high average velocity while those in m_2 have a lower average velocity (that is, $T_1^* > T_2^*$)—thus, there is a certain degree of order in the system. But, when the objects are placed in thermal contact, this order is decreased as the excess energy of the molecules in m_1 is shared with the molecules in m_2. Heat energy always flows in the direction that allows a decrease in the order of a system; in other words, molecular motions always tend to produce a condition that is maximally random, a condition in which the internal energy is shared as equitably as possible among the constituents.

ENTROPY

The degree of order in a system can be expressed in a quantitative way by using the concept of *entropy.* An *ordered* system has a *low* entropy; a *disordered* system has a *high* entropy. The second law of thermodynamics therefore states that in an isolated system, some portion of the system may actually experience a decrease in entropy, but such a decrease is always more than compensated by an increase in entropy in the remainder of the system

[14] Any housekeeper will be able to provide ample evidence in support of this statement.

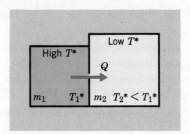

Fig. 7.24 *Heat energy can be transferred from a hotter to a colder body.*

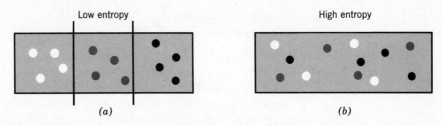

Low entropy   High entropy

(a)   (b)

Fig. 7.25 (a) *An ordered system with a low entropy.* (b) *Removal of the barriers in the box in* (a) *allows the particles to mix; the degree of order is lowered and the entropy is increased.*

so that the net effect for the system as a whole is an increase in entropy.

Biological systems are not exempt from the second law of thermodynamics. The metabolic processes in a cell increase the order in the cell by forming large molecules from small molecules as, for example, in photosynthesis. Although the entropy of the cell is decreased by these processes, the entropy of the surroundings is increased by an even greater amount. Any isolated system, and indeed the entire Universe, follows a course that continually increases its entropy. Therefore, we can look forward to the time (albeit in the distant future!) when the Universe will have reached a state of maximum entropy. Then, all objects will be at a common temperature so that no more work can be done by thermal energy and the Universe will die a "heat death."

TIME'S ARROW

The second law of thermodynamics, characterized by the statement that the entropy of the Universe (or of any isolated system) cannot decrease, makes a unique contribution to our understanding of the nature of the physical Universe. The fact that entropy is ever-increasing shows the *direction* of thermodynamic changes—that is, *time flows only in one direction.* The conclusion that time is unidirectional is consistent with our experiences (birth, growth, death, and so forth), but most of the fundamental laws of physics are not altered by the substitution of $-t$ for t. For these laws, the idea of the unidirectional flow of time is of no consequence. It is only the fact that entropy forever increases that provides us with a proper concept of the unidirectional flow of time.[15]

7.17 *Mass and Energy*

$$\mathcal{E} = mc^2$$

Thus far in our development of physical principles we have discovered several important quantities that remain constant during any physical process. These *conservation laws* relate to the following quantities:

1. Mass   (Section 2.3)
2. Linear momentum   (Section 5.9)

---

[15] Some experiments with elementary particles, when interpreted with a current theory, also show that time's arrow points only in one direction (see Section 16.5).

3. Angular momentum   (Section 5.10)
4. Electric charge   (Section 6.4)
5. Energy   (Section 7.7)

Although we have treated mass and energy as being separately conserved, the theory of relativity (Chapter 11) shows that there is an intimate connection between these two quantities and, in fact, that there can be interchanges between them. The mass of a system is related to the total energy of the system in a particularly simple way:

$$\mathcal{E} = mc^2 \qquad\qquad (7.53)$$

This is the famous Einstein mass-energy relation which shows that it is the *mass-energy* of a system (not the mass nor the energy separately) that is conserved in any physical process.

According to Eq. 7.53, there is a truly enormous amount of energy (*mass energy*) contained in even small amounts of mass. Suppose, for example, that 1 kg of matter could be converted entirely into energy. This process would yield 9×10^{23} ergs or about 3×10^{10} kW-hr! This is approximately the energy consumption of the United States per day.

THE LIMITATION ON EXTRACTING USEFUL MASS-ENERGY

Unfortunately, the entire mass-energy of a given substance is not available to be transformed into useful energy; we cannot destroy neutrons and protons—we can only release energy by rearranging them into forms that have different total mass. This is another conservation law: the total number of heavy particles (neutrons and protons, as distinct from light particles such as electrons and neutrinos, which are separately conserved) remains constant. We shall return to the conservation laws for elementary particles in Chapter 16.

One method of rearranging neutrons and protons within nuclei is the *fission* process in which a heavy nucleus (suitably jostled by a neutron) breaks into two fragments. The difference between the initial mass and the final mass is released as kinetic energy. (However, the total number of neutrons and protons remains the same.)

Example **7.18**

How much energy is released in the fission of 1 kg of U^{235}?

The amount of mass-energy that is converted to kinetic energy in the fission process is approximately 200 MeV per nucleus. (This is only about 0.1 percent of the total mass-energy of a uranium nucleus; the other 99.9 percent remains in the masses of the neutrons and protons and is therefore not available for conversion into kinetic energy.) In 1 kg of U^{235} there are approximately 2.5×10^{24} atoms. Therefore, the total energy release is

$\mathcal{E} = (200 \text{ MeV}) \times (2.5 \times 10^{24})$

$\quad = 5.0 \times 10^{26} \text{ MeV}$

$\quad = (5.0 \times 10^{26} \text{ MeV}) \times (1.6 \times 10^{-6} \text{ ergs/MeV})$

$\quad = 8.0 \times 10^{20} \text{ ergs}$

We can convert this into another popular unit by noting that the explosion of 1 ton of TNT releases approximately 4.1×10^{16} ergs. Thus, the fission of 1 kg of U^{235} releases an amount of energy

$$\varepsilon = \frac{8.0 \times 10^{20} \text{ ergs}}{4.1 \times 10^{16} \text{ ergs/ton TNT}} \cong 20 \text{ kilotons TNT}$$

This is approximately the size of the original atomic bomb of 1945.

The fission process in any heavy nucleus produces approximately 200 MeV; fission is therefore approximately 0.1 percent efficient in the conversion of total mass-energy into useful energy. On the other hand, chemical burning of fuels such as coal extract only about 10^{-10} of the total mass-energy as useful energy because there is relatively little energy stored in molecules in the form of binding energy. Therefore, fission is approximately 10^7 times more efficient than fuel burning in the generation of energy.

Summary of Important Ideas

Kinetic energy is the result of motion; *potential energy* is the result of work done against an opposing force.

The work required to move an object from one position to another against a *conservative force* is independent of the path taken. Gravitational and electrical forces are conservative; frictional forces are nonconservative.

The absolute value of energy has no physical meaning; only *changes* in energy are significant.

The total energy of an isolated system remains constant—energy is *conserved*. Energy may be changed from one form to another (for example, from potential energy to kinetic energy) without loss.

In an *elastic* collision, the internal energies of the colliding bodies remain constant; if the internal energies change, the collision is *inelastic*.

The *internal energy* of a body is that energy due to the motion of the constituent atoms or molecules.

The *first law of thermodynamics* expresses the principle of energy conservation when heat flow and internal energy are included.

The *second law of thermodynamics* expresses the fact that heat cannot be transferred from a colder body to a hotter body without work being done by an outside agent.

A system with a high degree of order has a low value of *entropy*, and conversely. Any physical process that takes place within an isolated system *increases* the entropy of the system.

Mass and *energy* can be interchanged but the mass-energy of a nucleus can be utilized to do useful work only insofar as protons and neutrons can be rearranged to change the mass of the nucleus; protons and neutrons cannot be destroyed.

Questions

7.1 Kinetic energy is imparted to the blood by the pumping action of the heart. What happens to this kinetic energy?

7.2 An object of mass m_1 collides "head on" with an object of mass m_2 which is initially at rest. If $m_1 < m_2$, will m_1 always reverse its direction of motion, independent of the magnitude of its initial velocity? What will happen if $m_1 > m_2$? What will happen if $m_1 = m_2$?

7.3 One often hears the statement that engines or machines are inefficient, that "they waste energy." Does this mean that the energy is lost? Explain.

7.4 A certain mercury thermometer has a nonuniform inside diameter. Explain how this will affect temperature measurements with this instrument. What procedures would be required to make this thermometer useful?

7.5 Examine the way in which your body has acquired thermal energy. Trace the history of energy transfer and show that the Sun is the ultimate source of this energy.

7.6 A certain volume of gas contains equal numbers of oxygen and nitrogen molecules. Is there any physical principle that dictates against the molecules arranging themselves with all of the oxygen molecules in one half of the volume and all of the nitrogen molecules in the other half? Explain.

7.7 Water evaporates from a salt solution and leaves behind salt crystals. Crystalline salt is a system with a high degree of order whereas the salt solution is a disordered system in which the atoms have random motion. Has the entropy law been violated? Explain.

7.8 The laws of physics dictate against the possibility of constructing perpetual motion machines on either of two grounds—the first law of thermodynamics (energy conservation) or the second law of thermodynamics. Explain the way in which one or the other of these laws will preclude the possibility of extracting useful work from a perpetual motion machine.

7.9 Explain why there is oxygen in the Earth's atmosphere but very little hydrogen.

7.10 A certain inventor claims to have constructed a machine that will produce 9×10^{19} J of useful energy from a ton of coal. Do you believe his claim? Explain.

Problems

7.1 When a certain loaded wagon is moving at constant speed, the resisting frictional force is 100 newtons. A man attaches a rope to the wagon in such a way that the rope makes an angle of 30° with the horizontal. He pulls the wagon at constant speed a distance of 100 m. A second man takes over the pulling job, but he attaches the rope at a higher point so that the rope is horizontal when he pulls. The second man also pulls the wagon for 100 m. With what force did each man pull on the rope? How much work did each do?

7.2* A man pushes on a large boulder with a force of 200 N for 1 min, but he finds that he cannot move it. How much work did he do on the boulder? Why is he tired by his exertion? (Comment on the process of muscular action involved.)

7.3 A man whose mass is 100 kg climbs stairs to a height of 10 m. How much work did he do? Is there a difference between the work required to climb stairs (which are slanted) to a given height and that required to climb a ladder (which is vertical) to the same height? Explain.

7.4 A spring with a force constant k is extended by a distance d in a time t. What average power was required to perform this act?

7.5* Eight identical boxes of mass 10 kg and height 10 cm all rest on a floor. How much work is required to arrange the boxes in a vertical stack? Calculate the potential energy of the stack by considering all of the mass to be concentrated at the center of mass. (Take the floor level to be the zero of potential energy.) Comment on the two results.

7.6 A certain spring has a force constant of 10^4 dynes/cm. One end of the spring is attached to an overhead support. On the other end is attached a 100-g mass. How far does the spring stretch? If the spring and mass are extended an additional 5 cm and then released, the mass will be set into motion. What maximum velocity will the mass attain?

7.7 A pile-driver is used to implant a stake in the ground. The resistive force of the ground for the particular stake is 2×10^6 N. The mass of the pile-driver head is 2000 kg and it is lifted to a height of 10 m above the stake. How far is the stake driven at each stroke? (Assume that all of the energy of the pile-driver is used in driving the stake.)

7.8 Refer to Example 5.8. How much work was required to shorten the radius of the orbit from 100 cm to 50 cm? If the string were played out from 100 cm to 200 cm, what would be the change in the ball's kinetic energy?

7.9 A water storage tank contains 2000 m³ of water and is at an average height of 40 m above ground level. How much work was required to fill the tank from a reservoir at ground level? How much work can be done by the water if it is piped to a place which is 20 m lower than ground level at the tank site? Is energy conservation violated here? Explain.

7.10* A conveyor belt lifts 20-kg blocks of material to a height of 10 m at a rate of 15 blocks per minute. At what rate is work being done? What is the minimum horsepower of the motor that drives the belt?

7.11 An object of mass 100 g rests on a frictionless horizontal plane. A certain constant force is applied to the object and the object is accelerated to a velocity of 100 cm/sec, moving a distance of 10 m in the process. At this point an additional, decelerating force is applied to the object and after moving an additional 10 m, the object is again at rest. How much work was done on the object in moving it the first 10 m? What was the net amount of work done on the object during the entire process?

7.12 A block of mass 100 g slides down an inclined plane from a height of 50 cm. The length of the plane is 100 cm. When the block reaches the bottom it is found to be traveling with a speed of 200 cm/sec. How much work was done against the frictional force? What was the average frictional force on the block while it was sliding?

7.13* If the gravitational potential energy of the water contained in the lake behind a dam can be converted into electrical energy by hydroelectric generators with an efficiency of 20 percent, how many cubic meters per day must fall through a distance of 30 m if the plant is to generate 10 megawatts of electrical power?

7.14 How long would it take a 2-hp motor to raise a 10^3-kg mass to a height of 100 m? (Neglect friction.)

7.15 What is the kinetic energy of an athlete ($m = 75$ kg) while running a 10-sec 100-m dash? (Assume constant speed.)

7.16 The engine of a certain automobile develops 100 hp while driving the automobile at a speed of 60 mi/hr over a level pavement. What is the retarding force (road resistance plus air resistance) acting on the automobile?

7.17 A 500-kg roller-coaster car starts from rest at a point 30 m above ground level. The car dives down into a valley 4 m above ground level and then climbs to the top of a hill that is 24 m above ground level. What velocity did the car have in the valley and at the top of the hill? (Neglect friction.)

7.18 In the preceding problem, the length of track from the starting position to the top of the hill is 100 m. If the car just reaches the top of the hill ($v = 0$), what was the average frictional force between the car and the track?

7.19 How much work is required to increase the speed of an automobile ($m = 1500$ kg) from 10 m/sec to 20 m/sec? (Neglect friction.)

7.20 An automobile ($m = 2000$ kg) is traveling on a level road at a speed of 20 m/sec. What is the automobile's kinetic energy? If friction were negligible, could the automobile *coast* to the top of a hill that is 15 m higher than the road? If so, what would be the speed at the top?

7.21* A baseball pitcher can throw a ball ($W = 5$ oz.) at a speed of 100 mi/hr. How much energy is absorbed by the catcher when he catches the ball? (Convert to metric units.)

7.22 A ball is dropped onto a steel plate from a certain height h. Each time the ball strikes the plate it loses 10 percent of its motional energy. How many collisions will be required before the ball fails to rise to a height greater than $\frac{1}{2}h$?

7.23* How much work is required to raise a 100-kg object from the surface of the Earth to a height of 1000 miles? If the object were dropped from such a height, what would be its impact velocity (neglecting air resistance)?

7.24 The galaxy Andromeda has a mass of approximately 4×10^{11} solar masses and a diameter of about 10^5 L.Y. What velocity is necessary to escape from Andromeda from a position near its outer boundary?

7.25 Two small spherical charges of $+3$ statC and -8 statC are initially separated by 4 cm. What is the change in electrostatic potential energy if the separation is increased by 2 cm? How much work has been done? What is the change in electrostatic potential energy if the separation is reduced to 2 cm? Is work done on the charges in the latter case?

7.26   How much work is required to move the charge $q = +2$ statC from A to B?

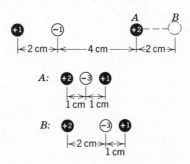

7.27   What is the difference in electrostatic potential energy between configurations A and B?

7.28   There is a potential difference of 10^5 V between a point at ground level and one at a height of 3 m. How much work is required to raise a 1-g object that carries a charge of $+10^3$ statC from ground level to the 3-m point? How much work would be required if the charge on the object were $q = -10^3$ statC?

7.29   Two protons are separated by 10^{-7} cm. An electron is on the line connecting the protons and is at a distance of 10^{-8} cm from one of the protons. How much work is required to move the electron to point B which is at a distance of 10^{-8} cm from the other proton? What is the change in potential energy if the electron is moved to point A, midway between the two protons?

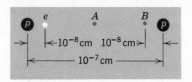

7.30   A negative helium ion (He$^-$, a helium nucleus with *three* electrons) starts from rest at point A and is accelerated toward a positively charged cylinder B. There is a potential difference of $+10^6$ V between B and A. On reaching B, the helium ion passes through a thin foil that removes all of the electrons and produces a completely ionized helium nucleus (He^{++}). This positive ion is now repelled by the positive charge on B and is accelerated toward C. There is zero potential difference between C and A. How much kinetic energy will the ion have at C? What happens to the electrons? (This is basically the principle of operation of the *Tandem Van de Graaff* accelerator.)

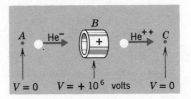

7.31 A 2-MeV neutron strikes a proton (initially at rest) and attaches to it to form a deuteron (the nucleus of "heavy hydrogen," consisting of a neutron and a proton bound together). Consider the collision to be perfectly inelastic and compute the velocity of the deuteron.

7.32 If a completely ionized carbon nucleus (6 protons and 6 neutrons) falls through a potential difference of 3×10^6 V, what will be its final kinetic energy and final velocity?

7.33 A 5-MeV proton makes an elastic, "head-on" collision with a helium nucleus. What are the velocities of the particles after collision?

7.34 An object of mass 1 kg is dropped from a height of 1 m onto a certain surface. The object is observed to rebound to a height of 80 cm. How many calories of heat were produced if all of the energy lost by the object was converted to heat?

7.35 A 1-kg mass of a soft material (such as clay) is dropped from a height of 40 m and sticks to the floor. If $c = 0.2$ cal/g-°C for the material, and if all of the heat is produced in the material, what temperature rise will the material experience?

7.36 An amount of work, 10^4 J, is done by hammering on each of two 1-kg bars, one made of copper and the other of gold. Assuming no heat losses, which bar will become hotter and by how much?

7.37 The daily food intake of a certain laborer is rated at 4000 Calories. During a day he performs 10^{13} ergs of work. What fraction of his food intake has been converted into useful work?

7.38 A kilogram of copper shot is placed in a bag. The bag is raised to a height of 10 m and dropped. The process is repeated until 10 drops have been made. Assuming that all of the heat is produced in the copper and that there is no heat loss from the copper, what is the temperature rise of the copper?

7.39* The Earth receives radiant energy from the Sun at the rate of 1.95 cal/min on an area of 1 cm^2 oriented perpendicular to the direction of the Sun. What total power (in ergs/sec) does the Earth receive from the Sun? During the lifetime of the Earth (4.5×10^9 yr) how much energy has it received?

7.40 A certain amount of gas is contained in a cylinder that has a movable piston of mass 1 kg. If 2 cal of heat is supplied to the gas and if the final temperature of the gas is the same as the initial temperature, through what *vertical* distance has the piston been moved?

7.41 A certain quantity of gas is confined in a cylinder with a movable piston. Heat is added to the gas and the temperature rises from its initial value of 27°C to a final value of 177°C. At the same time the volume is changed to $\frac{1}{2}$ the original value. What is the change in pressure?

7.42 A container has a volume of 1 m^3. The gas in the container has a temperature of 150°C and is at a pressure of 1 atm. If the temperature is lowered to -20°C, what is the new pressure if the volume remains constant? What change in volume will be necessary to maintain the pressure at 1 atm?

7.43 A gas, confined to a volume V, is heated from a temperature of 10°C (at which temperature the pressure is 2 atm) to a temperature of 400°C. What is the new pressure? What temperature decrease would be required to reduce the pressure to 1 atm?

7.44 The best laboratory vacuum apparatus can produce pressures of approximately 10^{-15} atm. How many molecules per cubic centimeter are there in a gas at this pressure?

7.45 A gas molecule of mass m impinges on a surface at an angle of 60° to the perpendicular and is "reflected" at the same angle. If the collision is elastic ($v = v'$), what momentum has been transferred to the surface?

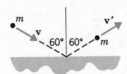

7.46 A gas is confined to a volume V. Initially, the pressure is 1 atm and the temperature is 100°C. What temperature is required to double the rms velocity of the gas molecules? At this new temperature, what is the pressure?

7.47 A certain volume of helium gas is at a temperature of 300°K. What is the average kinetic energy of a helium atom? (Express the result in electron-volts.) What is the rms velocity of an atom?

7.48 A certain volume of oxygen gas is at a temperature of 20°C. On the average, an oxygen molecule will move a distance of about 10^{-5} cm between collisions with other oxygen molecules. How many collisions per second, on the average, will a molecule make?

7.49 The interiors of most stars are at temperatures of approximately 10^{8}°K. What is the rms velocity of a hydrogen nucleus (that is, a proton; there are no *atoms* at these temperatures) in such a stellar interior? What is the average kinetic energy of such a proton in keV?

7.50 The nucleus Cf^{254} (*californium,* an artificially-produced element, $Z = 98$) has the property that it undergoes spontaneous fission, splitting into two fragments with about 200 MeV released as kinetic energy. Suppose one of the fission fragments has a total of 105 protons and neutrons. What is its recoil energy and velocity?

7.51* Suppose that in a nuclear reactor during a certain period of time, 2 kg of fuel (plutonium) is consumed by undergoing fission. How many kW-hr of electrical energy are produced if the conversion process from kinetic energy of the fission fragments to electrical energy is 25 percent efficient?

7.52 The energy of a 10-kiloton atomic weapon is released in approximately 1 microsecond. What is the horsepower rating of the weapon during the explosion period?

7.53* It has been estimated that by the year 2000 the annual consumption of electrical energy in the United States will be about 8×10^{12} kW-hr, of which about $\frac{1}{4}$ will be generated by nuclear reactors. According to this estimate, what will be the annual rate of use (in kg) of fissionable material? (Assume that 25 percent of the fission energy can actually be converted into electrical energy.)

Temperature field: Thermograph of Manhattan skyline

8 *Fields*

Most of the forces that we encounter in everyday experience are of the *contact* type—we push or pull on something or one object strikes another.[1] To ancient men, contact forces were the only real forces. For the Sun to exert a real force on the Earth seemed impossible because there was no contact between the two bodies. This belief persisted until relatively modern times.

An entirely new concept emerged when Newton established the theory of universal gravitation. According to this theory, the Earth, moon, Sun, and planets all exert forces on one another without contact or any material medium between them through which such forces could be propagated. The term "action at a distance" was used to describe the gravitational interaction (and later, electrical forces as well) because the propagation of the gravitational force through the vacuum of space was inconceivable.

Newton did not attempt to explain *why* the gravitational force acts through a vacuum. When discussing this remarkable property of gravity in his *Principia,* Newton wrote, "Hypotheses non fingo" ("I frame no hypotheses").[2] Newton sought only to interpret the observational evidence within a consistent and useful mathematical framework. This he could and did do without the necessity of crossing the boundary between physics and metaphysics and asking "why?"

The most popular solution to the dilemma posed by the action-at-a-distance forces was the invention of the *ether*—one of the most famous *ad hoc* assumptions in physics. The ether assumption explained that which it was invented to explain and did nothing else. The ether was conceived as possessing only one property—transmitting the action-at-a-distance forces. It was believed to be an invisible, weightless jelly—poke it at one point and the pressure induces a strain that is transmitted to other points. In a jelly, there are material particles that contact one another and thereby propagate the strain. The ether that was invented to explain action-at-a-distance forces was not material and contact forces (at least, in the ordinary sense) were not involved. But this did not seem to bother the proponents. They proceeded to work out elaborate theories of stresses and strains and vortices within the ether. This same ether was considered to be the medium through which light was transmitted, it was therefore sometimes called the *luminiferous ether.*

The ether concept was in vogue until the early 20th century when it was finally laid to rest by Einstein in his theory of relativity. The ether theory had been forced to include so many ad hoc assumptions to explain so many facts that it finally collapsed, mainly by its own weight. In its place came the *field theory* approach to all action-at-a-distance forces.

---

[1]See Section 7.10 where it is argued that on a *microscopic* scale, the term "contact force" loses its significance.

[2]"But hitherto I have not been able to discover the cause of those properties of gravity . . . and I frame no hypotheses . . . . To us it is enough that gravity does really exist, and act according to the laws which we have explained, and abundantly serves to account for all the motions of the celestial bodies . . . ."

8.1   *Scalar and Vector Fields*

WHAT IS A FIELD?

Any physical quantity that has a well-defined value at any point in space can be considered to be a *field* quantity. That is, in a two-dimensional area or a three-dimensional space, we can imagine measurements being made of a certain physical quantity (the field quantity) at any point. We must obtain a unique value of the field quantity at every point. And, furthermore, there must be a smooth variation of the field quantity from point to point within the space. There cannot be a discontinuous change (that is, a jump) in the value of a physical quantity from, say, -10 units at one point to $+20$ units at a neighboring point which is an arbitrarily short distance away. There must be a gradual change from -10 units to $+20$ units. It is this smooth variation from point to point in space that is the essential feature of a *field*.

PRESSURE AND TEMPERATURE FIELDS

A weather map (or meterological map) is actually a representation of the *pressure field* for the particular area. Such maps are prepared by measuring the atmospheric pressure at a large number of points throughout the country and then plotting curves (called *isobars*[3]) to connect points of equal pressure. These maps (see Fig. 8.1) show the slow variation of pressure across the country; particularly evident are the regions of high and low pressure, the movements of which generally determine our weather patterns. The pressure at a given point is specified by a single number. That is, pressure is a *scalar* quantity and the pressure field is a *scalar field*.

Weather maps also show the variation of *temperature* across the country. In this case, curves (called *isotherms*) are drawn connecting points at which equal temperatures have been measured. The temperature field is also a scalar field since temperature is a scalar quantity.

VELOCITY FIELDS

When water flows in a river or stream, generally the flow velocity is not the same at all points but varies in a smooth way from surface to bottom and from midstream to bank. The variation of flow rate depends on the contour of the banks and bed and may produce stagnant regions or eddy currents and even whirlpools. Because the flow velocity varies smoothly from point to point in the river, we can describe the situation in terms of a *velocity field*. A velocity field differs in an essential way from pressure or temperature fields because velocity requires both magnitude and direction for its specification. The velocity field is a *vector field*.

The flow of air can also be characterized by a vector field. Aeronautical engineers frequently use a *wind tunnel* to study the pattern of air flow over aircraft wings. In such experiments, a constant flow of air through a long, cylindrical tube is maintained by a high-speed fan. At a number of points

---

[3]From *iso-* (equal) and *-bar* (standing for "barometric pressure").

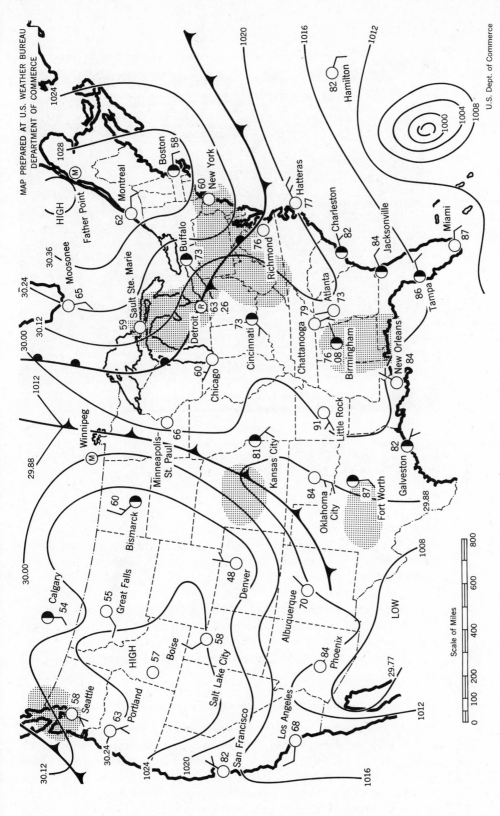

Fig. 8.1 *A weather map of the United States. The curves connect points at which measurements have shown the atmospheric pressure to be the same and so define the "pressure field" across the country. Each curve (or isobar) is labeled with two numbers—one of the numbers (near 1000) gives the pressure in millibars (1000 millibars = 1 atm), and the other number (near 30) gives the pressure in units of the number of inches of mercury that a vacuum will support. The temperature (°F) is given for various cities along with a symbol that designates the wind velocity and direction. The shaded regions represent areas of precipitation. Notice the hurricane just off the Florida coast.*

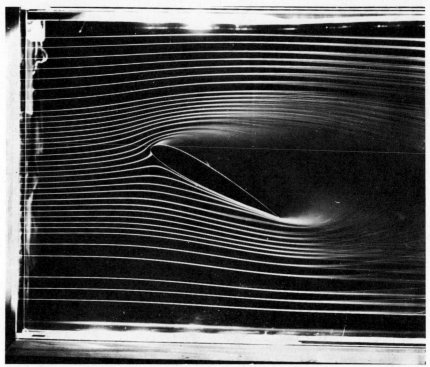

Rudolf Lehnert

Fig. 8.2 *Lines of smoke in a wind tunnel reveal the flow of air around a section of an aircraft wing. What is the significance of the fact that in certain regions the lines of smoke bunch together increasingly as they move along?*

within the chamber, smoke is injected into the air stream through narrow tubes. The smoke streams make visible the flow pattern of the air (Fig. 8.2). At every point, the smoke particles have a well-defined direction and speed; hence, the velocity of the air can be examined as it flows over the airfoil. At any particular point the velocity flow vector is *tangential* to the *flow line* (or *stream line*).

Any physical quantity that has a well-defined magnitude and direction at every point in space can be considered to be a vector field quantity. Most of the interesting field quantities that we encounter in physics are *vectors*— for example, the gravitational field, the electric field, the magnetic field, and others.

WHEN IS THE FIELD CONCEPT USEFUL?

Ideally, a quantity should vary in space in a mathematically *continuous* way in order to permit a field description. Some physical quantities, such as the gravitational force vector, do vary in a continuous fashion and field descriptions of these quantities are entirely proper. On the other hand, pressure and temperature, which we have argued can be described by fields, are not quantities that really have continuous variation. The reason is that these quantities are macroscopic manifestations of large numbers of microscopic effects. It makes no sense to inquire about the pressure or temperature of a single molecule or even a dozen molecules. Large numbers of molecules are required before pressure and temperature have meaning. But the volume occupied by even a million molecules is still small by ordinary standards so that for all practical purposes there is in effect a continuous variation of pressure and temperature from one group of a million molecules to the next. For the description of the gross behavior of a large amount of gas, the field concept is entirely appropriate.

Each particular case must be examined to determine if the physical quantity involved has an acceptable variation in space to make the field description a useful one.

8.2   *The Gravitational Field*

THE FIELD VECTOR

According to Newton's gravitation law, the magnitude of the gravitational force *on* a mass m_2 *due to* a mass m_1 a distance r away is

$$F_{G,21} = G\frac{m_1 m_2}{r^2} \tag{8.1}$$

We know that $\mathbf{F}_G$ is a vector quantity and that the force on m_2 is directed *toward* m_1 (see Fig. 8.3).

It proves convenient to describe this situation in the following way. The mass m_1 sets up a certain condition in space to which m_2 reacts and m_2 experiences a force directed toward m_1. This "condition" is the *gravitational field* of m_1. (Of course, m_1 also experiences a force directed toward m_2 due to the gravitational field of m_2, but let us continue to consider the effects due to the field of m_1.) Since m_1 in some way produces this gravitational field that attracts m_2, we say that m_1 is the *source* of the field and we refer to m_1 as the *source mass*. Any object (such as m_2) that is placed in this field, at any point, will experience a force that depends on the gravitational field set up by m_1 at that point.

Instead of writing a *force* law specifically for the case of a particular mass m_2, let us divide both sides of Eq. 8.1 by m_2:

$$\frac{F_{G,21}}{m_2} = G\frac{m_1}{r^2} \tag{8.2}$$

The right-hand side of this expression now depends only on the *distance* of m_2 from m_1 and not on the mass of m_2. That is, the right-hand side is

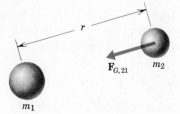

Fig. 8.3 *The source mass* m_1 *sets up a gravitational field in space. The mass* m_2 *experiences a force* $\mathbf{F}_{G,21}$ *due to that field.*

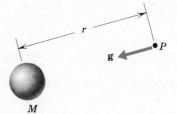

Fig. 8.4 *The source mass* M *sets up a gravitational field at the point* P. *This field is specified in magnitude and direction by the field vector* **g.**

a specification of the *gravitational field* at this distance due to the source mass and will be the same no matter what mass m_2 is placed at this position. Therefore, let us rewrite this expression in a way that emphasizes only the source mass. The new quantity, which is the right-hand side of the equation above and which characterizes the gravitational field of m_1, will be denoted by *g:*

$$g = G\frac{M}{r^2} \tag{8.3}$$

where the source mass m_1 is now designated by M. The dimensions of g are those of force divided by mass or *acceleration.*

Since the gravitational force $\mathbf{F}_G$ is a vector, the quantity **g** is also a vector. The complete specification of the gravitational field (magnitude and direction) due to the source mass M at any point P is given by **g,** the *gravitational field vector* (see Fig. 8.4).

The field vector **g** gives the *force per unit mass* on (the *acceleration* of) any object placed in the gravitational field of the source mass M. The gravitational force on a mass m is

$$\mathbf{F}_G = m\mathbf{g} \tag{8.4}$$

This equation is just the vector counterpart of the familiar scalar equation, $F = mg$, that we have used to calculate gravitational forces. In fact, the acceleration due to gravity g, as we have used it, is just the magnitude of the field vector **g.** Of course, the field vector **g** is a more general quantity and varies with position in space, but it does have the magnitude 980 cm/sec² at the surface of the Earth.

THE PRINCIPLE OF SUPERPOSITION

One of the facts that makes the field concept so useful for the gravitational case (as well as for the electrical case, as we shall see), is that the gravitational force vector and the gravitational field vector obey the *principle of superposition.* That is, if we wish to calculate the gravitational force on a given object due to many other objects (Fig. 8.5), the net force is the vector sum of all the individual forces; each of these individual forces can be calculated as if the other objects were not present. Thus,

$$\mathbf{F}_{G,\text{net}} = \mathbf{F}_1 + \mathbf{F}_2 + \mathbf{F}_3 + \cdots \tag{8.4a}$$

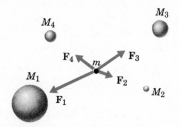

Fig. 8.5 *The net gravitational force on* m *is the (vector) sum of all the individual forces due to* M_1, M_2, M_3, *and* M_4.

Since the gravitational field vector is just the force per unit mass, it follows that **g** obeys a similar additive relation:

$$\mathbf{g}_{net} = \mathbf{g}_1 + \mathbf{g}_2 + \mathbf{g}_3 + \cdots \tag{8.4b}$$

To state that the gravitational force on an object is the vector sum of all contributing forces, each calculated without regard to the others, is not an empty or trivial statement. For example, consider the force on a mass m due to two other masses, M_1 and M_2, as in Fig. 8.6. The principle of super-position states that the force on m is

$$F_G = \frac{M_1 m}{r_1^2} + \frac{M_2 m}{r_2^2}$$

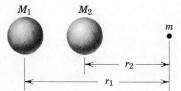

Fig. 8.6 *The force on* m *due to* M_1 *is not "screened" by the presence of* M_2.

The fact that M_2 lies *between* M_1 and m does not affect the calculation of the force due to M_1. That is, M_2 does not "screen" or "shadow" the force of M_1 on m—it is the same whether M_2 is present or not.[4] This result is obtained only through experiment. First, the superposition principle is hypothesized and the consequences are deduced. (For example, it is not just the surface of the Earth that attracts the moon, but the entire mass of the Earth that acts, each small portion performing its function independent of the other portions.) Then, these consequences are checked against experi-mental findings. Since no contradictions have ever been found, the super-position principle for gravitational forces is considered to be established.

LINES OF FORCE

A diagram or a map of a vector field, such as the gravitational force field of a source mass, is more complex than that of a simple scalar quantity, such as atmospheric pressure (Fig. 8.1), because both magnitude *and* direc-tion must be specified. Suppose that we begin to map the gravitational force field around a certain source mass M by measuring the force on a small

[4]Thus, it is extremely unlikely that a "gravitational shield" will ever be discovered.

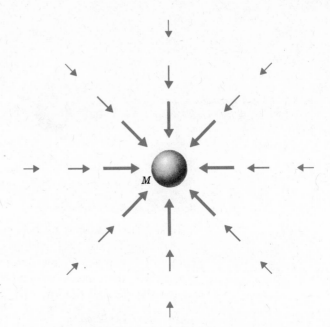

Fig. 8.7 *The mapping of the gravitational field of M. Each arrow represents the force on a test mass placed at the end of the arrow.*

test mass.[5] The results of such measurements can be represented by a series of arrows, as in Fig. 8.7. The length of each arrow is proportional to the gravitational force at the end of the arrow (that is, at the end opposite the point) and the direction of the force is given by the direction of the arrow. Alternatively, we can construct around the source mass a set of continuous lines, called *lines of force,* so that at any point the *direction* of the force is the direction of the line of force passing through that point. The *magnitude* of the force at any point in such a diagram is proportional to the *density* of the lines in the immediate vicinity of that point. At a distance r from the center of the mass M, the density of lines of force is proportional to $1/r^2$, as demanded by the radial dependence of the gravitational force law. Therefore, the simple inspection of a lines-of-force diagram reveals where the force is greatest (where the lines bunch together) and where the force is least (where the lines spread out to low density).

Although the lines-of-force scheme is useful in visualizing the force field surrounding an object, it is important to realize that this picture is only an *invention*—there are no rubber-band-like lines that extend through space and exert forces on other objects. Lines of force are not *real*—they serve only to provide a crutch for our thinking when we consider force-field problems.

For a simple spherical mass, the lines of force are all straight lines in the radial direction. But for objects with complicated shapes or for a group of bodies (even spherical bodies), the lines of force will generally be curved. For example, consider the case of two identical spherical objects that are located close together, as in Fig. 8.9. The lines of force can be plotted by

[5]A "test mass" is an hypothetical object whose mass is so small that it does not disturb the gravitational field of the source mass and whose dimensions are so small that the field of the source mass is essentially uniform throughout the volume of the object.

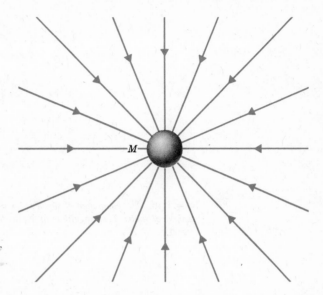

Fig. 8.8 *The lines of gravitational force around a spherical object are all straight lines. The density of lines decreases with the square of the distance from* M *because the force varies as* $1/r^2$.

measuring the force on a test mass at many places in the field or by calculating the vector sum of the two gravitational forces at every point. In the next section we show an additional significance of the lines-of-force picture.

8.3 *The Gravitational Potential*

POTENTIAL ENERGY PER UNIT MASS

In Eq. 7.20 we found that the gravitational potential energy of a test mass m that is a distance r from a source mass M is

$$PE_G = -G\frac{Mm}{r} \tag{8.5}$$

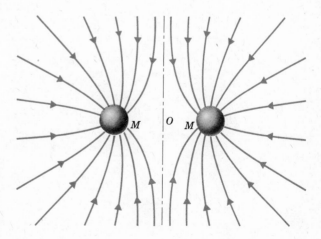

Fig. 8.9 *The lines of force for two identical, nearby objects of mass* M. *What is the force on a test mass placed on the plane of symmetry (represented by the dot-dash line in the figure)? How will a test mass react in the vicinity of* O?

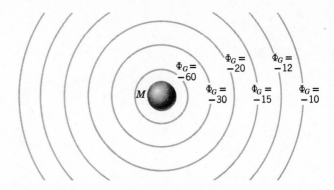

Fig. 8.10 *The equipotential surfaces for a uniform spherical source mass are all concentric spherical shells. The magnitude of* Φ_G *decreases inversely with the distance from the center of* M.

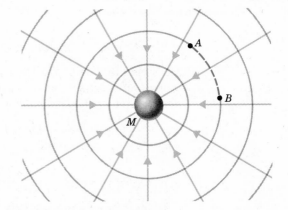

Fig. 8.11 *The lines of force and the equipotential surfaces for a uniform spherical source mass intersect at right angles. No work is required to move a test mass at constant velocity along an equipotential, as from A to B.*

As in the case of the gravitational force and the gravitational field vector, if we divide the gravitational potential energy by m we obtain a quantity that is characteristic of the source mass M and does not depend on the test mass. We call this new quantity the *gravitational potential* and assign it the

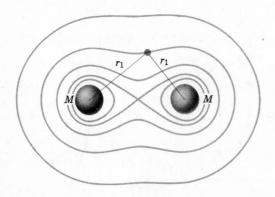

Fig. 8.12 *Equipotential surfaces for a pair of identical spherical masses. These equipotentials are all perpendicular to the lines of force which are shown in Fig. 8.9.*

symbol Φ_G:

$$\Phi_G = \frac{PE_G}{m} = -\frac{GM}{r} \tag{8.6}$$

This quantity is *potential energy per unit mass.*

Since the absolute magnitude of gravitational potential energy has no physical meaning, the same is true for gravitational potential. Since only *changes* in PE_G or Φ_G are significant we can always select, for convenience, an arbitrary position for the zero level. Equations 8.5 and 8.6 are written in the convention that PE_G and Φ_G are zero when $r = \infty$.

The gravitational potential Φ_G is a *scalar* quantity and obeys the superposition principle:

$$\Phi_{G,\text{net}} = \Phi_1 + \Phi_2 + \Phi_3 + \cdots$$

The quantity Φ_G clearly has a definite value at every point in space and satisfies all the requirements of a field quantity. Therefore, Φ_G represents the *scalar gravitational potential field* whereas **g** represents the *vector gravitational force field.*

EQUIPOTENTIALS

According to Eq. 8.6, the gravitational potential due to a uniform, spherical object depends only on the radial distance from that object. Therefore, the potential will be the same at every point on a spherical shell that has the source mass at its center. Such a shell is called an *equipotential surface.* For a uniform spherical source mass, the equipotential surfaces are a series of spherical shells (Fig. 8.10).

Recall that the lines of force for a spherical source mass are all straight, radial lines. Therefore, the lines of force intersect the equipotentials at *right angles* (Fig. 8.11). This is, in fact, a general result: *the lines of force and the equipotential surfaces for any source mass or group of source masses are always mutually perpendicular.*

We can prove this statement in the following way. We know that no work is required (in the absence of friction) to move an object at constant velocity perpendicular to the direction of the force acting on the body; furthermore, the perpendicular direction is the *only* direction for which this is true. If no work is done on or by a body, there can be no change in the potential energy of that body. Therefore, no work is required to move a body at constant velocity along an equipotential surface (starting at any point and proceeding in any direction, as from A to B in Fig. 8.11) since such a movement does not alter the potential energy. Because no work is done in the movement on an equipotential surface, this surface must everywhere be perpendicular to the lines of force.

Consider again the case of two identical masses separated by a certain distance. What are the equipotential surfaces for this situation? The gravitational potential for the pair of masses is

$$\Phi_G = -GM\left(\frac{1}{r_1} + \frac{1}{r_2}\right)$$

where Φ_G is the potential at a point that is at a distance r_1 from one of the masses and also at a distance r_2 from the other mass (see Fig. 8.12). All of the combinations of r_1 and r_2 that give to Φ_G the same value define one of the equipotential surfaces. Different sets of distances will give additional surfaces with different (but constant) values of Φ_G. Several such surfaces are shown in Fig. 8.12. (In three dimensions, the surfaces are generated by rotating these curves around the line that connects the two masses.)

Figures 8.9 and 8.12 have been drawn to the same scale and placed on opposite sides of the same page. Therefore, by holding this page up to the light, it will become apparent that the lines of force and the equipotentials are everywhere mutually perpendicular.

8.4   *The Electric Field*

THE FIELD VECTOR

Having discussed the gravitational field in detail, it is now a simple matter to apply the same reasoning to the electrical case.

If a test charge q_2 is placed a distance r from a source charge q_1, the force *on* q_2 *due to* q_1 is, according to Eq. 6.11,

$$F_{E,21} = -\frac{q_1 q_2}{r^2} \tag{8.7}$$

Dividing $F_{E,21}$ by q_2, we obtain a quantity characteristic of q_1:

$$\frac{F_{E,21}}{q_2} = -\frac{q_1}{r^2}$$

This new quantity, which is the *force per unit charge* is denoted by E and specifies the magnitude of the *electric field* due to q_1. Again, we change the notation and write the source charge as Q. Thus, the electric field of Q (a uniform spherical charge) at a distance r is

$$E = -\frac{Q}{r^2} \tag{8.8}$$

The quantity that characterizes the electric field is, of course, a vector: **E**. The direction of **E** is arbitrarily chosen to be the direction of the force exerted on a *positive* test charge in the field. Therefore, the field vector of a positive source charge is directed *away* from the source and the field vector of a negative source charge is directed *toward* the source (see Fig. 8.13).

The dimensions of **E** are

$$[E] = \text{statC/cm}^2 = \text{statV/cm} = \text{dyne/statC} \tag{8.9}$$

If we place a test charge q in this electric field, the charge will experience a force given by

$$F_E = q\mathbf{E} \tag{8.10}$$

The electric field vector obeys the superposition principle:

$$\mathbf{E}_{\text{total}} = \mathbf{E}_1 + \mathbf{E}_2 + \mathbf{E}_3 + \cdots$$

where $\mathbf{E}_1, \mathbf{E}_2, \mathbf{E}_3, \ldots$, are the field vectors at a given point for individual source charges calculated without regard for the other charges.

It is important to realize that the gravitational and electrical fields are *independent* quantities. The two fields can coexist at a particular point in space and neither field influences the other. The total force on a test particle (which possesses both mass and charge) is the vector sum of $\mathbf{F}_G$ and $\mathbf{F}_E$, but it makes no sense to sum the two field vectors, $\mathbf{g}$ and $\mathbf{E}$ (the *dimensions* are different). Only the *forces* are the measurable (and therefore the physically significant) quantities.

THE ELECTRIC POTENTIAL

According to Eq. 7.25, the potential energy of a charge q that is a distance r from another charge (now the *source* charge) Q is

$$PE_E = \frac{Qq}{r} \tag{8.11}$$

Following the procedure used in the preceding section, we divide this expression by q and call the new quantity the *electric potential* Φ_E:

$$\Phi_E = \frac{PE_E}{q} = \frac{Q}{r} \tag{8.12}$$

Φ_E is *potential energy per unit charge* and has the dimensions

$$[\Phi_E] = \text{statC/cm} = \text{statV} = \text{erg/statC} \tag{8.13}$$

The electric potential obviously obeys the superposition principle:

$$\Phi_{E,\text{total}} = \Phi_1 + \Phi_2 + \Phi_3 + \cdots$$

In Section 7.13 we discussed the work required to move a charge from one point to a second point in the electrostatic field of another charge. This work is just the potential energy difference ΔPE_E between the two points. The work *per unit charge* required to make this movement is just the change in potential $\Delta\Phi_E$ between the points. We have given this quantity the symbol V (see Eq. 7.28):

$$\frac{W}{q} = \frac{\Delta PE_E}{q} = \Delta\Phi_E = V \tag{8.14}$$

Fig. 8.13 (*a*) *The electric field vector for a positive source charge is directed* away *from the source.* (*b*) *The field vector for a negative source charge is directed* toward *the source. The direction of* **E** *is always the direction of the force exerted on a positive test charge in the field.*

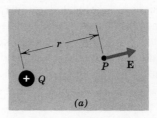

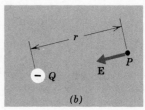

Table **8.1**   *Units of Electrical Quantities*

Quantity	CGS	MKS
F_E	dyne	N
Q	statC	C
E	statV/cm	V/m
Φ_E, V	statV	V

$$1 \text{ N} = 10^5 \text{ dynes}$$
$$1 \text{ statV} = 300 \text{ V}$$
$$1 \text{ C} = 3 \times 10^9 \text{ statC}$$

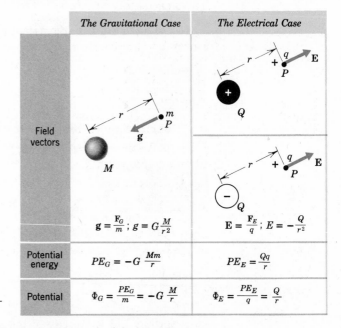

	The Gravitational Case	The Electrical Case
Field vectors	$\mathbf{g} = \dfrac{\mathbf{F}_G}{m}$; $g = G\dfrac{M}{r^2}$	$\mathbf{E} = \dfrac{\mathbf{F}_E}{q}$; $E = -\dfrac{Q}{r^2}$
Potential energy	$PE_G = -G\dfrac{Mm}{r}$	$PE_E = \dfrac{Qq}{r}$
Potential	$\Phi_G = \dfrac{PE_G}{m} = -G\dfrac{M}{r}$	$\Phi_E = \dfrac{PE_E}{q} = \dfrac{Q}{r}$

Fig. 8.14   *Comparison of the various field quantities for the gravitational and electrical cases.*

V is the *potential difference* or the *voltage* between the two points.

The units of the various electrical quantities that we have introduced are summarized in Table 8.1. A graphical comparison between the gravitational and electrical cases is made in Fig. 8.14.

Example **8.1**

Compute the electric field and the electric potential at point P midway between two charges, $Q_1 = Q_2 = +5$ statC, separated by 1 m.

Because P is located midway between two identical charges, any test charge placed at this point will experience equal but *oppositely directed* forces due to Q_1 and Q_2. The *net* force is therefore zero, so that $\mathbf{E} = 0$ at P.

Although the electric field at P is zero, this does *not* imply that the electric potential is also zero. The total potential $\Phi_{E,\text{total}}$ is the sum (the *algebraic* sum since potential is a *scalar*) of the potentials due to Q_1 and Q_2:

$$\Phi_{E,1} = \frac{Q_1}{r_1} = \frac{5}{50} = 0.1 \text{ statV}$$

$$\Phi_{E,2} = \frac{Q_2}{r_2} = \frac{5}{50} = 0.1 \text{ statV}$$

Therefore,

$$\Phi_{E,\text{total}} = \Phi_{E,1} + \Phi_{E,2} = 0.2 \text{ statV}$$

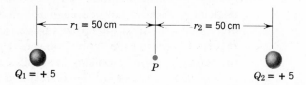

Notice that if either Q_1 or Q_2 is changed from $+5$ statC to -5 statC, the electric *potential* will vanish but the electric *field* will not. Therefore, the fact that either the field or the potential is zero in any particular case does not necessarily mean that the other quantity will also be zero; each quantity must be calculated separately.

Example **8.2**

In the Bohr model of the hydrogen atom, the electron is considered to move around the nuclear proton in a circular orbit that has a radius of 0.53×10^{-8} cm. In what electric field and in what potential does the electron move?

$$E = \frac{Q}{r^2}$$

$$= \frac{4.8 \times 10^{-10} \text{ statC}}{(0.53 \times 10^{-8} \text{ cm})^2}$$

$$= 1.7 \times 10^7 \text{ statV/cm}$$

which is a very large field indeed. Sparking usually occurs in air when a field strength of 100 statV/cm is reached.

$$\Phi_E = \frac{Q}{r}$$

$$= \frac{4.8 \times 10^{-10} \text{ statC}}{0.53 \times 10^{-8} \text{ cm}}$$

$$= 0.09 \text{ statV}$$

which is a rather small potential. The potential difference between the terminals of a flashlight is 1.5 V or 0.005 statV.

The electric field E depends on $1/r^2$, whereas the potential Φ_E depends on $1/r$; since r is extremely small (0.5×10^{-8} cm) in the case of the hydrogen atom, the field strength is large while the potential is small.

POINT SOURCES AND EXTENDED SOURCES

The expressions that have been given for the forces, fields, and potentials for both the gravitational and electrical cases are valid if the source mass or source charge is a uniform sphere or if it is a "point" (so small that its dimensions are negligible in comparison with all other dimensions in the problem and therefore for all practical purposes can be considered as a mathematical point). Of course, since real physical objects are generally not of such regular shape the equations as given are not adequate. However, because the field quantities all obey the superposition principle, any extended object can be considered to be composed of a large number of "point" objects and the field quantities can be calculated by summing the contributions from all of the "point" objects. Thus, any problem can be reduced to a series of "point" problems. We shall confine our attention here to "point" or spherical sources (and flat plates) and shall not be concerned with the sophisticated mathematical techniques that are required for dealing with the problems of extended sources with complicated shapes—there is no new physics in such cases, only new mathematics.

ELECTRIC FIELD LINES

Because a test mass placed in the gravitational field of a source mass always experiences an attractive force, the lines of force (or the gravitational *field lines*) are always directed *toward* the source mass. Because an electrical test charge will be either attracted or repelled by a source charge depending on the signs of the charges, we must adopt a convention regarding the direction ascribed to electric field lines. We choose to give to the electric field lines for a source charge of either sign the direction of the force that would be experienced by a *positive* test charge. Therefore, the field lines from a positive source charge are directed radially *outward,* whereas those from a negative source charge are directed radially *inward* (see Fig. 8.15). The convention for the electric field lines is therefore the same as that for the electric field vector.

Where do the field lines go? If we had an isolated charge, the field lines would go indefinitely far into space as straight lines. But, of course, it is not physically possible to have an isolated charge. All macroscopic matter (and, presumably, the entire Universe) is composed of equal numbers of elementary positive and negative charges and is therefore electrically neutral. (Objects can be charged, but this condition is brought about by separating the positive and negative charge of an originally neutral object.) Let us

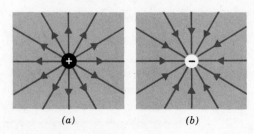

Fig. 8.15 *The electric field lines from a positive source charge (a) are directed radially outward whereas those from a negative source charge (b) are directed radially inward.*

(a)   *(b)*

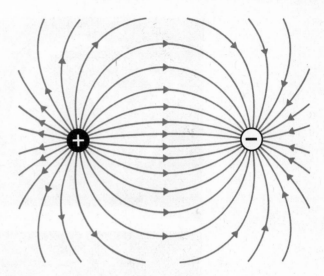

Fig. 8.16 *The electrical field lines originate on the positive charge and terminate on the negative charge.*

consider the case of two objects carrying equal and opposite charges (Fig. 8.16). As always, we can generate the field lines by measuring or calculating the magnitude and direction of the force on a positively-charged test body. If we do this, we find that the field lines emerge from the object carrying positive charge and go smoothly in curves that terminate on the negatively-charged object.

If we perform this kind of experiment or do the calculation for an arbitrary assembly of charges (but with zero *net* charge), we always find the same result: *the electric field lines originate on positive charges and terminate on negative charges.* This is one of the important results in the theory of electrostatics.

The conclusion regarding the origination and termination of electric field lines is different from that for gravitational field lines. In the latter case, the field lines have no definite point of origin but extend to infinity (at least in our nonrelativistic view).

FIELDS FOR DISTRIBUTIONS OF CHARGE

The shapes of the field lines for different geometrical configurations of charges can be obtained in a variety of ways that present graphic pictures of the fields. One such way is to suspend grass seeds in an insulating liquid (such as oil or glycerine) which surrounds the configuration of charges under investigation. The electric field induces a separation of the charge in the seeds: one end of a seed becomes negatively charged and the other end becomes positively charged, but the seed remains electrically neutral. This process is called *polarization*. The polarized seeds then orient themselves along the field lines, thereby rendering "visible" the shape of the field lines. Some photographs of field configurations made in this manner are shown in Fig. 8.17.

In Fig. 8.17a notice that the field lines for a pair of similarly charged objects are the same as those shown in Fig. 8.9 for the case of two identical masses. Notice also that in 8.17b the field lines are of the same shape as those in Fig. 8.16. There are two other general results illustrated in Fig. 8.17.

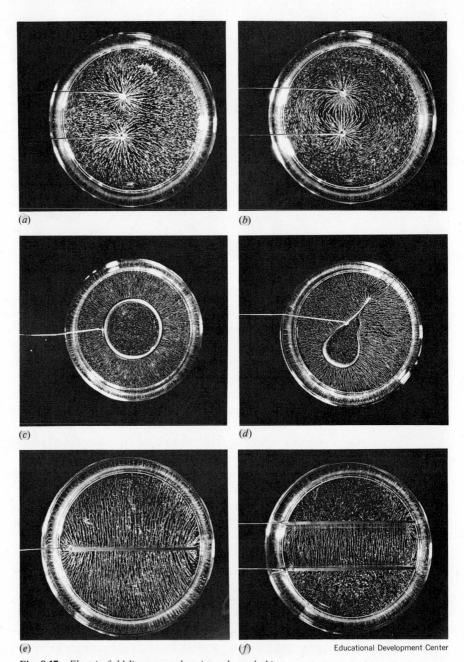

Fig. 8.17 *Electric field lines around various charged objects.*

(a) *Two charges of the same sign.*

(b) *Two charges with opposite signs.*

(c) *A charged ring (cross section of a charged sphere); the field inside is zero.*

(d) *A charged conductor of arbitrary shape; the field inside is zero.*

(e) *A charged plate.*

(f) *A pair of plates carrying equal and opposite charges, uniformly distributed.*

Educational Development Center

1. *The electric field inside a current-free conductor (solid or hollow) is zero.* (Figures 8.17c and 8.17d illustrate this point for the case of the hollow conductor.) First, consider a *solid* conductor. If an amount of charge is placed within such a conductor and if the charges are free to move, there will be mutual repulsion among all of the individual charges and they will rapidly migrate to the surface. Now, if this surface charge sets up an electric field *within* the conductor, then the conduction electrons will be induced to move and will then constitute a current, contrary to the stipulation that the conductor be current free. (In the next chapter we investigate the effects of electric current.)

Next, consider a *hollow* conductor. Let us take the simple case of a hollow sphere. If the sphere is charged, there will be a uniform distribution of charge on the surface.[6] A test charge placed at the center of the sphere will experience no net force; the field at the center must be zero. But what about other interior points? Figure 8.18 shows the geometry of the situation. We wish to determine the net force on a test charge at the point P. Two concentric cones, extending in opposite directions, are constructed with their vertices at P. These cones intercept two areas on opposite sides of the sphere. Since the charge is uniformly distributed on the surface of the sphere, the force on the test charge due to the charge on either intercepted segment of the shell is proportional to the area of that segment. The two forces are oppositely directed. But the larger area is farther away from P, and the increase in force due to the larger area ($\propto r^2$) is offset by the decrease in the force due to the greater distance ($\propto 1/r^2$). Therefore, the two forces are equal in magnitude and opposite in direction, giving no net effect. Because the same argument applies to the remaining area of the sphere, there is zero net force on the test charge. An identical result is found for any point within the sphere. Hence, the electric field *within* the spherical shell is everywhere *zero*. By using more complicated mathematics, the same result can be obtained for closed surfaces of arbitrary shape.

2. *The electric field between two uniformly charged parallel plates of equal area is uniform* (see Fig. 8.17f). First, consider the situation if the two plates are of infinite extent, one with a uniform positive charge and the other with a uniform and equal negative charge. The lines of force must originate on positive charge and terminate on negative charge. Because of the uniformity of the charge, no region between the plates can be different from any other region. Therefore, the lines of force must be uniform and parallel. The absence of any bunching of the lines of force means that the electric field is also uniform. In the real case of plates with finite size (Fig. 8.19), the field lines tend to curve at the edges. However, if the spacing between the plates is small compared to the linear dimensions of the plates, this so-called *edge effect* is unimportant for most calculations and is therefore neglected.

A pair of plates carrying charges, as in Fig. 8.19, is called a *condenser* or a *capacitor*. Such devices have important applications in electronic circuits.

Fig. 8.18 *The fact that the electrostatic force varies as $1/r^2$ insures that there is no force on a test charge at any point within a uniformly charged spherical shell; therefore, the electric field within the shell is zero.*

---

[6]This is insured by the *symmetry* of the sphere; no point on the surface is different from any other point and so, because the charges mutually repel each other, they distribute themselves uniformly over the surface.

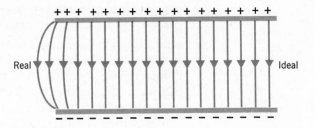

Fig. 8.19 *The actual curving of the field lines at the edges of a parallel-plate capacitor is usually neglected and only the ideal situation is treated. The field lines in an ideal parallel-plate capacitor are uniform both in magnitude and in direction.*

VOLTAGE

The difference in electrical potential between two points is called the *voltage* between those points: $\Delta\Phi_E = V$. Suppose we have a pair of parallel plates that are separated by a distance d, as in Fig. 8.20. If the plates are connected to a battery, the plate that becomes positively charged will have a potential $\Phi_{E,A}$ and the negatively-charged plate will have a potential $\Phi_{E,B}$, so that

$$\Phi_{E,A} - \Phi_{E,B} = \Delta\Phi_E = V \tag{8.15}$$

where V is the voltage of the battery.

The electric field between the plates is given by the electrical force on a test charge in the field:

$$E = \frac{F_E}{q} \tag{8.16}$$

In moving the test charge from B to A, we are working against the electrical force on q. The work required is

$$W_{B\to A} = F_E \times d = qEd$$

or,

$$E = \frac{W_{B\to A}}{qd} \tag{8.17}$$

The work done per unit charge $W_{B\to A}/q$ is just the voltage (see Eq. 8.14), so that

$$\boxed{E = \frac{V}{d}} \quad \text{(uniform field)} \tag{8.18}$$

The *direction* of the electric field is always given by the direction of the force on a positive test charge placed in the field.

Fig. 8.20 *The magnitude of the electric field between a pair of parallel plates connected to a battery is V/d.*

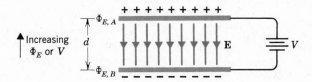

Example **8.3**

An electron is released from rest at one point in a uniform electric field and moves a distance of 10 cm in 10^{-7} sec. What is the electric field strength and what is the voltage between the two points?

For a uniform electric field, the force on the electron is constant; hence, the acceleration, $a = F_E/m_e$, is also constant. The distance traveled is $d = \frac{1}{2}at^2$, so

$$d = \frac{1}{2}at^2 = \frac{1}{2}\left(\frac{F_E}{m_e}\right)t^2$$

The field strength is the force divided by the charge:

$$E = \frac{F_E}{e} = \frac{2m_e d}{et^2}$$

$$= \frac{2 \times (9.1 \times 10^{-28}\,\text{g}) \times 10\text{ cm}}{(4.8 \times 10^{-10}\,\text{statC}) \times (10^{-7}\,\text{sec})^2}$$

$$= 3.8 \times 10^{-3}\,\text{statV/cm}$$

The voltage is

$$V = E \times d$$

$$= (3.8 \times 10^{-3}\,\text{statV/cm}) \times 10\text{ cm}$$

$$= 0.038\,\text{statV}$$

$$= 0.038 \times 300 = 11.4\text{ V}$$

THE DETERMINATION OF THE ELECTRONIC CHARGE e

In 1911 Robert A. Millikan performed a beautifully simple but highly significant experiment that established the discreteness of the charge on the electron; his experiment also obtained, for the first time, a precise value for this quantity. Millikan set up an electric field between a pair of parallel plates, as in Fig. 8.21. He sprayed a fine mist of oil droplets in this field. (Oil does not evaporate as readily as water.) Some of the droplets became negatively charged by friction in the process of spraying. Millikan viewed the droplets with a microscope and found that by adjusting the voltage V between the plates to a particular value he could suspend a given droplet in an equilibrium position. When suspended, the downward gravitational force, mg, equaled the upward electrical force, qE, where the magnitude

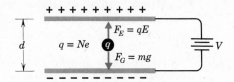

Fig. 8.21 *Schematic diagram of Millikan's oil drop experiment for the determination of* e.

of E is just V/d. Therefore, at equilibrium,[7]

$$mg = qE = \frac{qV}{d}$$

so that

$$q = \frac{mgd}{V} \tag{8.19}$$

The charge on the droplet q is given in terms of measurable quantities. The values of g, d, and V were measured directly but it was necessary to determine the mass m by an indirect method. First, the density of the oil was obtained by bulk measurements. Next, the rate of fall of the droplet in the absence of the electric field was measured. For the tiny droplet that Millikan used, the rate of fall was the *terminal velocity* (see Section 4.6) and it was known from the theory of motion of objects in a sluggish (or *viscous*) medium, such as air, that the terminal velocity is a sensitive function of the radius. Therefore, a measurement of the terminal velocity determined the radius (and, hence, the volume) and, when combined with the density measurement, provided a value for the mass.

Millikan found that the values of the charge on various droplets, determined in this way, were not of arbitrary sizes. Instead, he found that every charge was an integer number times some basic unit of charge; that is, $q = Ne$, $N = 1,2,3, \ldots$. This basic unit of charge is the charge on the electron, $e = 4.8 \times 10^{-10}$ statC. (Millikan's 1911 value of e was slightly in error due to the use of a faulty value for the viscosity of air. When the correction was made later, the resulting value was quite close to that accepted at present.)

8.5   *The Field of the Nuclear Force*

THE DIFFERENCE BETWEEN THE NUCLEAR FORCE AND $1/r^2$ FORCES

The gravitational field of a spherical object or the electrical field of a spherical charge has an exceedingly simple character. The strength of such a field decreases with distance in accordance with the same geometrical law that governs the variation with distance from the source of the intensity of a quantity, such as light, that is emitted uniformly in all directions into space. This $1/r^2$ variation of the field strength with distance means that a force will be experienced by a test object at any definite distance from the source. It is only at the mathematically well-defined but physically unrealizable distance, $r = \infty$, that the force decreases to zero.

The nuclear force, on the other hand, is of a decidedly different character. The effect of this force does not extend to infinity, but is instead confined to extremely small distances around the source. What is the nature of the *nuclear* field? Indeed, is the field concept of any validity in the description of these strange and complicated forces?

---

[7]Millikan actually used a *dynamic* instead of a *static* method for determining e in his oil drop experiment, but the distinction is not important here.

THE PION FIELD

The field concept has been applied to the strong nuclear force, but the nature of the problem in this case requires a departure from our previous reasoning regarding fields. The basic new idea that paved the way for our present (and still incomplete) understanding of the nuclear force was provided by the Japanese physicist Hideki Yukawa (1907–   ) in 1935. Yukawa hypothesized that two nucleons experience an attractive force at small distances because of the exchange between them of a new elementary particle (which had not yet been observed at the time) called a *meson*. A meson is a particle with mass intermediate between the mass of an electron and that of a proton. Several brands of mesons are now known, but the meson that is responsible for the strong nuclear force has been found to have a mass approximately 273 times that of an electron (or about 0.15 of the mass of a nucleon) and is called the π *meson* or *pion*.[8]

The *pion exchange force,* when treated in detail, requires complicated mathematics, but a qualitative description of an exchange force can be made. Suppose that two boys are grappling for control of a basketball. One boy grabs the ball from the other boy and then the second boy snatches it back again; the process is repeated over and over. This continual exchange of the ball results in each boy being pulled toward the other; that is, there is an attractive "basketball exchange force."

Suppose that a proton and a neutron collide. The collision process is controlled by the nuclear force (the pion exchange force) that acts between the two particles. Where does the pion that mediates the interaction come from? In a sense, the pion was always "there." It is very useful to picture the proton as consisting of a neutron and a positively-charged pion:

$$p \longleftrightarrow n + \pi^+$$

That is, a proton is equivalent to a neutron "core" surrounded by a "meson cloud." Or, in the field picture, we say that there is a pion field around the neutron and that the combination appears as a proton. Therefore, in the collision of a neutron and a proton, the π^+ meson is exchanged between the two neutrons and mediates the nuclear force between the particles. This exchange of the pion is an extremely rapid process, requiring only about 10^{-23} sec.

If the proton and the neutron differ in mass by only about 0.1 percent, how can the proton be considered to be a neutron combined with a pion when the latter has a mass of 15 percent of the proton? We cannot answer this question in detail until we have discussed the famous *uncertainty principle* of quantum mechanics (Section 12.6). In essence, this principle states that we can conceive of processes, such as $p \rightarrow n + \pi^+$, in which the mass does not balance, as long as the system reverts to its original state, that is, $n + \pi^+ \rightarrow p$, within a sufficiently short period of time. Since the complete exchange process takes only about 10^{-23} sec, the exchange of a pion between the colliding neutron and proton is allowed.

---

[8]The pion is the *principal* particle involved in the strong nuclear force, but we now know that there are also more massive mesons that contribute to this interaction (see Section 16.4).

By using the pion field concept of the strong nuclear force, a great deal of progress has been made, although we still have much to learn before we can claim to understand completely this basic force of Nature. The situation with regard to the weak force is even more bleak. Presumably, there is a similar elementary particle that mediates the force between electrons and neutrinos, but as yet we have no clear conception of the nature of this particle.

8.6   *Energy in the Field*

DOES AN OBJECT POSSESS POTENTIAL ENERGY?

When we raise an object to a height h above the surface of the Earth, we say that the object possesses a gravitational potential energy mgh relative to its initial position. But does the *object* really possess this potential energy? Or does the *Earth* share in the energy? According to our field description of the gravitational interaction we should not ascribe the increase in potential energy to *either* body. An amount of work mgh has been done *on the field* by changing the relative positions of the two bodies and it is the *gravitational field* that has acquired this energy. The energy can be recovered from the field by allowing the field to set the objects into motion. Similar comments also apply for the electrical and nuclear force fields.

Although it is well to keep in mind that the *field* possesses potential energy, we shall usually follow our previous custom and continue to refer to the potential energy "of a body" or the potential energy "of a system of bodies."

DOES THE FIELD REALLY POSSESS ENERGY?

We have argued that a field can possess energy. But is this just a fiction? Can a vacuum, completely devoid of any material particle, actually retain a physical quantity as real as energy? Consider the following situation. A space vehicle has landed on the moon and is telemetering information back to the Earth via radio waves (Fig. 8.22). Radio waves, as we shall see in Chapter 10, are a propagating disturbance in an electromagnetic field. The transmitter on the moon sends out a pulse that contains a certain amount

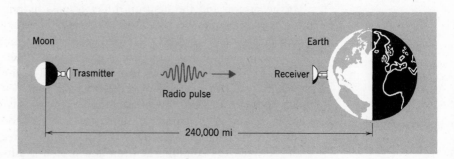

Fig. 8.22 *It requires approximately 1.3 sec for a radio signal to be transmitted from the moon to the Earth. During this interval, the transmitted energy resides in the electromagnetic field.*

of energy. This pulse propagates with the velocity of light and about 1.3 sec later it can be detected by a receiver on the Earth. Energy has been transmitted from an instrument on the moon and detected by another instrument on the Earth. Where was the energy during the 1.3-sec interval between transmission and reception? The energy conservation principle insures that the energy was *somewhere;* it could only have been contained in the *electromagnetic field.*

THREE TYPES OF ENERGY

If we choose to group together all types of potential energy under the heading "field energy," we would have three types of energy:

1. Kinetic energy
2. Field energy
3. Mass energy

The field theoretical approach to physical problems has been so successful and productive that we might expect future developments to permit the unification of our concepts and the classification of *all* types of energy as *field energy.* This would be a step in the direction toward the goal of simplifying physical theory because it would allow energy to be treated on a single basis, no matter what its source.

ARE FIELDS "REAL"?

We have continually emphasized the importance of the concept of energy in physics. If a field can contain energy, then must not we conclude that the field is indeed a *real* entity? In physics we attribute *reality* exclusively to those quantities that are *measurable.* Distance, mass, velocity, and momentum are surely *real.* But we never *measure* the electric field vector **E** or the gravitational field vector **g.** We always measure the *effect* of these field quantities; that is, we always measure a *force.* Therefore, the fields are only mathematical constructions that enable us to interpret in a consistent and experimentally verifiable way the actions of the gravitational and electrical forces. Are fields *real?* It is almost too fine a distinction to make. Whether or not they are real, the concept of the field has been one of the most fruitful ideas in physics, and it is clear that we shall continue to reap the benefits of field theories in many areas of science and technology.

Summary of Important Ideas

In the broadest sense, any physical quantity that is smoothly varying and well-defined at all points in space can be considered a *field* quantity.

Field quantities can be either *scalars* or *vectors.*

The *gravitational field quantity* is the force per unit mass (that is, the *acceleration*) acting on a test mass.

Gravitational field lines are always directed *toward* the source mass.

Field lines and equipotential surfaces always intersect at *right angles.*

Movement along an equipotential involves *no work*.

The *electric field quantity* is the force per unit charge acting on a test charge.

Gravitational (electrical) *potential* is the potential energy per unit mass (charge).

Electric field lines always *originate* on positive charges and *terminate* on negative charges.

The electric field inside a hollow conductor (or inside a current-free solid conductor) is *zero*.

The electric field between a pair of parallel, uniformly charged plates is *uniform*.

Fields can possess *energy*. There are *three* basic types of energy: kinetic energy, field energy, and mass energy.

Questions

8.1 Under what conditions can the following be usefully considered to be fields? Specify whether the field is *scalar* or *vector*.
(a) The mass density distribution in the Earth.
(b) The population density in a country.
(c) The population density in a city block.
(d) The density of stars in a galaxy.
(e) The flow of air masses in the atmosphere.

8.2 How could one extract useful work by moving an object across a region in which there exists pressure differences, as in Fig. 8.1? (Your method need not be a *practical* one.)

8.3 If a body is released in a force field to which it is sensitive, its motion does not necessarily follow a line of force. Explain why. Under what special conditions would the motion be exactly along a line of force?

8.4 Two concentric, hollow spheres carry equal and opposite charges. Argue whether an electric field exists in each of the regions *A, B,* and *C.* Reexamine the situation if both spheres carry equal charges of the same sign.

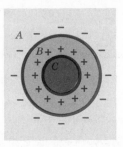

8.5 What is the gravitational potential inside a hollow spherical shell of matter?

Problems

8.1 Make a tracing of the map outline in Fig. 8.1. On this map show a set of isotherms (label these with values) that depict the following: low temperature areas in the Pacific northwest and in the northeastern states; high temperature areas in the southwest and in southern Florida; intermediate temperatures in the midwest.

8.2 Water is flowing in a cylindrical tube. Sketch the flow vectors along a diameter across the tube. Sketch the flow vectors if the fluid is oil with the same volume of flow per unit time. (The flow of oil is more sluggish or *viscous* than that of water and therefore has a tendency to flow less rapidly along the inner surface of the tube.)

8.3 In a certain oval-shaped pool, the water inlet is at the bottom near one end of the pool. The outlet is at the bottom near the opposite end. Sketch flow diagrams that represent water flow in this pool as seen from above and from one side.

8.4 What is the difference in gravitational potential between the base and the top of the Washington Monument ($h = 555$ ft)?

8.5 What is the gravitational potential due to the Earth at the position occupied by the moon?

8.6 Consider two spherical masses, one of which has twice the mass of the other, that are separated by a certain distance. Sketch the lines of gravitational force and the equipotentials (as in Figs. 8.9 and 8.12).

8.7 Calculate the ratio of the gravitational potential of the Sun to that of the Earth at the position of the moon during a lunar eclipse. Calculate the same ratio during a solar eclipse and compare the results. What can you say about the variation of the gravitational potential in which the moon finds itself?

8.8 What is the gravitational potential due to the Earth at the surface of the Earth?

8.9* Three charges are placed on the vertices of an equilateral triangle. One charge is $+2$ statC and the other two are each -1 statC. Sketch the electric field lines.

8.10 A charge $Q = -50$ statC is located at the origin of a coordinate system. What is the electric field vector and the potential at the point $x = 4$ cm, $y = 4$ cm?

8.11 At a certain point P in space a source charge produces an electric field of 30 statV/cm in the $+x$ direction. At this same point another source charge produces a field of 60 statV/cm in the $+y$ direction. What force will a proton experience at P?

8.12 An electron is placed in a uniform field of 100 statV/cm. What force does the electron experience?

8.13* At a distance of 100 km from a small, electrically charged asteroid it is found that there is an electric field of 0.1 statV/cm. What charge does the asteroid carry? If the asteroid is spherical with a radius of 1 km, what is the charge per unit area on its surface. (The total charge is uniformly distributed.)

8.14 Suppose that the surface of the Earth carries a uniform surplus of electrons amounting to 1 electron per cm². What is the charge on the Earth? What is the electric potential at the surface of the Earth?

8.15 Sketch the equipotentials for two charges of equal magnitude but opposite sign that are separated by a certain distance.

8.16 Three charges are located in a straight line as follows: $+2$ statC at $x = -10$ cm; -4 statC at $x = 0$ cm; $+2$ statC at $x = +10$ cm. Sketch the electric field lines.

8.17 Calculate the gravitational potential and the electrical potential at a distance of 10^{-8} cm from a proton.

8.18 A completely ionized helium nucleus is in a uniform electric field of 50 statV/cm. What acceleration does the helium nucleus experience?

8.19 Consider the proton to be a uniformly charged sphere with a radius of 10^{-13} cm. (This is a dubious model of the proton.) What is the electric field at the surface of the proton?

8.20 A sphere of radius 1 m carries a uniform surface charge of 0.3 statC/cm². What is the electric field and the potential at the surface of the sphere? What are the values 10 cm above the surface?

8.21 Two charges, $Q_1 = +5$ statC and $Q_2 = -3$ statC, are placed at opposite corners of a square whose sides are 50 cm in length. What is the potential at each of the unoccupied corners?

8.22 Four charges are situated at the corners of a square with 10-cm sides. The charges are, reading clockwise, $+3$ statC, -8 statC, -5 statC, and $+10$ statC. Graphically construct the picture and show the net force vector acting on the $+10$ statC charge. Change the sign of the -8 statC charge and show the new net force vector.

8.23 Six equal charges, $Q = +10$ statC, are positioned at equal intervals on the circumference of a circle of 1 m radius. What is the electric field and the electric potential at the center of the circle? Next, change alternate charges from $Q = +10$ to $Q = -10$ statC (so that there are now three charges with $+10$ and three with -10 statC, each set being at equal intervals) and recompute the field and the potential at the center of the circle.

8.24 At a certain position in space the electric potential is $\Phi_E = +800$ statV. What is the potential energy at this point of (a) an electron and (b) a proton?

8.25 Two points, A and B, have electric potentials of $+100$ statV and -150 statV, respectively. How much work is required to move an electron from A to B? Is the same amount of work required to move a proton from A to B?

8.26 Two parallel plates are separated by 2 cm. A battery is used to put a potential difference of 600 V across the plates. What electrical force will an oil droplet carrying a charge of $4e$ experience in the field between the plates?

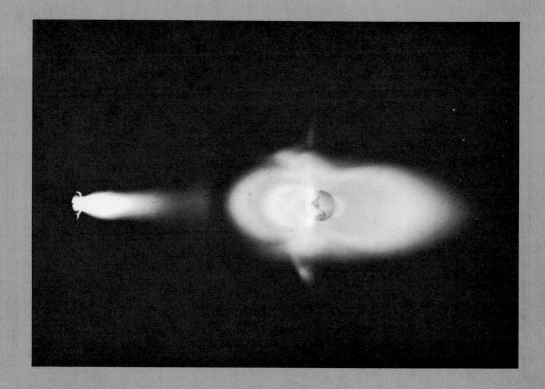

Simulation of the effect of the solar wind on the Earth's magnetosphere.

NASA

9 *Electric Charges in Motion*

The motion of electric charges is a fundamental aspect of much that takes place in the Universe. Not only do moving charges influence the dynamics of the Sun and other stars as well as provide the Earth with its magnetism, but moving charges in the form of electric current perform countless tasks in our present-day electrically-oriented world by operating all manner of electrical motors, radio circuits, and other devices. There is almost no area of modern society that does not depend in a crucial way on the effects of moving electric charges.

In this chapter we are concerned, not with the numerous electrical devices that are applications of charges in motion, but with the fundamental aspects of electric current and its relationship to magnetic fields. An important part of the discussion will be concerned with the way in which charged particles behave when moving in magnetic fields. The high energy particles that are used to probe nuclei and to investigate the properties of elementary particles are usually confined in the accelerating machines and guided to their targets by magnetic fields. The charged particles of the cosmic rays interact with the magnetic field of the Earth and give rise to many interesting phenomena, such as the northern lights. The Earth's field also *traps* charged particles and produces the radiation belts that encircle the Earth. The fact that charged particles trapped in magnetic fields may one day make feasible the inexpensive generation of electrical power from fusion reactors is of great economic importance.

We shall conclude this chapter by discussing the four Maxwell equations that provide a complete description of electromagnetic field phenomena. In the next chapter we continue the study of electromagnetism by treating the important topic of electromagnetic waves.

9.1 *Electric Current*

THE TRANSPORT OF ELECTRONS

The movement of electric charge from one position to another constitutes a *current*. In almost all practical situations, the positively charged nuclei of atoms remain essentially stationary compared to the movement of the much less massive electrons. Thus, in almost all of the cases with which we shall be concerned, electric current is due to the motion of *electrons*.

In conductors, a fraction of the electrons are not attached to specific atoms but are free to move about in the material. In the absence of an applied electric field, these *free electrons* (or *conduction electrons*) move at random, colliding with the stationary atoms at frequent intervals and thereby changing their directions of motion. Across any given surface in the material, just as many electrons will move in one direction as in the other (Fig. 9.1). Therefore, there is no *net* transport of electrons across the surface and the electric current is *zero*.

If the ends of the conductor are attached to the terminals of a battery (Fig. 9.2), an electric field is established within the conductor. There will now be a net transport of electrons across any surface that is perpendicular

Fig. 9.1 *If no electric field is present, the motion of the conduction electrons is random.*

to the field lines. That is, in Fig. 9.2, there will be a few more electrons crossing the surface from *right* to *left* than will be crossing from *left* to *right*.

The direction in which the electrons drift is opposite to that of the field lines. However, just as we arbitrarily elected to consider the direction of electric field lines to be the direction of the force on a *positive* charge, it is traditional to refer to the direction of *current* flow as the direction in which *positive* charges would move (even though it is only the electrons that actually move). A negative current (or an electron current) in one direction is equivalent to a positive current in the opposite direction. We shall always use the terms *current* or *current flow* to mean this so-called *conventional* current which flows in the same direction as the electric field lines. The terms *electron flow* or *electron current* will be used to refer explicitly to the motion of the electrons. The advantage in retaining this seemingly awkward convention will soon become apparent.

THE DEFINITION OF CURRENT

Electric current is defined quantitatively as the *rate* at which electric charge flows across a surface. When 1 statC of positive charge crosses a given surface in 1 sec, a current of 1 *statampere* (statA) is flowing. In general, the current I is given by the net charge Q crossing a surface during a time t (Fig. 9.3):

$$I = \frac{Q}{t} \tag{9.1}$$

The unit of current is

$$[I] = \frac{\text{statC}}{\text{sec}} = \text{statA} \qquad \text{(CGS)} \tag{9.2}$$

Of course, if we wish to define the *total* current flow in a wire, the surface must be one that is a complete cross section of the wire.

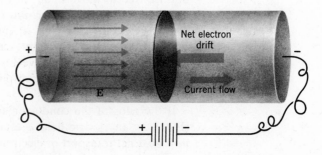

Fig. 9.2 *The application of an electric field to a conductor causes a net motion of the electrons in the direction opposite to the electric field lines. By convention, the current is said to flow in the same direction as the field lines.*

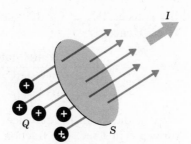

Fig. 9.3 *If 1 statC of charge Q flows across the surface* S *in 1 sec, the current* I *is 1 statA.*

In the MKS system of units, the unit of current is the *ampere* (A):

$$1 \text{ A} = 1 \text{ C/sec} = 3 \times 10^9 \text{ statC/sec}$$
$$= 3 \times 10^9 \text{ statA} \tag{9.3}$$

Because the unit is of a convenient size, the *ampere* is used to specify household current. Household circuits are usually capable of carrying currents of 15–20 A and household appliances usually require currents of a few amperes.

ELECTRON DRIFT VELOCITIES

In ordinary conductors, the free electrons move with velocities of about 10^8 cm/sec. In the absence of an electric field, as already pointed out, this motion is *random* and there is no net current flow. If only a small net drift velocity is imposed on these electrons by an electric field, a sizable current can result. Suppose that we have a copper conductor with a cross sectional area of 1 cm^2 (Fig. 9.4). What net drift velocity is necessary to constitute a current of 1 A?

Consider a 1-cm length of a copper rod (shown by the volume *AB* in Fig. 9.4). We have previously calculated (Example 6.3) that there are approximately 0.92×10^{22} atoms in 1 g of copper. The density of copper is 8.92 g/cm^3; therefore, 1 cm^3 of copper contains $8.92 \times (0.92 \times 10^{22}) = 8.2 \times 10^{22}$ atoms. There is one free electron per copper atom, so the volume element *AB* contains 8.2×10^{22} free electrons. The effect of the electric field on the motion of the free electrons in the volume *AB* is the same as if all 8.2×10^{22} of these electrons were to move forward by a distance d to the position $A'B'$.

In order for a current I to flow, a certain number N of electrons must move across the cross sectional surface at B. That is,

$$I = \frac{Q}{t} = \frac{Ne}{t}$$

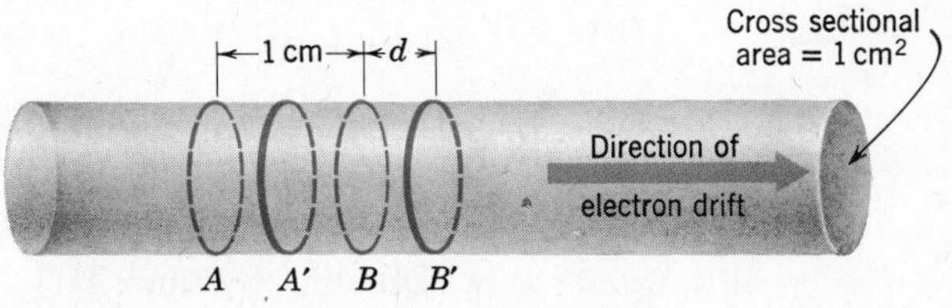

Fig. 9.4 *The free electrons in the 1-cm^3 volume AB drift to the position A'B' which constitutes a net movement through the distance* d.

‹—1 cm—›‹d ›

Cross sectional area = 1 cm^2

Direction of electron drift

A A' B B'

If I is to be 1 A = 3 × 10⁹ statA, then Ne must be 3 × 10⁹ statC for t = 1 sec. Hence,

$$N = \frac{(Ne)}{e} = \frac{3 \times 10^9 \text{ statC}}{4.8 \times 10^{-10} \text{ statC/electron}}$$

$$= 6.2 \times 10^{18} \text{ electrons}$$

That is, the result is equivalent to moving 6.2×10^{18} electrons across the surface B in 1 sec. The fraction f of the free electrons in the 1-cm³ volume that cross the surface is

$$f = \frac{6.2 \times 10^{18}}{8.2 \times 10^{22}} = 7.6 \times 10^{-5}$$

Since the length of the volume AB is 1 cm, the distance of net movement d must be this same fraction of 1 cm; that is, $d = 7.6 \times 10^{-5}$ cm.

The drift velocity is therefore

$$v_{\text{drift}} = \frac{7.6 \times 10^{-5} \text{ cm}}{1 \text{ sec}} = 7.6 \times 10^{-5} \text{ cm/sec}$$

If the cross sectional area were smaller, the drift velocity would have to be larger in order for the same number of electrons to pass a given surface per unit time. Common No. 18 household wire has a diameter of 0.040 in. or a cross sectional area of 0.0083 cm². Hence, the drift velocity in such a wire required to produce a current of 1 A would be

$$v_{\text{drift}} = \frac{1 \text{ cm}^2}{0.0083 \text{ cm}^2} \times (7.6 \times 10^{-5} \text{ cm/sec})$$

$$= 9.2 \times 10^{-3} \text{ cm/sec}$$

or approximately 0.1 mm/sec.

Because conductors contain such enormous numbers of free electrons, only very small drift velocities are necessary to produce sizable currents. It should be remembered that this net drift velocity of a fraction of a cm/sec is superimposed on the large random electron velocities of ~10⁸ cm/sec.

ELECTRICAL POWER AND ENERGY

The amount of electrical power required to operate a device (such as a motor or a radio) is equal to the product of the voltage applied and the current that flows:

$$P_E = VI \tag{9.4}$$

In the MKS system,

$$[P_E] = [V] \times [I] = \text{volts} \times \text{amps}$$

$$= \text{watts (W)} \tag{9.5}$$

In the CGS system, the unit of power is the *erg/sec;* 1 W = 10^7 ergs/sec.

We can easily verify that the product VI has the correct dimensions: power = work per unit time. According to Eq. 7.31, *voltage × charge = work*. Dividing by time,

$$\text{voltage} \times \frac{\text{charge}}{\text{time}} = \frac{\text{work}}{\text{time}} = \text{power}$$

and *charge/time* is just *current*.

The electrical *energy* consumed by a device (and transformed into some other form of energy such as kinetic energy or heat) is equal to the product of the power and the time during which the power was supplied:

$$\mathcal{E}_E = P_E \times t$$
$$= VIt \tag{9.6}$$

Example **9.1**

An electric motor, rated at 1 hp, operates from a 100-V line for 1 hr. How much current does the motor require and how much electrical energy was expended?

The current is

$$I = \frac{P_E}{V} = \frac{1 \text{ hp}}{100 \text{ V}} = \frac{746 \text{ W}}{100 \text{ V}}$$

$$= 7.46 \text{ A}$$

The electrical energy is

$$\mathcal{E}_E = P_E \times t = 1 \text{ hp} \times 1 \text{ hr}$$

$$= 746 \text{ W-hr}$$

$$= 0.746 \text{ kW-hr}$$

or, in CGS units,

$$\mathcal{E}_E = (746 \text{ W-hr}) \times \left(\frac{10^7 \text{ ergs/sec}}{1 \text{ W}}\right) \times \left(\frac{3600 \text{ sec}}{1 \text{ hr}}\right)$$

$$= 2.68 \times 10^{13} \text{ ergs}$$

9.2 *Magnetism*

LODESTONES AND THE COMPASS

Ancient men were acquainted with the fact that certain natural stones (*lodestones*) attracted one another and also attracted bits of iron. These stones were found in Magnesia in Asia Minor and they became known as *magnets*. The property of these stones that causes attraction (that is, the *magnetism*) cannot be cancelled as can the electricity formed by friction charging (which was also known to these ancient people). Somehow, magnetism is a permanent aspect of the material.

Lodestones, when suspended freely or floated in still water on pieces of wood, were found to take up a definite direction with respect to the Earth— long, narrow lodestones orient themselves in a north-south direction. This observation led to the introduction of a practical magnetic compass for purposes of navigation.

The compass has provided us with the standard by which we define

Fig. 9.5 *Like magnetic poles repel and opposite magnetic poles attract.*

directions in describing magnetic effects. The end of a compass magnet which is *north-seeking* is called the *north pole* or *N pole* of the magnet. Similarly, the south-seeking end is the south or S pole of the magnet.

Magnets have the familiar property (similar to that of electric charges) that *opposite poles attract* and *like poles repel* (Fig. 9.5). Therefore, the north-seeking N pole of a compass magnet actually is attracted to and points toward the Earth's S magnetic pole which is located near (but not at) the geographic north pole (see Fig. 9.7).

MAGNETIC FIELDS

We know that at every point in the vicinity of the Earth a compass will assume a definite direction. That is, there is a *magnetic field* due to the Earth's magnetism just as there is a gravitational field due to the Earth's mass. The magnetic field of the Earth or that of any magnetic material, can be mapped by using a compass. (This instrument will indicate the *direction* of the field at any point but it will not directly provide information regarding the *strength* of the field.) Compass measurements of the field of a simple bar magnet show that the magnetic field lines are as indicated in Fig. 9.6. By convention, we take the direction of the field lines to be the direction in which the N pole of a compass magnet points; that is, the field lines in the region external to the bar magnet run *from the N pole to the S pole,* as in Fig. 9.6. Similar measurements of the Earth's magnetic field show that the Earth's magnetism is practically the same as that of a bar magnet (Fig. 9.7).[1]

Magnetic field lines can be made "visible" by a simple technique. If a sheet of paper is placed over a bar magnet and iron filings are sprinkled on the paper, the tiny pieces of iron will take up positions with their long dimensions along the field lines. (That is, the pieces of iron act as small compass magnets.) Figure 9.8 shows the magnetic field map of a bar magnet obtained in this way.

ELEMENTARY MAGNETS

If a bar magnet is cut into two pieces, as in Fig. 9.9, we find that the two halves are themselves complete magnets with N and S poles in the same orientation as the original magnet. Further division of the magnet produces additional magnets, again with N and S poles oriented in the same direction as the original magnet.

What will happen if we continue this process of division down to the

[1]The Earth's magnetism is not static; instead, it changes slowly with time. Measurements made on archeologically dated samples of magnetic materials have shown that the intensity of the Earth's magnetic field fluctuates with a time interval of about 5000 years between successive maxima and minima. Furthermore, during the last 10–20 million years, the field has actually *reversed* its polarity every 300,000 years or so.

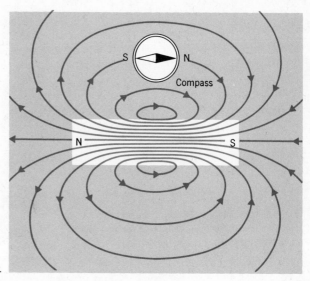

Fig. 9.6 *Magnetic field lines for a simple bar magnet.*

atomic level? Can we ever separate the N pole from the S pole? As we shall discuss in detail in later sections, even individual atoms can behave as microscopic but *complete* magnets with N and S poles. In fact, because we cannot take an atom apart and retain in its components the complete magnetic effects of the atom as a whole, we must conclude that the N and S poles of a magnet have no independent existence. Although there are theoretical speculations that individual magnetic poles (magnetic *monopoles*) may exist, no experiment has yet given any evidence for their existence (see Problem 9.27).

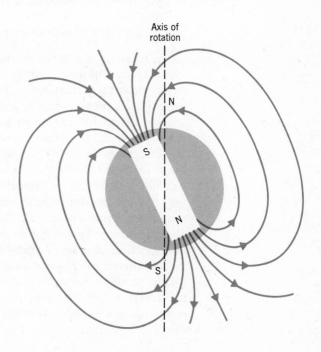

Fig. 9.7 *The Earth's magnetic field is like that of a bar magnet with the S pole near the north geographic pole.*

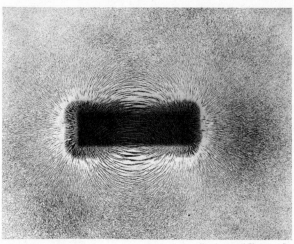

Fig. 9.8 *The magnetic field of a bar magnet obtained by sprinkling iron filings on a piece of paper covering the magnet.*

Fundamental Photographs

Fig. 9.9 *Cutting a magnet produces two magnets with N and S poles in the same orientation as the original magnet.*

9.3 *Magnetic Fields Produced by Electric Currents*

THE FIELD OF A CURRENT-CARRYING WIRE

The magnetism of a bar magnet appears to be a completely static affair. The magnetic material is electrically neutral and there does not appear to be any electrical current flowing in the magnet. What, then, is the connection between electrical currents and magnetism? Until early in the 19th century, electricity and magnetism were thought to be two independent phenomena. In 1820, Hans Christian Oersted (1777–1851), a Danish physicist, discovered (quite by accident) that a current-carrying wire influenced the orientation of a nearby compass magnet. Once this discovery established a definite connection between electrical current and magnetism, it required only a short time for experiments to reveal many of the details of *electromagnetism*. Foremost among the scientists who contributed to the rapid accumulation of knowledge of this new branch of physics was the French mathematical physicist André-Marie Ampère (1775–1836), for whom the unit of electric current is named.

One result of these early experiments was the discovery that a current-carrying wire produces a magnetic field (which influenced Oersted's compass); it was found that the magnetic field lines are *circles* centered around and perpendicular to the current-carrying wire. Figure 9.10 shows the circular field lines as revealed by the iron-filing technique.

RIGHT-HAND RULE

We can establish the *direction* of the magnetic field lines due to a current-carrying wire by observing the orientation of a compass magnet when placed in the vicinity of the wire. The results of such an experiment are summarized in the so-called *right-hand rule:*

If a current-carrying wire is grasped with the right hand in such a way that the thumb is in the direction of conventional current flow, then the fingers encircle the wire in the same direction as the magnetic field lines (see Fig. 9.11).

The magnetic field vector (analogous to the electric field vector **E**) is given the symbol **B**.

We see here the advantage of having adopted the convention of always using the term *current flow* to mean the (equivalent) flow of *positive* charge, namely, that it permits us to use a *right*-hand rule for the direction of the field lines and the field vector **B**. We have already defined the direction of the angular momentum vector (Section 5.10) in terms of the direction

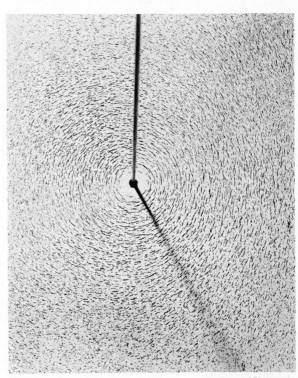

Fig. 9.10 *The circular magnetic field lines surrounding a current-carrying wire are revealed by using the iron-filing technique.*

Fundamental Photographs

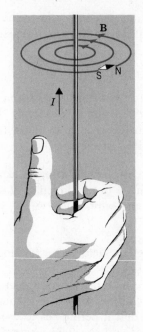

Fig. 9.11 *Illustration of the right-hand rule for determining the direction of the magnetic field lines due to a current flowing in a wire.*

of advance of a right-hand screw and we shall later have additional conventions regarding *right* hands. Therefore our choice allows us always to specify the directions of these vector quantities in terms of rules using the *right* hand.

UNIFORM MAGNETIC FIELDS

In the discussion of electric fields in the preceding chapter, it was apparent that there are many situations for which it is desirable to have uniform electric fields. It is also true that many experiments require uniform *magnetic* fields. We know that uniform electric fields can be produced, for example, by parallel, uniformly charged plates. How can we produce magnetic fields that are uniform? One of the simplest ways is to bend a bar magnet until the N and S pole faces are close together and parallel, as in Fig. 9.12. This so-called "C" magnet has a uniform **B** field between the poles.

Although the "C" magnet does produce a uniform field, the magnet is "permanent" and therefore the strength of the field cannot easily be changed.

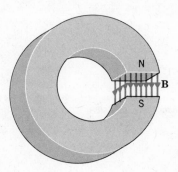

Fig. 9.12 *A "C" magnet, formed by bending a bar magnet, produces a uniform magnetic field between the poles.*

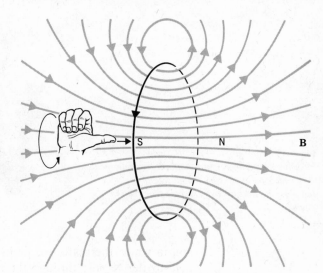

Fig. 9.13 *The field lines for a current-carrying loop of wire. Notice that the field resembles that of a bar magnet.*

Another type of right-hand rule is in operation here. If the fingers of the right hand are curled in the same sense that current flows in the loop, the thumb points in the same direction as the magnetic field lines that pass through the loop. (Check that this is consistent with the application of the previous right-hand rule.)

Therefore, "C" magnets are of limited usefulness in many types of experiments; a uniform magnetic field that can be set at a strength appropriate for a particular experiment (just as the electric field between a pair of capacitor plates can be changed by adjusting the voltage across the plates) would be most useful.

One way of producing a variable, uniform magnetic field is to pass a current through a wire wrapped around a piece of iron that has been bent into the shape of a "C." Each loop of current-carrying wire that encircles the iron yoke produces a magnetic field as shown in Fig. 9.13. In the center of the loop, where the field lines merge, the magnetic field is strong. The iron yoke concentrates the field lines and they are carried by the yoke to the gap where a uniform field is produced between the pole tips. When the current in the wire is varied, the magnetic field in the gap is also varied. Such an arrangement is called an *electromagnet*.

An even more efficient design for an electromagnet is the so-called "H" magnet (Fig. 9.14) in which current windings are placed on each of the poles

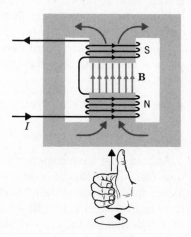

Fig. 9.14 *Schematic design of an "H" electromagnet.*

Fig. 9.15 *Photograph of an "H" magnet used for laboratory experiments with uniform magnetic fields. By turning the cranks on the outside of the yoke, the separation distance of the pole faces can be changed.*

Alpha Scientific Laboratories

and in which there are two paths for the field lines to return from the S pole to the N pole through the iron. Most uniform-field electromagnets found in laboratories are of this general design (see Fig. 9.15).

In all electromagnets the strength of the **B** field between the poles depends not only on the current in the windings but also on the geometry of the yoke and the poles, and the detailed composition of the iron alloy that is used.

WHERE DO MAGNETIC FIELD LINES BEGIN AND END?

We have previously seen (Section 8.4) that electric field lines originate on positive charges and terminate on negative charges. The lines that specify the magnetic field are distinctly different in character: *magnetic field lines have no beginning and no end.* Thus, the field lines surrounding a straight current-carrying wire are circles; the field lines of a current loop are not circles but nevertheless the lines have no point of origination or termination. Even the field lines of a bar magnet or electromagnet do not begin at the N pole and end at the S pole; the lines extend through the interior of the bar or core without termination.

The fact that magnetic field lines have neither a beginning nor an end is equivalent to the statement that there are *no magnetic monopoles.* Electric field lines originate and terminate on electric monopoles (charges) but there are no magnetic monopoles to terminate the magnetic field lines. The continuity of magnetic field lines and the nonexistence of magnetic monopoles is one of the important facts of electromagnetism.

9.4   *Effects of Magnetic Fields on Moving Charges*

THE STRENGTH OF THE MAGNETIC FIELD

Although we have completely specified the *direction* of the magnetic field lines (or, equivalently, the direction of the magnetic field vector **B**), we have as yet made no quantitative statement regarding the *strength* of the field (that is, the *magnitude* of **B**). In the case of the electric field, the magnitude of **E** was defined in terms of the *force* on a stationary test charge in the field. Similarly, we can define the magnitude of **B** in terms of the force exerted by the field on a test charge. But a test charge that is *stationary* in a magnetic field experiences no force. Only in the event that the test charge is in motion is there a magnetic force on the charge.

In a given magnetic field it is found that the magnetic force is directly proportional to both the charge and the velocity of the test particle. The proportionality factor that connects the magnetic force F_M with the charge and velocity (in units of c, the velocity of light) of the test particle is the magnetic field strength B:

$$F_M = q\frac{v}{c}B \qquad \text{(for } \mathbf{v} \perp \mathbf{B}\text{)} \tag{9.7}$$

The magnetic field unit is named for Karl Friedrich Gauss (1777–1855), the great German mathematician:

$$[B] = \text{gauss} \quad \text{(CGS)} \tag{9.8}$$

Notice that the dimensions of B are *force per unit charge.* Thus, in terms of fundamental quantities, the unit of B is equivalent to that for E in the CGS system (dyne/statC), but the special name *gauss* is given to B to prevent confusion.

THE DIRECTION OF THE MAGNETIC FORCE

The quantities v and B that appear on the right-hand side of Eq. 9.7 are the magnitudes of the vectors **v** and **B**. And, of course, F_M is the magnitude of the force vector $\mathbf{F}_M$. What is the relationship among the directions of these three vectors?

The case of the magnetic force is distinctly different from that of the electric force. As shown in Fig. 9.16, the electric force vector $\mathbf{F}_E$ that acts on a positive test charge has the *same* direction as the electric field vector **E** and is independent of the direction of the velocity vector **v**. On the other hand, it has been found experimentally that when a charged particle enters a magnetic field, the direction of the magnetic force is *perpendicular* to both **v** and **B** (Fig. 9.17).

The direction of $\mathbf{F}_M$ relative to **v** and **B** is given by another right-hand rule: The vector $\mathbf{F}_M$ has the same direction as that of the advance of a

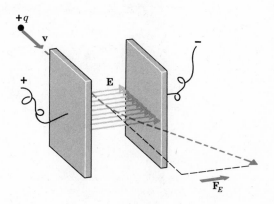

Fig. 9.16 *The electric force vector* $\mathbf{F}_E$ *has the same direction as the electric field vector* **E.**

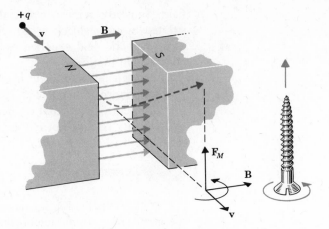

Fig. 9.17 *The magnetic force is perpendicular to both **v** and **B**. Application of the right-hand rule shows that the direction of* $\mathbf{F_M}$ *is up.*

right-hand screw when rotated in the sense that moves the vector **v** toward the vector **B** (see Fig. 9.18a). Alternatively, we can state the rule in the following way: Point the fingers of your right hand in the direction of **v** and curl the fingers toward the direction of **B**; the thumb then points in the direction of $\mathbf{F_M}$ (see Fig. 9.18b). Note that the rules apply to the case of a particle carrying a *positive* charge; for a negatively charged particle the direction is *opposite* to that given by the rules.

The maximum force exerted on a moving charged particle by a magnetic field occurs when the velocity vector of the particle is perpendicular to the field vector **B**, as in Fig. 9.17. The magnitude of this maximum force is given by Eq. 9.7. If **v** is not perpendicular to **B**, the force is less and becomes zero when **v** is parallel to **B.** In general, if θ is the angle between **v** and **B,** the magnetic force is (in the CGS system)

$$F_M = q\,\frac{v}{c}\,B\,\sin\theta \tag{9.9}$$

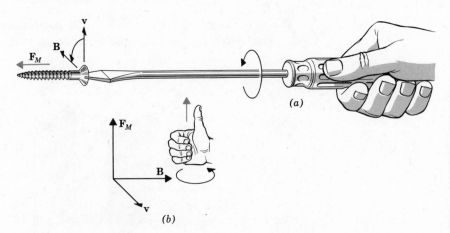

Fig. 9.18 *Illustration of the right-hand rules for determining the direction of the magnetic force* $\mathbf{F_M}$ *on a moving positive charge.*

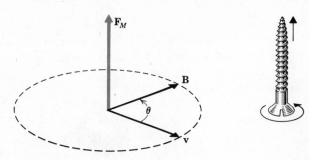

Fig. 9.19 *The magnitude of* **F_M** *depends on the sine of the angle θ between* **v** *and* **B**. *The force vector* **F_M** *is always perpendicular to the plane defined by* **v** *and* **B**.

If the angle between **v** and **B** is not zero, the two vectors define a plane and the force vector **F_M** is always perpendicular to this plane, as in Fig. 9.19.

Another way of stating the result expressed by Eq. 9.9 is the following: The magnetic force on a moving charged particle depends only on the component of the velocity vector that is perpendicular to **B**. The perpendicular component is (Fig. 9.20)

$$v_\perp = v \sin \theta \tag{9.10}$$

which is just the factor that appears in Eq. 9.9.

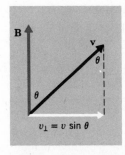

Fig. 9.20 *The component of* **v** *that is perpendicular to* **B** *is* $v_\perp = v \sin \theta.$

THE LORENTZ FORCE

In a combination electric and magnetic field the *total* force exerted on a moving charged particle is

$$F_L = F_E + F_M \tag{9.11}$$

where

$$\left. \begin{array}{l} F_E = qE \\[2mm] F_M = q\dfrac{v}{c} B \sin \theta \end{array} \right\} \tag{9.12}$$

The total electromagnetic force **F_L** is called the *Lorentz force* after Hendrik Antoon Lorentz (1853–1928), the Dutch physicist whose studies of electromagnetism paved the way for Einstein's development of relativity theory.

Although the total electromagnetic force **F_L** is written as the sum of an electric and a magnetic term, it is important to realize that the magnetic force is not a *new* force at all and that we do not have to add it to our list of the basic forces of Nature. As pointed out earlier, magnetic fields have no existence independent of electric charges; a magnetic field arises solely by virtue of the motion of a charged particle relative to the point at which the field is measured.

Suppose that we have a charge q and a meter that is sensitive to a magnetic field; suppose that both are at rest in some coordinate system, as in Fig. 9.21a. Clearly, the meter will show zero field, $B = 0$. (There is, of course, an *electric* field at the position of the meter.) However, if the charge is in motion, as in Fig. 9.21b, we know that this is equivalent to a current and that there is produced a magnetic field which will be registered by the meter. But it is only the *relative* motion of the charge and the meter that is impor-

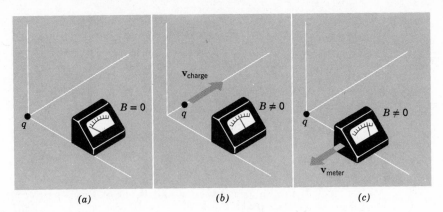

Fig. 9.21 *A magnetic field is produced solely as the result of relative motion between the charge* q *and the "B meter."*

tant. (The laws of physics are the same in all inertial reference frames.) Therefore, if q remains at rest in the coordinate system and the *meter* moves, as in Fig. 9.21c, this is entirely equivalent to the situation in Fig. 9.21b, and the presence of a magnetic field will again be shown by the meter. *A magnetic field is produced only by a changing or moving electric field.*

THE MAGNETIC FIELD OF A MOVING CHARGED PARTICLE

The strength of the electric field due to a charged particle decreases with distance away from the source as $1/r^2$. Similarly, the strength of the magnetic field due to a moving charged particle has a $1/r^2$ dependence on distance. In the plane perpendicular to the velocity vector $\mathbf{v}$ of a charged particle (plane A' in Fig. 9.22), the magnetic field at the point P' is given by

$$B' = \frac{q}{r'^2} \times \frac{v}{c}$$

That is, the magnetic field at P' is a factor v/c *smaller* than the electric field at that point. Also shown in plane A' of the figure is the circular magnetic field line that passes through P'.

The plane A in Fig. 9.22 is also perpendicular to $\mathbf{v}$ but is farther along the line of motion of q. At the point P, the field B is proportional to $1/r^2$ but it is also reduced by the sine of the angle ϕ between $\mathbf{v}$ and the line connecting q with P. Therefore, the general expression for the magnetic field of a moving charged particle is (in the CGS system)

$$\boxed{B = \frac{qv}{cr^2}\sin\phi} \qquad \text{(moving charged particle)} \qquad (9.13)$$

Figure 9.22 shows the field line in the plane A that passes through the point P. This line is also circular but the strength of the field on this line is smaller than that on the field line in the plane A' (that is, $B < B'$). Because of the factor $\sin\phi$ in Eq. 9.12, the moving charge produces no magnetic field on its line of motion (where $\sin\phi = 0$).

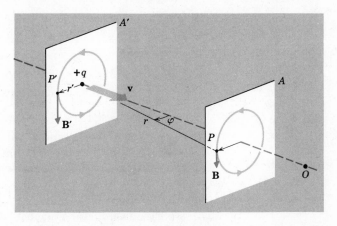

Fig. 9.22 *The magnetic field due to a moving charged particle is proportional to sin φ and to the inverse square of the distance from the particle. The field at the point O is zero.*

THE FIELD OF A CURRENT-CARRYING WIRE

In order to calculate the magnetic field at any point due to a steady current flowing in a long, straight wire, we could do so by computing the net force on a test charge resulting from the force exerted by each of the charges moving in the wire. This is a straightforward calculation but the necessary mathematics is above the level of this book and so we only state the result. At a distance r from a long, straight wire that carries a steady current I, the strength of the magnetic field is (see Fig. 9.23)

$$B = \frac{2I}{cr} \qquad \text{(long, straight wire)} \qquad (9.14)$$

Although the field of a single moving charge decreases as $1/r^2$ (Eq. 9.13), the field of a *line* of moving charge (that is, a current in a long, straight wire) decreases only as $1/r$.

According to our right-hand rule, the direction of the magnetic field vector at point P in Fig. 9.23 is *into* the plane of the page. The field lines are circular, of course, as shown in Figs. 9.10 and 9.11.

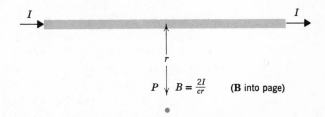

Fig. 9.23 *The magnetic field at a distance r from a long, straight wire carrying a current I is 2I/cr.*

$P \downarrow \quad B = \frac{2I}{cr} \qquad$ (**B** into page)

Example **9.2**

What is the magnetic field 10 cm from a long straight wire carrying a current of 10 amp?

$$10 \text{ amp} = (10 \text{ A}) \times (3 \times 10^9 \text{ statA/A}) = 3 \times 10^{10} \text{ statA}$$

$$B = \frac{2I}{cr}$$

$$= \frac{2 \times (3 \times 10^{10} \text{ statA})}{(3 \times 10^{10} \text{ cm/sec}) \times 10 \text{ cm}}$$

$$= 0.2 \text{ gauss}$$

For comparison, this is approximately equal to the magnitude of the Earth's magnetic field at the surface of the Earth and at middle latitudes.

MAGNETIC EFFECTS ARE IMPORTANT IN SPITE OF v/c

We have seen that the magnetic field of a moving charge is smaller than the electric field of the particle by a factor v/c. We have also seen (Section 9.1) that electron velocities in current-carrying wires are of the order of 10^{-2} cm/sec. Thus, the factor v/c is of the order of 10^{-12}. Why, then, are magnetic effects of any practical importance if the electric field so completely dominates the magnetic field? The reason lies in the fact that current-carrying wires contain equal numbers of positive and negative charges and are therefore electrically neutral and produce no electric fields. But within these wires the electrons are drifting (albeit with small velocities) whereas the positive charges are stationary. Thus, the electrostatic effects of the two types of charges are cancelled although the magnetic effects of the moving electrons are still important.

9.5   *Torque on a Current Loop—Magnetic Moments*

THE MAGNETIC FORCE ON A CURRENT ELEMENT

When electric charges move through a wire that is located in a magnetic field, a magnetic force is exerted on each of the charges. Because the moving charges constitute a current, it is usually more convenient to discuss the magnetic force on the wire in terms of the net current than to consider the charges individually. In order to derive an expression for the magnetic force on a current-carrying wire, we consider a *current element,* that is, a short length of wire in which flows a steady current. Of course, an isolated length of wire cannot carry a steady current (why?)—we must have a complete loop in order for the current to flow uniformly. But, keeping this point in mind, we can consider the effect of the magnetic field on a *segment* of a complete loop of wire. Such a current element, of length *l,* which carries a steady current *I,* is shown in Fig. 9.24; the segment makes an angle ϕ with the direction of the magnetic field **B.**

The current in the wire segment is the amount of charge passing a given point per unit time. The total amount of charge moving in the segment is Q. If each individual charge requires a time t to travel the length of the segment, then the velocity is $v = l/t$. Therefore, the current flowing in the segment is

$$I = \frac{Q}{t} = \frac{Q}{l} \times \frac{l}{t} = \frac{Q}{l} \times v$$

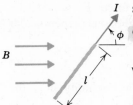

Fig. 9.24 *A current element is a short segment of wire (which must actually be a portion of a complete loop) that carries a steady current.*

so that

$$Qv = Il \qquad (9.15)$$

Now, the magnetic force on an individual charge q moving with a velocity **v** in a field **B** is (Eq. 9.8)

$$F_M = \frac{qv}{c} B \sin \phi$$

where the appropriate angle, as shown in Fig. 9.24, is ϕ. The magnetic force on the current element in which a total charge Q is moving is given by this expression for F_M with Qv from Eq. 9.14 substituted for qv. Thus,

$$F_M = \frac{1}{c} IlB \sin \phi \qquad \begin{array}{l}\text{(force on a current}\\ \text{element of length } l)\end{array} \qquad (9.16)$$

A CURRENT LOOP IN A UNIFORM MAGNETIC FIELD

We now consider the effect of a magnetic field on a wire loop that carries a steady current. As shown in Fig. 9.25a, the loop is rectangular with a width a and a height b; the perpendicular to the plane of the loop makes an angle θ with the direction of the magnetic field vector **B.** The loop carries a steady current I.

What forces are exerted on the loop by the magnetic field? Consider first the top and bottom segments of the loop. The magnetic forces **F**$_M$ on these segments clearly have equal magnitudes and opposite directions; the forces are *colinear* (that is, they are directed along the same straight line) and so exactly cancel.

The forces on the side elements of the loop are easier to visualize by referring to Fig. 9.25b, which is a top view of the situation shown in Fig. 9.25a. Again, it is clear that the forces on the two segments have equal magnitudes, but here the vectors are *not* colinear. The result is that there is a *torque* exerted by the magnetic field on the current loop. (See the discussion of torque in Section 5.10.)

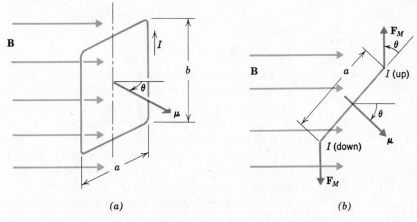

(a)   *(b)*

Fig. 9.25 *(a) A current loop in a uniform magnetic field. (b) The magnetic field exerts a torque on the loop in such a direction that the magnetic moment vector* **μ** *tends to align with the field vector* **B.**

The magnitude of the torque due to each of the forces $\mathbf{F}_M$ is given by the product of the distance to the pivot point ($\frac{1}{2}a$) and the perpendicular component of the force (compare Eq. 5.12):

$$T = (\tfrac{1}{2}a) \times F_M \sin \theta \tag{9.16}$$

The value of F_M is given by Eq. 9.15 where $\sin \phi$ is now unity since the side segments of the loop are perpendicular to the field:

$$F_M = \frac{1}{c} IbB \tag{9.17}$$

Substituting Eq. 9.17 into Eq. 9.16 and *doubling* the result since the force on each of the two side segments contributes equally to the torque, we have for the *total* torque,

$$T_{\text{total}} = \frac{1}{c} IabB \sin \theta \tag{9.18}$$

Since the product ab is just the *area A* of the loop, we can write the result as

$$T_{\text{total}} = \frac{1}{c} IAB \sin \theta \tag{9.19}$$

MAGNETIC MOMENTS

The current loop that we have been considering has only two properties of interest from the standpoint of magnetic effects: the area of the loop[2] and the current flowing in the loop. Since the product of these two quantities appears in Eq. 9.19, it proves convenient to define this product as a new quantity that completely specifies the properties of the loop. We call this quantity the *magnetic moment* of the loop and denote it by μ:

$$\mu = \frac{IA}{c} \tag{9.20}$$

where we include the constant c in the definition of μ in order to use CGS units in a consistent way throughout. Equation 9.19 can now be expressed as

$$T_{\text{total}} = \mu B \sin \theta \tag{9.21}$$

Notice that torque has the dimensions of (*force*) $\times$ (*distance*) = *work* or *energy*. Therefore, μB also has the dimensions of energy and so the unit of magnetic moment in the CGS system is the *erg/gauss*.

The magnetic moment of a current loop is actually a *vector*. The direction of μ is determined in exactly the same way as we determine the direction of the angular momentum vector, namely, μ has the direction of advance of a right-hand screw when turned in the same sense in which the (positive) current flows in the loop. Therefore, the magnetic moment vector for the current loop in Fig. 9.25 has the direction shown in Fig. 9.26.

---

[2]It is clear that any rectangular loop with an area A will experience the same torque in the field **B.** It is also true, although we shall not prove the result here, that the torque is the same for *any* loop with an area A, regardless of its shape.

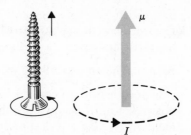

Fig. 9.26 *The direction of the magnetic moment vector μ is the direction of advance of a right-hand screw when turned in the same sense in which* I *flows.*

The torque on any magnetic system (current loop, bar magnet, etc.) when placed in a magnetic field can always be expressed in terms of the magnetic moment of that system.

ELEMENTARY MAGNETS

The smallest units of magnetism are found at the atomic level. In a highly simplified model of atomic structure, electrons are considered to move around the atomic nucleus in definite orbits. A single electron that executes a circular orbit around a stationary positively-charged nucleus is shown in Fig. 9.27. The motion of this electron is equivalent to a current loop (but with the current flowing in the direction opposite to the electron velocity). Therefore, a magnetic field is produced with the same configuration as that shown in Fig. 9.13. Furthermore, this orbiting electron possesses a definite magnetic moment.

Because of quantum effects (discussed in Chapter 13), the magnetic moment associated with the orbital motion of an electron in an atom cannot have any arbitrary value. Instead, the electron *orbital magnetic moment* can only take on values that are integer multiples (including zero) of a basic

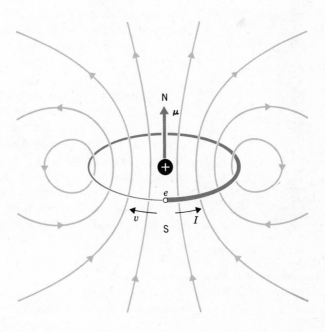

Fig. 9.27 *An electron moves around a stationary atomic nucleus and produces a magnetic field equivalent to that of a current loop. Notice that the direction of current flow is opposite to that of the electron's velocity because the electron carries a negative charge.*

unit called the *Bohr magneton* in honor of Neils Bohr (1885–1962), the Danish theoretical physicist whose early studies of atomic structure were the first steps toward a quantum mechanical description of matter.

$$\text{Bohr magneton: } \mu_0 = 9.27 \times 10^{-21} \text{ erg/gauss} \tag{9.22}$$

In Section 5.11 we pointed out that elementary particles such as electrons possess an intrinsic angular momentum. (We can picture this angular momentum as rising from a spinning of the particle, but actually no mechanical motion is involved.) Since a spinning charge is equivalent to a current, there is a magnetic field associated with electron spin. The magnetic moment of the electron due to this spin is also one Bohr magneton.[3]

ATOMIC MAGNETISM

Why are not all atoms elementary magnets with magnetic moments proportional to the number of orbiting electrons? The answer lies in the fact that the electrons that make up the outer portions of all atoms are ordered according to the rules of quantum mechanics. In most atoms these rules lead to the *pairing* of electron spins so that their magnetic effects cancel. The so-called *ferromagnetic* group of elements (iron, cobalt, and nickel) all have four unpaired electrons that give these atoms relatively large magnetic moments. When the fields of ferromagnetic atoms in solid matter are all predominantly oriented in the same direction, an appreciable net field results and a permanent, bar-type magnet can be formed.

9.6   Orbits of Charged Particles in Magnetic Fields

CIRCULAR ORBITS

Consider a uniform magnetic field **B** and a charged particle of mass m moving in the field with a velocity **v**, where $\mathbf{v}_\perp$ **B**, as in Fig. 9.28. The magnetic field exerts on the particle a force of constant magnitude:

$$F_M = \frac{qvB}{c}$$

The direction of this magnetic force is always perpendicular to the instantaneous direction of motion of the charged particle, that is, $\mathbf{F}_M \perp \mathbf{v}$. Hence, there is no component of the force in the direction of motion and *no work is done on the particle by the magnetic field*. Although the direction of motion is continually changing as a result of the magnetic force, the *speed* of the particle (that is v) is constant and the kinetic energy remains always the same.

The magnetic force produces a centripetal acceleration of constant magnitude that is always perpendicular to **v**:

$$a_c = \frac{F_M}{m} = \frac{qvB}{mc} \tag{9.23}$$

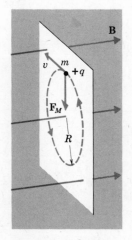

Fig. 9.28 *A charged particle moves in a circular orbit in a uniform magnetic field if* **v** ⊥ **B.**

[3] Actually, the value is slightly greater, 1.00115 μ_0, because of certain corrections that are required by the quantum nature of the electromagnetic field (see Section 13.4).

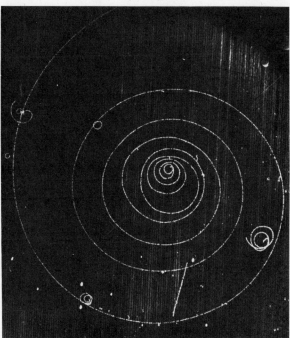

Fig. 9.29 *The path of a charged particle in a*
bubble chamber (*consisting of liquid hydrogen*) *is*
made visible by the many tiny bubbles that are
formed in the wake of the particle. This photograph
shows the orbit of a fast electron in a bubble chamber
which is in a strong magnetic field. The electron
loses energy through collisons with the hydrogen
atoms and so the radius of the orbit decreases,
causing the electron to move in a spiral path. The
tracks of some secondary electrons released in
encounters with hydrogen atoms can be seen near
the main track.

Lawrence Radiation Laboratories

Thus, the particle moves in a *circular* orbit. In terms of the velocity and
the orbit radius, the centripetal acceleration is given by (see Eq. 4.31)

$$a_c = \frac{v^2}{R} \tag{9.24}$$

Equating these two expressions for a_c and solving for R, we have

$$\boxed{R = \frac{mvc}{qB}} \tag{9.25}$$

Therefore, a charged particle moving at right angles with respect to a
uniform magnetic field executes a *circular* orbit in the field with a radius
that is directly proportional to its momentum (*mv*) and inversely proportional to the field strength.

It is important to realize that a charged particle executing an orbit in
a static magnetic field in vacuum will neither gain nor lose energy.[4]

Example **9.3**

What is the radius of the orbit of a 1-MeV proton in a 10^4-gauss field?
We have

$$m = 1.67 \times 10^{-24} \text{ g}$$

$$q = e = 4.8 \times 10^{-10} \text{ statC}$$

---

[4]In the event that the magnetic field changes with time, the energy of the particle will, in
general, be altered; see Section 9.8.

Also, from Example 7.12 we know that the velocity of a 1-MeV proton is $v = 1.38 \times 10^9$ cm/sec

Therefore,

$$R = \frac{mvc}{eB}$$

$$= \frac{(1.67 \times 10^{-24} \text{ g}) \times (1.38 \times 10^9 \text{ cm/sec}) \times (3 \times 10^{10} \text{ cm/sec})}{(4.8 \times 10^{-10} \text{ statC}) \times (10^4 \text{ gauss})}$$

$$= 14.4 \text{ cm}$$

In essentially every accelerator laboratory, magnets are used to steer the beam of charged particles to the various stations at which experiments are carried out. Figure 9.30 shows the "magnetic switch yard" at the National Bureau of Standards electron accelerator laboratory. The five large magnets (which are all of the "H" type shown in Figs. 9.14 and 9.15) are used to direct the electron beam into three separate experimental rooms.

THE CYCLOTRON

One type of device that is often used in the acceleration of charged particles (protons, deuterons, α particles, etc.) to high velocities is the *cyclotron*. The basic idea of cyclotron operation is to accelerate charged particles

National Bureau of Standards

Fig. 9.30 *The "magnet switch yard" at the National Bureau of Standards electron accelerator laboratory. The electron beam enters from the left and the five large "H" magnets are used to steer the beam to any of three experimental rooms located to the right of the area pictured.*

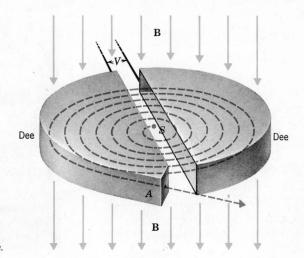

Fig. 9.31 *Schematic of a cyclotron.*

by means of electric fields while confining the particles with a magnetic field.

A schematic representation of a cyclotron is shown in Fig. 9.31. The essential elements of a cyclotron are a hollow, cylindrical cavity which is split along a diameter to form two "dees" (so-called because of their shape), an electromagnet (not shown in the figure) which produces a magnetic field perpendicular to the plane of the "dee" structure, and high voltage apparatus which produces a potential difference V between the "dees." Neutral gas atoms (for example, hydrogen) are ionized at the source S, located near the center of the "dees," to produce charged particles that are injected into the left-hand "dee." These particles move in a circular orbit under the influence of the field $\mathbf{B}$ until they emerge from the left-hand "dee." The particles are then accelerated across the "dee" gap by the electric field and are increased in energy by an amount qV. The electric field exists only *between* the "dees;" the interiors of the conducting "dees" have no electric field. Therefore, when the particles enter the right-hand "dee" they again move in a circular orbit but now of increased radius corresponding to their greater velocity. By the time the particles reach the "dee" gap again, the high-voltage apparatus has switched the polarity of the voltage on the "dees" so that the particles experience another accelerating voltage at the gap. This process is continued for many passages through the gap; each passage increases the energy by the amount qV and the particles spiral outward to greater and greater radii. Near the outer wall of the "dee" structure the particles pass into an *extractor* (usually a pair of plates across which is placed a high voltage) and emerge as a beam of high energy particles at A.

It is essential for the operation of a cyclotron that the voltage across the "dee" gap always be of the correct sign to accelerate the particles rather than to retard them. This is relatively easy to accomplish because of the important fact that a charged particle moving in a given uniform magnetic field requires a *fixed* time to execute an orbit *independent of its velocity*[5] (see Problem 9.20). Therefore, the particles require the same time to complete each half revolution in the "dees" and arrive at the gap *in phase* with

---

[5]This statement is only true for velocities sufficiently low that relativistic effects can be ignored.

Fig. 9.32 *One of E. O. Lawrence's early cyclotrons constructed at the University of California in the 1930s.*

those particles executing orbits with different radii, ready to accept the next accelerating voltage. For a given type of particle and for a given magnetic field there is a single frequency (called the *cyclotron frequency*) at which the polarity of the voltage must be alternated to provide continuing acceleration.

The first cyclotron (only 11 inches in diameter) was constructed by E. O. Lawrence and M. S. Livingston at the University of California in 1930. (Lawrence received the 1939 Nobel Prize in physics for this work.) One of Lawrence's early cyclotrons is shown in Fig. 9.32. The University of Maryland cyclotron, one of the most sophisticated in existence, is shown in Fig. 9.33. This accelerator, completed in 1969, is now being used in an extensive research program devoted to studies of nuclear structure and reactions.

Modifications of the basic cyclotron principle have been made to permit the acceleration of particles to ultra-high (relativistic) energies. These machines are known as *synchrocyclotrons* and *synchrotrons;* a view of the beam lines emerging from one such machine is shown in Fig. 4.38.

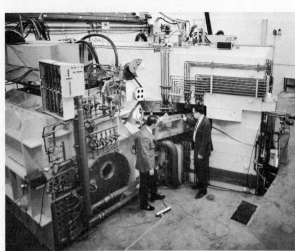

Fig. 9.33 *Photograph of the University of Maryland cyclotron which can produce protons with energies up to 140 MeV. The diameter of the largest orbit is 90 in.*

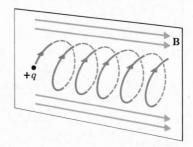

Fig. 9.34 *If the velocity vector is not perpendicular to* **B**, *the orbit is a helix.*

HELICAL ORBITS

If a charged particle moves in a magnetic field with its velocity vector at an angle other than 90° with respect to the field direction, the orbit will not be circular. Recall that only that component of the velocity that is *perpendicular* to **B** contributes to the magnetic force (Eqs. 9.9 and 9.10); the parallel component is unaffected by the field. Therefore, if a particle moves in a field with velocity components both parallel and perpendicular to the field, the total motion will be a combination of circular motion (the field action on $v_\perp$) and a steady drift along the field direction ($v_\parallel$ unaffected by the field). The combination of circular motion and a steady drift produces a *helical orbit,* the axis of which coincides with the direction of the magnetic field, as shown in Fig. 9.34.

TRAPPED ORBITS

The radius of the orbit of a charged particle in a magnetic field is inversely proportional to the strength of the field; if B increases, then, for a given velocity, the orbit radius of the particle must decrease. Figure 9.35 shows a magnetic field which increases in strength from left to right. (The bunching together of the field lines indicates a high field strength.) If a charged particle, for example, an electron, enters this field with its velocity vector at an angle of less than 90° with respect to the field lines, the electron will begin to move in a helical orbit. In the first stages of the motion shown in Fig. 9.35, the electron moves in a uniform field that is directed toward the right, and the motion is in the form of a simple helical orbit. But where the field intensity increases (that is, where the field lines bunch together),

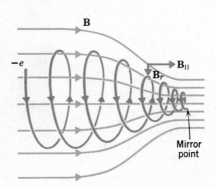

Fig. 9.35 *An electron "spirals down" the lines of increasing magnetic field strength and is reflected back at the mirror point.*

B

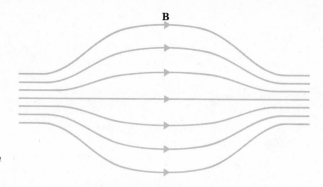

Fig. 9.36 *A magnetic bottle in which charged particles can be confined by reflections back and forth between the mirror points in the regions of high magnetic field strength.*

the electron begins to "spiral down" the field lines. In this region the field vector has a component toward the right ($\mathbf{B}_\parallel$) and also a component directed toward the center of the bundle of field lines ($\mathbf{B}_r$, the *radial* component). If we calculate the force on the electron due to $\mathbf{B}_r$, using the right-hand rule, we find that this force is always toward the *left*, regardless of the position of the electron in its helical orbit. That is, the bunching together of the field lines produces a force on the electron that retards its motion into the high-field region. At a certain point (called the *mirror point*), the motion of the electron into the high-field region will be completely stopped and the electron will be reflected back into the region of weaker magnetic field. The action at the mirror point is similar to the elastic collision of a ball at a wall—the electron "strikes" the magnetic wall at the mirror point and is reflected.

By increasing the magnetic field strength (that is, constricting the field lines) at two different points, as in Fig. 9.36, a *magnetic bottle* is produced in which charged particles can be trapped by successive reflections from the high-field regions. Such devices show great promise of being able to confine high energy particles (such as deuterons) for times that are sufficient[6] to allow useful power to be extracted from the nuclear reactions that take place when the particles collide (see Section 15.5). Perhaps *fusion reactors*

[6]Perfect magnetic bottles cannot be constructed and all are "leaky" to some degree.

Fig. 9.37 *The "Stellarator," a device for confining charged particles with magnetic fields, in a laboratory at Princeton. Studies with such devices may lead to economical fusion power plants.*

Princeton University

of this general type will be in use before the end of the century. Experiments toward this end are now being pursued with a variety of instruments that confine charged particles by means of magnetic fields. One such device, the "Stellarator" at Princeton, is shown in Fig. 9.37.

9.7 *Charged Particles in the Earth's Magnetic Field*

COSMIC RAYS

The Earth is under continual bombardment by energetic charged particles from outer space. Some of these particles originate outside the solar system and consist primarily of protons (about 85 percent) and α particles (about 14 percent) with the remainder being heavier nuclei. The term *galactic cosmic rays* is applied to these particles to emphasize the fact that they are not of local origin.[7] In addition there are *solar cosmic rays* that are also primarily protons and that are continually ejected from the Sun but most copiously during periods when violent disturbances (magnetic storms) take place on the Sun's surface.

Solar cosmic rays have energies that extend up to a few GeV (10^9 eV), but galactic cosmic rays with energies up to 10^{19} eV have been observed. (This latter energy is approximately 1 joule!) When these cosmic ray particles approach the Earth, they are affected by the Earth's magnetic field. If the kinetic energy of a particle is less than a certain minimum value, the particle will be deflected by the field and will never reach the surface of the Earth. For protons entering the Earth's magnetic field along a path that lies in the *geomagnetic equatorial plane,* the cut-off energy is approximately 15 GeV. That is, protons with energies less than 15 GeV will be deflected back into space while those with energies greater than 15 GeV can penetrate to the Earth's surface. At higher geomagnetic latitudes, the cut-off energy is smaller and at the magnetic poles even particles with low energies can spiral down the field lines and reach the Earth's surface. Figure 9.38 shows the paths of several charged particles in the Earth's magnetic field.

AURORA

Another interesting phenomenon that is associated with charged particles in the Earth's magnetic field is the occurrence of the *northern lights* or *aurora.* Auroral displays (see Fig. 9.39) occur most frequently at high northern latitudes but occasionally they can be observed as far south as the middle of the United States. Auroral light is generated by solar protons that penetrate the Earth's field down to altitudes of about 100 km. At these altitudes the atmosphere is quite tenuous but there are present sufficient numbers of oxygen and nitrogen atoms that collisions between the protons and the atoms produce enough light to be prominently visible. Although auroral activity takes place continuously, the amount of light generated is usually insufficient to be noticeable. On those occasions when there are pronounced solar disturbances in the form of Sun spots and magnetic storms, the number of protons injected into the upper atmosphere is greatly increased and spectacular auroral displays occur.

---

[7] Most of these particles probably originate within our own Galaxy.

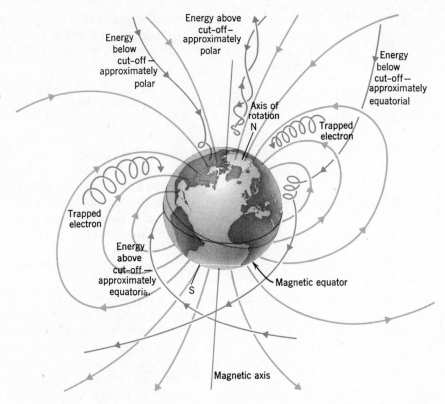

Fig. 9.38 *The motion of charged particles in the Earth's magnetic field* (*the magnetosphere*). (*Adapted from Alonso and Finn.*)

THE EARTH'S RADIATION BELTS

During the flight of the satellite Explorer I in 1958, evidence was obtained by James A. Van Allen and his co-workers that there are regions in space near the Earth that contain large numbers of charged particles. These regions are known as the *Van Allen radiation belts* and are regions in which charged particles are trapped in the Earth's magnetic field. Figure 9.40 shows an early map of the radiation belts constructed from satellite data. The various curves represent the surfaces on which the electron density is the same; the numbers are the counts per second that were registered by Van Allen's geiger counters. There are two regions (indicated by the shaded areas) in which the electron density is quite high and these are known as the *inner* and *outer* radiation belts.

It is now known that there are protons as well as electrons trapped in the Earth's field. Just as for the electrons, there is a variation in the proton density with position. The distribution of low energy protons (energies up to about 10 MeV) are shown in Fig. 9.41. The three belts that are shown are actually not well separated; instead, there is a continuous distribution of protons. The particles with the lowest energies are concentrated in the outer belt and the density distribution of the highest energy protons is a maximum much closer to the Earth.

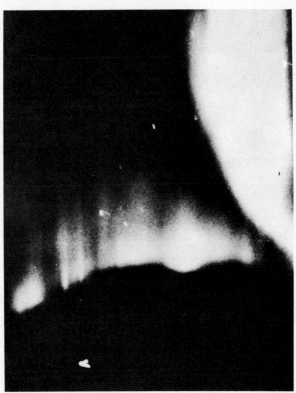

NASA

Fig. 9.39 *Photograph of an aurora taken in northern Canada from a NASA aircraft flying above the cloud cover. The V-shaped spot on the lower portion of the photograph is Jupiter—the strange shape is caused by the motion of the camera during the 20-sec time exposure.*

One of the mechanisms that seems to be responsible for at least a portion of the electrons and protons that are trapped in the Earth's magnetic field is the following. High energy protons that are ejected from the Sun are incident on the upper atmosphere. Collisions between the protons and the nuclei of gas atoms produce nuclear reactions in which neutrons are emitted. Neutrons carry no electric charge, and thus they are unaffected by the Earth's magnetic field. Therefore, neutrons leave the sites of the nuclear reactions and travel out of the atmosphere in straight paths to regions of the magnetosphere where the field strength is smaller (the regions of the radiation belts). The neutron is an unstable particle and it decays within an average time of 12.8 min into a proton, an electron, and a neutrino. The electrons and protons from neutron decays can be trapped in the magnetic field and contribute to the radiation belts. (The neutrinos are uncharged and thus escape.) This process is shown schematically in Fig. 9.42.

Although the neutron decay mechanism is clearly responsible for a portion of the electrons and protons in the inner radiation belt, the particles in the outer belt probably are injected directly from solar radiations. A great deal of information regarding the radiation belts has been obtained in recent years from satellite and rocket experiments, but, in fact, we do not yet have any really satisfactory theory that describes in detail the data regarding charged particles in the magnetosphere.

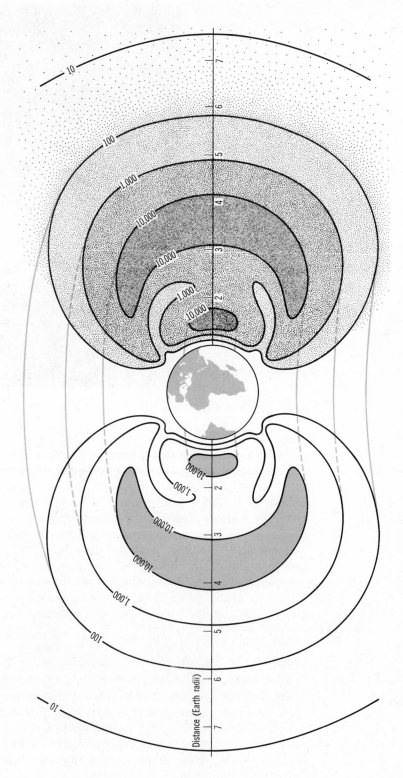

Fig. 9.40 *Map of the Van Allen radiation belts constructed from satellite data. The two shaded areas are the inner and outer radiation belts in which the electron density is high.*

Distance (Earth radii)

Fig. 9.41 *Three low-energy proton belts. The belts actually overlap and there are no gaps. The belts shown here indicate the regions in which protons of various energies have the maximum density. The lowest energy protons are concentrated in the outer belt and the highest energy particles are predominately in the inner belt.*

THE EFFECT OF THE SOLAR WIND ON THE MAGNETOSPHERE

The Earth's magnetic field is not a static affair nor does the simple bar-magnet type field (Fig. 9.7) extend great distances into space. The magnetosphere is profoundly influenced by the so-called *solar-wind,* a continuous stream of protons and electrons that are ejected by the Sun and, at the position of the Earth, have velocities of about 400 km/sec. (For protons, this velocity corresponds to an energy of 170 keV.) The density in space (near the Earth) of the solar wind particles is in the range from 10–100 per cm^3; the value varies with time and is a function of the surface activity of the Sun. The magnetic fields produced by these moving charged particles must be vectorially added to the Earth's magnetic field in order to obtain the net field in space. The result is that the Earth's magnetosphere is

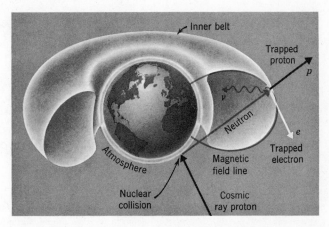

Fig. 9.42 *Trapping of electrons and protons from neutron decay. The neutrons are produced when cosmic ray protons interact with the nuclei of gas atoms in the upper atmosphere.*

cancelled in certain regions (or "blown back") by the solar wind. The distortion of the Earth's field lines is so great in the direction toward the Sun that the magnetosphere ends abruptly at a distance of 8–10 Earth radii; in the direction away from the Sun the field lines trail off into space. A photograph of a laboratory simulation of the "blowing back" of the magnetosphere by the solar wind is shown in the illustration at the beginning of this chapter.

9.8   Fields that Vary with Time

INDUCED CURRENTS DUE TO THE RELATIVE MOTION OF WIRES AND FIELDS

Thus far we have considered only the effects that are produced when moving charges and currents interact with *static* magnetic fields. We now turn to a discussion of *time-dependent* phenomena, including cases in which there is relative motion between current-carrying wires and magnetic fields or in which a magnetic field varies with time.

For the same reason that a moving charge experiences a force in a magnetic field, a current-carrying wire (equivalent to a line of moving charges) that passes through a magnetic field, as in Fig. 9.43, will also be acted on by a magnetic force. If we disconnect the external source (for example, a battery) that supplies current to the wire, then, of course, the magnetic force will disappear. Now, suppose that we move the current-free wire through the magnetic field in the manner illustrated in Fig. 9.44. This motion will produce a magnetic force on the charges in the wire and they will begin to move along the wire. As long as the wire is in motion in the

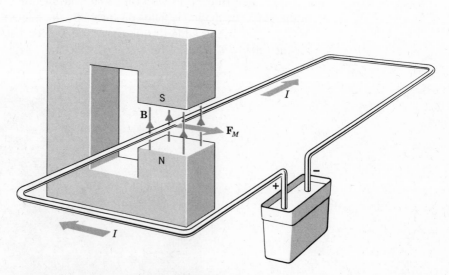

Fig. 9.43 *A static magnetic field exerts a force on a current-carrying wire. As long as the wire is in motion in the field, these charges will continue to flow; that is, a current has been* induced *in the wire. The phenomenon of* induction *was discovered in 1831 by the great English physicist, Michael Faraday (1791–1867).*

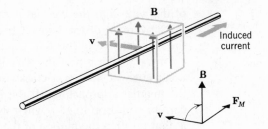

Fig. 9.44 *The motion of a wire through a magnetic field induces a current to flow in the wire. The section of wire shown is only a portion of the complete loop of wire that is necessary in order for a steady current to flow; compare Fig. 9.43.*

field, these charges will continue to flow; that is, a current has been *induced* in the wire. The phenomenon of *induction* was discovered in 1831 by the English physicist, Michael Faraday (1791–1867).

In order for a current to be induced in a wire it is not crucial that the field be stationary while the wire moves. Clearly, it is only the *relative* motion of wire and field that is important. Therefore, the situation shown in Fig. 9.45, in which a magnetic field moves to the *right* across a stationary wire is entirely equivalent to the case in which the wire moves to the *left* through a stationary field, as in Fig. 9.44. In both cases the induced current flows in the same direction.

Another situation in which a current is induced by the relative motion between a magnetic field and a wire is shown in Fig. 9.46. Here a bar magnet is thrust into a wire loop that is connected to a current-measuring device (an *ammeter*). Notice that the velocity vector, which describes the motion

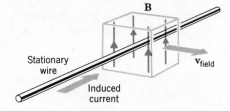

Fig. 9.45 *Current is induced by the relative motion of wire and field. This situation is entirely equivalent to that shown in Fig. 9.44. Only a section of the complete loop of wire is shown; compare Figs. 9.43 and 9.44.*

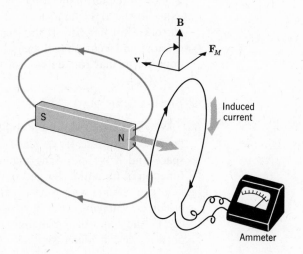

Fig. 9.46 *The relative motion between the field of the bar magnet and the wire loop induces a current in the wire which is detected by the ammeter.*

of the wire (and the charges it carries), has a direction opposite to that of the movement of the magnet because this vector describes the velocity of the *wire* relative to the *field.* Application of the right-hand rule for F_M (for example, at the top of the wire loop, as indicated in Fig. 9.46) shows that a current will flow in a clockwise sense when viewed with the N pole of the magnet approaching. If the magnet continues its motion, the induced current will drop to zero when the magnet is centered in the loop because at that position the motion of the wire (and the charges it carries) is directly *along* the field line; when v is parallel to **B,** the magnetic force vanishes. As the S pole of the magnet passes through the loop, an induced current will flow again, but now the direction of flow will be opposite to that shown in Fig. 9.46. (Why?)

CURRENTS INDUCED BY TIME-VARYING FIELDS

Consider a loop of wire that is connected to a battery through a switch. If the switch is open, no current flows and there is no magnetic field. Closing the switch causes the current to flow. But the current does not instantaneously attain its final steady value. Instead, the current is zero at the exact instant that the switch is closed and builds up to its final value during a certain short interval of time. Similarly, the magnetic field that is due to the flow of current starts at zero when the switch is closed and builds up to its final value just as does the current. Therefore, at any particular position in the vicinity of the wire, the magnetic field increases with time during the interval required for the current to attain its final steady value.

We can describe this situation in a pictorial way by saying that the circular magnetic field lines originate at the wire (beginning at the instant that the switch is closed) and spread out into space until the final steady field configuration is attained. The "movement" of the field lines in this outward expansion is similar in its effect to the physical movement of a magnet. Therefore, a current will be induced in a wire that lies in the path of the "moving" field lines. Figure 9.47 shows such a situation; the field that expands from the right-hand loop when the switch is closed induces a current to flow in the left-hand loop. As soon as the current reaches its final steady value in the right-hand loop, the field ceases to expand and the induced current drops to zero. If the switch is now opened, the magnetic field will collapse and a current will be induced in the left-hand loop but in a direction opposite to that for the case of the expanding field.

A CHANGING MAGNETIC FIELD PRODUCES AN ELECTRIC FIELD

We can summarize in the following way all of the results discussed thus far in this section. If we place a charged particle at a certain position in space and if we allow the magnetic field at that position to change as a function of time, the charged particle will experience a force. The magnetic field can change with time because of the motion of a permanent magnet or current-carrying wire that gives rise to the field or the change with time can be the result of the changing current in a wire. If, instead of a charged particle, we consider a conductor (such as a wire) that contains free electrons,

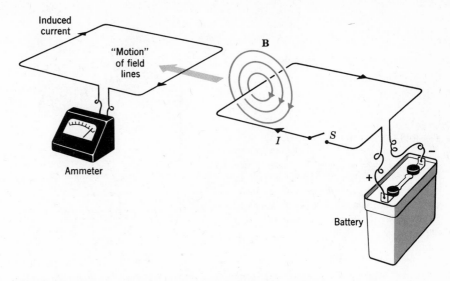

Fig. 9.47 *When the switch S is closed, a current begins to flow in the right-hand loop of wire. A magnetic field expands from this wire as the current builds up; the "moving" field lines induce a current in the left-hand loop.*

the force exerted on these charges by the changing field will induce a current to flow.

From the expression for the Lorentz force on a charged particle (Eqs. 9.11 and 9.12), we know that when the particle is at rest ($v = 0$), then $F_M = 0$, but $F_E \neq 0$. That is, a charged particle at rest in a static magnetic field will remain at rest; only an electric field can exert a force on a *stationary* charge. This must mean that *a changing magnetic field produces an electric field* and it is the electric field that causes the charge to move. Symbolically, we can express this statement as

$$\frac{\Delta \mathbf{B}}{\Delta t} \propto \mathbf{E} \tag{9.26}$$

A CHANGING ELECTRIC FIELD PRODUCES A MAGNETIC FIELD

We know that a magnetic field is produced in the vicinity of moving charged particles. But is there any other way of producing a magnetic field? Just as a changing magnetic field produces an electric field, it is also true that *a changing electric field produces a magnetic field*.

Consider the situation illustrated in Fig. 9.48. Initially, the two plates carry charges of $+q$ and $-q$. With the switch in the circuit *open*, the equal and opposite charges on the two plates produce an electric field in the region between them. No current flows in the wire, so there is no magnetic field at any point around the circuit. When the switch is closed, a current begins to flow in the wire and a magnetic field is produced around the wire in the manner that we have previously discussed. But in the region between the plates, there is no flow of ordinary current—there is only a *change* in the *electric field.* In spite of the fact that there is no movement of electric charge in this region, nevertheless there is a magnetic field produced as

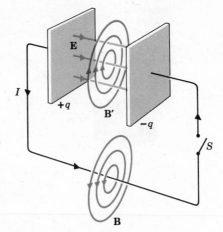

Fig. 9.48 *When the switch* S *is closed, a current flows in the wire from the positively charged plate to the negatively charged plate. This current produces the magnetic field* **B**; *also, the changing electric field between the plates produces the magnetic field* **B′**. *Notice that the field lines of* **B** *and* **B′** *have the same direction relative to the direction of current flow.*

a result of the changing electric field. Since we find experimentally that magnetic fields are produced as the result of current flow as well as changing electric fields, we can summarize the situation symbolically as

$$\frac{\Delta \mathbf{E}}{\Delta t} + \text{current} \propto \mathbf{B}$$ (9.27)

The strengths of the magnetic fields that are due to changing electric fields are, in general, considerably smaller than those due to current flow. Nonetheless, these fields can be detected in favorable situations and the detailed equation relating $\Delta \mathbf{E}/\Delta t$ and **B** can be completely verified.

We have previously stated that magnetic fields have no existence apart from moving electric charges. The fact that magnetic fields can be produced in regions of space where **E** is changing, *even though there are no charged particles to move,* does not invalidate this statement because the electric field cannot change unless there is a flow of current in another part of the circuit. Therefore, the origin of the magnetic field lies ultimately in the motion of charge.

LENZ' LAW

Because of the magnetic force on the current-carrying wire in Fig. 9.43, the wire will begin to move. As soon as the wire has a velocity relative to the field, there will be a new magnetic force, $\mathbf{F}'_M$, on the charged particles due to their motion in the direction of the original $\mathbf{F}_M$. Application of the right-hand rule shows that $\mathbf{F}'_M$ is in the direction *opposite* to the direction of current flow. That is, there is an induced current that tends to oppose the original current flow. This is a general result—if any electromagnetic change A causes an effect B, then B will always induce a reaction C that tends to oppose A. This principle was discovered by Heinrich Lenz (1804–1865), a German physicist, and is known as Lenz' law. This law can always be used to predict the direction of current flow induced in a circuit due to external changes.

Although Lenz' law is reminiscent of Newton's third law, it is, in fact, a completely distinct statement. Actually, Lenz' law is simply a statement of energy conservation applied to induced currents. (If induced currents *aided* one another, in opposition to Lenz' law, then the currents would grow without the benefit of any energy input to the system. This is a clear violation of energy conservation.)

MAXWELL'S EQUATIONS

Just over a hundred years ago, in 1865, the great Scottish theoretical physicist James Clerk Maxwell (1831–1879) showed that it was possible to base a complete description of electromagnetic field phenomena on a set of only four equations. Much as Newton had drawn on the previous work of others in devising his famous equations of dynamics, Maxwell leaned heavily on the formulations of electric and magnetic phenomena that had been made by others, especially Michael Faraday, in a long series of experimental and theoretical investigations. Thus, Maxwell did not *invent* the equations that now bear his name (in fact, he was responsible for setting down for the first time only *one* of these equations); his important contribution to the subject was to show in a definitive way that these equations form the basis of interpreting all manner of electromagnetic effects (including electromagnetic *waves,* which we shall treat in the next chapter).

Maxwell's four equations can be summarized in words in the following way:

1. Coulomb's law; the electric field of a point charge.
2. Magnetic field lines are continuous and have neither beginning nor end; there are no magnetic monopoles.
3. A changing magnetic field produces an electric field (electromagnetic induction).
4. A magnetic field can be produced by current flow as well as by a changing electric field.

Bettmann Archive   New York Public Library

Fig. 9.49 *The giants of electromagnetism—Michael Faraday (left) and James Clerk Maxwell. Faraday conducted thousands of experiments on all manner of electrical and magnetic phenomena; he discovered electromagnetic induction and was responsible for the concept of field lines. Maxwell incorporated Faraday's results into his all-encompassing theory of electromagnetic fields.*

Summary of Important Ideas

The net motion of electric charge constitutes a *current*.

Like magnetic poles *repel;* unlike magnetic poles *attract*.

A freely-suspended magnet aligns itself from S pole to N pole *along* the magnetic field lines.

The magnetic field lines in the space surrounding a magnet have the direction from the N pole to the S pole.

The S pole of the Earth is near the *north* geographic pole.

The poles of a magnet have no independent existence; N and S poles *always* occur together.

The direction of the magnetic field lines surrounding a current-carrying wire is determined by the *right-hand rule*.

Magnetic field lines are *continuous;* they have no beginning and no end. This is equivalent to the statement that *magnetic monopoles do not exist*.

Magnetic fields have no existence independent of electric charges.

The magnetic force $\mathbf{F}_M$ on a moving positively-charged particle is *perpendicular* to the plane defined by $\mathbf{v}$ and $\mathbf{B}$. The direction of $\mathbf{F}_M$ is the same as the direction of advance of a right-hand screw when turned in the sense that rotates $\mathbf{v}$ toward $\mathbf{B}$.

Any magnetic system that possesses a magnetic moment will experience a *torque* (but no net force) when placed in a uniform magnetic field.

The magnetic force on a charged particle *does no work* (unless the magnetic field changes with time).

The magnetic force involves only the component of a particle's velocity that is *perpendicular* to $\mathbf{B}$.

Charged particles can be *trapped* in nonuniform magnetic fields.

Elementary atomic magnets are produced by the fields of *circulating electrons*.

Elementary charged particles are themselves tiny magnets by virtue of the fields produced by *spinning charges*.

A changing magnetic field produces an *electric* field; a changing electric field produces a *magnetic* field.

Maxwell's four equations are a complete description of the electromagnetic field.

Questions

9.1 Explain carefully how a *steady* current can flow in a wire when the electron drift velocity is only 0.1 mm/sec and the thermal speeds of the electrons are of the order of 10^8 cm/sec.

9.2 Electrical current can be conducted through certain liquids (called *electrolytes*) by the movement of *ions* rather than electrons. Even water is normally ionized to a small degree into H^+ and OH^- ions. Sodium sulfate, Na_2SO_4, forms two Na^+ ions and an SO_4^{--} ion. Explain the qualitative differences between the conduction of electrical current in an electrolyte and the conduction in a copper wire.

9.3 If you were given two iron bars of identical appearance, one of which is magnetized and one of which is not, how would you decide which is magnetized without using any additional bars or magnets? Would it be possible to make the determination without taking advantage of the Earth's magnetic field?

9.4 An electron is projected into a current loop exactly along the axis. Describe the motion of the electron. What difference will there be if the electron's velocity vector is at a slight angle with respect to the axis of the loop?

9.5 It has been proposed that the Earth's magnetic field is due to a ring of electron current that flows in the molten metallic interior of the Earth. In what direction would the electrons have to flow in order to give the correct polarity for the Earth's field?

9.6 Two identical cardboard tubes are wound with wire in exactly the same way. The tubes are placed end-to-end and equal currents are passed through the wires. The currents circulate about the tubes in the same way. Will there be attraction or repulsion between the tubes?

9.7 A current-carrying wire lies in a north-south direction. A compass is placed immediately above the wire and the N pole points eastward. In what direction are the electrons in the wire moving?

9.8 An electron moves in an eastward direction near the equator. In what direction does the Earth's magnetic field exert a force on the electron?

9.9 A wire lies in a north-south direction and a current flows north in the wire. A positively-charged particle moves in the vicinity of the wire. In what direction will $\mathbf{F}_M$ act if (a) the particle is over the wire and moves north, (b) the particle is east of the wire and moves toward the wire, and (c) the particle is west of the wire and moves away from the wire?

9.10 Describe what will happen to the current loop shown in Fig. 9.25 if the direction of current flow is reversed.

9.11 Use the fact that a bar magnet is composed of atoms that have magnetic moments and give an atomic explanation of why such a magnet tends to align itself with a magnetic field.

9.12 If you wished to produce a proton beam with an energy of 100 MeV, why would a cyclotron be a suitable accelerator but a Van de Graaff generator would not?

9.13 Suppose that a number of electrons were released, with various velocities, at a high altitude above the Earth. Describe qualitatively what would happen to those electrons that move initially in directions more-or-less along the magnetic field lines. (Such experiments have been performed by detonating nuclear devices carried aloft by rockets.)

9.14 One end of a bar magnet is thrust into a wire loop. The induced current in the wire flows in the clockwise direction as viewed by looking along the direction of motion in the magnet. Which pole of the magnet was thrust into the loop? In what direction will the current flow if the magnet is *withdrawn* from the loop?

9.15 A steady current I flows in the wire loop (1) in the direction shown. If loop (1) is moved toward loop (2), in what direction will the induced current flow in loop (2)?

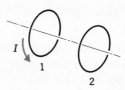

9.16 A *short length* of wire is originally at rest in a static magnetic field. A mechanical force is suddenly applied to the wire, causing it to move through the field at a constant velocity. Explain why current will flow in the wire for a short time but will then stop even though the wire continues to move.

9.17 A bar magnet is dropped through a horizontal loop of wire with the N pole entering first. Describe the induction of current in the wire. Use Lenz' law to determine whether the magnet will experience an acceleration greater than or less than g while passing through the loop.

Problems

9.1 A certain copper wire has a diameter of 2 mm. When a current of 10 amps is flowing in this wire, how long will it take an electron, on the average, to drift a distance of 1 m?

9.2 A copper wire has a cross sectional area of 0.5 mm^2 and is formed into a circular loop with a radius of 60 cm. When a certain current flows in the wire it requires 10 hours, on the average, for an electron to flow completely around the loop. What is the value of the current?

9.3 Sketch the magnetic field lines for the two pairs of bar magnets shown in the diagram.

9.4 Two long wires lie parallel and carry equal currents in opposite directions. Sketch the lines of **B** in a plane that is perpendicular to the wires. Will the wires be mutually repelled or attracted?

9.5 An electric heater requires 15 amps at 110 volts for its operation. If the price of electrical power is 2 cents per kilowatt-hour, what is the daily cost of operating the heater?

9.6 What is the maximum force that a 10^4-gauss magnetic field can exert on an electron whose energy is 10 keV? What is the minimum force and under what conditions would it be attained?

9.7 At a distance of 1 m from a current-carrying wire the magnetic field is 0.5 gauss. How many electrons are passing through a cross section of the wire each second?

9.8 An electron is at the position shown in the figure and is moving with a velocity of 8×10^9 cm/sec. What is the magnitude and direction of the magnetic field at each of the points A, B, and C?

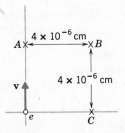

9.9 There are sharp boundaries between a field-free region of space and a region containing a uniform magnetic field with the dimensions shown in the diagram. A charged particle enters the field region (from the field-free region) and moves perpendicular to the field lines. Describe the subsequent motion of the particle for the cases in which the orbit radius, R, has the values: (a) $R < \frac{1}{2}l$, (b) $\frac{1}{2}l < R < l$, and (c) $R > L$.

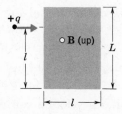

9.10 A certain wire carries a current of 100 A. A 1-MeV proton moves parallel to the wire at a distance of 5×10^{-7} cm. What is the magnetic force on the proton? If the proton moves in the same direction as the conduction electrons in the wire, in what direction is the force?

9.11 An electron is accelerated, starting from rest, by falling through a potential difference of 1000 volts. The electron then enters a magnetic field and is found to execute a circular orbit with a radius of 20 cm. What is the strength of the magnetic field?

9.12* A proton moves in a helical path in a uniform magnetic field of 1000 gauss, as shown in Fig. 9.34. It requires 0.3 μsec for the proton to drift a distance of 30 cm along the field direction. The radius of the helix is 10 cm. What is the *speed* of the proton?

9.13 Two particles have the same momentum but one particle carries twice the charge of the other. What will be the ratio of their orbit radii in the same uniform magnetic field?

9.14 A singly-charged carbon ion (C^{12+}) is found to have the same orbit as a 2-MeV proton in a certain magnetic field. What is the energy of the carbon ion?

9.15* A 1-MeV proton is moving horizontally in an eastward direction near the Earth's equator where the magnetic field strength is 0.3 gauss. By how much will the proton be deflected from its original line of motion after traveling 10 m? In what direction is the deflection?

9.16 A proton and a deuteron fall through the same potential difference and enter a uniform magnetic field. What is the ratio of the radii of the orbits?

9.17 The *magnetic rigidity* of a charged particle is BR where R is the radius of the orbit that the particle would follow in a magnetic field of strength B. Compare the magnetic rigidity of a 5-MeV proton and a 5-MeV α particle (He^{4++}).

9.18 A 1-g mass is moving eastward at the equator (where $B = 0.3$ gauss) with a velocity $v = 0.1c$. What charge must the mass have if the upward magnetic force cancels the downward gravitational force?

9.19 A cosmic ray proton with an energy of 10^{18} eV behaves as if its mass were approximately 10^9 times its mass when at rest (because of the relativistic increase of mass with velocity). The velocity of such a proton is essentially the velocity of light. Calculate the radius of the orbit that a 10^{18}-eV proton would execute in galactic space where the average magnetic field is about 3×10^{-6} gauss. Compare the result with the size of the local Galaxy.

9.20 A particle of mass m and charge q moves in a circular orbit in a magnetic field of strength B. Show that the time required to complete an orbit does not depend on the velocity of the particle.

9.21 It is desired that 40-MeV α particles in a cyclotron should execute an orbit with a radius of 1 m. What magnetic field strength is required?

9.22* A certain cyclotron can produce 10-MeV protons. If the field strength is 10^4 gauss, how much time is required for each revolution? If the accelerating voltage across the "dee" gap is 40 kV, how many revolutions are necessary to accelerate protons to the final energy? How long does the acceleration process take?

9.23 In a certain region of space there are uniform electric and magnetic fields. The **E** and **B** field vectors are at right angles with respect to one another. A charged particle with a velocity **v** enters this field in a direction perpendicular to both **E** and **B**. Show that the particle will proceed through the field region *undeflected* if $v = cE/B$. (Such a device is called a *crossed-field analyzer* and is useful for selecting particles with a given velocity.) If $E = 10^4$ volts/cm, what value of B is required to allow 1-MeV protons to pass undeflected?

9.24 A current of 1 amp is flowing in a circular ring whose radius is 1 cm. What is the magnetic moment of the loop? Express the result in Bohr magnetons.

9.25 A 1-MeV proton is moving in a circular orbit with a radius of 1 meter. What is the magnetic moment associated with the motion?

9.26 Calculate the orbital magnetic moment of an electron in a hydrogen atom. (Refer to Example 6.6 for the details of the orbit.) Express the result in Bohr magnetons.

9.27* For certain theoretical reasons it has been proposed that magnetic monopoles might exist. (A magnetic monopole would be the magnetic analog of the fundamental electric monopole, the electron.) No experiment has yet detected such a magnetic monopole, but if it existed, the magnetic monopole should have a strength f that is approximately 70 times the electron charge: $f \cong 70e$. A magnetic monopole placed in a magnetic field of strength B would experience a force $\mathbf{F}_M = f\mathbf{B}$ (just as a charge q in an electric field of strength E experiences a force $\mathbf{F}_E = q\mathbf{E}$). If a magnetic monopole were to be accelerated by the force $\mathbf{F}_M$ for a distance of 10 cm in a field of 10^3 gauss, what would be its energy in MeV? What would be the advantages of producing high energy particles by accelerating magnetic monopoles instead of accelerating electric monopoles?

"The Florescent Sea"

10 *Oscillations, Waves, and Radiation*

Oscillatory phenomena play an extremely important role in all realms of science as well as in our everyday lives. A pendulum bob oscillates to and fro on its suspension string; the piston in an engine oscillates up and down as it drives an automobile along; and we ourselves (in a manner of speaking) "oscillate" back and forth from our homes to our jobs. In fact, many of the natural processes in the Universe occur in repetitive cycles and therefore bear a strong resemblance (at least mathematically) to the simple oscillatory motion of a vibrating pendulum.

Wave motion is also an oscillatory phenomenon. As a water wave travels across the sea, water molecules vibrate up and down; as a sound wave travels through the air, air molecules vibrate back and forth. In each case, the particles of the medium move cyclically in a small region while the *wave* moves forward. Electromagnetic radiation (radio waves, light, X rays, etc.) is oscillatory in character but in this case particle motion is not required for the wave to progress—it is the *electromagnetic field* that oscillates.

In this chapter we begin by treating in detail one of the simplest types of oscillatory motion—the vibration of a small mass attached to a spring. We shall then use the results of this discussion in analyzing wave and radiation phenomena. As we progress to the later chapters we shall see the great importance of wave phenomena in the areas of modern and contemporary physics. We shall even see that effects displayed by *material particles* can be described in terms of *waves*.

10.1 *Simple Harmonic Motion*

THE EQUATION OF MOTION

The oscillatory motion of a mass attached to a coiled spring is the prototype of a large and important class of oscillatory phenomena called *simple harmonic motion*. Figure 10.1a shows such a mass at rest in its equilibrium position on a frictionless surface. If we apply an external force to displace the mass to the right, and then remove the external force, there will be a restoring force **F** exerted on the mass by the spring and directed to the left (Fig. 10.1b). In Example 7.2 we found that such an elastic force is directly

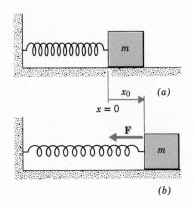

Fig. 10.1 (a) *A mass* m *rests on a frictionless surface and is attached to a spring; the mass and spring combination is in its normal (*equilibrium*) condition.* (b) *If* m *is displaced to the right an amount* x_0 *(by an external force), there will be a restoring force to the left given by* $F = -kx_0$, *where* k *is the force constant characteristic of the particular spring.*

proportional to the magnitude of the displacement, if the displacement is not too great. In general, for any displacement x, the restoring force is[1]

$$F = -kx \qquad (10.1)$$

where the negative sign means that the direction of the force is opposite to the direction of displacement. If the initial displacement were to the *left*, the spring would be compressed and the restoring force would be to the *right*.

We know from Newton's equation of motion that the net force on an object must equal the product of its mass and its acceleration. Therefore, at any extension x,

$$F = ma = -kx$$

so that

$$a = -\frac{k}{m}x \qquad (10.2)$$

This is the basic *equation of motion* for an object undergoing *simple harmonic motion*.

A GRAPHICAL SOLUTION

If the mass is released from its position of initial extension ($x = x_0$), it will be accelerated to the left by the restoring force. At $x = x_0$ the acceleration is a maximum; as the mass moves toward $x = 0$ the velocity increases while the acceleration decreases. When the mass reaches $x = 0$, the restoring

[1] In Example 7.2 we wrote $F = kx$, where F was the *external* applied force; here, F is the force exerted by the *spring* and is in the opposite direction (Newton's third law).

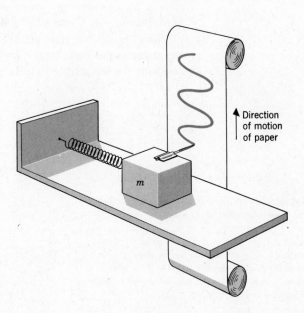

Direction
of motion
of paper

m

Fig. 10.2 *A simple method for recording the motion of an oscillating mass as a function of time.*

Table **10.1**   *The Displacement of an Oscillating Mass as a Function of Time*

t	$\dfrac{2\pi}{\tau}t$	$\cos\dfrac{2\pi}{\tau}t$	$x(t) = x_0\cos\dfrac{2\pi}{\tau}t$
0	0 or 0°	1	x_0
$\frac{1}{4}\tau$	$\frac{1}{2}\pi$ or 90°	0	0
$\frac{1}{2}\tau$	π or 180°	-1	$-x_0$
$\frac{3}{4}\tau$	$\frac{3}{2}\pi$ or 270°	0	0
τ	2π or 360° $= 0°$	1	x_0

force (and, hence, the acceleration) will have decreased to zero, but the velocity of the mass will have been increased to its maximum value at this point and the inertia of the mass will carry it into the region of negative x (to the *left* of $x = 0$). In this region the restoring force (and, hence, the acceleration) is directed to the *right* and the mass will be slowed down. At $x = -x_0$, the motion will stop[2] and the acceleration (which is still toward the right) will cause the mass to reverse its motion and move toward $x = x_0$ again. The entire process is one of *cyclic* (or *oscillatory* or *periodic*) motion, with the mass vibrating back and forth between $x = x_0$ and $x = -x_0$.

We can obtain a record of the motion of the mass as a function of time in the following simple way. As shown in Fig. 10.2, we attach a pen to the mass and allow it to touch a roll of paper that is moved uniformly in a direction perpendicular to the direction of motion of the mass. In this way we obtain a displacement-time graph of the motion. Examination of the graph shows that it is a cosine curve[3] of the form[4]

$$x(t) = x_0\cos\frac{2\pi}{\tau}t \tag{10.3}$$

where τ is the *period* of the motion; that is, after every interval of time τ, the motion repeats itself (see Fig. 10.3 and Table 10.1). The quantity x_0 is the *amplitude* of the motion, the maximum excursion from the equilibrium position experienced by the mass. A sinusoidal function (sine or cosine) varies in a simple and regular way (that is, the variation is *harmonic*), and

[2] This is insured by the conservation of energy, as we shall shortly see.
[3] Or a *sine* curve, depending on what point we elect to designate $t = 0$.
[4] $x(t)$ means "the displacement x as a function of time t."

Fig. 10.3 *The displacement as a function of time for a mass undergoing simple harmonic motion. There is a time interval τ (the period) between any two successive corresponding points on the curve.*

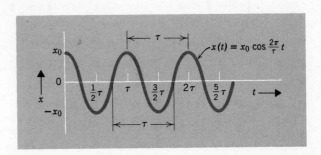

therefore motion described by such functions is termed *simple harmonic motion.*

Equation 10.3 describes an oscillatory motion that continues indefinitely without any change in amplitude. This is the ideal case in which frictional losses are ignored. Of course, in a real situation, friction will be present and the motion will not persist forever unless energy is continually supplied to the system. Friction causes the amplitude of the motion to decrease with time (this is called *damping*), and eventually the oscillations will cease.

DISPLACEMENT, VELOCITY, AND ACCELERATION

Let us now see how the fact that the displacement-time graph is a cosine function compares with the equation of motion (Eq. 10.2), which states that the acceleration of the mass is directly proportional to the negative of the displacement. In order to do this we use the fundamental definitions of instantaneous velocity (equal to the slope of the displacement-time graph at the particular instant of time) and instantaneous acceleration (equal to the slope of the velocity-time graph).

Figure 10.4a shows the displacement *versus* time curve and is the same as Fig. 10.3. The slope of this curve is zero at $t = 0$ and is downward or *negative* for all times between $t = 0$ and $t = \frac{1}{2}\tau$; therefore, the velocity is negative in this interval. The maximum negative slope and, hence, the maximum negative velocity, $-v_m$, is realized at $t = \frac{1}{4}\tau$. Physically, this means that the mass, after release from $x = x_0$, moves to the *left* and therefore has a negative velocity; the maximum negative velocity is attained when the mass passes through $x = 0$. At $t = \frac{1}{2}\tau$, the mass has reached the farthest point to the left, $x = -x_0$, and the velocity is zero again. For times between $t = \frac{1}{2}\tau$ and $t = \tau$, the motion is to the right with positive velocity. The maximum positive velocity, v_m, is realized at $t = \frac{3}{4}\tau$. The complete velocity-time curve is shown in Fig. 10.4b. This curve is simply an inverted sine curve; that is,

$$v(t) = -v_m \sin \frac{2\pi}{\tau} t \tag{10.4}$$

We can construct the acceleration-time curve in exactly the same way by measuring the slope of the velocity-time graph at each instant of time. If

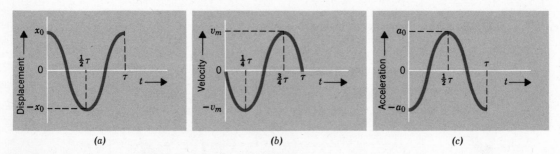

(a)   (b)   (c)

Fig. 10.4 (a) *The displacement-time graph for the oscillating mass.* (b) *The velocity-time graph obtained by measuring the slope of the curve in* (a) *at every instant of time.* (c) *The acceleration-time graph obtained by measuring the slope of the curve in* (b) *at every instant of time. Curve* (c) *is just the negative of curve* (a)*, as demanded by the equation of motion (Eq. 10.2).*

we do this, we find the curve shown in Fig. 10.4c. This curve is just the negative of a cosine curve such as that shown in Fig. 10.4a; that is,

$$a(t) = -a_0 \cos \frac{2\pi}{\tau} t \qquad (10.5)$$

where $-a_0$ is the initial acceleration and, according to Eq. 10.2, must be equal to $-(k/m)x_0$. Therefore,

$$a(t) = -\frac{k}{m} x_0 \cos \frac{2\pi}{\tau} t \qquad (10.5a)$$

Comparing this equation with Eq. 10.3 shows that $a = -(k/m)x$, which is just the equation of motion, Eq. 10.2. This result is not accidental, the *only* function that can satisfy the equation of motion (Eq. 10.2) is a sinusoidal function (sine or cosine).

ENERGY IN THE SIMPLE HARMONIC OSCILLATOR

In the ideal simple harmonic oscillator there are no frictional losses and so energy is conserved for the system. At any instant of time, we have

$$\mathcal{E} = PE + KE = \text{const.} \qquad (10.6)$$

From the result of Example 7.2, we know that an amount of work $\frac{1}{2}kx^2$ (supplied by an outside agency) is required to extend the spring by an amount x from its equilibrium position. The potential energy of the simple harmonic oscillator is just the energy stored in the spring, so that

$$PE = \frac{1}{2}kx^2 \qquad (10.7)$$

The kinetic energy of the oscillator is the motional energy of the mass, so that

$$KE = \frac{1}{2}mv^2 \qquad (10.8)$$

Therefore,

$$\mathcal{E} = \frac{1}{2}kx^2 + \frac{1}{2}mv^2 = \text{const.} \qquad (10.9)$$

At the position of maximum extension of the spring, $x = x_0$, the velocity of the mass is zero and so the total energy is just $\frac{1}{2}kx_0^2$. Similarly, at $x = 0$, the potential energy is zero, the mass has its maximum velocity, $v = v_m$, and the total energy is $\frac{1}{2}mv_m^2$. That is,

$$\mathcal{E} = \frac{1}{2}kx_0^2 = \frac{1}{2}mv_m^2 \qquad (10.10)$$

Solving for the relationship between x_0 and v_m, we find

$$v_m = \pm \left(\sqrt{\frac{k}{m}} \right) x_0 \qquad (10.11)$$

where a positive or a negative sign must be chosen for the square root in order to indicate the direction of the velocity relative to the direction of the initial displacement. (The mass can pass through $x = 0$ moving in *either* direction; the magnitude of v_m will be the same for each case, but the *signs* will be different.)

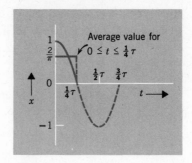

Fig. 10.5 *The average value of cos* $(2\pi t/\tau)$ *in the interval from* $t = 0$ *to* $t = \frac{1}{4}\tau$ *is* $2/\pi$.

THE PERIOD OF MOTION

Consider the motion of the mass from the point of release, $x = x_0$, to the position $x = 0$. This motion starts at $t = 0$ and is completed at $t = \frac{1}{4}\tau$. That is, it requires $\frac{1}{4}$ of the period of the oscillation to first reach the position $x = 0$. The velocity at $t = \frac{1}{4}\tau$ is v_m and is given by the average acceleration during this time interval multiplied by the time interval; that is,

$$v_m = \bar{a} \times \tfrac{1}{4}\tau \tag{10.12}$$

The acceleration $a(t)$ is given by Eq. 10.5a. Since the coefficient $(-kx_0/m)$ does not depend on the time, only $\cos(2\pi t/\tau)$ must be averaged from $t = 0$ to $t = \frac{1}{4}\tau$. This averaging process is shown graphically in Fig. 10.5; analytical methods show that the average value is $2/\pi$. If we use this result and Eq. 10.11 for v_m,[5] Eq. 10.12 for the period becomes

$$\tau = \frac{4v_m}{\bar{a}}$$

$$= \frac{4 \times \left(-\sqrt{\dfrac{k}{m}}x_0\right)}{\left(-\dfrac{k}{m}x_0 \times \dfrac{2}{\pi}\right)}$$

so that

$$\boxed{\tau = 2\pi\sqrt{\frac{m}{k}}} \tag{10.13}$$

Notice that the period is independent of x_0, the amplitude of the oscillations. This fact is characteristic of simple harmonic motions.

FREQUENCY

The *period* of a vibratory motion is the time interval required for one complete oscillation of the system. The *frequency* of a vibratory motion is the number of complete oscillations that the system makes in unit time. It is customary to use the symbol ν to denote frequency. The unit of frequency

---

[5]In this case we must use the negative sign for the square root since the velocity at $t = \frac{1}{4}\tau$ is in the direction *opposite* to the initial displacement.

is *cycles/sec* (or simply *sec*$^{-1}$), the abbreviation for which is *Hertz* (*Hz*), in honor of Heinrich Hertz (1857–1894) who made great contributions to the study of electrical oscillations. If the period of a system is τ, it is clear that the system will experience $1/\tau$ vibrations per second. That is, the period and the frequency are reciprocally related:

$$\boxed{\nu = \frac{1}{\tau}}$$

(10.14)

Example **10.1**

Suppose that a mass of 8 grams is attached to a spring that requires a force of 1000 dynes to extend it to a length 5 cm greater than its natural length. What is the period of the simple harmonic motion of such a system?

The force constant is

$$k = \frac{F}{x} = \frac{1000 \text{ dynes}}{5 \text{ cm}}$$

$$= 200 \text{ dynes/cm}$$

Therefore, the period is

$$\tau = 2\pi\sqrt{\frac{m}{k}}$$

$$= 2\pi\sqrt{\frac{8 \text{ g}}{200 \text{ dynes/cm}}}$$

$$= 2\pi \times \sqrt{\frac{4}{100} \text{ sec}^2}$$

$$= 2\pi \times 0.2 \text{ sec}$$

$$= 1.26 \text{ sec}$$

and the frequency is

$$\nu = \frac{1}{\tau} = \frac{1}{1.26 \text{ sec}}$$

$$= 0.8 \text{ Hz}$$

THE SIMPLE PENDULUM

The results we have obtained for the oscillations of a mass attached to a spring are not unique to this particular case. There are many other situations in which the motion is simple harmonic. In fact, almost all oscillatory motions are approximately simple harmonic, *if the amplitude of the motion is sufficiently small.*

The oscillatory motion of a simple pendulum[6] will be *simple harmonic* if we can find a force constant for the system that relates the restoring force and the displacement. According to Eq. 10.1, the force constant is given by

---

[6]A *simple* pendulum is one that we can consider to consist of a point mass suspended by an extensionless and massless string.

$$k = -\frac{\text{restoring force}}{\text{displacement}} \qquad (10.15)$$

As shown in Fig. 10.6, there are two forces acting on the pendulum bob—the downward gravitational force, $F_G = mg$, and the tension in the suspension string, T. The net force acting on the bob is $F_{net} = F_G + T = mg \sin \theta$; this is the *restoring force*. The *displacement* x is the distance through which the bob moves along the arc starting at A; that is, $x = l\theta$ (compare Eq. 1.10). Therefore, Eq. 10.15 becomes

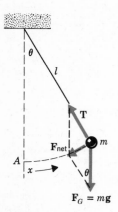

Fig. 10.6 *A simple pendulum of length* l. *The net force on the pendulum bob is the vector sum of the gravitational force* F_G *and the tension in the string,* T.

$$k = \frac{mg \sin \theta}{l\theta} = \frac{mg}{l}\left(\frac{\sin \theta}{\theta}\right) \qquad (10.16)$$

where we have written only the magnitude of F_{net} and where we have cancelled the negative sign by taking into account the fact that the displacement and the restoring force have *opposite* directions.

Now, if θ is small, $\sin \theta \cong \theta$ (see Eq. 4.18). Then, $(\sin \theta)/\theta \cong 1$ so that

$$k \cong \frac{mg}{l} \qquad (10.17)$$

Using this expression for k in Eq. 10.2, we have

$$a = -\frac{k}{m}x \cong -\frac{g}{l}x \qquad (10.18)$$

Since the factor g/l in the equation of motion for the simple pendulum corresponds to the factor k/m in the equation for the mass attached to a spring, we can immediately write down the expression for the period of a simple pendulum (compare Eq. 10.13):

$$\tau = 2\pi\sqrt{\frac{l}{g}} \qquad (10.19)$$

Notice that the period is *independent* of the mass of the bob and is directly proportional to the square root of the length of the pendulum.

The approximation that we have used, namely, $\sin \theta \cong \theta$, introduces an error of only 0.1 percent if θ is 7° and only 1 percent when θ is as large as 23°.

Example **10.2**

What must be the length of a simple pendulum that will have a period of 1 sec at the surface of the Earth?

Squaring Eq. 10.19 and solving for l, we have

$$l = \frac{\tau^2 g}{4\pi^2}$$

$$= \frac{(1 \text{ sec})^2 \times (980 \text{ cm/sec}^2)}{4\pi^2}$$

$$= 24.8 \text{ cm}$$

10.2   *Wave Motion*

THE PROPAGATION OF WAVE PULSES

The oscillatory motions we have just discussed involve the motions of only a single particle or mass. We now turn our attention to situations in which the motion of any given particle influences and is influenced by the motion of its neighbors. An important case of such a cooperative phenomenon is that of *wave motion.*

Examples of wave motion are to be found virtually everywhere. Water waves travel across the seas. Sound is propagated by waves in the air. A piano or a violin produces its characteristic sounds by the wave motions of strings. The air (and even empty space) is permeated by electromagnetic waves in the form of radio waves and light. In all cases of mechanical waves (as distinct from electromagnetic waves, which are due to the periodic variations of the electromagnetic field), the vibratory motion of particles is involved. But even though these waves propagate through the air or along a string or across the ocean, the individual motions of the particles are never very large. For example, Fig. 10.7 shows a wave disturbance propagating along a coiled spring. A ribbon, tied to one of the coils, provides a marker for observing the motion of a particular portion of the spring. It is evident that no portion of the spring moves very far in the horizontal direction and yet the pulse travels along the spring by virtue of the fact that the particles at the front of the disturbance are forced to move upward and those at the rear are forced to move downward to their original positions (see Fig. 10.8).

By noting the position of the wave pulse at various instants of time, we can determine the propagation velocity of the pulse in the conventional way (see Fig. 10.9). If the pulse travels a distance Δx in a time interval Δt the propagation velocity is

$$v = \frac{\Delta x}{\Delta t} \tag{10.20}$$

SUPERPOSITION

Wave motion is another physical phenomenon that obeys the principle of superposition (see Sections 8.2–8.4). If two separate pulses travel toward one another along the same spring, their motions are completely independent.[7] Figure 10.10 is a photographic record of two pulses on a coiled spring that approach and pass one another. After passing, they continue in opposite directions without having suffered any change in shape or size (or velocity). When the two pulses pass one another, the displacement of the spring is the *sum* of the two individual displacements. If these individual displacements are of the same shape and size but are of opposite signs, then a *cancellation* will result at the moment of passing (see the 5th frame of Fig. 10.10 and Fig. 10.11c). If the pulses are of the same sign, then they will *add* at the moment of passing.

---

[7] Both pulses will travel with the same velocity because the propagation velocity depends on the properties of the spring and not on the particular shape or size of the pulse (if frictional losses are ignored).

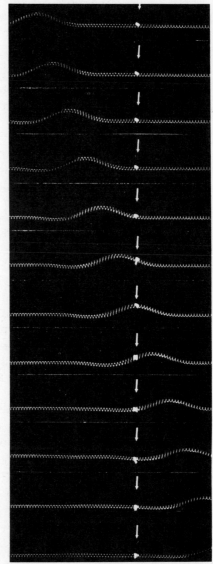

Fig. 10.7 *A wave pulse travels along a coiled spring from left to right. The various pictures are individual frames from a film of the motion recorded by a movie camera. A ribbon is tied to one of the coils (indicated by the arrow) in order to illustrate the motion of a particular small portion of the spring.*

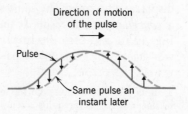

Fig. 10.8. *The relation between the motion of the portions of the spring and the forward motion of a pulse.*

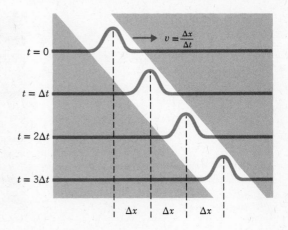

$t = 0$

$t = \Delta t$

$t = 2\Delta t$

$t = 3\Delta t$

$v = \dfrac{\Delta x}{\Delta t}$

$\Delta x \quad \Delta x \quad \Delta x$

Fig. 10.9 *The wave pulse travels toward the right with a propagation velocity,* $v = \Delta x/\Delta t.$

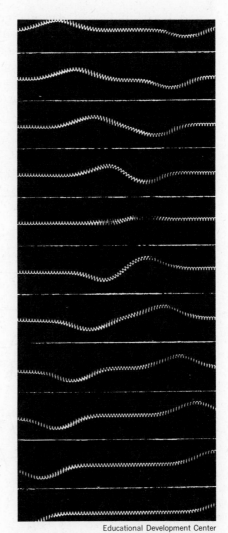

Fig. 10.10 *The superposition of two almost identical wave pulses traveling along a coiled spring. At the moment of passing (fifth frame), they almost cancel each other. Notice, however, that the blurring of the photograph indicates that, although the spring has almost no net displacement, there is a substantial vertical velocity at two positions.*

Educational Development Center

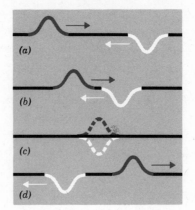

Fig. 10.11 *A schematic representation of the super-position of two identical pulses as shown photo-graphically in Fig. 10.10. In (c) the two pulses exactly cancel.*

TRAVELING WAVES

Figure 10.12 represents the generation of a *traveling wave*. A string or spring is initially taut with no displacement at any point along its length. At $t = 0$ an external force acts on one end of the string and begins to drive it up and down in a periodic motion. This motion of the end portion of the string is propagated to the adjacent portion, which in turn acts on the next portion, and so on along the string. The wave therefore propagates (or "travels") along the string.

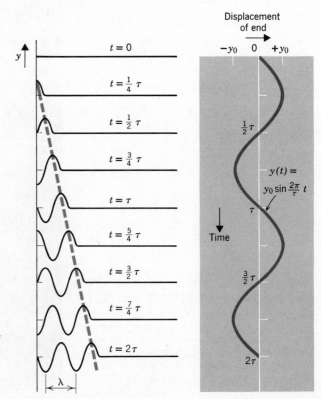

Fig. 10.12 *The generation of a traveling wave in a length of string. The left end of the string is driven up and down in a sinusoidal manner (right-hand graph) and the motion is propagated along the string, forming a sinusoidal traveling wave.*

The velocity of the wave propagation can be calculated from the standard expression, $v = \Delta x/\Delta t$. The period τ of the driving oscillation is also the period of the wave motion. During a time interval τ the wave moves forward by an amount called the *wavelength* λ of the wave. Therefore, if $\Delta t = \tau$, then $\Delta x = \lambda$, so that the important characteristics of the wave are related according to

$$v = \frac{\lambda}{\tau} = \lambda\nu \qquad\qquad (10.21)$$

At a given position in space, the variation of the wave amplitude with time is

$$y(t) = y_0 \cos\frac{2\pi}{\tau}t \qquad\qquad (10.22)$$

Also, the distance x that any particular portion of the wave moves forward in a time t is

$$x = vt = \frac{\lambda t}{\tau}$$

where Eq. 10.21 has been used for v. Hence,

$$\frac{t}{\tau} = \frac{x}{\lambda}$$

and making this substitution in Eq. 10.22 we have the variation of the wave amplitude with *position* at any instant of time:

$$y(x) = y_0 \cos\frac{2\pi}{\lambda}x \qquad\qquad (10.23)$$

These results for $y(t)$ and $y(x)$ are shown in Fig. 10.13.

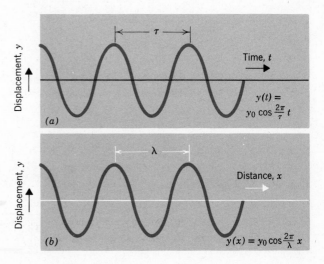

Fig. 10.13 (a) *The displacement of a given point of the string as a function of time.* (b) *The displacement of the string as a function of distance for a given instant. Both curves are sinusoidal curves.*

TRANSVERSE AND LONGITUDINAL WAVES

In the wave motions we have been discussing, the particles in the strings or springs move at right angles to the direction of propagation of the wave. Such waves are therefore called *transverse* waves. Wave motion is also possible in which the particles move back and forth along the direction of wave propagation. Such waves are called *longitudinal* waves. An example of this type of wave motion is the *compressional* waves that can be propagated in a spring (see Fig. 10.14a). At any instant, portions of the spring are alternately compressed and extended, and the variation is sinusoidal along the length of the spring. A similar situation exists in the *sound* waves that can be propagated in a column of gas (see Fig. 10.14b). In this case the *density* of the gas molecules (and the gas pressure) varies sinusoidally along the column. Since some physical property (such as the density of particles or the pressure) in longitudinal waves varies with distance and time in exactly the same way as in transverse waves, these two different types of wave motion are described by identical mathematical expressions. For example, in Eq. 10.23 we need only call $y(t)$ the pressure in a column of gas in order to represent the propagation of a sound wave with a frequency ν.

SOUND WAVES

The propagation of sound waves through the air (at standard conditions) takes place with a velocity of approximately 3.3×10^4 cm/sec or 1100 ft/sec, which, within wide limits, is independent of the frequency of the wave.[8] When such a pressure wave reaches our ears, it produces vibrations in the ear's membranes which provoke a nervous response and we *hear* the sound. But a hearing sensation is produced in the human nervous system only if the frequency of the wave is between ~ 16 Hz and $\sim 20,000$ Hz. For these extreme frequencies, the corresponding wavelengths are

$$\lambda_1 = \frac{v}{\nu_1} = \frac{3.3 \times 10^4 \text{ cm/sec}}{16 \text{ sec}^{-1}} \cong 2 \times 10^3 \text{ cm} = 20 \text{ m}$$

$$\lambda_2 = \frac{v}{\nu_2} = \frac{3.3 \times 10^4 \text{ cm/sec}}{2 \times 10^4 \text{ sec}^{-1}} \cong 1.6 \text{ cm}$$

---

[8] Aircraft speeds are frequently given in units of the speed of sound. These units are called *Mach numbers;* Mach 1 corresponds to the speed of sound (approximately 730 mi/hr at sea level), Mach 2 corresponds to twice the speed of sound, etc.

Fig. 10.14 (a) *Longitudinal compressional waves in a coiled spring are initiated by the application of a periodic driving force at one end.* (b) *Longitudinal waves (sound waves) in a column of gas are initiated by the application of a periodic force to a piston located in one end.*

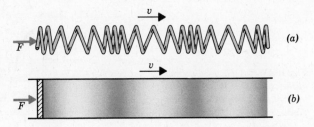

Table **10.2** *Sound Velocities in Some Materials*

Substance	Velocity (cm/sec)
Granite	$\sim 6 \times 10^5$
Iron	5.13×10^5
Sea water	1.53×10^5
Lead	1.23×10^5
Air	3.31×10^4

Sound waves in air with wavelengths longer or shorter than these are possible, but they are inaudible to the human ear. (There are also variations in the hearing range from individual to individual. In general, as a person grows older the upper limit of audible frequencies tends to decrease due to the decreasing flexibility of the ear drum.)

Sound propagates in solid materials not only by longitudinal compressional waves but by transverse waves as well. The density of solids is much higher than that of gases and the velocity of propagation of sound waves is correspondingly greater in solids. Velocities approaching 10^6 cm/sec are found for some materials (see Table 10.2).

Example **10.3**

What is the frequency of a 2-cm sound wave in sea water?

From Table 10.2, $v = 1.53 \times 10^5$ cm/sec in sea water, so

$$\nu = \frac{v}{\lambda}$$

$$= \frac{1.53 \times 10^5 \text{ cm/sec}}{2 \text{ cm}}$$

$$= 7.6 \times 10^4 \text{ Hz} = 76 \text{ kHz}$$

which is an *ultrasonic* wave (that is, above the human audible range). A 2-cm sound wave in *air* would be audible.

MODULATED WAVES

Wave motion, in the form of sound waves or radio waves, is frequently used to communicate information. For this purpose a simple wave that is propagated with a single frequency and a uniform amplitude (Fig. 10.15a) is not suitable. Information is impressed on waves by various methods of changing the wave form (*modulation*). Broadcast radio waves are high frequency waves (*carrier* waves), modulated in amplitude by the voice or music signal that is to be transmitted (Fig. 10.15b). This technique is called *amplitude modulation* (AM). Alternatively, the amplitude can be maintained constant and the *frequency* modulated, as in FM broadcasting (Fig. 10.15c). A typical sound wave is modulated both in frequency and amplitude (Fig. 10.15d).

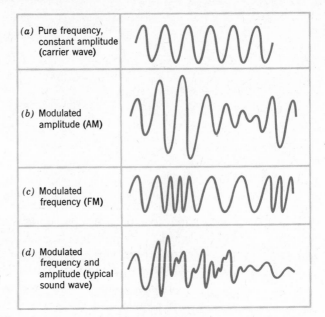

(a) Pure frequency, constant amplitude (carrier wave)

(b) Modulated amplitude (AM)

(c) Modulated frequency (FM)

(d) Modulated frequency and amplitude (typical sound wave)

Fig. 10.15 *Various methods of modulating the uniform carrier wave shown in (a). (b) Amplitude modulation. (c) Frequency modulation. (d) A typical sound wave is modulated in both frequency and amplitude.*

THE DOPPLER EFFECT

We are all familiar with the fact that the frequency of the sound from a siren on a moving vehicle changes dramatically when the vehicle passes. When the sound source is approaching, the apparent frequency is high, and, when the source passes and is moving away, the apparent frequency is low. The dependence of the frequency of a wave disturbance on the relative motion of the source and the observer is termed the *Doppler effect,* after the Austrian physicist, Christian Johann Doppler (1803–1853), who extensively studied this phenomenon.

Figure 10.16 shows the spherical sound waves (of constant frequency and amplitude) that are emitted by a source S, which moves at constant velocity v_S toward a listener L. (We are interested here, not in the three-dimensional character of the waves, but only in those portions of the waves that move directly from the source to the listener.)

In the reference frame of the source, the frequency of the waves is ν_S. Therefore, in a time t, the source will emit $\nu_S t$ waves. Consider the time interval from $t = 0$ until $t = t$. The wave emitted at $t = 0$ will travel a distance vt during this interval, where v is the velocity of the waves in the particular medium. (The velocity of waves in a medium depends only on the mechanical properties of that medium and does not depend in any way on the velocity of the source.) When the last of the $\nu_S t$ waves has been emitted, the source will have traveled a distance $v_S t$. Hence, the $\nu_S t$ waves occupy the distance $vt - v_S t$ and so their wavelength λ_L as determined by the listener, is

$$\lambda_L = \frac{\text{distance}}{\text{no. of waves}} = \frac{vt - v_S t}{\nu_S t} = \frac{v - v_S}{\nu_S}$$

The frequency ν_L, again as determined by the listener, is

$$\nu_L = \frac{v}{\lambda_L} = v \times \frac{\nu_S}{v - v_S}$$

or

$$\boxed{\nu_L = \nu_S \left(\frac{v}{v - v_S} \right) = \frac{\nu_S}{1 - v_S/v}}$$ (S moving toward L)   (10.24)

This expression shows that the frequency of the sound heard by L is *increased* over that which he would hear if the source were at rest. If the source moves *away from* the listener, the sign of v_S is changed and the negative sign in Eq. 10.24 becomes a positive sign. Thus, the frequency is *lowered.* Similar results are obtained if the source is stationary (relative to the medium) and the listener moves toward or away from the source.

The Doppler effect exists for all types of wave motion—sound waves in air, elastic waves on springs or in solids, water waves, etc. Even though the propagation of electromagnetic waves (for example, *light*) differs in an essential respect from the propagation of mechanical waves (see Section 11.1), electromagnetic waves also exhibit the Doppler effect. Indeed, we have already seen (Sections 3.1 and 3.2) that measurements of the Doppler shifts in the light from stars and galaxies constitute one of the most important methods for determining the motions of and the distances to astronomical objects.

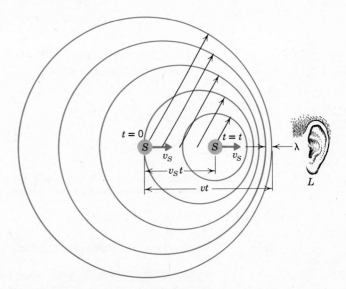

Fig. 10.16 *The source of sound waves* S *moves toward the listener* L *with a velocity* v_S. *The waves become "bunched up" and the listener perceives a* higher *frequency of sound than he would if the source were at rest. If the listener were at the* left, *he would hear a* lower *frequency. (The arrows directed upward and to the right indicate the radii to which the waves emitted at various points have propagated by the time* t.)

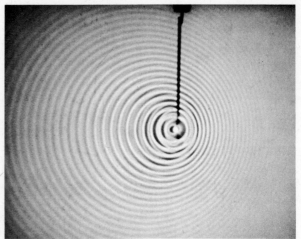

Fig. 10.17 *Photograph of water waves produced by a moving source. The dark line in the photograph is a vibrating wire that is moved through the water from left to right. An observer on the right would measure a shorter wavelength (higher frequency) than if the source were at rest.*

Example **10.4**

An airplane is flying at Mach 0.5 and carries a sound source that emits a 1000-Hz signal. What frequency sound does a listener hear if he is in the path of the airplane?

$$\nu_L = \frac{\nu_S}{1 - v_S/v}$$

But a speed of Mach 0.5 means that $v_S/v = 0.5$; therefore,

$$\nu_L = \frac{\nu_S}{1 - 0.5} = 2\nu_S$$

so that the listener hears a 2000-Hz sound. What frequency does the listener hear after the airplane has passed?

$$\nu_L = \frac{\nu_S}{1 + v_S/v}$$

$$= \frac{\nu_S}{1 + 0.5} = \frac{2}{3}\nu_S$$

so that the listener hears a 667-Hz sound.

10.3 Standing Waves

PULSES ON TERMINATED STRINGS

Consider a string or a spring that is terminated at one end; that is, the end of the string is attached to a rigid wall and cannot be moved. Suppose that we send a wave pulse along such a terminated string, as in Fig. 10.18a. When the pulse reaches the terminated end, the wave motion will not simply stop (because energy must be conserved); rather, the pulse will be *reflected* and will proceed back along the string in the opposite direction (Fig. 10.18b).

Fig. 10.18 (a) *A wave pulse travels along a string toward the terminated end.* (b) *After reaching the end, the pulse is reflected and travels back along the string in the opposite direction with the opposite sign of the displacement. The motion of the wave pulse is very similar to that of a ball rebounding from a wall.*

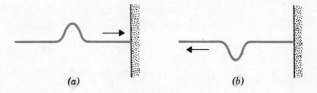

(a)   (b)

Notice that the reflected wave pulse has the same size and shape as the initial pulse, but the *sign* of the displacement is reversed. This reversal in the sign of the displacement is due to the fact that the wall exerts a *downward* force on the string as a reaction to the *upward* force that the incident pulse exerts on the wall. Since the wall does not move, this reaction force manifests itself in a downward displacement of the reflected pulse. Figure 10.19 is a photographic record of a pulse traveling along a coiled spring; at the

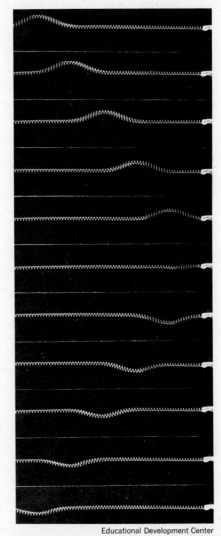

Fig. 10.19 *Photographic record of a wave pulse reflected from the terminated end of a coiled spring. The reflected pulse is slightly smaller than the initial pulse because of losses that occur at the termination.*

Educational Development Center

terminated end the pulse is reflected with the attendant change in sign of the displacement. In this case, the reflected pulse is slightly smaller than the initial pulse because there are frictional losses that occur when the pulse is reflected at the termination.

If a pulse travels along a string that is terminated at both ends (Fig. 10.20), reflection will occur at each end and the pulse will travel back and forth with a positive displacement while moving to the right and a negative displacement while moving to the left. We can represent this situation in another entirely equivalent way. Suppose that we have a long string that we set into motion in such a way that a series of positive-displacement pulses travel to the right and a series of negative-displacement pulses travel to the left, as in Fig. 10.21. We arrange the distance between successive pulses to be L (equal to the distance between the termination points in Fig. 10.20). Now, focus attention on the behavior of the pulses at the points A and B in Fig. 10.21. As the positive pulse b moves past A toward the right, the negative pulse b' moves past A toward the left. Similarly, as the positive pulse c moves past B toward the right, the negative pulse c' moves past B toward the left. Because of the uniform spacing of the pulses, it is clear that the behavior of the pulses in the region AB will continue to repeat in a regular way. Furthermore, the behavior of the pulses in this region is exactly the same as if the string were terminated at A and B and a pair of pulses were being reflected at the ends. That is, pulse c is reflected at B and becomes pulse c'; pulse b' is reflected at A and becomes pulse b. We conclude, therefore, that the behavior of wave pulses on a string that is terminated at both ends can be represented by the superposition of two trains of pulses (traveling waves) moving in opposite directions along the string.

SINUSOIDAL WAVES ON TERMINATED STRINGS

Let us now extend the reasoning in the preceding paragraph and consider the case of two sinusoidal waves traveling in opposite directions along a stretched string. The situation is shown in Fig. 10.22, where the wave moving to the right is indicated by the solid curve (Fig. 10.22a) and the wave moving to the left is indicated by the dotted curve (Fig. 10.22b). Both waves have the same amplitude and each wave has a wavelength λ equal to L, the

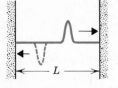

Fig. 10.20 *A pulse traveling along a string that is terminated at both ends will be reflected back and forth.*

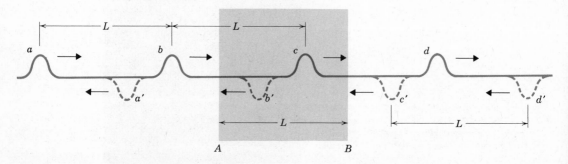

Fig. 10.21 *The two trains of pulses,* a b c d *and* a′ b′ c′ d′, *combine to produce, in the interval* AB, *exactly the same effect as a pair of pulses reflected from terminations at* A *and* B.

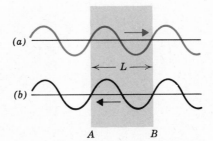

(a)

←— L —→

(b)

A   B

Fig. 10.22 *To waves of the same amplitude and wavelength* (λ = L) *travel along a stretched string in opposite directions.*

distance between two points, *A* and *B,* that eventually will represent termination points.

We arrange the two traveling waves in such a way that at the instant $t = 0$ the displacements of the waves are the same at every point along the string, with each wave having zero displacement at the points *A* and *B*. Since we know that the superposition principle applies, the net effect of the two wave motions is that the displacement is everywhere exactly twice the displacement of a single wave (Fig. 10.23a). (The dashed curve is the net displacement.) In this situation we say that the waves are *in phase* and reinforce one another. A short time later, at $t = \Delta t$, each wave will have moved a small distance along its direction of motion. Therefore, when the two wave displacements are added to obtain the net displacement at every point along the wave, the result is again a sine curve but now with a smaller amplitude (Fig. 10.23b). Notice also that, even though the waves have moved slightly from their positions at $t = 0$, the displacements at *A* and *B* are equal and opposite so that the net displacement at these points is zero. At the time $t = \frac{1}{4}\tau$, each wave has moved a distance equal to $\frac{1}{4}\lambda$ so that the displacements are everywhere equal and opposite. The net effect is everywhere zero and the resultant curve is a straight line (Fig. 10.23c). In this situation, we say that the waves are *out of phase* and cancellation results. At the time $t = \frac{1}{2}\tau$, each wave has progressed a distance of $\frac{1}{2}\lambda$ and the waves are again *in phase* but now with displacements opposite to those at $t = 0$. Hence, the net displacement is a sine curve, as at $t = 0$, but with opposite sign (Fig. 10.23d).

There are two important points that emerge from this analysis: (1) At every instant of time the displacements of the two traveling waves at points *A* and *B* are equal and opposite so that there is always *zero* net displacement at these two points. (2) At every instant of time the net displacement along the string is described by a *sine curve*. The amplitude of this curve varies from a maximum in the positive sense at $t = 0$ (Fig. 10.23a), to zero at $t = \frac{1}{4}\tau$ (Fig. 10.23c), and to a maximum in the negative sense at $t = \frac{1}{2}\tau$ (Fig. 10.23d); at $t = \tau$, the amplitude is again a maximum in the positive sense and the situation is the same as at $t = 0$.

Because there is always zero displacement at the points *A* and *B,* if we wish to consider the wave motion only in this interval, we can dispense with the long string and consider the string to be terminated at *A* and *B*. That is, the superposition of two traveling waves on a long string duplicates exactly the motion of reflected waves on a terminated string whose length *L* is equal to the wavelength λ of the waves. A sinusoidal wave that maintains its overall shape between two termination points (but that changes its amplitude as a function of time) is called a *standing wave.*

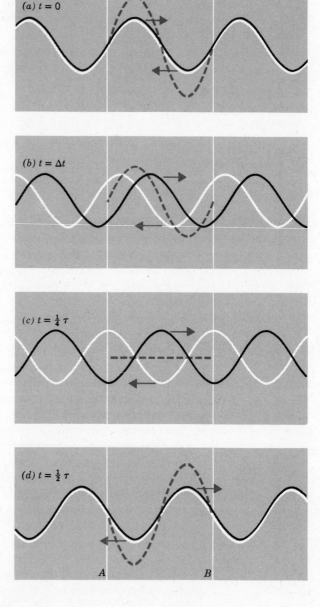

Fig. 10.23 *Two sine waves traveling in opposite directions along a stretched string combine to produce a standing sinusoidal wave. The standing wave pattern in the interval AB will be the same whether the string is very long or whether the string is terminated at A and B.*

THE WAVELENGTH RELATION FOR STANDING WAVES

We have just demonstrated that a standing wave of wavelength λ can be set up on a stretched string that is terminated at two points separated by a distance $L = \lambda$. But examination of Fig. 10.23 will reveal that the only feature of the particular case represented that is essential for the conclusion is the occurrence of zero displacement for the wave at the points A and B. (Points that have zero displacement for all times are called *nodes.*) Therefore, a string of length L can support a standing wave if the wavelength is such that there are nodes at the termination points. Figure 10.24 shows

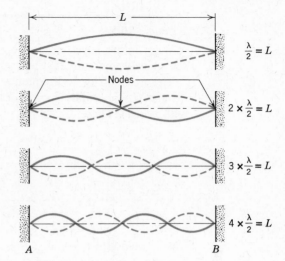

Nodes

$\frac{\lambda}{2} = L$

$2 \times \frac{\lambda}{2} = L$

$3 \times \frac{\lambda}{2} = L$

$4 \times \frac{\lambda}{2} = L$

Fig. 10.24 *Standing waves on a stretched string. All such waves must have nodes at the termination points.*

four such waves; it is evident that the condition for standing waves of wavelength λ on a string of length L is

$$n\frac{\lambda}{2} = L, \qquad n = 1, 2, 3, \ldots \tag{10.25}$$

Standing waves occur for situations other than transverse vibrational waves on stretched strings. Standing sound waves can be set up between reflecting walls, and electromagnetic waves (including light) can be reflected from a pair of surfaces to form standing waves. Standing waves can even be set up in automobile tires when the velocity of propagation of a disturbance in the rubber of the tire just equals the forward velocity of the tire along the road (Fig. 10.25). In this situation the stresses in the tire at the bulging points are extremely large and a blow-out is likely if the motion is sustained.

Fig. 10.25 *Standing waves are produced in an automobile tire when driven at high speeds. The distortion is quite evident in this tire, driven in place at the equivalent of 140 mi/hr.*

Example **10.5**

What is the lowest frequency of the standing sound wave that can be set up between walls that are separated by 25 ft?

The wave of lowest frequency has the longest wavelength. For a given distance L, the standing wave of longest wavelength has $\lambda = 2L$ (Eq. 10.25 with $n = 1$). Therefore,

$$\lambda = 2L = 2 \times 25 \text{ ft} = 50 \text{ ft}$$

The frequency of the wave is

$$\nu = \frac{v}{\lambda} = \frac{1100 \text{ ft/sec}}{50 \text{ ft}}$$

$$= 22 \text{ Hz}$$

which is close to the lowest frequency that can be heard by a human ear. Therefore, a room somewhat larger than 25 ft is necessary in order to set up standing waves of the lowest audible frequency (for example, organ notes of 16 Hz).

10.4   *Huygens' Principle and Diffraction*

WAVES IN TWO AND THREE DIMENSIONS

The waves we have been discussing move along straight lines guided by strings or springs; such waves are called *one-dimensional waves*. There are also important types of wave motions that take place in two or three dimensions. For example, if we touch a vibrating rod to the surface of an expanse

Fig. 10.26 *Circular waves are produced by touching the tip of a vibrating rod to the surface of an expanse of water.*

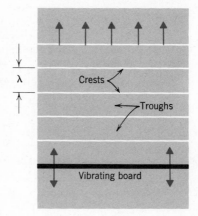

Fig. 10.27 *A vibrating board in an expanse of still water produces two-dimensional plane waves.*

of still water, a circular, two-dimensional wave will be propagated outward from the source (Fig. 10.26). Or, if a source of sound is located within a certain volume of air, spherical (three-dimensional) sound waves will be generated.

Another important type of wave motion can be generated in the following way. Suppose we place a long, straight board in still water, with its longest dimension horizontal, so that part of the board is submerged. Now we oscillate the board back and forth with a regular periodic motion that is transverse to the board's longest dimension (Fig. 10.27). Each time the board moves, it piles up the water in front of it and pushes the water forward. As a result, a series of crests are propagated across the water with troughs between them. This type of movement constitutes a wave motion and the distance between successive crests (or troughs) is the wavelength of the disturbance. Waves that propagate in this manner are called two-dimensional *plane waves*. Sound waves or electromagnetic waves that are propagated in such a way that the status of the disturbance is the same at all points on a series of planes that move in the direction perpendicular to the planes are called *three-dimensional plane waves*.

HUYGENS' PRINCIPLE

Suppose that a plane wave of wavelength λ is incident on a panel into which is cut a slot of width d, where d is small compared to λ (see Fig. 10.28). Will the wave that emerges from the slot be confined to the narrow region of width d? The answer to this question is "no," and the reason is that waves exhibit the phenomenon of *diffraction*. A convenient way to view this effect is to use a clever construction invented by the Dutch mathematical physicist, Christiaan Huygens (1629–1695), who first formulated the wave theory of light. According to *Huygens' principle,* the manner in which a wave front of arbitrary shape will advance can be determined by considering every point on a given wave front at any instant to be the source of a circular wave (or spherical wave in the case of three dimensions). Therefore, by drawing a series of circular waves emanating from a given wave front and then constructing the envelope of these waves, the shape and position of the entire wave at a later time can be found. Huygens' method is illustrated

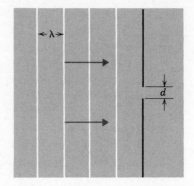

Fig. 10.28 *A plane wave is incident on a panel into which is cut a slot. Will the emerging waves be confined to the narrow region of width* d?

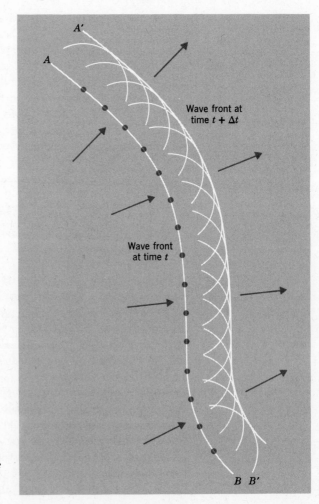

Fig. 10.29 *Huygens' construction for determining the advance of a wave front of arbitrary shape. Every point on the wave front AB at time* t *is considered to be the source of a circular wave. By drawing these circular waves and constructing the envelope, the wave front* A'B' *at time* t + Δt *is determined.*

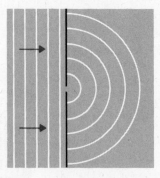 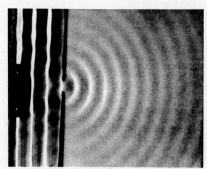

Physics I by Elisha Huggins Benjamin

Fig. 10.30 *Pattern of waves produced by a plane wave incident on a slot. Even though the width of the slot is actually not much smaller than the wavelength of the incident wave, nevertheless, a circular wave pattern results. (The dark rectangle at the left of the photograph is a part of the mechanism that produces the plane wave.)*

in Fig. 10.29. This type of construction is clearly correct for the cases of plane or circular waves (try the constructions), and it can be shown to be valid in general.

We are now in a position to predict the form of the wave motion on the right-hand side of the panel in Fig. 10.28. Every point of the slot is to be considered as a source of circular waves propagating to the right.[9] If the slot width is small compared to the wavelength of the incident plane wave ($d \ll \lambda$), then the slot is essentially a point source. Therefore, to the right of the panel we would expect only circular waves emanating from the slot. Figure 10.30 shows the construction of the wave pattern according to Huygens' principle and a photograph of water waves that exhibit the predicted diffraction effect.

If the width of the slot is large compared to the wavelength of the incident disturbance ($d \gg \lambda$), then the Huygens' construction shows that the wave pattern to the right of the panel is essentially a plane wave, with curved ends (Fig. 10.31). Thus, the wave is largely unaffected by the presence of the panel but the circular wave effect persists near the edges of the slot.

10.5 *Interference*

TWO-SOURCE INTERFERENCE

By combining the principle of superposition with Huygens' principle, we are able to explain a variety of important and interesting *interference* effects that occur in wave motions of all sorts. Any wave motion in which the amplitudes of two or more waves combine will exhibit *interference*. Wave pulses of opposite signs that are traveling along a string will *cancel* when they pass (Figs. 10.10 and 10.11); we call this effect *destructive interference*. If the pulses are of the same sign, they will *add* when passing; this is *constructive interference*.

[9] A more detailed analysis is required to show that there are no backward-going waves even though the statement of Huygens' principle would allow such waves.

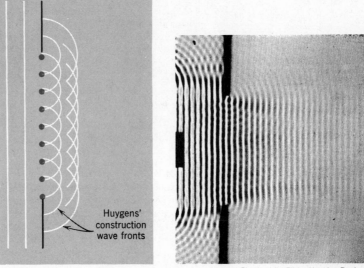

Fig. 10.31 *If the width of the slot is large compared to the wavelength of the incident plane wave, the wave passes through the slot almost unaffected. There is a noticeable diffraction effect only near the edges of the slot.*

In two or three dimensions, a pattern of constructive and destructive interference will be developed throughout a region that supports the propagation of waves from two or more sources. For example, consider two sources of circular water waves of the same wavelength, each similar to that shown in Fig. 10.26, and separated by a certain distance. At certain definite positions on the surface of the water, the amplitudes of the waves from the two sources will have the same sign (the waves will be *in phase*) and constructive interference will result; that is, at these positions the disturbance of the water will be enhanced. At other positions the waves will have opposite signs (the waves will be *out of phase*) and destructive interference will result; that is, at these positions the water will remain calm. Figure 10.32 is a photograph of the interference pattern produced by two sets of circular water waves.

If we allow a plane wave to strike a panel in which there are two slots (each with $d \ll \lambda$), Huygens' principle tells us that these slots will act as separate sources of circular waves. Therefore, the situation should be exactly the same as in Fig. 10.32, and a similar interference pattern should result. Figure 10.33 shows the geometry of such a case and the photograph in Fig. 10.34 demonstrates the validity of Huygens' principle in predicting interference effects.

THE INTERFERENCE CONDITION

Two separate sets of circular waves emanate from a pair of source slits, S_1 and S_2, as in Fig. 10.35. How can we determine whether constructive or destructive interference will occur at a point P that is at a distance L_1 from S_1 and at a distance L_2 from S_2? The nature of the interference depends

Fig. 10.32 *The pattern of constructive and destructive interference produced by two sets of circular water waves.*

Physics I by Elisha Huggins Benjamin

on whether the waves arrive at *P in phase* or *out of phase.* The wave that emanates from S_2 must travel a greater distance than the wave from S_1; this path difference is $L_2 - L_1$. Now, if there are exactly an integer number of wavelengths in the distance $L_2 - L_1$, the two waves will arrive at *P* with their maxima together (that is, *in phase*), and constructive interference will result. (Refer to Fig. 10.33 and verify that at any point along the dashed lines the difference in the distances from S_1 and S_2 is always *one* wavelength; for the central maximum, the distances are equal.) Therefore, the condition for constructive interference is

$$\frac{L_2 - L_1}{\lambda} = N, \qquad N = 0, 1, 2, \ldots \qquad \text{(constructive)} \qquad (10.26a)$$

Similarly, if the distance $L_2 - L_1$ contains an *odd* number of *half* wavelengths (that is, an integer number of wavelengths plus one-half wavelength), the waves will arrive at *P* with their maxima displaced from one another by $\frac{1}{2}\lambda$. Hence, the waves will be *out of phase* and destructive interference

Fig. 10.33 *A plane wave is incident on a panel containing two slots (S_1 and S_2, each with $d \ll \lambda$). The circular waves emanating from these slots produce an interference pattern similar to that shown in Fig. 10.32. The dot-dash line is the central maximum and the dashed lines show the positions of secondary constructive interference.*

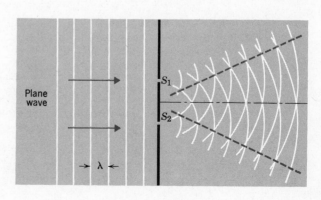

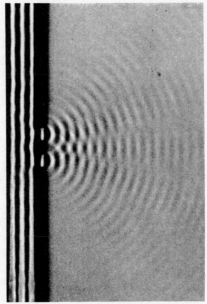

Fig. 10.34 *Photograph of the interference pattern in water waves produced in a geometry similar to that shown in Fig. 10.33.*

Physics I by Elisha Huggins Benjamin

will result. This condition is

$$\frac{L_2 - L_1}{\lambda} = N + \tfrac{1}{2}, \qquad N = 0, 1, 2, \ldots \qquad \text{(destructive)} \qquad (10.26b)$$

If $L_2 - L_1$ is between $N\lambda$ and $(N + \tfrac{1}{2})\lambda$, there will be oscillations of the medium but the amplitudes will be smaller than the maximum amplitude.

YOUNG'S DOUBLE-SLIT EXPERIMENT

The first clear demonstration that light is a wave phenomenon (and therefore exhibits interference effects as mechanical waves are known to do) was made by the English physicist, Thomas Young (1773–1829), early in the 19th century. Young allowed sunlight (essentially a plane wave) to illuminate a panel into which had been cut two very narrow slots or pinholes.[10] Thus, his experimental arrangement was the same as that shown

[10] For example, two pinholes in a sheet of black paper or two scribe marks on a smoked glass plate.

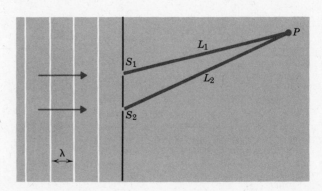

Fig. 10.35 *The condition for constructive interference at* P *is that* $(L_2 - L_1)/\lambda$ *is an integer number.*

in Figs. 10.34 and 10.35 except that the slots were much narrower. On a screen placed some distance away, Young found alternating bright and dark lines corresponding to the positions of constructive and destructive interference.[11]

We can account for the variation of the light intensity on the viewing screen in the following way. Refer to Fig. 10.36. Light waves with a definite wavelength λ are incident on the pair of narrow slits, A and B, which are separated by a distance D. The interference pattern is to be viewed on a screen that is a distance L away. (The diagram has been distorted for clarity; we must actually have L much greater than D.) At a position P on the screen, there will be a bright line, resulting from constructive interference, if the path difference Δ between the waves arriving from A and B is an integer number of wavelengths of the incident light. That is, for a bright line,

$$\Delta = N\lambda, \quad N = 0, 1, 2 \ldots$$

Now, if $L \gg D$, the angle $\angle \overline{ACB}$ will be essentially a right angle and $\overline{AC}$ will be approximately perpendicular to $\overline{PQ}$; therefore, the angle $\angle \overline{BAC}$ will be approximately equal to θ, the angle that specifies the position P on the screen. Consequently, the path difference Δ is given by

$$\Delta \cong D \sin \theta$$

Therefore, bright lines on the screen will be observed at positions for which θ follows the relation

Bright lines: $\boxed{N\lambda \cong D \sin \theta, \quad N = 0, 1, 2 \ldots}$ (10.27a)

There is a central maximum for $N = 0$ (that is, there is no path difference, $\Delta = 0$) and there are secondary maxima on either side that correspond to larger values of N.

[11]Interference can be observed if sunlight (which consists of light with many frequencies) is used, but the effect is much more striking if monochromatic light (that is, light with a *single* frequency) is used. See Question 10.9 and Problem 10.21.

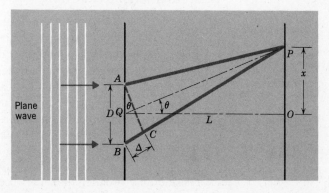

Fig. 10.36 *The geometry of Young's double-slit experiment. The viewing screen is located at a distance that is large compared to the slit separation, i.e.,* L $\gg$ D.

Dark lines in the interference pattern occur when Δ is equal to an odd number of half wavelengths. That is,

Dark lines: $(N + \tfrac{1}{2})\lambda \cong D \sin \theta, \qquad N = 0, 1, 2 \ldots$ (10.27b)

Photographs of the interference patterns obtained for slits with three different separations are shown in Fig. 10.37.

The spacing between adjacent bright lines on the screen can be calculated as follows. Because $L \gg D$, the distance from the midpoint between the slits (Q) to P is approximately equal to L. Therefore, $\sin \theta \cong x/L$, and Eq. 10.27a becomes

$$x \cong \frac{N\lambda L}{D}$$ (10.28)

Since N increases by unity in going from one bright line to the next, the spacing between adjacent bright lines (or dark lines) is just

$$\Delta x \cong \frac{\lambda L}{D}$$ (10.29)

Example **10.6**

In a double-slit experiment, $D = 0.1$ mm and $L = 1$ m. If yellow light is used, what will be the spacing between adjacent bright lines?

The wavelength of yellow light is approximately 6×10^{-5} cm (see Section 10.6). Therefore, the spacing is

$$\Delta x \cong \frac{\lambda L}{D}$$

$$= \frac{(6 \times 10^{-5} \text{ cm}) \times (100 \text{ cm})}{10^{-2} \text{ cm}}$$

$$= 0.6 \text{ cm}$$

Thus, the spacing between lines is about 6 mm or $\tfrac{1}{4}$ of an inch.

SINGLE-SLIT DIFFRACTION

If we allow a plane wave to illuminate a *single* slit of narrow width, we shall also find a regular pattern of constructive and destructive interference. Figure 10.38 shows a photograph of water waves incident on a slit which has a width only slightly larger than the wavelength of the incident wave. It is apparent that there is a broad central maximum in the pattern of transmitted waves. Closer examination will reveal that there are also weak secondary maxima with interference minima between them.

Light waves exhibit a similar effect. Figure 10.39 shows a photograph of the diffraction pattern on a screen located behind a single slit illuminated with light. This photograph clearly shows that the central maximum is considerably broader than the secondary maxima that lie on either side. In fact, a detailed analysis shows that the spacing between successive dark lines on either side of the central maximum is $\lambda L/d$, where d is the width

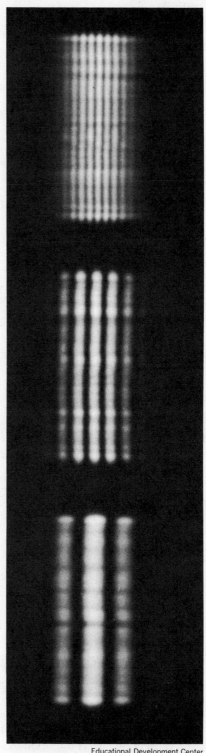

Fig. 10.37 *Double-slit interference patterns for slits with three different separations. Which case corresponds to the wide spacing and which to the narrow spacing?*

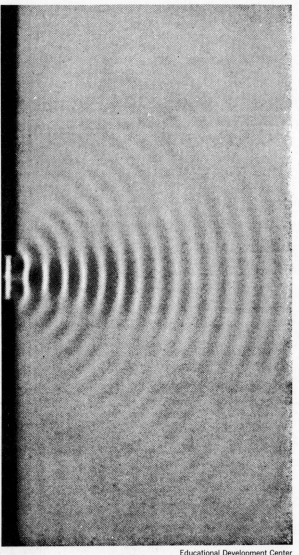

Fig. 10.38 *Interference pattern produced by water waves incident on a single slit. There is a broad central maximum with weaker secondary maxima on either side.*

Educational Development Center

of the slit, whereas the two central dark lines are separated by a distance $2\lambda L/d$. The graph above the photograph gives, in a schematic way, the distribution of light intensity along the screen. The intensities shown for the secondary maxima have been enlarged to show the pattern in more detail; these secondary maxima are actually quite weak.

MULTIPLE-SLIT DIFFRACTION

The shape of the diffraction pattern that results when light is passed through slits depends on the frequency of the light. Because of this fact, diffraction effects are often used instead of refraction effects in glass prisms (Fig. 3.8) to analyze the frequency spectrum of light from a source. The difficulty with using single or double slits for this purpose is that the various

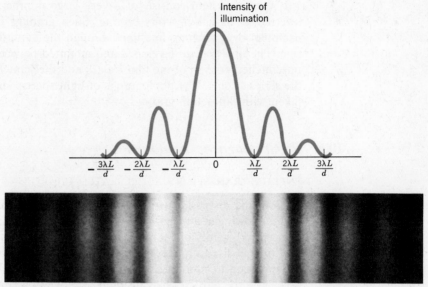

Fig. 10.39 *The diffraction pattern produced by a single slit. The photograph has been over-exposed in order to reveal the secondary maxima; therefore, the central maximum is "washed out" and does not appear in its full intensity. The intensity graph at the top is only schematic; the secondary maxima are actually considerably less intense than shown.*

intensity maxima for a given frequency of light are relatively broad so that the maxima for one frequency blend into those for another, not-too-different frequency. That is, the ability of a single or double slit to distinguish between (or to *resolve*) two nearby frequencies is poor. This defect can be overcome by using a diffraction panel into which are cut many slits; as the number of slits is increased, the resolution improves. Figure 10.40 shows a comparison of the intensity maxima for a double slit and for a set of 20 slits. There is a dramatic increase in the sharpness of the lines in the latter case.

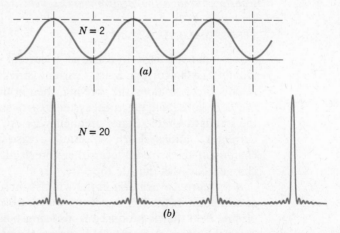

Fig. 10.40 *Intensity patterns for light of a single frequency passed through (a) a double slit and (b) a diffraction grating consisting of 20 slits.*

If the slit system consists of a very large number of slits, it is usually referred to as a *diffraction grating*. Such gratings can be prepared, for example, by scratching fine lines through the emulsion of a photographic film. The film is then developed and mounted between pieces of glass; the unscratched portion of the film is dark and the scratches are clear, forming the series of slits. The photographs of light spectra shown in Fig. 3.7 were taken with diffraction gratings.

10.6   *Electromagnetic Radiation*

ENERGY TRANSFER BY ACCELERATED CHARGES

From the discussions in the preceding chapter, we know that a steady current flowing in a wire will produce a static (that is, unchanging) magnetic field in the vicinity of the wire. If a charged particle moves in such a field, the force exerted on the particle by the field is always at right angles to the direction of motion of the particle. Therefore, although the field can accelerate the particle (by causing it to move in a circular orbit), the *speed* of the particle is not changed. A static magnetic field can do no *work* on the particle. That is, there can be no *energy* transferred from the moving charges in the wire to the particle via the intermediary of the field.

Now let us consider the case in which the current in the wire is allowed to vary so that the magnetic field is no longer static. A changing current produces a changing magnetic field. We know that a changing magnetic field produces an electric field and that this electric field can act on a charged particle to change its energy. Therefore, energy can be transferred from a wire carrying a changing current to a charged particle in the vicinity of the wire, although no such transfer can occur for the case of a wire carrying a steady current. The essential difference in the two situations is that the charges that move in the wire in the case of the changing current undergo accelerations, whereas there is no acceleration of the moving charges in the case of the steady current. *Only accelerating charges can produce energy transfers through the electromagnetic fields that they generate.*

ELECTROMAGNETIC WAVES

It is relatively easy to construct a current source that provides a sinusoidally varying current. Such a current varies in a regular way, first flowing in one direction along the wire and then in the opposite direction. (In fact, ordinary household electrical systems carry such *alternating current* or AC; the standard United States' frequency for AC current is 60 Hz.) Sinusoidal current variations clearly constitute a case of accelerated motion of the charges in the wire and can therefore lead to the transfer of energy to charged particles outside the wire.

If the current variation in the wire is sinusoidal, then the magnetic field at any point in the vicinity of the wire is also sinusoidal. Furthermore, the electric field that is produced is also sinusoidal. Therefore, at any point in space the electromagnetic field vectors, **E** and **B,** can be represented by the expressions

$$\left.\begin{array}{l} \mathbf{E} = \mathbf{E}_0 \cos \dfrac{2\pi}{\tau} t \\[2mm] \mathbf{B} = \mathbf{B}_0 \cos \dfrac{2\pi}{\tau} t \end{array}\right\} \qquad (10.30)$$

where $\mathbf{E}_0$ and $\mathbf{B}_0$ are the (vector) amplitudes of the varying field vectors and where τ is the period of the current oscillations that produce the field.

The generation of an electromagnetic field by the changing current in a wire is similar to the production of a traveling wave in a string by applying an oscillation to one end, as in Fig. 10.12. Just as a sinusoidal displacement in the string is propagated along the string, the sinusoidal variation of the field vectors, $\mathbf{E}$ and $\mathbf{B}$, is propagated through space (Fig. 10.41).

The similarity between the mechanical and electromagnetic cases goes further. For the mechanical wave motion shown in Fig. 10.13, the displacement as a function of time at any position is sinusoidal, and the displacement as a function of distance at any instant of time is also sinusoidal. The same is true for the electromagnetic case. That is, there is an *electromagnetic wave* that is propagated through space with a sinusoidal variation in space and in time.[12] At a given instant of time, there will be a sinusoidal variation of the field vectors, $\mathbf{E}$ and $\mathbf{B}$, along a particular direction in space away from the wire that carries the changing current (Fig. 10.42). A length of wire that carries a changing current and therefore generates electromagnetic waves is called an *antenna*. A simple antenna is shown in Fig. 10.41 and a large radio broadcast antenna array is pictured in Fig. 10.43.

In a region of space that is far from the antenna (located in the negative x-direction in Fig. 10.42), and at a given instant of time, there will be essentially no variation of the field vectors over the y-z plane. That is, the condition of the electromagnetic field is uniform over a plane that is perpendicular to the direction of propagation of the wave away from the antenna. The propagating electromagnetic wave is therefore a *plane wave*. Electromagnetic waves have the following important properties:

1. The electromagnetic field vectors, $\mathbf{E}$ and $\mathbf{B}$, in a *plane wave* are everywhere mutually perpendicular:

    $$\mathbf{E} \perp \mathbf{B} \qquad \text{(plane wave)} \qquad (10.31)$$

---

[12] But only at large distances from the wire that carries the changing current; near the wire the electromagnetic field is quite complicated.

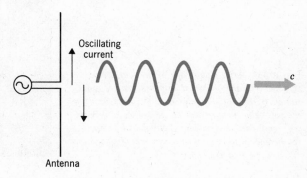

Fig. 10.41 *The sinusoidal variation of current flowing in a wire produces sinusoidal variations in* $\mathbf{E}$ *and* $\mathbf{B}$ *which are propagated through space with the velocity* c. *Here, the source of the electromagnetic wave is a simple antenna.*

Oscillating current

Antenna

c

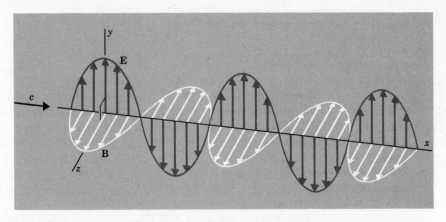

Fig. 10.42 *At a large distance from the antenna (located in the negative x-direction), the variation of* **E** *and* **B** *with distance at a given instant is sinusoidal. Furthermore, the electromagnetic wave is a plane wave and has* **E** ⊥ **B** *and* E = B *(in CGS units) at every point and for all values of the time.*

2. The magnitudes of the field vectors in a *plane wave* are everywhere identical:[13]

$$E = B \quad \text{(plane wave)} \tag{10.32}$$

Of course, **E** and **B** vary with position in space and with time, but at any particular position in space and for all values of the time, properties (1) and (2) are true.

3. The direction of propagation of an electromagnetic wave is given by a right-hand rule: the direction of propagation is the same as the

[13] That is, the magnitude of **E** (in statV/cm) is equal to the magnitude of **B** (in gauss).

Fig. 10.43 *Aerial photograph of the four towers that make up the broadcast antenna array of radio station WMAL in Washington, D.C. The reason for the particular arrangement of the towers is to beam the signal only in directions that will not produce interference with stations in other locations operating on the same frequency (630 kHz).*

Radio Station WMAL

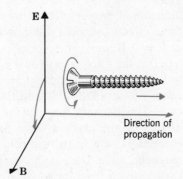

Direction of
propagation

Fig. 10.44 *The direction of propagation of an electromagnetic wave is the same as the direction of advance of a right-hand screw when it is turned in the sense that carries* **E** *into* **B.**

direction of advance of a right-hand screw when turned in the sense that carries **E** into **B** (Fig. 10.44). Even though the directions of the vectors **E** and **B** change with time and with position, the wave always propagates in the same direction. Application of the right-hand rule to the various portions of the wave in Fig. 10.42 shows this to be the case.

4. Electromagnetic waves are *transverse waves*. In Fig. 10.42 notice that the vectors **E** and **B** are always *perpendicular* to the direction of propagation of the wave; at large distances from the source, the field vectors never have any component in the direction of propagation.

5. The velocity of propagation in empty space of all types of electromagnetic waves (light, radio waves, X rays, etc.) is $c = 3 \times 10^{10}$ cm/sec.

ELECTROMAGNETIC ENERGY AND MOMENTUM

When electric charges undergo accelerations, a time-varying electromagnetic field is produced and electromagnetic waves are propagated outward from the source. It is easy to see that this *electromagnetic radiation* carries both energy and momentum. Consider a charged particle ($+q$) that is in the path of an electromagnetic wave (Fig. 10.45). If the particle is initially at rest, the magnetic field **B** will have no influence on the particle, but the electric field **E** will exert a force on the particle and will give it an acceleration in the direction of **E**. As soon as the particle has acquired a velocity in this direction, the magnetic field will exert a force on the particle. Application of the right-hand rule for the magnetic force (turn **v** into **B** and advance as a right-hand screw) shows that this force is in the

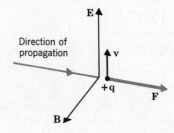

Direction of
propagation

Fig. 10.45 *An electromagnetic wave always exerts a force on a charged particle that is in the direction of propagation of the wave. (Does the direction of the force depend on the sign of the charge?)*

direction of propagation of the wave. Furthermore, the force will be in the same direction during the next half-cycle of the wave when the directions of **E** and **B** are both reversed. (Try the right-hand rule again.) Now, it is true that the force exerted on the particle by the electric field is perpendicular to the direction of propagation of the wave, but the direction of this force reverses every half-cycle and thus cancels when averaged over a complete cycle. The magnetic force, on the other hand, is always in the same direction. Therefore, the average force on the particle is the magnetic force. If a charged particle, initially at rest, can be given an acceleration by an electromagnetic wave, then both energy and momentum have been transferred to the particle from the wave. We must conclude that electromagnetic waves carry both energy and momentum.

Since an electromagnetic wave exists throughout a region of space and changes with time, how do we specify the *energy* of the wave? It proves convenient to speak in terms of the *density* of energy flow in the wave by stating the electromagnetic energy that crosses a unit area (perpendicular to the direction of propagation of the wave) in unit time. To this quantity we give the symbol S. A calculation, which is too involved to give here, shows that

$$S = \frac{c}{8\pi}(E_0^2 + B_0^2) \tag{10.33}$$

where E_0 and B_0 are the magnitudes of the vector amplitudes, $\mathbf{E_0}$ and $\mathbf{B_0}$. In the CGS system $[S] = \text{erg/cm}^2\text{-sec}$. Since $E = B$ in a *plane wave,* half of the energy associated with the wave is electric and half is magnetic. For purposes of calculation, we can substitute $2E_0^2$ or $2B_0^2$ for $E_0^2 + B_0^2$ in Eq. 10.33 for such waves. That is,

$$S = \frac{c}{4\pi}E_0^2 = \frac{c}{4\pi}B_0^2 \quad \text{(plane waves)} \tag{10.34}$$

The momentum associated with an electromagnetic wave can be obtained by a calculation similar to that for the energy flow in the wave, but a simple argument will suffice for our purposes here. Consider a portion of an electromagnetic wave[14] that has an energy $\mathcal{E}$. According to the Einstein mass-energy relation (Eq. 7.53), a mass m has a mass-energy $\mathcal{E} = mc^2$. Or, conversely, an entity that possesses a total energy $\mathcal{E}$ has an equivalent mass, m. Therefore, electromagnetic radiation that has an energy $\mathcal{E}$ has an equivalent mass $m = \mathcal{E}/c^2$. And since electromagnetic radiation travels with a velocity c, the momentum of the radiation (that is, equivalent mass $\times$ velocity) must be

$$\boxed{p = mc = \frac{\mathcal{E}}{c}} \tag{10.35}$$

That electromagnetic radiation carries energy should come as no surprise. After all, energy from the Sun carried to the Earth by electromagnetic radiation supplies the light and heat necessary to sustain life on Earth. But

[14]As we shall see later, electromagnetic radiation actually exists in discrete units called *photons.* We consider here such a photon with an energy $\mathcal{E}$.

it is more difficult to appreciate the fact that electromagnetic waves have *momentum.* Why is it that the momentum which accompanies the enormous amount of energy supplied to the Earth by the Sun does not push the Earth out of its orbit? Why do we see the light and feel the heat from a light bulb but we have no sensation of being *pushed* by the momentum of the light? The answer, of course, lies in the fact that the momentum associated with radiation of energy $\mathcal{E}$ is equal to $\mathcal{E}/c$ and c is a very large number. Therefore, the momentum transferred by electromagnetic radiation is always very small, and, in fact, it is quite difficult to measure the effects of electromagnetic momentum in the laboratory.

There is, however, a striking effect, readily observable, that is the direct result of the transfer of momentum by electromagnetic radiation. Comets are members of the solar system which are composed of swarms of solid particles, rocky material, and frozen gases.[15] Cometary orbits are highly elongated,[16] and so most of the time a comet is far from the Sun. When it approaches the Sun, however, the solar radiation vaporizes and "boils off" some of the cometary material. This material is subject to the effects of electromagnetic momentum carried by the solar radiation (*radiation pressure* or *light pressure*). As a result, the vaporized material is forced away from the comet and is visible as the cometary *tail* (Fig. 10.46). The effect of radiation pressure is always to force the tail of a comet to point away from the Sun (Fig. 10.47), so that the tail can even precede the head of the comet through space. This behavior of cometary tails has long been known (even in Kepler's time), but the explanation was lacking until it was recognized to be the result of momentum carried by electromagnetic radiation.

AMPLITUDE AND INTENSITY

The electromagnetic field at any position in space is completely specified by the vectors **E** and **B** at that point. In almost all cases of interest, however, we are concerned not with the field vectors themselves but with how much work can be done by the field on a charged particle or a group of charged particles. That is, we are interested in the *energy* content of the field. The reception of a radio signal by a receiver antenna or the action of a light wave on the eye both depend on transfers of energy. In Eq. 10.11 we found that the energy of a vibrating mass is proportional to the *square* of the amplitude of the vibration. And in Eq. 10.33 we stated that the energy flow in an electromagnetic wave is proportional to the *squares* of the amplitudes of the field vectors.

When we speak of the *intensity* of a wave, we refer to the *energy* content of the wave, and so *intensity* is always proportional to the square of the *amplitude* of the wave:

$$\left.\begin{array}{l} \text{Amplitude} = \mathbf{E}_0 \\ \text{Intensity} \quad \propto |\mathbf{E}_0|^2 = E_0^2 \end{array}\right\} \tag{10.36}$$

---

[15] The astronomer Fred Whipple has said that a comet resembles a "dirty iceberg."

[16] The orbits are ellipses but, generally, the elongation is so great that the orbits are essentially indistinguishable from parabolas.

Fig. 10.46 *Comet Seki-Lines photographed from Frazier Mountain, California, August 9, 1962.*

Photograph by Alan McClure

THE SPECTRUM OF ELECTROMAGNETIC RADIATION

In 1862, James Clerk Maxwell, predicted on the basis of his theory of electromagnetism that electromagnetic *waves* should exist. His calculations showed that these electromagnetic waves should propagate with a velocity that was the same as that previously found for the propagation of light in air (or empty space). This fact immediately suggested that light is just a particular form of an electromagnetic wave.

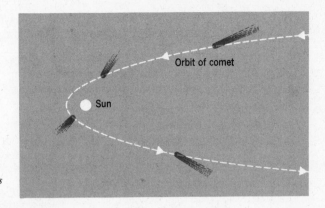

Fig. 10.47 *Because of radiation pressure (transfer of electromagnetic momentum), cometary tails always point away from the Sun.*

Electromagnetic waves (apart from *light*) were not observed until 1887 when Heinrich Hertz produced waves with wavelengths of 10–100 m by forcing a charged sphere to spark to a grounded sphere. Hertz's experiments indicated that these waves were identical in every respect to light waves, except that the wavelength was much longer.

Following the pioneering experiment of Hertz, electromagnetic waves were generated with an ever-increasing range of frequencies. Indeed, it seemed that waves with *any* frequency could be produced if some method of driving electric charges with the appropriate oscillation frequency could be found. This is, in fact, the situation today: there appears to be no physical limitation on the frequency of electromagnetic waves—all we require is a suitable source. Electronic methods have been used to generate electromagnetic

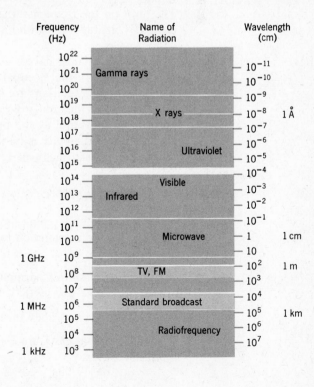

Fig. 10.48 *A part of the spectrum of electromagnetic radiation. The classification terms used for electromagnetic radiations are not well-defined; there is actually considerable overlapping of adjacent categories.*

Table **10.3**   *Some Typical Radiations in and near the Visible Spectrum*

Name	Wavelength (cm)	Wavelength ($\mathring{A}$)	Frequency (Hz)
Near infrared	1.0×10^{-4}	10,000	3.0×10^{14}
Longest visible red	7.6×10^{-5}	7,600	3.9
Orange	6.1×10^{-5}	6,100	4.9
Yellow	5.9×10^{-5}	5,900	5.1
Green	5.4×10^{-5}	5,400	5.6
Blue	4.6×10^{-5}	4,600	6.5
Shortest visible blue	4.0×10^{-5}	4,000	7.5
Near ultraviolet	3.0×10^{-5}	3,000	10
X ray (long wavelength)	3.0×10^{-6}	300	100

waves with frequencies up to about 10^{12} Hz. In this range of frequencies we classify the radiation as either *radiofrequency* (RF) waves or as *microwaves* (see Fig. 10.48). In the former category are the standard broadcast, FM, TV, air and marine, and amateur broadcast bands. Radar and point-to-point relay signaling use microwaves.

In order to generate radiation with frequencies above the microwave range, direct electronic methods are no longer useful, and we employ *atomic* radiations. *Infrared* or *heat radiation* lies in the frequency range between microwaves and the narrow band of frequencies that constitute *visible* radiation. At still higher frequencies are *ultraviolet* radiation and *X rays*. The limit on the frequency that can be generated by atomic systems lies near 10^{20} Hz; radiation with higher frequencies (*gamma rays*) are produced within *nuclei*.

The fact that names have been assigned to these various bands of radiation must not obscure the essential feature of electromagnetic waves, namely, that all of these radiations are *identical* in character and differ only in their *frequency*. Thus, the radio waves that are broadcast from a 500-ft radio tower are exactly the same as the penetrating gamma rays that originate in a nucleus whose diameter is only 10^{-12} cm; the only difference between the two radiations is a factor of 10^{15} or so in frequency.

Summary of Important Ideas

For a particle undergoing simple harmonic motion, the displacement, the velocity, and the acceleration can all be represented by *sinusoidal* functions of the time.

The *period* of simple harmonic motion is independent of the amplitude of the vibration.

Waves of all types obey the principle of *superposition*.

Sound waves in gases are *longitudinal* waves; electromagnetic waves are *transverse* waves.

The observed *frequency* of a wave depends on the relative motion of the source and the observer (the *Doppler effect*).

Standing waves can be represented as the superposition of *traveling waves*.

Standing waves are produced on strings or in enclosures when the wave disturbance always has *nodes* at the end positions.

The propagation of a wave in space can be determined by considering every point on a given wave front to be the source of out-going spherical waves (*Huygens' principle*).

Diffraction causes waves to "bend around" obstacles and to spread out upon passing through narrow apertures.

When waves from two (or more) sources arrive at a particular point, *constructive interference* results if the waves are *in phase* and *destructive interference* results if the waves are *out of phase.*

Electromagnetic waves are produced only by electric charges that are undergoing *accelerations.*

In a *plane electromagnetic wave* the field vectors, **E** and **B,** are mutually perpendicular.

An electromagnetic wave possesses and can transfer both energy and momentum.

The *intensity* of a wave (that is, the *energy* content) is proportional to the *square* of the *amplitude.*

All electromagnetic waves have *identical* properties; they can differ only in *frequency.*

Questions

10.1 What information would be required to calculate the maximum velocity experienced by the bob of a simple pendulum? Outline the method that you would use to make the calculation.

10.2 Sound waves do not propagate for great distances through air; eventually, the waves "die out." What happens to the *energy* in the sound wave?

10.3 Compare the average velocity of molecules in air (Table 7.5) with the velocity of sound in air (Table 10.2). Why should these two numbers be so close? Why should the sound velocity be *less* than the molecular velocity? (Consider how sound waves are propagated through air.)

10.4 Transverse waves are possible in solids but not in gases. What is the physical differences in the two situations that accounts for this fact?

10.5 Explain why small rooms are not generally suited for listening to high-fidelity music.

10.6 What do you think would be your bodily reaction to an intense sound wave of wavelength $\lambda = 50$ m? (What is the frequency of the wave?)

10.7 Explain how the string in Fig. 10.11c, which has no displacement at any position, can have a displacement at a later time. Describe the physical situation that allows a displacement to be generated from a condition of no displacement.

10.8 Is there a Doppler effect when the source moves perpendicular to the line connecting the source and the observer?

10.9 Suppose that a double slit is illuminated with *white light* (that is, light containing *all* frequencies) instead of *monochromatic light* (light of a *single* frequency). Describe the interference pattern that would be produced.

10.10 A sound wave in air and an electromagnetic wave in air each have a wavelength of 10 cm. Classify the two waves (audible, inaudible; light, radio wave, etc.). Why are two waves of the same wavelength so different in their properties?

10.11 The Sun radiates huge amounts of electromagnetic energy. We know that such radiation carries momentum. Does the momentum of the Sun change with time? Explain.

10.12 Argue that if a particle in the interplanetary region of the solar system is sufficiently small, radiation pressure will overcome the solar gravitational attraction and the particle will be accelerated *away* from the Sun. (Consider the particle to be spherical. What properties of the particle determine the magnitude of the gravitational force? What properties determine the magnitude of the force exerted by radiation pressure? Examine the way in which these forces change as the size of the particle is reduced.) Particles with radii less than about 2×10^{-4} cm are actually expelled from the solar system by radiation pressure.

10.13 Huge antennas are required to generate radio waves, but X rays are produced by atoms and gamma rays are produced by nuclei. Why are electromagnetic waves of the highest frequency generated by the smallest systems?

Problems

10.1 A force of 500 dynes applied to a 20-g mass that is attached to a spring displaces the mass 5 cm from its equilibrium position. What maximum velocity will the mass attain after release? What maximum acceleration will the mass experience? What will be the period of the motion?

10.2 A mass attached to a spring vibrates with a period of 0.5 sec. When the mass passes through its equilibrium position, it is moving with a velocity of 20 cm/sec. What is the amplitude of the motion?

10.3 A 10-g mass attached to a spring vibrates with a period of 2 sec. How much force is required to stretch the spring by 10 cm starting from its equilibrium position? (Use the approximation that $\pi^2 \cong 10$.)

10.4 A 20-g mass vibrates at the end of a spring. When the mass passes through the equilibrium position, the velocity is 40 cm/sec. What amount of work was required initially to stretch the spring?

10.5 A simple pendulum of length 24.8 cm will have a period of 1 sec at the surface of the Earth (see Example 10.2). What will be the period of such a pendulum at the surface of the moon?

10.6 A pendulum has a period of 2.00 sec when vibrating near the surface of the Earth at the equator. What will be the period of this pendulum at the North Pole? (See Fig. 4.16.)

10.7 A flash of lightning is observed and 8 sec later a clap of thunder is heard. About how far away was the lightning?

10.8 A stick of dynamite is exploded on the surface of the sea. The sound is propagated through the water as well as the air. At a distance of 3 km, which signal will be heard first? What will be the time interval between the arrival of the two signals?

10.9 What is the wavelength of a 10-kHz sound wave in iron? What would be the wavelength of the same wave in air?

10.10 What is the speed (in km/hr) of a submerged submarine that moves at Mach 0.0025? (Mach numbers are never actually used to specify such speeds because the values are so small.)

10.11* Sketch a sinusoidal wave which has a wavelength λ and another of the same amplitude which has a wavelength $\lambda/4$. Add the two waves graphically to produce the wave that results from the superposition of one on the other and show that a "modulated" wave is obtained.

10.12 An organ note ($\lambda = 22$ ft) is sustained for 1 sec. How many full vibrations of the wave have been emitted?

10.13 Waves travel with a velocity of 20 m/sec on a certain taut string, which is attached to two supports that are separated by 2 m. What are the frequencies of the first four standing waves (starting with the longest wavelength) that can be set up in the string? Which of these modes of vibration will produce an audible sound?

10.14 A train approaches you at a speed of 60 mi/hr and its whistle emits a 2-kHz note. What frequency do you hear? When the train passes, what frequency do you hear?

10.15 As a train passes you, you hear the frequency of its whistle drop from 1000 Hz to 800 Hz. What is the speed of the train in mi/hr.

10.16 Show that the wavelength of a particular spectral line in the light from a distant galaxy as observed on Earth (λ_E) is given in terms of the wavelength of that line from a laboratory source (λ_S) by $\lambda_E = \lambda_S \times (1 + v_G/c)$, where v_G is the recessional velocity of the galaxy.

10.17 Show by means of constructions that Huygens' principle yields the correct results for the propagation of *plane* waves and *circular* waves (in *two* dimensions).

10.18 A measurement will show that the waves in Fig. 10.32 have a wavelength that is $\frac{1}{3}$ of the distance between the two sources. Use a compass and construct two sets of circular waves that duplicate the conditions of Fig. 10.32. Show that the regions of constructive and destructive interference are the same as those in the figure.

10.19* In Fig. 10.32, suppose that the two sources do not vibrate exactly together, but that the upper source is at its maximum displacement when the lower source is going through its equilibrium position. (That is, the two waves produced by the sources are shifted in time by $\frac{1}{4}\tau$.) Use a Huygens' construction similar to that shown in Fig. 10.33 and identify the lines of constructive and destructive interference. Is there a central maximum?

10.20 Two narrow slits are so close together that a direct measurement of their separation is difficult to make. By illuminating these slits with light ($\lambda = 5 \times 10^{-5}$ cm) it is found that on a screen 4 m away adjacent bright lines in the interference pattern are separated by 2 cm. What is the separation of the slits?

10.21* Blue light ($\lambda_B = 4 \times 10^{-5}$ cm) and yellow light ($\lambda_Y = 6 \times 10^{-5}$ cm) simultaneously illuminate a pair of slits whose separation is 0.02 mm and bright diffraction lines (blue and yellow) are formed on a screen that is 2 m away. If the central lines of the two colors are numbered "zero," what will be the numbers of the lines at the first point beyond the central lines where a blue line falls at exactly the same place as a yellow line? How far is this point from the central lines?

10.22 Light waves have the same *frequency* whether they travel through air (essentially the same as empty space as far as light waves are concerned) or through some medium such as glass or water. But in glass or water, the velocity of light is only about 0.7 c and so the *wavelength* of the light is different. Suppose that the double-slit apparatus of Example 10.6 were immersed in water. What would be the spacing between adjacent bright lines on the screen?

10.23 Yellow light ($\lambda = 6 \times 10^{-5}$ cm) illuminates a single slit whose width is 0.1 mm. What is the distance between the two dark lines on either side of the central maximum if the diffraction pattern is viewed on a screen that is 1.5 m from the slit?

10.24 Microwaves ($\lambda = 0.5$ cm) are incident on a pair of slits that are separated by 25 cm. Describe the intensity pattern that would be found by moving a microwave detector along a screen that is 15 m from the slits.

10.25 What is the wavelength of the 25-MHz radiation that WWV uses to broadcast time signals? What is the frequency of radiation in the 10-m shortwave broadcast band?

10.26* The maximum electric field strength in a certain electromagnetic plane wave is 300 microvolts/meter. (This is a typical value for radio waves.) How much electromagnetic energy flows across an area of 1 cm² in 1 sec? (It is necessary to convert from *volts* to *statvolts* in order to use CGS units throughout.)

10.27 A proton oscillates back and forth across the diameter of a nucleus (about 10^{-12} cm) with a velocity $v \cong 0.05c$. What is the approximate frequency of the emitted radiation? (The frequency of the radiation is equal to the frequency of the oscillation.) How would you classify this radiation?

Relativity M. C. Escher

11 *Relativity*

During the latter part of the 19th century, it became clear that physical theory was in serious trouble. Newtonian dynamics was well established; it was recognized that this theory is valid in any inertial reference frame and that all such frames are equivalent. Maxwell's theory of electromagnetism was also well established and it was understood that light is a wave phenomenon correctly described by Maxwell's equations. An integral part of the theory of the propagation of electromagnetic waves was the concept of the *ether*.[1] Because of the mechanistic view of electromagnetism that was in vogue, it was considered essential that an ether exist in order to provide a medium in which the waves could propagate. Maxwell's equations were considered valid in a reference frame at rest with respect to the ether. Unlike Newton's equations, which were known to be valid in *all* reference frames, Maxwell's equations seemed to demand a *preferred* reference frame.

By continuing to force mechanical solutions on the problems of electromagnetism, physicists worked themselves deeper and deeper into a hole. It was necessary to make numerous *ad hoc* assumptions in order to account for an increasing number of experimental facts. Eventually, almost all of the pertinent equations that we now accept as correct were, in fact, worked out. But these results were not at all satisfying; they were mixed in a morass of assumptions and tied indissolubly to the elusive *ether*.

The ether theory entered finally into a state of terminal shock when three separate experiments required the following conclusions: (a) the ether is dragged along by the Earth so that laboratory experiments are always carried out *at rest* with respect to the ether;[2] (b) the Earth moves freely through the ether which remains at rest with respect to the "fixed" stars; and (c) a moving material medium (such as water), through which light can propagate, drags along the ether but with only *half* the velocity of the medium. Faced with these contradictions, the ether theory, at last, collapsed.

In 1905, a crucial new idea was contributed by Albert Einstein (1879–1955). In a single bold stroke, Einstein swept away the ether theory and all of the *ad hoc* assumptions and replaced them with only two postulates. Using these postulates as a foundation, he was able to construct a beautiful theory which is a model of logical precision. Einstein's relativity theory provided the key link between mechanics and electromagnetism—it unified the two great theories of classical physics.

Einstein's solution to the problem required that we discard the previous belief that space and time are distinct and unrelated concepts. In Einstein's view we do not exist in a three-dimensional space on which we superimpose the concept of time; instead, he proposed that space and time coordinates exist together on an equal basis in a four-dimensional world of *space-time*.

When we first encounter them, the ideas of relativity seem somewhat strange and forced. But relativistic effects are important only when velocities

---

[1] See the introductory section of Chapter 8 for additional comments on the *ether*.

[2] This was the conclusion based on the famous experiment carried out by Michelson and Morley in the 1890s.

Fig. 11.1 *Albert Einstein as a young man (1905).*

Lotti Jacobi

approaching the velocity of light are encountered, and our intuition is based on our everyday experience in which we almost never meet situations involving such high velocities. Perhaps if we were reared in a much faster-moving world, relativistic concepts would be natural and easy to accept. Nevertheless, we must adhere to the tenet that if experimental facts conflict with our preconceived notions, we cannot change the facts—only our ideas. After all, it was this same brand of "common sense" that once supported the view that the Earth is flat.[3]

11.1   *Light Signals—The Basis of Relativity*

THE GALILEAN TRANSFORMATION AND ITS FAILURE

The principle of *Galilean* (or *Newtonian*) *relativity* states that the laws of mechanics are the same in any inertial reference system. Newton's equations of dynamics satisfy this principle—if $\mathbf{F} = m\mathbf{a}$ is valid in an inertial system K, then the equation also holds in a system K' that moves at *constant velocity* with respect to K (Fig. 11.2). The reason for this equivalence of the two systems is that acceleration is the *rate of change* of velocity; this rate of change is not influenced by the relative motion of K and K' and so $\mathbf{a}$ is the same in both systems (see Section 5.4).

If we wish to express the position of an object in one of these reference frames in terms of the coordinates in the other frame, we make use of the so-called *Galilean transformation equation*. Consider an object P (Fig. 11.3) that lies at a distance x from the origin O of system K. To an observer in

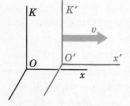

Fig. 11.2 *The coordinate system* K *is an inertial system.* K' *moves with constant velocity relative to* K *and so* K' *is also an inertial system. The laws of mechanics are the same in both systems.*

[3] Einstein once remarked that "common sense" is the layer of prejudices built up before the age of 18.

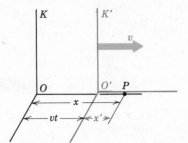

Fig. 11.3 *The Galilean transformation equation,* x′ = x − vt, *connects the coordinates of* P *as viewed by observers in* K *and in* K′.

K' the position of P is changing with time according to

$$x' = x - vt \quad \text{(Galilean)} \tag{11.1}$$

(Primed quantities always refer to the K' system.)

In Newtonian dynamics, *time* is considered to be an *absolute* quantity; that is, the specification of time is unique and its value is the same in all inertial reference frames, regardless of their motions; that is, $t = t'$, always.

Although the principle of Newtonian relativity, embodying the Galilean transformation equation and the concept of absolute time, is adequate for the description of all ordinary mechanical processes, the principle fails when applied in the realm of electrodynamics. This failure is demonstrated in the situation pictured in Fig. 11.4. A long, straight, uniformly charged wire lies parallel to the x-axis in K and is at rest in this system (Fig. 11.4a). There is a stationary charge q a distance r away from the wire. Clearly, q experiences a repulsive force $\mathbf{F}_E$ due to the charged wire. Now, in Fig. 11.4b, we have the same wire and charge as viewed by an observer in the system K' which is in motion relative to K. Since K' moves to the *right* with respect to K, the observer in K' sees the wire and the charge moving to the *left*. Since the wire is long and uniformly charged, the K' observer calculates the repulsive electrical force that acts on q and obtains $\mathbf{F}_E$, the same as that calculated by the observer in K. The moving charged wire constitutes a uniform current and since the charge q is also in motion, the K' observer knows that there is a *magnetic* force on q in addition to the electrical force. Application of the rule for determining the direction of a magnetic force shows that $\mathbf{F}_M$ is directed *opposite* to $\mathbf{F}_E$. Therefore, the observer in K' concludes that the net force on q is *smaller* than the value obtained in the K system.

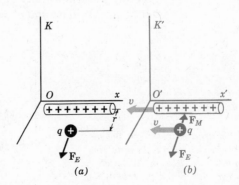

Fig. 11.4 *Using the principle of Newtonian relativity, the observers in the systems* K *and* K′ *reach different conclusions regarding the net force on the charge* q.

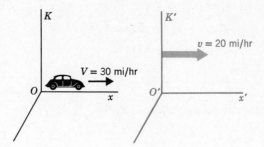

Fig. 11.5 *According to Galilean theory, an observer in K' would measure the velocity of the automobile to be 10 mi/hr.*

This is an intolerable situation. Taken at face value it means that there is a fundamental difference between the laws of *mechanical* dynamics (which are the same in all inertial reference frames) and the laws of *electro*-dynamics (which are *not* the same). But how do we distinguish between *mechanical* and *electrical* systems? All matter is composed of charged particles and electrical forces are responsible for holding matter together. Thus, all mechanical systems involve *charged* particles and all electrodynamic systems involve the motion of particles that have *mass*. We can accept no alternative to the statement that *all* physical laws must be the same in *all* inertial reference frames. We are forced to conclude that some essential feature of the problem has escaped us. The resolution of this dilemma (and others) has led to the development of the *theory of relativity*.

THE VELOCITY OF LIGHT IS CONSTANT

Suppose that an automobile is moving along the x-axis of system K with a velocity of 30 mi/hr (Fig. 11.5). If K' moves relative to K with $v = 20$ mi/hr, then to an observer in K' the automobile would appear to be traveling with a velocity V' where

$$V' = V - v \tag{11.2}$$

or $V' = 10$ mi/hr. If the automobile reversed its direction of motion in K, the K' observer would measure the new velocity to be -50 mi/hr.

Suppose that we now replace the automobile with a pulse of light that travels with a velocity c in the K system (Fig. 11.6). The K' observer can determine the velocity of this light pulse in his system by measuring the transit time t' between a pair of light detectors, A' and B', that are separated by a precisely known distance l'; that is, $c' = l'/t'$. What result will the K' observer obtain? Will he find $c' = c - v$? In view of the fact that essentially

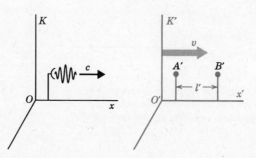

Fig. 11.6 *The K' observer can determine the velocity of the light pulse emitted in K by measuring the transit time between the detectors A' and B'. The result is that light travels with the velocity c in K' as well as in K.*

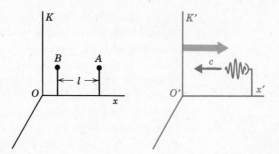

Fig. 11.7 *The* K *observer measures the velocity in his system of the light pulse emitted in* K' *to be* c, *not* c − v.

all of our everyday experience is governed by Newtonian principles, the answer to this question is somewhat surprising: in K', the velocity of the light pulse is just c! Similarly, if a light pulse were emitted by a source in the moving system K' (Fig. 11.7), the K observer could also determine the velocity in his system by measuring the transit time between the detectors A and B. The K observer also finds the velocity in his system to be c.

The results of these experiments can be simply stated: *The velocity of light is independent of any relative motion between the light source and the observer.* Modern experiments have shown that the velocity of light (in a vacuum) is $c = (2.997\,925 \pm 0.000\,010) \times 10^{10}$ cm/sec. In this book, we shall usually use the approximate value, $c = 3 \times 10^{10}$ cm/sec.

Because it is extremely difficult to achieve velocities exceeding 10^5 cm/sec in the laboratory, it is not practical to test the constancy of the velocity of light with experiments on the Earth that are of the type just described. (If $v = 10^5$ cm/sec, extreme precision would be necessary to distinguish between c and $c - v$.) However, there are astronomical observations that can be made, which clearly show that the velocity of light is constant and independent of the motion of source or observer. Figure 11.8 shows a *binary star system,* a pair of stars that revolve in orbits around one another. Such systems are by no means rare—probably half of the stars in our Galaxy exist in binary systems. The orbital velocities of binary stars are quite large, often exceeding 3×10^6 cm/sec.

In Fig. 11.8 we suppose that both stars of the binary system are of equal size and equal brightness and that we view the system in a direction that

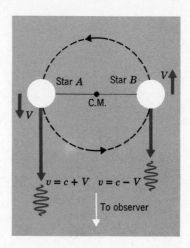

Fig. 11.8 *If the velocity of light depended on the motion of the source, the light from the stars in a binary system would travel toward us with velocities* c + V *and* c − V.

passes through the plane of the orbit. Therefore, when the stars are in the positions shown, one is approaching with the velocity V while the other is receding with the same velocity. If the velocity of light depended on the motion of the source, the light from the approaching star (A) would travel toward us with a velocity $c + V$, while the light from the receding star (B) would have a velocity $c - V$. If $V = 30$ km/sec $= 3 \times 10^6$ cm/sec and if the stars are 100 light-years away (only 1 percent of the size of our Galaxy), then the difference in the arrival times for the light signals traveling with velocities of $c + V$ and $c - V$ would be approximately *one week*. Therefore, the news from star B, transmitted by light signals, would arrive at the Earth a week later than the news from A. However, if the rotation period of the stars is 12 days, then it takes 6 days for star B to move into the position occupied by star A. Therefore, if the star-B people are interested in speedy transmittal of their news, it behooves them to wait 6 days until they have moved into the position where the light velocity will be $c + V$ and thereby save one day in transmission time. Notice also that the variation in light velocity would cause each of the stars to be apparently in *two* positions at the same time—a ludicrous situation indeed!

Apart from comparing the datelines on the news dispatches from star A and star B, what observations can be made to ascertain the velocity of light emitted by the two stars? The simplest measurement that we can make is to record as a function of time the *intensity* (that is, the *amount*) of light received from the system as a whole. (Thus, we do not even require that our instruments be able to give separate images of the stars.) If the velocity of light is independent of the motion of the source, then the so-called *light curve* that we measure will be that shown in Fig. 11.9a. The intensity will be constant except for the brief intervals when one star passes behind (*eclipses*) the other star and the intensity drops to one-half its normal value. On the other hand, if the velocity of the source must be added to (or subtracted from) the velocity of light, then by the time the light has reached the Earth, some of the "fast" light will have overtaken some of the "slow" light and the intensities will add. That is, the light will be "bunched up,"

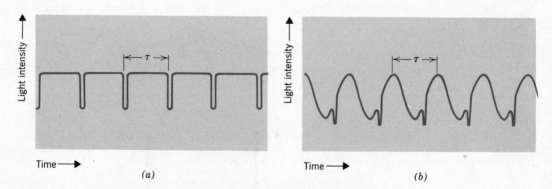

(a)   (b)

Fig. 11.9 (a) *The light curve for a binary star system that would be obtained if the velocity of light is independent of the motion of the source.* (b) *The light curve for a binary star system if the velocity of light were to depend on the motion of the source. For all binary star systems observed in the geometry of Fig. 11.8, the light curve is of the form* (a). τ *is the period of the motion and contains an eclipse of each of the two stars.*

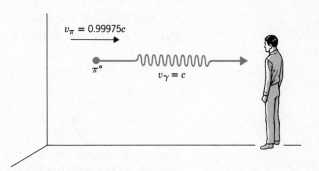

Fig. 11.10 *A neutral pion* ($\pi°$) *is traveling with a velocity* $v_\pi = 0.99975$ c *as measured by the observer in the laboratory. When the pion decays, it emits gamma radiation (the same as high-frequency light), the velocity of which in the laboratory is measured by the observer to be* c. *This experiment is the most exacting laboratory test so far performed of the postulate that the velocity of light is always* c, *independent of any relative motion between the light source and the observer.*

with periods of greater than average intensity followed by periods of less than average intensity. The resulting light curve will have the shape shown in Fig. 11.9b. There will be a sinusoidal variation in light intensity with the eclipse effect superimposed. (The time interval between an intensity maximum and the next drop in intensity due to eclipse depends on the orbital velocity, the period of the orbit, and the distance from the Earth.) Notice that the time between successive maxima in Fig. 11.9b and between successive eclipses in Fig. 11.9a is the same and is equal to one-half of the orbit period τ.

No light curve of the shape shown in Fig. 11.9b has ever been observed; all light curves from binary star systems have the form of Fig. 11.9a. (Of course, if the line of observation does not pass through the plane of orbit, the shape will be different but the interpretation of such curves is always consistent with the constancy of the velocity of light.)

An elegant demonstration that the velocity of light is constant, independent of the source velocity, has been made by measuring the velocity of the light emitted in the decay of a beam of neutral pions which had a velocity of $0.99975c$. The experiment showed that the light velocity was c to within 1 part in 10^4 (see Fig. 11.10).

Although many different types of experiments have been performed, none has ever shown a result in contradiction with the statement that the velocity of light is the same for all observers.

THE VELOCITY ADDITION RULE

How can the statement that the velocity of light is independent of the motion of the source be consistent with the fact that all ordinary mechanical velocities simply *add* algebraically, as in Fig. 11.5? Einstein showed that the simple addition formula for mechanical velocities is not correct and must be modified. If two velocities v_1 and v_2 are to be added, the sum is

$$V = \frac{v_1 + v_2}{1 + \dfrac{v_1 v_2}{c^2}} \tag{11.3}$$

In the event that v_1 and v_2 are small compared to the velocity of light (the case for all ordinary mechanical situations), the term $v_1 v_2/c^2$ is much less than unity and can be neglected. Then, the sum velocity is $V = v_1 + v_2$, identical to the result that would be obtained by applying Newtonian reasoning.

If one of the velocities is the velocity of light, $v_1 = c$, then

$$V = \frac{c + v_2}{1 + \dfrac{cv_2}{c^2}} = \frac{c + v_2}{1 + \dfrac{v_2}{c}} = \frac{c + v_2}{\left(\dfrac{c + v_2}{c}\right)} = c$$

This result insures that the velocity of light is the same for all observers because no matter what velocity v_2 is added to c, the addition rule always yields c. In particular, if $v_1 = c$ and $v_2 = c$, we still have $V = c$.

Example **11.1**

A spaceship moving away from the Earth at a velocity $v_1 = 0.75\,c$ with respect to the Earth, launches a rocket (in the direction *away* from the Earth) that attains a velocity $v_2 = 0.75\,c$ with respect to the spaceship. What is the velocity of the rocket with respect to the Earth?

$$V = \frac{v_1 + v_2}{1 + \dfrac{v_1 v_2}{c^2}} = \frac{0.75c + 0.75c}{1 + \dfrac{(0.75c)(0.75c)}{c^2}} = \frac{1.5c}{1 + 0.5625} = 0.96c$$

Therefore, in spite of the fact that the simple sum of the two velocities exceeds c, the actual velocity relative to the Earth is slightly less than c.

THE MOMENTUM OF LIGHT

If light signals are received by an observer from two identical sources, one stationary and one in motion with respect to the observer (Fig. 11.11), the result of the analysis just given requires that the observer measure equal velocities for the two signals. But there is clearly a physical difference in the two cases—one source is moving and the other is not. How will this difference be manifest in the two signals?

The answer to this question lies in taking account of the *Doppler effect* for moving sources. As was discussed in Section 10.2, the *frequency* of the sound wave or light wave emitted by a source that is approaching an observer is *increased* over that from an identical stationary source. Therefore,

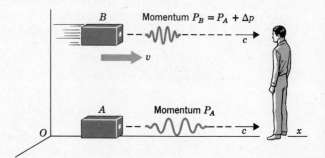

Fig. 11.11 *As determined by the observer, the frequency (and, hence, the momentum) of the light from the moving source (B) is greater than that of the light from the stationary source (A). The velocity of each light signal is the same.*

the light reaching the observer from source *B* in Fig. 11.11 will have a frequency higher than that of the light from source *A,* although both signals will travel toward the observer at the *same velocity.* The relationship between the two frequencies is (see Eq. 10.24):[4]

$$\nu_B = \frac{\nu_A}{1 - v/c}$$

We shall see in a later chapter that the *energy* $\mathcal{E}$ associated with an electromagnetic or light wave is proportional to the frequency of the light. We have already seen (in Section 10.6) that the *momentum* of such a wave is $p = \mathcal{E}/c$. Thus,

$$p_A \propto \nu_A$$

$$p_B \propto \nu_B = \frac{\nu_A}{1 - v/c} \cong \nu_A \left(1 + \frac{v}{c}\right)$$

where the last, approximate equality is valid if v is small compared to c.[5] Therefore,

$$p_B \cong p_A + \Delta p$$

where

$$\Delta p \propto v$$

We conclude, therefore, that the motion of the source transfers additional momentum (and energy) to the light wave but that the motion does not affect the velocity of the light.

The fact that light waves emitted by moving sources are Doppler shifted is crucial in the determination of the recessional velocities of distant galaxies (Section 3.2).

THE DIFFERENCE BETWEEN LIGHT WAVES AND WAVES IN MATERIAL MEDIA

Suppose that we again have two coordinate systems, *K* and *K'*, that have a relative velocity v. Suppose also that there is a body of water that is stationary with respect to system *K.* At the instant that *K'* passes *K* and the two origins coincide (Fig. 11.12a), a stone is dropped into the water and waves begin to spread out in circles from the origin *O.* Figure 11.12b shows the situation a short time later. Of course, the *K* observer sees the waves spreading out from his origin. But what does the *K'* observer see? Because the water is a *material* medium and defines a specific reference frame for any observer, moving or stationary, the observer in *K'* will also see waves spreading out from *O* (not from his own origin *O'*). Everyone has observed this or an equivalent phenomenon and the reason for this effect is clear.

Let us repeat this experiment using light waves instead of water waves.

---

[4]According to relativity theory, even this expression needs modification, but this fact is not important for the argument here.

[5]We use the expression (letting $v/c = x$), $1/(1 - x) = 1 + x + x^2 + x^3 + \cdots$. Then, if $x \ll 1$, the terms x^2, x^3, etc., can be neglected.

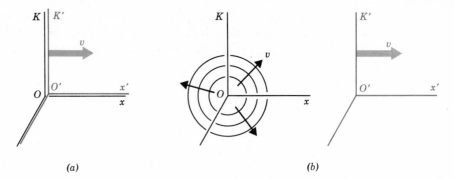

Fig. 11.12 *Water waves are seen by both observers (in* K *and in* K') *as propagating outward from the origin* O.

A light source is located at O in the system K and as K' moves past K, a light pulse is emitted the instant O and O' coincide (Fig. 11.13a). Because the source is at rest in his system, it is clear that the K observer will see a spherical light wave radiating outward from O. The K' observer sees a light source approaching him at a velocity v and at the instant the source is at O', a light pulse is emitted. Since the velocity of light is independent of the velocity of the source, the situation, as observed in K', is exactly the same as if the source were located in a stationary position at O'. That is, the K' observer also sees a spherical light wave radiating from his origin O', not from O (Fig. 11.13b). Therefore, each observer sees exactly the same thing—a spherical light wave spreading out uniformly in all directions from his origin with the velocity c.

At first thought this seems to be a surprising (if not fantastic) result. What is the crucial difference between water waves and light waves that causes such a remarkable disparity in the results of the two similar experiments? The point is actually quite simple. In the case of the water-wave experiment there is a tangible, material medium (the water) that is stationary in the K system. The water wave propagates by virtue of the motions of the water molecules. These material particles "belong" to K and both observers perceive this fact. In the light-wave experiment, however, there is no material

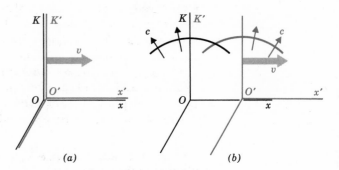

Fig. 11.13 *A light pulse is emitted from the common origin of* K *and* K' *as they pass. Each observer then sees a spherical light wave radiating outward from his origin with the velocity* c.

medium involved—the light wave propagates through empty space. Since there is no matter to be identified as "belonging" to one system or the other, each observer views the light wave relative to his own coordinate system and therefore sees light waves radiating outward with spherical wave fronts.

SIMULTANEITY

In our everyday experience we have become accustomed to consider that all events proceed in time in an orderly and regular way—there is a past, a present, and a future, and we can always establish whether one event preceded or followed another event or whether the two events occurred simultaneously. Einstein showed, however, that in the relativistic world there is no clean-cut distinction between past and future—events that appear to occur in a certain sequence according to one observer may appear to occur in quite a different sequence to another observer who is in motion with respect to the first observer. This is perhaps the most startling result of the Einstein theory, but it is easy to show that this conclusion follows in a direct and simple way from the constancy of the velocity of light.

In order to demonstrate that time is a relative concept, consider the following example (which is due to Einstein). In Fig. 11.14a an observer K sees two lightning bolts strike the ends of a moving railway car just as the midpoint of the car passes him. Since the ends of the car are equidistant from him, K sees the light from the flashes *simultaneously*. Observer K' stands in the middle of the car. Now, K knows that K' is moving toward the flash of light that originates at B and *away* from the flash that originates at A. Therefore, K concludes that the B flash will reach K' *before* the A flash reaches K'. But K' is a stationary observer in an inertial reference frame (the railway car) and he knows that both flashes of light travel with a velocity c in his reference frame. Since K' is equidistant from the two ends of the car and since the flash from B reaches him first (Fig. 11.14b), he concludes that the B flash must have occurred *before* the A flash. Thus, two events that appear to be simultaneous in the K system do *not* appear to be simultaneous in the K' system because the two systems are in relative motion.

If K had seen the lightning strike A *slightly* before the strike at B occurred,

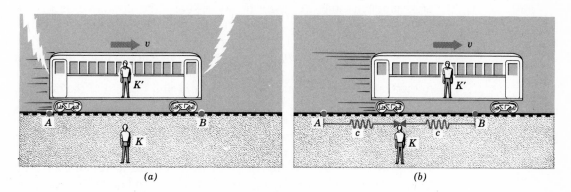

(a)　　　　　　　　　　　　　　　　(b)

Fig. 11.14　*Observer* K *sees two lightning bolts strike the ends of the railway car simultaneously. But observer* K′, *who is moving toward the right-hand bolt with a velocity* v, *sees the right-hand bolt strike first.*

he would have seen *A* precede *B*, whereas *K'* would have still have seen *B* precede *A.* Hence, the two observers would have seen the *opposite* time ordering of events; "past" and "future" would have been interchanged. But notice that *K* cannot inform *K'* of the event that will be in his future (the flash from *A* which *K'* sees *after* the *B* flash) because information can be transmitted at a maximum velocity of *c.* Therefore, *K*'s message would be received only *after K'* perceived the occurrence of the event.

Although the time sequence of events as seen by different observers depends on their relative velocity, the physical law of *cause and effect* must still be valid in the relativistic world; *no observer (whatever his motion) can perceive an event which is an effect prior to an event which is the cause of the first event.*

11.2   *Lorentz Transformations*

EINSTEIN'S POSTULATES

In 1905 Einstein showed that it was possible to remove all apparent discrepancies between the dynamics of mechanical and electromagnetic systems by basing a theory on only two postulates:

I.   *All physical laws are the same in all inertial reference frames.*
II. *The velocity of light (in vacuum) is the same for any observer regardless of the relative motion between the light source and the observer.*

The theory that is based on these postulates and that applies to all nonaccelerating systems is called the *special theory of relativity.* (The more complicated situation of accelerating systems is the subject of the so-called *general theory of relativity,* which is described briefly in Section 11.5.) It is indeed remarkable that such a far-reaching theory—one that forced a complete reexamination of the traditional views concerning the fundamental concepts of space and time and that has had such a profound effect on the interpretation of atomic, nuclear, and astrophysical effects—can be built on only two postulates as simple as those given by Einstein. If it is the goal of physical theory to formulate the laws of Nature with brevity and economy of assumptions, then relativity theory is surely the showpiece of science.

On the basis of Einstein's two relativity postulates, it is not difficult to develop (although we shall not give the details here) the equations that relate the space coordinates and the time in two reference systems that are in *uniform* motion with respect to one another. These equations are similar to the Galilean transformations, $x' = x - vt$ and $t' = t$, but they contain essential differences when the velocity of the relative motion is appreciable compared to the velocity of light.

These transformation equations were first developed by Hendrick Antoon Lorentz (but on the basis of several *ad hoc* assumptions that Einstein later swept away with his simplifying postulates) and are therefore called the *Lorentz transformation equations.* If the relative motion of the two reference frames is along their respective *x* axes, as in Fig. 11.2, the space coordinates and the time in the two frames are related by

$$x' = \frac{x - vt}{\sqrt{1 - \beta^2}}; \qquad y' = y; \qquad z' = z \tag{11.4a}$$

$$t' = \frac{t - \dfrac{\beta}{c}x}{\sqrt{1 - \beta^2}} \tag{11.4b}$$

where we use the customary notation,

$$\beta = \frac{v}{c} \tag{11.5}$$

In using these equations we understand that the K observer has a meter stick with which he measures distances x, y, z and a clock with which he measures the time t. The K' observer is equipped with similar instruments for the corresponding measurements in his system and he has verified that his calibrations are the same as the K observer's when the two systems are at rest relative to one another. Then, when the two systems are in relative motion, the two clocks can then be simultaneously set to zero and started at the instant when the two origins coincide.

The meaning of Eqs. 11.4 is that when the K observer determines that a certain event took place at the point x, y, z and at the time t in the K system, the K' observer sees the same event at the point x', y', z' and at the time t' in the K' system.

Notice that the space coordinates *transverse* to the relative motion of the two systems, y and z, are unaffected by the motion and have the same values in both systems.

If the relative velocity v is small compared to c so that $\beta \cong 0$, the factor $\sqrt{1 - \beta^2}$ is essentially equal to unity and the factor $(\beta/c)x$ becomes negligibly small. Therefore, when $v \ll c$, the Lorentz equations become indistinguishable from the Galilean equations. Because in our everyday experience we rarely encounter velocities that are at all comparable with the velocity of light, our world is essentially Newtonian and directly observable relativistic effects are generally absent.

Two important consequences of the Lorentz equations are the *contraction of length* and the *dilation of time,* which will now be described.

THE LORENTZ CONTRACTION OF LENGTH

Consider a rod of length l that lies along the x-axis of the system K with one end at the origin (Fig. 11.15). What will be the length of this rod as measured by an observer in the system K'? The K' observer can make such a measurement by determining the time required for his origin O' to travel the length of the rod. The time interval starts at the instant when O and O' coincide; at this position $t_1 = 0$, $t_1' = 0$. The origin O' moves with a velocity v and when O' reaches the end of the rod, the K' clock reads t_2' and the K clock reads t_2. The K observer sees O' travel a distance l with a velocity v, so

$$t_2 = \frac{l}{v}$$

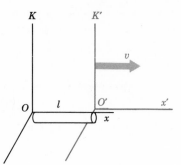

Fig. 11.15 *The* K' *observer determines the length of the rod as viewed in his system by measuring the time interval necessary for* O' *to travel from one end of the rod to the other.*

The time interval measured by the K' observer is

$$\Delta t' = t_2' - t_1' = \frac{t_2 - \frac{\beta}{c} l}{\sqrt{1 - \beta^2}}$$

since $t_1' = 0$ and $x_2' = l$. Substituting l/v for t_2 we have

$$\Delta t' = \frac{\frac{l}{v} - \frac{v}{c^2} l}{\sqrt{1 - \beta^2}} = \frac{\frac{l}{v} - \frac{v^2}{c^2} \frac{l}{v}}{\sqrt{1 - \beta^2}}$$

$$= \frac{\frac{l}{v}(1 - \beta^2)}{\sqrt{1 - \beta^2}} = \frac{l}{v} \sqrt{1 - \beta^2}$$

If we multiply through by v and note that $v\Delta t'$ is just l', the length as viewed by the K' observer, we find

$$l' = l\sqrt{1 - \beta^2} \tag{11.6}$$

We therefore conclude that the observer who is in motion with respect to the rod finds a length that is *smaller* (that is, *contracted*) in comparison to the length determined by the observer at rest with respect to the rod. The situation is symmetrical between the two moving systems. The K' observer sees a contraction of a rod in K, and the K observer will also see a contraction of a similar rod that is at rest in the K' system.

The contraction of the length of a moving rod is a *real* effect because the two observers use identical meter sticks (which they have compared when at rest) and actually *measure* different lengths for the rod. The only physically meaningful quantities (that is, *real* quantities) are those that we can *measure*.

Example **11.2**

An observer moves past a meter stick at a velocity that is one-half the velocity of light. What length does he measure for the meter stick?

$$l' = l\sqrt{1 - \beta^2}$$
$$= (100 \text{ cm}) \times \sqrt{1 - (0.5)^2}$$
$$= (100 \text{ cm}) \times \sqrt{0.75}$$
$$= 86.6 \text{ cm}$$

In order to obtain numerical results for problems that involve the relativistic factor $\sqrt{1 - \beta^2}$, the following approximate expressions can be used when the velocity v is either very small compared to c or comparable to c:

If $\beta \ll 1$ $(v \ll c)$: $\sqrt{1 - \beta^2} \cong 1 - \frac{1}{2}\beta^2$ (11.7a)

$$\frac{1}{\sqrt{1 - \beta^2}} \cong 1 + \frac{1}{2}\beta^2 \qquad (11.7b)$$

If $\beta \cong 1$ $(v \cong c)$: $\sqrt{1 - \beta^2} \cong \sqrt{2(1 - \beta)}$ (11.7c)

Example **11.3**

Suppose that the velocity of the observer relative to the meter stick in the previous example is reduced to 30 m/sec (about 67 mi/hr). What length does he now measure for the meter stick?

$$l' = (100 \text{ cm}) \times \sqrt{1 - \left(\frac{3 \times 10^3 \text{ cm/sec}}{3 \times 10^{10} \text{ cm/sec}}\right)^2}$$
$$= (100 \text{ cm}) \times \sqrt{1 - 10^{-14}}$$
$$\cong (100 \text{ cm}) \times (1 - 0.5 \times 10^{-14})$$
$$= 99.9999999999995 \text{ cm}$$

It is easy to see that the relativistic contraction of length is of little practical consequence in everyday matters!

TIME DILATION

Not only do moving and stationary observers find a difference in their measurements of the length of an object, but they also disagree on the rate at which clocks run in the two systems. Let us prepare a "standard clock" in the following way. At a distance L from the origin along the y-axis we place a mirror M, as in Fig. 11.16a. At the origin we place a light flasher and a light detector. The standard unit of time will be the interval required for light to travel from the flasher to the mirror and back to the detector. If the K observer operates his flasher at $t = 0$, he finds that the light pulse makes the round trip and returns to the origin at the time

$$t = \frac{2L}{c} \qquad (11.8)$$

Now, an identical clock is installed in the K' system and the K *observer views its operation.* The K' system moves past the K system at a velocity

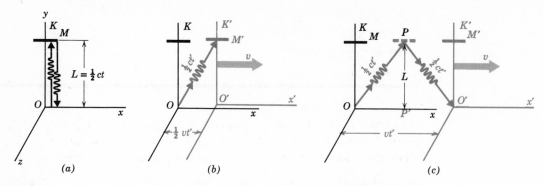

Fig. 11.16 *According to the observer in* K, *it requires a shorter time for the round trip* OMO *of the light signal in* K *than for the round trip* OPO′ *in* K′. *The* K *observer concludes that the clock in* K′ *runs* slower *than the clock in* K.

v along the x-axis. (Notice that we have been careful to place the mirrors on the y axes so that there will be no Lorentz contraction effects.) At the instant when O and O' coincide ($t' = 0$), the flasher of the K' clock is operated. Since the K' system is moving relative to K, the K observer notices that the light flash must travel from O' to M' on a slanted path that is longer than the path traveled by the light flash in the K clock. When the light flash reaches M', this time corresponds to one half of a standard interval of the K' clock, that is, $\frac{1}{2}t'$ (Fig. 11.16b). The complete standard interval t' ends when the reflected light flash again reaches O' having traveled the path OPO' (Fig. 11.16c). During this time, the origin O' has moved a distance vt' from O.

In order to compare the intervals t and t', we use the Pythagorean theorem for the triangle $OM'O'$ (Fig. 11.16b) or OPP' (Fig. 11.16c). Thus,

$$(\tfrac{1}{2}ct')^2 = (\tfrac{1}{2}vt')^2 + L^2$$

or, using Eq. 11.8 for L,

$$(\tfrac{1}{2}ct')^2 = (\tfrac{1}{2}vt')^2 + (\tfrac{1}{2}ct)^2$$

from which

$$(c^2 - v^2)t'^2 = c^2t^2$$

so that

$$t'^2 = \frac{c^2t^2}{c^2 - v^2} = \frac{t^2}{1 - \dfrac{v^2}{c^2}}$$

or, finally,

$$\boxed{t' = \frac{t}{\sqrt{1 - \beta^2}}}$$

(11.9)

The standard time interval t' of the K' clock, *as viewed by the K observer,* is *longer* than the interval of the K clock. (The K observer must draw this conclusion since he views the light flash in the K' clock traveling a greater distance than the light flash travels in his own clock.) Therefore, the observer in K finds that the K' clock runs *slower* than his clock. Of course, if the K' observer views the K clock, he concludes that the K clock runs slower than his. Therefore, we can state that *any observer will find that a moving clock runs slower than an identical clock that is stationary in his reference frame.*

We used a light-flash clock for this argument; will the same result be found if we use some other type of clock, such as a mechanical clock? The answer is that indeed it must. Consider the alternative. Suppose that we have a clock that does not slow down when in motion relative to our particular reference frame. We could then set this clock to agree with our standard clock and then (at least in principle) send this clock on an expedition to all manner of moving reference frames for the purpose of adjusting their clocks to agree with ours. By using these synchronized clocks we would then be able to determine unambiguously the time ordering of events as viewed in any reference frame. But we have already concluded that the finite velocity of light prevents any such determination of the absolute sequence of events in moving systems. We are therefore forced to concede that it is impossible to construct such a "perfect" clock.

TIME DILATION IN PION DECAY

The Lorentz contraction of length and time dilation have a definite and intimate relationship and the appreciation of this relationship makes it easier to understand both effects. We can illustrate this point by considering the motion of the short-lived elementary particles, the π mesons (or *pions*). When viewed at rest, pions have an average lifetime[6] of $\tau_\pi = 2.6 \times 10^{-8}$ sec before decaying into other elementary particles. Pions are produced in copious quantities by the interaction of high-energy protons with matter and therefore are relatively easy to study.

If pions move with a velocity of $0.75c$, the average distance they would travel before decay is $l_\pi = v\tau_\pi = 0.75 \times (3 \times 10^{10}$ cm/sec$) \times (2.6 \times 10^{-8}$ sec$) = 5.85$ m. At the Columbia University cyclotron a beam of pions with $v = 0.75c$ was produced and it was found that the average distance these particles traveled before decay was not 5.85 m but 8.5 ± 0.6 m. We can account for this difference in terms of the time dilation effect. Because the pions are moving in the laboratory system (corresponding to our customary K system), the laboratory observer sees any clock in the system moving with the pions (the K' system) running slowly. But the decay rate of the pions is a type of clock and so the laboratory observer will find the average lifetime of the pions to be longer than τ_π. In fact,

---

[6] The *half-life* of a particle or system that decays spontaneously (as in radioactive decay—see Section 3.4) is equal to 0.69 times the quantity that we call the *average* or *mean* life. Thus, the half-life of the pion is $\tau_{1/2} = 1.8 \times 10^{-8}$ sec.

$$\tau_{lab} = \frac{\tau_{\pi}}{\sqrt{1 - \beta^2}}$$

$$= \frac{2.6 \times 10^{-8} \text{ sec}}{\sqrt{1 - (0.75)^2}}$$

$$= 3.9 \times 10^{-8} \text{ sec}$$

Therefore, the average distance in the laboratory that the pions will travel before decay is

$$l_{lab} = v\tau_{lab}$$

$$= 0.75 \times (3 \times 10^{10} \text{ cm/sec}) \times (3.9 \times 10^{-8} \text{ sec})$$

$$= 8.8 \text{ m}$$

which agrees with the measured value of 8.5 ± 0.6 m.

Let us now view the situation from the standpoint of the pions. By their clock, the pions will live an average of 2.6×10^{-8} sec and will travel an average distance of 5.85 m according to a measurement made with a meter stick in their frame of reference. But the pions see the Columbia laboratory moving past them at a velocity of $0.75c$. Therefore, the laboratory dimensions are contracted and a distance of 8.8 m in the laboratory appears to be only $(8.8 \text{ m}) \times \sqrt{1 - \beta^2} = 5.85$ m to the pions.

This example shows us that time dilation and length contraction are just reciprocal results of the same basic relativistic effect.

THE TWIN PARADOX

One of the results of relativity theory that has been much discussed (and misunderstood) in recent years is the so-called "twin-paradox." Suppose that there are twins, Al and Bob, and that Bob is an astronaut. Bob embarks on a space journey to a star that is 10 light years distant; Al remains on Earth. If Bob's space ship travels at a velocity of $0.99c$ relative to the Earth, according to Al the trip will require a time[7]

$$\Delta t = \frac{10 \text{ L.Y.}}{0.99c} \cong 10 \text{ years}$$

An equal time will be required for the return journey, so Bob will arrive back on Earth when Al is 20 years older than when Bob departed.

In Bob's space ship, however, the Earth and the star appear to be moving with a velocity of $0.99c$ relative to Bob. Therefore, the Earth-star distance is contracted to

$$l' = (10 \text{ L.Y.}) \times \sqrt{1 - (0.99)^2} = 1.4 \text{ L.Y.}$$

According to Bob's clock, the trip will require only 1.4 years and he will return to Earth after having aged by 2.8 years. When he again greets his brother, Bob discovers that his twin is $20 - 2.8 = 17.2$ years *older* than he is! But we know that all motion is relative. Therefore, if the trip is viewed from Bob's reference frame, he sees Al (and the Earth) go on a round trip journey. Hence, Al's clock should run more slowly than Bob's and when

---

[7]Since 1 L.Y. is the distance traveled by light in 1 year, the quantity 1 L.Y.$/c$ is just *1 year*.

Al returns (along with the Earth), Bob should find that his twin is *younger* than he is. Thus, the paradox.

The "paradox" rests on invoking the symmetry of the situation. It should not matter which twin takes the trip and which remains at home. But it *does* matter, because *Al* (the stay-at-home) *is always in an inertial reference frame whereas Bob* (the traveler) *has undergone accelerations.* In leaving the Earth, Bob was accelerated to $0.99c$; he was accelerated when he turned around at the star; and he was accelerated again when he returned to Earth and landed. Therefore, the situation is *not* symmetric between Al and Bob. Because inertial reference frames are not involved throughout, the analysis must be carried out quite carefully. A proper calculation (which can be made within the context of special relativity if appropriate care is exercised) does in fact show that Bob ages less rapidly than his twin.

Because of the time dilation effect, we can imagine the exciting possibility of traveling to distant stars. If the trip is made at a velocity sufficiently close to the velocity of light, the traveler can easily cross vast distances of space within a time short compared to his lifetime. But he would return to a different Earth—one that has progressed (?) by hundreds or even thousands of years during his absence. There is, of course, a difficulty in this fanciful picture; we now have absolutely no conception of how to generate sufficient energy to accelerate a space ship to velocities that approach c!

It should be emphasized that the "twin paradox" is a real effect; the traveling twin will age less rapidly than his Earth-bound brother. On the other hand, the traveler cannot take advantage of his longevity because all of his biological processes progress at a slower rate (compared to the Earth rate) and he must function, think, and perform at this reduced pace.

11.3 *Variation of Mass with Velocity*

A COLLISION EXPERIMENT

The first postulate of relativity theory is that all physical laws must be the same in all inertial reference frames. One of these laws is the conservation of linear momentum and we shall now make use of the invariability of this law to assess the effect of motion on mass.

Consider two observers who are stationed in two reference frames, K and K', that are in relative motion with the velocity v, as in Fig. 11.17. In each reference frame there is a stationary mass m_o. (That the two masses are in

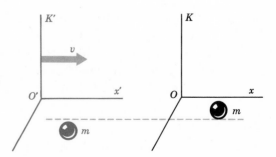

Fig. 11.17 *A mass* m_o *is stationary in each reference frame. The relative velocity is* v.

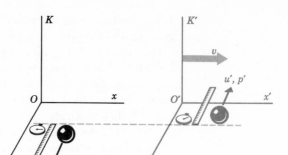

Fig. 11.18 *The grazing collision between the masses produces for each mass a velocity and a momentum transverse to the direction of relative motion.*

fact identical can be established beforehand by a balance comparison when the masses are at rest relative to one another.) The positions of the masses are such that when the reference frames pass one another, a grazing collision of the masses takes place. That is, each mass receives a small velocity at right angles (that is, *transverse*) to the direction of the relative velocity of K and K'. (In such a collision neither mass will receive any appreciable longitudinal velocity relative to its own reference frame.) Therefore, after collision, the situation is that shown in Fig. 11.18. The mass in K has a transverse velocity u and a transverse momentum p as measured by the observer in K; similarly for the mass in K' (but with primed quantities). Each observer uses a meter stick and a clock to measure the transverse velocity of his mass in his reference frame. Each observer obtains a numerical result for the velocity of his mass, which result he then communicates to his colleague in the other reference frame. They are both happy to note that the results are identical and congratulate themselves on having verified the conservation of linear momentum in the collision. In order to check the results, they decide to repeat the experiment twice again—once, K will observe K' making his measurements, and then K' will observe K making his measurements.

On the first rerun, K confirms that the meter stick used by K' is properly calibrated (transverse dimensions are unaffected by relative motion; the Lorentz contraction takes place only *along* the direction of relative motion), but that his clock runs *slowly*. Therefore, when K' reported that his mass traveled 1 meter in T seconds, K concludes that by *his* clock it required a time *greater* than T seconds for the 1-meter trip. Thus, K calculates that the velocity of the mass in K' is *smaller* than the value u' reported by K'—in fact, smaller by the time-dilation factor $\sqrt{1 - \beta^2}$. If the velocity is smaller and if conservation of linear momentum is still to hold, then the mass used by K' must be (so argues K) *larger* than that used by K—in fact, larger by the amount $1/\sqrt{1 - \beta^2}$.

Of course, during the second rerun of the experiment, K' draws exactly the same conclusions about the measurements of K. *Both* observers therefore agree that the mass of an object in motion is greater than the mass of an identical object which is at rest. The increase of mass with velocity (just as length contraction and time dilation) is symmetrical between the two reference frames in relative motion.

The mass of an object as measured in a reference frame at rest with respect to the object is denoted by m_0 and is called the *rest mass* or *proper mass*. Then, the mass m as measured by an observer moving with a velocity v relative to the object is

$$m = \frac{m_o}{\sqrt{1 - \beta^2}}$$

(11.10)

We must conclude from this equation that no material particle can attain or exceed the velocity of light because if $v = c$, the term $\sqrt{1 - \beta^2}$ vanishes and m becomes infinite. An infinite mass is a meaningless concept and we are therefore forced to accept the conclusion that material particles *always* move with velocities that are less than the velocity of light. And, because of the velocity addition rule (Eq. 11.3), the velocity of a material particle is less than c in *any* reference frame.

Example **11.4**

What is the velocity of an elementary particle whose mass is 10 times its rest mass?

The mass m is $10\ m_o$, so

$$10\ m_o = \frac{m_o}{\sqrt{1 - \beta^2}}$$

$$\sqrt{1 - \beta^2} = \frac{1}{10}$$

Squaring, we have

$$1 - \beta^2 = 0.01$$
$$\beta^2 = 1 - 0.01 = 0.99$$
$$\beta = \sqrt{0.99} = 0.995$$

Therefore,

$$v = 0.995\ c$$

PARTICLES WITH VARYING MASS

The difference between the mass m and the rest mass m_o is quite small unless the relative velocity v is greater than a few percent of the velocity of light. Therefore, the relativistic increase of mass with velocity is undetectable for all everyday velocities; it is only when we deal with elementary particles that have been given high velocities in accelerators that we encounter appreciable mass increases. Figure 11.19 shows Eq. 11.10 in graphical form; indicated on the curve are points corresponding to one everyday object (an automobile traveling at 50 mi/hr) and three high-velocity elementary particles. Notice that a 1-GeV proton has *twice* the mass of a proton at rest, whereas the mass increase of a 50-mi/hr automobile is insignificantly small.

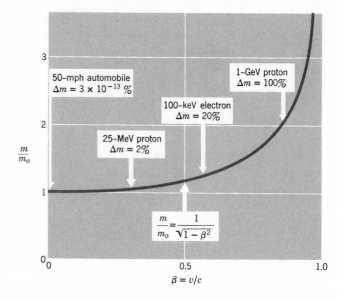

Fig. 11.19 *The relativistic increase of mass with velocity. (See also Fig. 1.11.) The energies given for the elementary particles are the kinetic energies.*

The increase in mass of an object in motion with a velocity *small* compared to the velocity of light is

$$\Delta m = m - m_o = m_o \left[\frac{1}{\sqrt{1 - \beta^2}} - 1 \right]$$

Since $v \ll c$ so that $\beta \ll 1$, we can use Eq. 11.7b to write

$$\Delta m \cong m_o \left[(1 + \tfrac{1}{2}\beta^2) - 1 \right] = m_o \times \tfrac{1}{2}\beta^2$$

Therefore, the fractional increase in mass is

$$\frac{\Delta m}{m_o} \cong \tfrac{1}{2}\beta^2 \qquad (v \ll c) \tag{11.11}$$

Example **11.5**

What is the fractional increase of mass for a 600-mi/hr jetliner?

$v = 600$ mi/hr $\cong 2.7 \times 10^4$ cm/sec

$$\beta = \frac{v}{c} = \frac{2.7 \times 10^4 \text{ cm/sec}}{3 \times 10^{10} \text{ cm/sec}} \cong 10^{-6}$$

Therefore,

$$\frac{\Delta m}{m_o} \cong \tfrac{1}{2}\beta^2 \cong 0.5 \times 10^{-12}$$

so that the mass is increased by only a trivial amount.

EXPERIMENTAL TESTS OF RELATIVISTIC MASS INCREASE

There are two ways to demonstrate in striking fashion that the relativistic formula for mass increase is correct. First, consider a particle, such as an electron, that falls through a certain potential difference and acquires a high velocity. If we project this electron into a magnetic field, we find that the radius of the orbit is *greater* than that calculated from the simple formula based on Newtonian dynamics (Eq. 9.25) if m_o is substituted for the mass. Therefore, a measurement of the orbit radius can be used to determine the mass. In fact, the results of measurements made by just this technique are shown in Fig. 1.11; the theoretical prediction and the experimental results are clearly in excellent agreement.

Next, consider an elastic collision between a moving particle and an identical particle at rest. From the discussion in Example 7.8, we know that Newtonian dynamics predicts that the velocity vectors of the particles after the collision should be at *right angles* (see Fig. 7.7). However, when a fast electron or proton collides with a similar particle at rest, it is found that the angle between the velocity vectors after collision is *smaller* than 90°, indicating that the incident particle had a mass *greater* than the struck particle (Fig. 11.20).

11.4 *Mass and Energy*

EINSTEIN'S MASS-ENERGY RELATION

If $v \ll c$, Eq. 11.10 can be written approximately as

$$m \cong m_o(1 + \tfrac{1}{2}\beta^2) \quad (v \ll c)$$

Multiplying both sides of this equation by c^2 and noting that $c^2\beta^2 = v^2$, we find

$$mc^2 \cong m_oc^2 + \tfrac{1}{2}m_ov^2 \quad (v \ll c) \tag{11.12}$$

The term $\tfrac{1}{2}m_ov^2$ is just the Newtonian result for the *kinetic energy*. The term m_oc^2 is clearly some *intrinsic* aspect of the object because it depends only on the *rest* mass. We call this quantity the *rest energy* of the object. The sum of the *rest* energy and the *moving* energy (that is, the kinetic energy) is the *total energy* of the object:

$$\underset{\text{(Total energy)}}{mc^2} = \underset{\text{(Rest energy)}}{m_oc^2} + \underset{\text{(Kinetic energy)}}{KE} \tag{11.13}$$

Fig. 11.20 *Photograph of the tracks left in a nuclear emulsion by a high-velocity electron colliding with an electron in an atom of the emulsion material. The angle between the outgoing particles is less than 90°, in agreement with the relativistic prediction.*

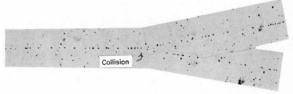

Collision

"The Study of Elementary Particles by the Photographic Method" by Powell

If v is not small compared to c, there are additional terms on the right-hand side of Eq. 11.12 that result from the expansion of $1/\sqrt{1-\beta^2}$. Nonetheless, the difference between the total energy and the rest energy is still the kinetic energy and Eq. 11.13 is correct. This equation is just the expression of the Einstein mass-energy relation:

$$\boxed{\mathcal{E} = mc^2}$$ (11.14)

where $\mathcal{E}$ is the *total* energy (rest energy + kinetic energy) of the object.

Example **11.6**

What is the mass of an electron that has a kinetic energy of 2 MeV? First, we calculate the rest energy of an electron:

$$m_o c^2 = (9.11 \times 10^{-28} \text{ g}) \times (3 \times 10^{10} \text{ cm/sec})^2$$

$$= (8.2 \times 10^{-7} \text{ ergs}) \times \left(\frac{1 \text{ MeV}}{1.6 \times 10^{-6} \text{ ergs/MeV}}\right)$$

$$= 0.511 \text{ MeV}$$

The kinetic energy expressed in units of $m_o c^2$ is

$$KE = 2 \text{ MeV} = (2 \text{ MeV}) \times \left(\frac{m_o c^2}{0.511 \text{ MeV}}\right)$$

$$\cong 4 \; m_o c^2$$

Therefore,

$$mc^2 = m_o c^2 + KE$$

$$\cong m_o c^2 + 4 \; m_o c^2 = 5 \; m_o c^2$$

Hence, the mass of a 2-MeV electron is approximately 5 times the mass of an electron at rest.

The relativistic increase in mass is appreciable for an electron even when the kinetic energy is rather low because the electron rest energy is only 511 keV. Therefore, the relativistic equations must be used whenever the electron KE is greater than a few tens of keV. Protons, on the other hand, have a rest energy of 938 MeV and so even a 10-MeV proton is "nonrelativistic" because its mass increase is only about 1 percent of its rest mass.

Example **11.7**

10 calories of heat are supplied to 1 g of water. How much does the mass of the water increase?

The increase in total energy of the water is

$$\Delta\mathcal{E} = 10 \text{ cal} = 10 \times (4.19 \times 10^7 \text{ ergs})$$

$$= 4.19 \times 10^8 \text{ ergs}$$

$$\Delta m = \frac{\Delta \mathcal{E}}{c^2}$$

$$= \frac{4.19 \times 10^8 \text{ ergs}}{(3 \times 10^{10} \text{ cm/sec})^2}$$

$$= 4.7 \times 10^{-13} \text{ g}$$

So that the mass of the water increases from 1 g to 1.00000000000047 g, a negligible increase indeed! But the mass *has* increased. Where does this additional mass come from? It is just the mass associated with the increase of kinetic energy that has been given to the water molecules by the addition of thermal energy.

Example **11.8**

What is the mass-energy conversion factor that relates AMU and MeV? From Eq. 3.7 we have

$$1 \text{ AMU} = 1.6605 \times 10^{-24} \text{ g}$$

Therefore,

$$(1 \text{ AMU}) \times c^2 = (1.6605 \times 10^{-24} \text{ g}) \times (3 \times 10^{10} \text{ cm/sec})^2$$

$$= (1.4945 \times 10^{-5} \text{ erg}) \times \left(\frac{1}{1.6022 \times 10^{-12} \text{ erg/MeV}} \right)$$

$$= 931.5 \text{ MeV}$$

Or, using more precise values for the constants, we can express c^2 as

$$c^2 = 931.481 \text{ MeV/AMU}$$

11.5 *General Relativity*

THE RELATIVITY OF ACCELERATED MOTION

Thus far we have considered only motions that take place with constant velocity. If we wish to treat the case of accelerated motion, then the special theory of relativity is no longer adequate and we must turn to the *general theory of relativity*. As we shall see, the general theory is more than a relativistic description of abstract accelerated motion, it is a theory of *gravitation*.

Although the special theory is supported by experimental tests of a wide range of predictions, the general theory enjoys much less in the way of experimental verification. In fact, there are only a few predictions that the theory in its current form can make and for none of these has a really definitive experimental test been carried out at the present time. Nevertheless, the issues that are raised by the general theory are so profound that a brief survey of the theory is warranted.

The first postulate of special relativity is that all physical laws are the same in all inertial reference frames. The general theory makes a much more sweeping statement:

All physical laws can be formulated in such a way that they are valid for any observer, no matter how complicated his motion.

If we allow an observer to undergo a complicated accelerated motion, it is clear that the mathematical expression of physical laws in his reference frame will also be complicated. Indeed, the general theory involves the use of an awesome arsenal of mathematical techniques. In spite of this fact, we can still gain some appreciation of the general theory without sophisticated mathematics.

THE PRINCIPLE OF EQUIVALENCE

The first important aspect of the general theory has to do with the equivalence of gravitational fields and accelerated motion. If we are in a laboratory on the Earth, as in Fig. 11.21a, a mass that is released will accelerate downward due to the gravitational attraction of the Earth. Now, let us move this laboratory into space, away from the gravitational influence of the Earth or any other body, and attach it to a rocket that is accelerating, as in Fig. 11.21b. If the magnitude of the rocket's acceleration a is equal to the acceleration due to gravity g and if the rocket pushes on the floor of the laboratory, the floor will be accelerated toward a mass that is released. Insofar as observations of the motion of the mass relative to the floor are concerned, the accelerated motion in the two cases will be exactly the same. If the laboratory has no windows, the observer can never distinguish between an acceleration due to gravity and an acceleration due to a push by a rocket.

Einstein incorporated this reasoning into his general theory by postulating the *principle of equivalence:*

In a closed laboratory, no experiment can be performed that will distinguish between the effects of a gravitational field and the effects due to an acceleration with respect to the "fixed" stars.

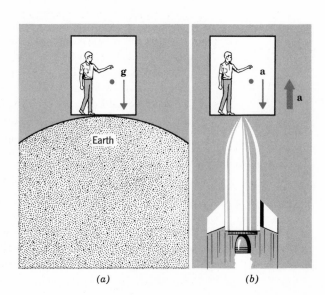

Fig. 11.21 *If* a = g, *the observer cannot distinguish between the acceleration produced by gravity and that produced by the push of the rocket. (The observer's box laboratory has no windows.)*

(a)   (b)

Notice that in the "experiment" of Fig. 11.21 we deal with *gravitational* mass in the first case and with *inertial* mass in the second case. (Refer to Section 5.5 for a discussion of these two types of mass.) If there were any difference between these two types of mass, then sufficiently precise measurements would permit the observer to decide whether he is in the Earth's gravitational field or accelerating in space. Therefore, the equivalence principle requires $m_G = m_I$. As pointed out in Section 5.5, experiments verify this equality to within 1 part in 10^{11}. In spite of the high precision of this result, an even sharper confirmation of the assertion of the equivalence principle would be desirable in order to provide greater support for this fundamental postulate of the general theory.

When Einstein first formulated his general theory of relativity (more than 50 years ago), he proposed two experimental tests of the theory. These tests deal with anomalies in the motions of the inner planets, particularly Mercury, and with the behavior of electromagnetic waves, particularly light, in the vicinity of massive objects such as the Sun. Next we shall briefly describe these two tests and the modern refinements.

THE PRECESSION OF THE PERIHELION OF MERCURY'S ORBIT

In Section 6.2 it was mentioned that the perihelion of Mercury's orbit has been observed to move in space (that is, to *precess*) at a rate that is larger than that predicted on the basis of Newtonian dynamics. After subtraction of the calculable perturbations due to the other planets, there remains a net precession of 43.11 ± 0.45 seconds of arc per century. If the special theory of relativity is used to calculate the effects of time dilation and mass increase with velocity, one finds a precession that is only one-half the observed value. Einstein was able to obtain, on the basis of his general theory, the value 43.03 seconds of arc per century. The excellent agreement between the calculated and the observed values is the most outstanding success of the general theory.

The corrected experimental precessional rate is based on calculations predicated on the assumption that the Sun is spherical in shape and therefore has gravitational field lines that are *straight*. Some recent high precision measurements of the shape of the Sun indicate that the Sun may be slightly *oblate*, that is, that there is a slight bulge at the Sun's equator. The interpretation of these measurements is not yet clear but they could mean that it is necessary to apply a further correction of about 4 seconds of arc per century to the observed value of the precessional rate for Mercury's orbit. The application of such a correction would, of course, destroy the agreement between experiment and the prediction of general relativity. If it is found that this new correction is indeed valid, then it may be necessary to make a fundamental modification of the general theory. On the other hand, the contention has been made that other astronomical data used in the perturbation calculations may be in error by sufficient amounts to cancel the oblateness correction. A new method is now being developed for improving the oblateness measurement and modern radar-ranging techniques are currently being used to obtain more precise measurements of planetary distances. Therefore, within the next few years we shall probably have a much improved value for the precessional rate with which to confront the general theory and to test its prediction.

Precession experiments with Earth satellites are also underway. These measurements should be free of many of the uncertainties that hamper the interpretation of the results for Mercury and may therefore eventually provide a sharper test of the theory.

THE BENDING OF LIGHT RAYS BY THE SUN

The general theory predicts that a light ray passing close to a massive object will follow a path that is slightly curved. We can understand this result qualitatively if we recall that electromagnetic radiation, including light, has *energy* and that energy has equivalent *mass*. Therefore, a gravitational field will affect a light ray and cause it to be bent in much the same way that a fast particle would bend when passing by a massive object. Because light travels at such an enormous velocity there is only a brief time during which the "attraction" is effective, and hence the deflection is small even for a passage near such a massive object as the Sun. (As in the case of Mercury's precession, the special theory, through the relation $\mathcal{E} = mc^2$, predicts for the deflection only one-half the value that results from the general theory.)

This prediction can be tested by observing the shift in the apparent position of a star when its light passes close to the Sun, as shown in Fig. 11.22. Because of the brightness of the Sun, measurements of the effect are carried out by comparing the apparent positions of stars during a solar eclipse with the positions six months later when the Sun is not in that part of the sky and ordinary night photographs of the stars can be taken. Such measurements are clearly difficult to make. (Not the least of the problems is the fact that there is a sudden change in temperature at the onset of the eclipse with resulting thermal contractions in the photographic apparatus!) However, the apparent shifts in position of hundreds of stars have been made by this technique and the average result is approximately 2 seconds of arc for the deflection; the general theory predicts 1.75 seconds of arc. Unfortunately, the uncertainty in the measurements is about 10 percent and there are some conflicting results, so that we cannot view this test as definitive.

New experiments are planned for the near future that will permit measurements to be made under daylight conditions, obviating the necessity of

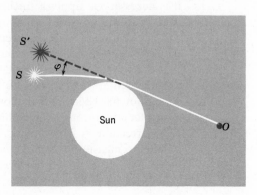

Fig. 11.22 *The light from a star S is bent upon passing close to the Sun, thus causing a shift in the apparent position of the star to S'.*

waiting for eclipses. These new measurements should decrease the uncertainty to about 1 percent and therefore will provide a stringent test of the general theory.

THE GRAVITATIONAL RED SHIFT

When a mass is released in a gravitational field, it accelerates downward and gains kinetic energy. Similarly, a light pulse—because it has an equivalent mass—will gain energy by falling in a gravitational field. The increase in kinetic energy of a massive particle results from the increase in its velocity. Since light always travels with the velocity c, an increase in energy can occur only by virtue of an increase in the *frequency* of the light wave. The converse is also true—if a light pulse is projected *opposite* to the gravitational field vector, it will *lose* energy and the frequency will decrease. Visible light emerging from the Sun will have its frequency *lowered* or, equivalently, the wavelength will be *increased* or shifted to the *red* part of the spectrum. The amount of the shift is small, but by comparing the wavelengths of certain solar radiations with the wavelengths of the radiations from the same types of atoms in an Earth laboratory, the predicted shift has been verified to a precision of about 10 percent.

We can derive the expression for the fractional change in frequency due to the gravitational red shift from Newtonian principles combined with the relativistic mass-energy relation. Since the energy of a light wave is proportional to its frequency, the fractional frequency change is equal to the fractional energy change:

$$\frac{\Delta \nu}{\nu} = \frac{\Delta \mathcal{E}}{\mathcal{E}}$$

Now, $\mathcal{E} = mc^2$ and the *change* in energy due to a change in position by an amount h in a uniform gravitational field g is $\Delta \mathcal{E} = mgh$. Therefore,

$$\frac{\Delta \mathcal{E}}{\mathcal{E}} = \frac{mgh}{mc^2} = \frac{gh}{c^2}$$

so that

$$\frac{\Delta \nu}{\nu} = \frac{gh}{c^2} \tag{11.15}$$

A verification of this expression has been carried out in the laboratory by allowing electromagnetic radiations from nuclei to fall through a distance of 74 ft $= 2.26 \times 10^3$ cm. The predicted fractional frequency change is

$$\frac{\Delta \nu}{\nu} = \frac{(980 \text{ cm/sec}^2) \times (2.26 \times 10^3 \text{ cm})}{(3 \times 10^{10} \text{ cm/sec})^2} = 2.5 \times 10^{-15}$$

This incredibly small frequency change has been measured to a precision of 1 percent (!) using a particular kind of nuclear frequency selector (based on the *Mössbauer effect*). In order to do as well in the calculation of the national debt ($\sim$300 billion dollars), your figures would have to be accurate to 1/1000 of a cent!

These experiments do not constitute a test of general relativity because the results can be accounted for entirely on the basis of the principle of equivalence (which is one of the *postulates* of the general theory) and the mass-energy relation, $m = \mathcal{E}/c^2$.

The periodic nature of an electromagnetic or light wave constitutes a kind of *clock*. Consider two observers, one in the strong gravitational field near the Sun and the other in a weak field in a space laboratory. Both observers have identical atomic light sources and standard clocks.[8] Each observer counts the number of oscillations in his source for a predetermined interval of time according to *his* standard clock. When the results are compared, it is found that the observer in the strong field has measured a greater number of oscillations (that is, a greater light frequency) than the observer in the weak field. They conclude that the clock in the strong field must have run more slowly in order for more oscillations to have occurred in the standard time interval. Notice that, unlike the case of time dilation in special relativity, *both* observers agree that the clock in the strong field is slow—there is no symmetry in the results.

Since the effects of gravity and acceleration are indistinguishable, we must conclude that an accelerating clock runs more slowly than a clock in an inertial reference frame.

NEW TESTS OF THE THEORY

During the past few years, several proposals for new tests of the general theory have been made. The theory predicts, for example, that a radar signal echoed from Mercury when the planet is on the far side of the Sun and close to one edge (as viewed from the Earth) should require a slightly longer transit time than that calculated on the basis of classical electromagnetic theory. Measurements of this effect have been made and preliminary results indicate a verification of the theoretical prediction of the general theory to within 10 percent. A considerably more precise experimental value of the transit time should be available in the near future.

The theory also predicts that the frequency of the radar pulse reflected from Mercury should be shifted and that the direction of the shift (that is, whether to a higher or a lower frequency) should depend on whether Mercury is entering or leaving the region behind the Sun (as viewed from the Earth). It is considerably more difficult to measure the frequency shift than to measure the time delay of the reflected signal, and so this particular test of the theory has not yet been made.

GRAVITATIONAL WAVES

In the previous chapter we found that an accelerating electric charge produces electromagnetic radiation. By analogy, then, should an accelerating massive object produce *gravitational* radiation? According to the general theory, the answer is *yes,* but the amount of energy radiated is so small that extraordinarily delicate equipment is required to detect even the radiation from an exploding star (a *nova* or *supernova*). Nevertheless, experiments

---

[8] Things are not quite as straightforward as implied here. The discussion of clocks (even *light* clocks) in the context of general relativity is extremely complicated.

Fig. 11.23 *The gravitational wave detection apparatus of Professor Joseph Weber and his colleagues at the University of Maryland. The heart of the system is a 1400-kg cylinder of aluminum which is suspended in vacuum in a tank. Sensitive instruments can detect motion of the end of the cylinder (caused by a gravitational wave) corresponding to an average displacement of only 10^{-14} cm!*

Professor J. Weber of University of Maryland

of this type are being conducted with equipment of the type shown in Fig. 11.23, and current results indicate the presence of signals at a rate of about one per week that are distinct from local effects (such as earthquakes) that would disturb the apparatus. This rate of occurrence of signals is approximately that expected for the explosions of stars in our Galaxy. It appears, therefore, that the general theory has been confirmed in another important respect—the prediction of the existence of gravitational waves.

GEOMETRY AND GRAVITY

Consider a pair of physicists who are *two-dimensional* men. That is, they appreciate length and breadth but have no comprehension of height. These physicists operate in a world that to them is a *plane*. Suppose these men are on the surface of the Earth at the equator—positions *A* and *B* in Fig. 11.24. They start out on a journey that takes them due north in parallel paths. After traveling a certain distance *d*, they discover that their separation

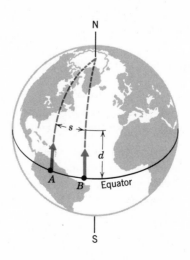

Fig. 11.24 *The two-dimensional physicists start from* A *and* B *and travel northward in parallel paths. They conclude that they are attracted together by some force.*

Fig. 11.25 *Albert Einstein in his later years.*

s is less than when they started. They conclude that they have been attracted together by some "force" and they give this "force" a name—*gravity.*

But, of course, there is no "force"—they have been deceived by the fact that their geometry is *curved* whereas they had assumed only plane Euclidean geometry in describing their positions. Thus it is with our real world. If we insist that the Universe be described by Euclidean geometry, then there is a mysterious force—gravity—for which we have no fundamental explanation. However, all of the effects of gravity result, in the general theory of relativity, from the non-Euclidean nature of the geometry (a four-dimensional space-time geometry) of the Universe. The presence of matter distorts the geometry and manifests itself in terms of the "gravitational force."

Einstein died (in 1955) still searching after his unproven vision that not only gravitation but all of the physical Universe can be completely described in terms of geometry alone. Current research by the disciples of Einstein has not succeeded in establishing this view of the Universe but the efforts continue.

Summary of Important Ideas

The *special theory of relativity* rests on two postulates:

I. The laws of physics are the same in all inertial reference frames.

II. The velocity of light in vacuum is the same for all observers.

In formulating the special theory of relativity, Einstein abandoned three basic ideas of Newtonian theory: (1) the concept of absolute space and time, (2) the principle of the addition of velocities, and (3) the law of conservation of mass (which he generalized to the conservation of mass-energy).

No *material particle* can have a velocity with respect to any reference frame that is equal to or greater than the velocity of light; no *signal* can be transmitted at a velocity greater than *c*.

The *time sequence of events* as perceived by two observers depends on the relative motion of the observers. But no observer, regardless of his motion, can perceive an *effect* before he perceives the *cause* of that effect.

A measurement by an observer of the length of an object that is in motion relative to the observer will give a *smaller* result than a measurement carried out by an observer at rest with respect to the object (*length contraction*). (The contraction is only in the dimension *along* the direction of relative motion; the transverse dimensions are unaffected.)

An observer who is in motion relative to a clock will find that clock to run more *slowly* than an identical clock at rest in his frame of reference (*time dilation*).

A particle moving with respect to an observer has a mass *greater* than that of an identical object at rest with respect to the observer.

The *total energy* of an object is the sum of its *rest energy* $m_o c^2$ and its *kinetic energy;* this total energy is $\varepsilon = mc^2$.

The *principle of equivalence* states that the effects of gravity and accelerated motion are indistinguishable.

The predictions of the *general theory of relativity* have been verified for (a) the precession of the perihelion of Mercury's orbit, (b) the bending of light rays that pass near the Sun, (c) the time delay of radar signals that pass near the Sun, and (d) the existence of gravitational waves. None of these tests is definitive, but the techniques used are being improved and new measurements are to be made in the near future.

The *gravitational red shift* is an immediate consequence of the principle of equivalence and the fact that light has "mass." That clocks run slowly in a gravitational field is due to the same effect.

Questions

11.1 The "writing speed" of a cathode-ray oscilloscope is the speed with which an electron beam can trace a line on the screen. A certain manufacturer claims that the writing speed of his oscilloscopes is 6×10^{10} cm/sec. Can his claim be true? Explain.

11.2 Using the results of relativity theory, comment on the anonymous limerick:

There was a young lady named Bright,
Who could travel much faster than light.
She departed one day,
In a relative way,
And returned on the previous night.

11.3 Suppose that the velocity of light suddenly became 30 mi/hr. Describe a few of the effects this would have on everyday events.

11.4 Discuss the implications for relativity theory if the velocity of light were infinite.

11.5 Refer to Fig. 11.14. Suppose that another observer K'' is moving towards the *left* with a velocity $-v$ and is at the position of K when the situation is that shown in Fig. 11.14b. Does K'' see the light flashes simultaneously or does one of the flashes precede the other? Explain.

11.6 A bicycle rider pedals past you at a velocity of 2.5×10^{10} cm/sec. Make a sketch of the way the bicycle would appear to you. Would the rider think you were your usual self?

11.7 In Section 7.15 we considered the connection between the *average velocity* of molecules in a quantity of matter and the *temperature* of the material. In this chapter we learned that there is an upper limit (viz., c) to the velocity of any material object. Should we conclude, then, that there is an upper limit for allowable temperatures?

11.8 Imagine a shaft drilled into the Earth and reaching to a depth of 1000 mi. Suppose that an observer in a closed box falls freely down this shaft. Argue that the principle of equivalence does not apply to this case. (Is the gravitational field *uniform?* What difference does this make?)

11.9 Should binary stars radiate gravitational waves? Why?

Problems

11.1 A space ship is moving away from the Earth with a velocity of 1.5×10^{10} cm/sec and launches a small rocket toward the Earth with a velocity of 2.5×10^{10} cm/sec with respect to the mother ship. What is the velocity of the rocket as determined by an observer on the Earth? (Notice that the *signs* of the two velocities are different.)

11.2 An observer sees one spaceship moving away from him at a velocity $0.9\,c$ and another spaceship moving in the *opposite* direction at the same velocity. What does he conclude is the relative velocity of the two spaceships? (Consider this carefully; the Einstein velocity addition rule is *not* involved in the answer. Why?) Does this result agree with that obtained by the occupants of the spaceships? Why?

11.3* A rocket ship traveling with a velocity of 1.5×10^{10} cm/sec with respect to the Earth fires a small rocket in the direction of its motion with a velocity of 10^7 cm/sec relative to the rocket ship. With what precision would a measurement on the Earth have to be carried out in order to distinguish between the Einstein velocity addition rule and that based on Newtonian mechanics?

11.4 A man sets up equipment to detonate dynamite charges by sending electrical signals along a wire. (The signals travel with the velocity of light.) 1 μsec after he pushes his firing button, a charge explodes at a distance of 400 m. After an additional 1 μsec, a second charge explodes at a distance of 500 m. Could the pushing of the button have been responsible for both explosions?

11.5 An observer measures the length of a moving meter stick and finds a value of 0.5 m. How fast did the meter stick move past the observer?

11.6 A highway billboard is in the form of a square, 5 m on a side, and stands parallel to the highway. If a traveler passes the billboard at a speed of 2×10^{10} cm/sec, what will be the dimensions of the billboard as viewed by the traveler?

11.7 The star nearest the Earth is *Proxima Centauri* (one of the three stars in the Alpha Centauri cluster); the distance, as measured by parallax, is approximately 1.3 parsec (or 4.3 light years). If a space traveler were to make the trip from Earth to Proxima Centauri at a uniform speed of $v = 0.95\ c$, how long would it take according to an Earth clock? How long would it take according to the space travelers clock?

11.8 *Muons* are elementary particles that are formed high in the atmosphere by the decay of pions which result from the interaction of cosmic rays with the gases in the atmosphere. The velocities of these muons are near the velocity of light ($v \cong 0.998\ c$). From laboratory experiments we know that the average lifetime of a muon is 2.2×10^{-6} sec (in its own rest frame). Show that muons formed at an altitude of 8,000 m can reach the surface of the Earth in spite of their short lifetime but that they would not do so without relativistic effects.

11.9* Could an astronaut, with a remaining life expectancy of 40 years, make a trip (at least in principle) to a galaxy that is 10^{10} light years distant? (This galaxy would be near the limit of the Universe that is visible from Earth.) If so, at what velocity would his spaceship have to travel?

11.10* In a paper published in 1905, Einstein stated: "We conclude that a balance clock at the equator must go more slowly, by a very small amount, than a precisely similar clock situated at one of the poles under otherwise identical conditions." Neglect the fact that the equator clock is actually rotating with the Earth and consider the velocity to be uniform (that is, consider the problem within the context of *special* relativity). Show that after a century the clocks will differ by approximately 0.0025 sec.

11.11 A distant star is receding from us at a speed of 0.8 c. The light intensity from this star is observed to vary with a period of 5 days. What is the time between intensity maxima in the reference frame of the star?

11.12 An observer measures the mass m and the length l of a rod that is stationary. He therefore concludes that the linear mass density of the rod is $\rho = m/l$. Now, suppose that he repeats the measurements while the rod is moving past him with a velocity v. (The motion is along the length of the rod.) What will the observer find for the linear mass density? What would he find if the motion were perpendicular to the length of the rod?

11.13* Electrons emerge from the Stanford Linear Accelerator (SLAC) with velocities only 1.5 cm/sec slower than the velocity of light. How long does a 1-km flight path in the laboratory appear to these electrons? What is the mass of these electrons? Compare this mass with that of an iron nucleus.

11.14 A 100-watt light bulb operates for 1 year. What is the mass equivalent of the radiated energy?

11.15 What is the velocity of a particle that has a kinetic energy equal to its rest energy?

11.16 How much energy is required to double the velocity of an electron which moves initially with a velocity of 10^{10} cm/sec?

11.17 At what velocity is the mass of an object equal to three times its rest mass?

11.18 A meter stick is in motion in the direction along its length with a velocity sufficient to increase its mass to twice the rest mass. What is the apparent length of the meter stick?

11.19* Complete the following table:

	KE (MeV)	v/c
Electrons	0.1	
		0.99
Protons		0.1
	1000	

(Use the approximate value 1000 MeV for the rest energy of the proton.)

11.20 Through what potential difference must an electron fall (starting from rest) in order to achieve a velocity of $0.5c$?

11.21* The rate at which radiation from the Sun reaches the Earth is approximately 0.14 watt/cm^2. At what rate is the Sun losing mass? If this rate were to continue, how long would the Sun last?

11.22* Calculate the orbit radius of a 2-MeV electron in a magnetic field of 1500 gauss.

11.23 A light pulse is directed parallel to the Earth's surface. After traveling 10 km, how far will it have fallen? (Recall that the result given by general relativity is twice the Newtonian result.)

11.24* What must be the magnitude of a uniform gravitational field in which the frequency of a light pulse changes by 1 part in 10^6 by falling through a distance of 100 m? Is a field of this size available in the solar system? Suppose that we carry out the experiment in the Earth's field over a distance of 1 km. What will be the fractional frequency shift?

11.25* By following the procedure used to obtain Eq. 11.15, derive the expression for the gravitational red shift of a light pulse that travels from the surface of the Sun to infinity. For $\Delta \mathcal{E}$ calculate the difference in potential energy of a mass that is moved from the surface of the Sun to infinity by using Eq. 8.5. Show that the shift is $\Delta \nu / \nu = GM/c^2 R$, where M is the mass of the Sun and R is the radius of the Sun. What is the numerical value of the fractional frequency shift?

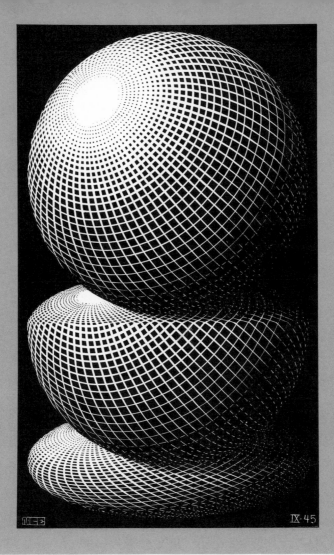

Mathematical Models

12 *The Foundations of Quantum Theory*

At the end of the 19th century many scientists viewed physics as a closed subject. What else that was significant and fundamental could be learned about the physical aspects of Nature? The laws of mechanics and the theory of universal gravitation had been established for more than 200 years. Maxwell's theory of electromagnetism was complete. It was understood that matter consisted of atoms. Thermodynamics had recently been placed on a firm foundation with the development of the statistical approach to systems with large numbers of particles. The great conservation principles—of energy, linear momentum, angular momentum, mass, and electrical charge—were well established and appreciated. What else of *real* importance could be discovered?

In spite of the general complacency among 19th-century physicists concerning the status of the subject, nevertheless, there *were* problems lurking about. It was soon to become clear that these were not all trivial problems and that the very heart of 19th-century physics was coming under violent attack. First, it was Einstein's relativity theory that forced a new way of thinking about such fundamental concepts as space and time. Even before this revolution could be digested, a new series of equally far-reaching questions was being asked about the nature of radiation and matter, how did they differ and in what ways were they the same—what was the inner structure of atoms—what was the origin of the newly discovered *radioactivity?* The answers to these questions began to emerge in the early years of the 20th century and culminated with the development of modern *quantum theory.*

In this chapter we sketch some of the origins of quantum theory. We describe the experiments that led to the discovery of the electron and the quantum of radiation. We shall then see that electrons and light quanta have remarkably similar properties—a crucial fact that led directly to the first formulation of the concepts of quantum theory. In the next chapter we shall see how the study of atomic radiations contributed to the development of the theory and in succeeding chapters we shall examine the consequences of the modern theory in describing the structure of molecules and bulk matter as well as the nuclei of atoms. Therefore, in this chapter we begin the study of the single most important theory in contemporary physics—quantum theory—the theory around which is centered the vast majority of present-day research in physics.

In beginning this chapter, we retrace some ground we have already covered—namely, the existence and the properties of the electron. We have previously used the fact that the electron is the elementary carrier of negative electricity to discuss the flow of electric current and we have studied Millikan's measurement of the electronic charge. We have become accustomed to the use of the concept of the electron, but in beginning the study of modern physics—and, in particular, quantum theory—it is appropriate to examine first the experiments that paved the way for the opening of the modern era—Thomson's discovery of the electron.

12.1   *Electrons*

CATHODE RAYS

Modern atomic physics begins essentially with the discovery of the electron. In the late 1880s many experimenters had observed the fact that an electrical discharge takes place in a partially evacuated tube containing electrodes which are connected to a source of high voltage.[1] But the nature of the discharge was unknown. In 1897 Sir Joseph John Thomson (1856–1940) finally established the atomistic character of the carrier of negative electricity by examining the properties of the particles involved in electrical discharges.

Thomson's experiments were performed with evacuated discharge tubes similar to that shown schematically in Fig. 12.1. (A photograph of one of Thomson's discharge tubes is shown in Fig. 12.2.) When a high voltage is connected between electrode C (the negative electrode or *cathode*) and electrode A (the positive electrode or *anode*), electric current flows through the circuit and across the space, A to C, in the discharge tube. This current is registered on the current meter M. Some of the current is due to the flow of negative charge from C to A. This fact is readily apparent because the residual gas atoms in the evacuated tube produce light when excited by the current flow. Thus, the path of the flowing charge is made visible. When a hole is cut in the anode A, some of the negative charge passes through, is further defined by another hole in a similar ring B, and emerges into the enlarged section of the tube to the right of B. Upon striking the fluorescent screen F, the flowing charges produce a bright spot. By deflecting these *cathode rays* in electric and magnetic fields, Thomson concluded that the rays are negatively-charged particles of matter. He was also able to measure the charge-to-mass ratio e/m for these particles, obtaining $e/m = 6.7 \times 10^{17}$ statC/g. This result, in spite of the crudeness of the measurements, compares

[1] A familiar example of this effect is the neon tube frequently used in illuminated signs.

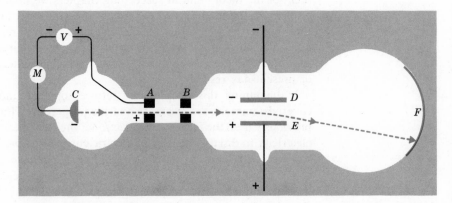

Fig. 12.1 *Schematic diagram of a discharge tube of the type used by Thomson in his investigations of cathode rays (electrons). A photograph of one of Thomson's tubes is shown in Fig. 12.2.*

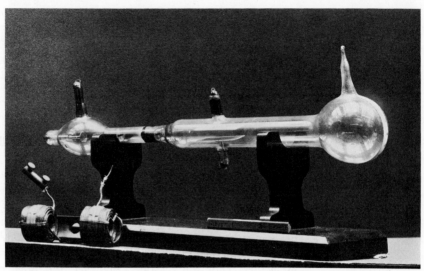

Science Museum, London

Fig. 12.2 *Photograph of Sir J. J. Thomson's* e/m *apparatus.*

quite favorably with the value for electrons that is accepted at present:

$$\frac{e}{m} = 5.27 \times 10^{17} \text{ statC/g} \tag{12.1}$$

Thomson found, in addition, that the value of e/m for the cathode rays was independent of the composition of the gas in the discharge tube as well as independent of the material from which the cathode (the origin of the rays) was made. From these experiments, Thomson concluded that the cathode rays are bits of matter, different from atoms, that are carriers of electricity and that are identical regardless of the details of their origin. These cathode rays are simply *electrons.*

POSITIVE RAYS

Studies with discharge tubes showed that in addition to the cathode rays there were also *positive rays,* streams of particles that moved in the direction *opposite* to that taken by the negative cathode rays. These positive rays were characteristic of the gas in the tube; in fact, they were found to be positively-charged atoms (or molecules) of the gas material. The e/m ratio for the hydrogen positive rays was measured to be about 1/1000 of that for electrons. (We now know that e/m for a hydrogen ion, that is, a proton, is 1/1836 of e/m for an electron.)

The information regarding the properties of the positive rays, coupled with the results of Thomson's experiments with cathode rays, led to several important conclusions. Cathode rays are electrons that originate in matter, that is, they come from atoms, but they are not characteristic of the originating atom. Positive rays, on the other hand, *are* atoms. When the electrons in the cathode rays strike the gas atoms in the discharge tube, electrons are removed from the atoms (and add to the electron current) leaving the atoms

Table **12.1** *Properties of the Electron*

Charge, e	$\begin{cases} 4.8032 \times 10^{-10} \text{ statC} \\ 1.6022 \times 10^{-19} \text{ C} \end{cases}$
Mass, m_e	9.1096×10^{-28} g
Rest-mass energy, $m_e c^2$	0.511004 MeV
Charge-to-mass ratio, e/m_e	5.2728×10^{17} statC/g

with a positive charge. These charged atoms are called *ions*. The simplest ion is the hydrogen ion, or *proton*.

Thomson concluded his 1897 paper with an argument regarding the size of the electron. Molecular ions (positive rays) were known to be able to penetrate only a very small amount of air whereas electrons were found to be capable of traveling a considerably greater distance before losing all of their kinetic energy. The passage of these ions and electrons through air (or any material) is governed by the frequency with which they collide with molecules. Thomson reasoned that electrons must be much smaller than molecular ions and therefore suffer collisions at a much lower rate; with fewer collisions to degrade the motion, electrons can penetrate a greater distance than can ions.

12.2 *Blackbody Radiation and Quanta*

THE SPECTRUM OF RADIATION

At about the time that Thomson was firmly establishing the existence of the electron and was determining some of its properties, other physicists were wrestling with another problem that was also to assist in setting the stage for the development of quantum theory. This problem concerned the distribution of wavelengths of the radiation from incandescent solids.

When a solid material is heated it becomes incandescent and emits radiation with a continuous range of wavelengths. Some of this radiation falls in the visible region of the electromagnetic spectrum; therefore, we can perceive the "glow" of a heated piece of metal. Depending on the temperature of the object, more or less of this radiation will be visible and the remainder will be distributed throughout the regions of longer and shorter wavelengths.

Consider now a hollow cavity within a certain material; a small hole connects this cavity with the outside world (Fig. 12.3). We bring the material

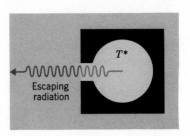

Fig. 12.3 *An ideal blackbody. The only radiation emitted is that which escapes through the small hole. At thermal equilibrium the escaping radiation is characteristic of the temperature of the blackbody.*

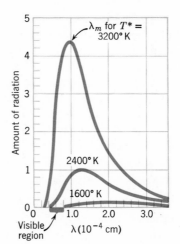

Fig. 12.4 *The blackbody spectrum for several different temperatures. As the temperature is increased, the wavelength of the maximum of the curve shifts to smaller values. (The visible part of the spectrum extends approximately from 0.4×10^{-4} cm to 0.7×10^{-4} cm.)*

to equilibrium at a certain absolute temperature T^*. In the ideal situation the only radiation to leave the object does so by escaping through the hole. (It is as though we were observing the radiation from a blast furnace by looking through an open door.) Therefore, except for the small hole, the object is black—*completely* black. Such an ideal object is called a *blackbody*.

If we measure the wavelength of the radiation escaping through the aperture of the ideal blackbody at some equilibrium temperature,[2] we find a *continuous spectrum* of wavelengths; that is, *all* wavelengths are present and for each wavelength there is a definite probability of occurrence that is independent of the way in which the cavity is maintained at the equilibrium temperature. Figure 12.4 illustrates the shape of the spectrum for several equilibrium temperatures. Each curve has a definite maximum at a wavelength labeled λ_m; as the temperature is increased, λ_m becomes smaller. Thus, a body that glows yellow is hotter than one that is dull red. The total *amount* of radiation increases as the temperature is increased (the increase is proportional to T^{*4}).

The value of λ_m at a particular temperature has been found experimentally to obey the relation

$$\lambda_m = \frac{0.29}{T^*} \text{ cm} - {}^\circ\text{K} \tag{12.2}$$

where T^* is the equilibrium temperature in degrees absolute.

In spite of the fact that they appear quite different from the ideal blackbody shown in Fig. 12.3, many heated solids (for example, electrically heated wire filaments) nevertheless radiate essentially as blackbodies.

THE INADEQUACY OF THE CLASSICAL THEORIES

Many attempts were made to explain the *shape* of the blackbody spectrum in terms of classical electromagnetic theory. Two of these attempts were those

[2]Of course, if we are to maintain a constant temperature as radiation escapes, we must continually replenish the energy; this can be accomplished, at least in principle, with a small electrical heater embedded in the cavity block.

made by Max Carl Wien (1866–1938) and by Lord Rayleigh (1842–1919). The predictions of these theories are compared with an experimental spectrum in Fig. 12.5. It is clear that neither is a satisfactory representation of the experimental data: Wien's result fails at long wavelengths whereas Rayleigh's result agrees with experiment only in the long wavelength region. Thus, the classical theories were found inadequate to explain the blackbody spectrum, and so remained the situation until a bold and imaginative step forward was made by the German physicist Max Karl Ernst Ludwig Planck (1858–1947).

PLANCK'S QUANTUM HYPOTHESIS

In 1900 Planck undertook to accomplish what Wien and Rayleigh and others had failed to do—to explain in detail the shape of the blackbody spectrum for *all* wavelengths, not just in the long- or short-wavelength regions. Planck found a formula that did in fact reproduce the complete shape of the spectrum, but there was no theoretical justification for the formula whatsoever. In order to provide a basis for his formula, Planck was forced to make an unorthodox assumption. He concluded that energy exchange between the radiation and the cavity material does not take place in a continuous way. Instead, Planck hypothesized, energy exchange takes place in discrete steps by *quanta*. The amount of energy associated with a quantum of frequency v is a certain constant times v; that is,

$$\varepsilon = hv \qquad (12.3)$$

where h is the proportionality constant, now known as *Planck's constant*. The value of the constant h is extremely small:

$$h = 6.625 \times 10^{-27} \text{ erg-sec} \qquad (12.4)$$

Therefore, each quantum has associated with it only a very small amount of energy. It is not surprising, then, that when large amounts of radiation are involved, the discrete nature of the energy is not evident since a change of only a few quanta is completely negligible.

In formulating his theory of blackbody radiation, Planck did not draw on any *direct* evidence for radiation quanta; he only inferred the existence of quanta because this concept appeared necessary to explain the shape of the blackbody radiation spectrum. Consequently, Planck's method of reproducing the blackbody spectrum was regarded by most physicists as an

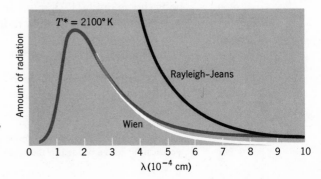

Fig. 12.5 *Comparison of an experimental blackbody spectrum (red curve) with the results of the theories of Wien and Rayleigh (the latter as modified by James Jeans).*

Fig. 12.6 *Max Planck, winner of the Nobel Prize in 1918 for his quantum explanation of the black-body radiation spectrum.*

interesting trick of no fundamental importance. In 1900 it was still generally believed that all physical processes were *continuous,* and even Planck did not go so far as to suggest that all electromagnetic radiation was quantized; his hypothesis applied only to the exchange of energy between radiation and the cavity. Thus, the one single great idea that was necessary to under-stand the nature of radiation—the quantum hypothesis—was incompletely developed and was generally ignored.

Planck's idea remained in limbo for several years before Einstein found the quantum hypothesis necessary for the explanation of the photoelectric effect (described in the next section). Einstein extended Planck's idea and postulated that *all* electromagnetic radiation occurs in the form of quanta (or *photons*). Finally, Planck's great contribution was realized and he was awarded the 1918 Nobel Prize in physics for his introduction of the quantum concept and his explanation of the blackbody spectrum.

Example **12.1**

A 1-gram mass falls through a height $H = 1$ cm. If all of the energy acquired in the fall were converted to yellow light ($\lambda = 6 \times 10^{-5}$ cm), how many photons would be emitted?

The energy acquired in the fall is just the potential energy:

$$PE = mgH$$
$$= (1 \text{ g}) \times (980 \text{ cm/sec}^2) \times (1 \text{ cm})$$
$$= 980 \text{ ergs}$$

The energy of each yellow photon is

$$\varepsilon = h\nu = \frac{hc}{\lambda}$$
$$= \frac{(6.6 \times 10^{-27} \text{ erg-sec}) \times (3 \times 10^{10} \text{ cm/sec})}{6 \times 10^{-5} \text{ cm}}$$
$$= 3 \times 10^{-12} \text{ ergs/photon}$$

Therefore, the number of photons is

$$\text{No. of photons} = \frac{980 \text{ ergs}}{3 \times 10^{-12} \text{ ergs/photon}}$$

$$= 3 \times 10^{14} \text{ photons}$$

Thus, only a minute amount of energy is associated with an individual photon.

THE FUNDAMENTAL ATOMIC CONSTANTS

With Planck's constant h we have completed the introduction of the fundamental atomic constants: c, m_e, e, and h. All the other important numbers that occur in atomic theory are either multiples or combinations of these four constants. Because of the largeness of c and the smallness of m_e, e, and h, the physical phenomena that depend on these quantities do not reveal themselves in everyday occurrences. Therefore, the true nature of these phenomena was not appreciated until the modern era when sufficiently delicate instruments were developed for measuring effects taking place at the atomic level.

Table **12.2**   *The Fundamental Atomic Constants*

Velocity of light	$c = 2.9979 \times 10^{10}$ cm/sec
Mass of electron	$m_e = 9.1096 \times 10^{-28}$ g
Charge of electron	$e = 4.8032 \times 10^{-10}$ statC
Planck's constant	$h = 6.6262 \times 10^{-27}$ erg-sec

12.3   *The Photoelectric Effect*

EJECTION OF ELECTRONS BY LIGHT

As early as 1887 it had been observed by Heinrich Hertz, the discoverer of electromagnetic waves, that the intensity of the spark between two high-voltage electrodes was increased when the electrodes were exposed to ultra-violet (UV) light. Almost immediately, this effect was pursued by others and it was found, for example, that a clean zinc plate acquires a positive electric charge when irradiated by UV light (Fig. 12.7). These two effects can both

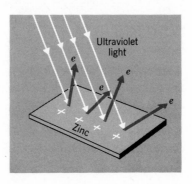

Fig. 12.7 *The photoelectric effect. Ultraviolet light incident on a metal plate, such as zinc, ejects electrons and the plate acquires a positive charge.*

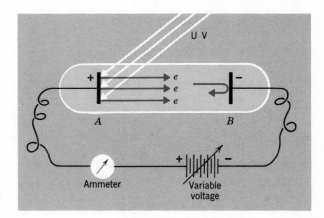

Fig. 12.8 *Apparatus for the study of the photoelectric effect. The motion of the energetic electrons emitted from plate* A *is retarded by the negative potential of plate* B *relative to* A.

be explained in terms of the ejection of electrons from the materials by the incident light. In Hertz's experiment, the ejected electrons assisted in the conduction of the spark between the electrodes and therefore, the act of irradiating the electrodes with UV light increased the intensity of the spark. In the zinc-plate experiment the removal of electrons by the action of UV light leaves behind a positively charged plate. This phenomenon of electron ejection by light is called the *photoelectric effect.* Thus, Hertz, who believed the wave theory of light was a "certainty," accidentally uncovered the first piece of evidence for the particle-like aspect of light.

The qualitative explanation of the photoelectric effect in terms of the removal of electrons by the action of light was in no way revolutionary. Light was known to consist of oscillating electric and magnetic fields that carry electromagnetic energy. It was entirely consistent with classical theory that these waves could transfer energy to electrons in the metal and when an electron had acquired sufficient energy from the wave it could escape from the metal. But further experiments showed that several aspects of the photoelectric effect were at variance with the predictions of classical electromagnetic theory.

THE RESULTS OF PHOTOELECTRIC EXPERIMENTS

Quantitative information regarding the photoelectric effect was obtained by using apparatus similar to that shown schematically in Fig. 12.8. Two plates, *A* and *B*, are contained within an evacuated tube and are connected to a source of variable voltage and a sensitive current-measuring instrument (a galvanometer or an ammeter). Plate *A* is the photoemissive surface and electrons are ejected from this plate when it is exposed to ultraviolet radiation; plate *B* is the collector plate. With this apparatus the current of photoelectrons can be measured as a function of the frequency and the intensity of the UV radiation and of the voltage between plates *A* and *B*. Also, the effect of using different materials for the photoemissive surface can be studied. Certain materials, such as lithium, sodium, and potassium, were found to eject photoelectrons under irradiation by visible blue light, but most other elements must be exposed to higher frequency UV radiation before photoemission occurs.

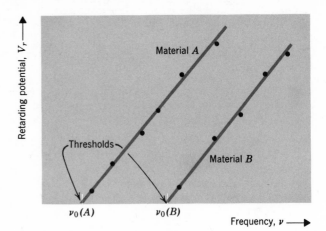

Fig. 12.9 *The maximum retarding potential that can be overcome by photoelectrons is directly proportional to the UV frequency for all frequencies above the threshold at $v = v_0$. The dots show the results of various measurements made with discrete UV frequencies for two different materials. The two lines have the same slopes but different thresholds.*

One of the important measurements that can be made in the investigation of the photoelectric effect is the determination of the maximum retarding (or stopping) potential V_r that can be overcome by the energetic photoelectrons ejected when plate A is irradiated with UV light of a definite frequency. By making plate B negative with respect to plate A, the electrons are slowed down as they move from A to B. If the retarding potential is sufficiently great, no electrons can reach plate B and the ammeter will register no current flow. Therefore, by increasing the negative potential of B and observing the current flow, the voltage V_r necessary to just stop the electrons from reaching plate B can be determined. Measurements of this type led to the important result that V_r increases in direct proportion to the *frequency* of the incident UV radiation. Furthermore, there was found to be a *threshhold frequency* v_0 below which no photoelectrons could be produced *regardless of the intensity of the incident radiation*. It was also observed that the number of photoelectrons ejected is directly proportional to the *intensity* of the UV radiation for any frequency of radiation greater than v_0.

The dependence of V_r on the UV frequency is illustrated in Fig. 12.9. A satisfactory representation of all the experimental points for a given material is a straight line, and this line intersects the axis at $v = v_0$. Therefore, the equation that represents the relationship between V_r and v is

$$V_r = a(v - v_0) \tag{12.5}$$

where a is a constant and is equal to the slope of the straight line. It was found that photoemissive surfaces of different materials all produced straight-line relationships *with the same slope* but that the threshold frequency is characteristic of the material (see Fig. 12.9). If an electron with kinetic energy KE is just stopped when the potential difference between the plates is V_r, then $KE = eV_r$. That is, eV_r is the *maximum* kinetic energy of the ejected photoelectrons. (Electrons liberated below the surface of the material can lose kinetic energy before emerging from the surface.)

The experimental facts regarding the photoelectric effect that must be explained are these:

1. The intensity of the incident UV radiation determines the *number* of photoelectrons ejected, but the intensity does *not* determine the retarding potential required to stop the electrons (that is, the *energy* of the electrons does not depend on the *intensity* of the UV radiation).

2. The *maximum energy* of the photoelectrons depends only upon the *frequency* of the UV radiation.

3. For each material there exists a definite threshold frequency ν_0, and UV radiation with a lower frequency will *not* eject photoelectrons no matter what the intensity of the radiation. However, if $\nu > \nu_0$, then even the most feeble UV radiation will cause photoelectrons to be ejected (albeit in small numbers) with *no time delay* between the turning on of the UV source and the appearance of the first photoelectron.

EINSTEIN'S EXPLANATION OF THE PHOTOELECTRIC EFFECT

According to the classical theory, UV radiation delivers electromagnetic energy to a photoemissive surface and, when sufficient energy has been concentrated in one electron, this electron is ejected. An increase in the UV intensity causes more electrons to be ejected. Therefore, the classical theory is capable of explaining the first part of fact No. 1, above. But this theory is completely inadequate to account for the remaining facts. For example, if the UV intensity is low, the classical theory would predict that it would require a certain time to concentrate sufficient energy in any one electron to cause its ejection; this prediction is contrary to the experimental result (fact No. 3).

In 1905 Einstein offered a theory that provides an explanation of the entire set of facts concerning the photoelectric effect. Einstein's photoelectric theory was beautifully simple, with the same brevity and elegance that characterized his relativity theory (proposed in the same year).[3] Einstein extended Planck's hypothesis of quantized energy exchange in a cavity and proposed that *all* electromagnetic radiation was in the form of discrete bundles of electromagnetic energy called *quanta* or *photons*. He further proposed that when a photon interacts with matter it behaves as a *particle* and delivers its energy, not to the material as a whole or even to an atom as a whole, but to an *individual* electron. The occurrence of a threshold is due to the fact that a certain amount of energy must be supplied to the electron in order to free it from the material (even if it receives no kinetic energy). Furthermore, different materials have different values for the threshold energy.

According to Einstein, then, the kinetic energy of a photoelectron must be the difference between the energy of the incident UV photon and the minimum energy necessary to free the electron from the material (called the *work function* of the material). That is,

$$(\text{electron } KE) = (\text{photon energy}) - (\text{work function}) \tag{12.6}$$

When $h\nu$ is used for the photon energy (Eq. 12.3), this equation becomes

---

[3] Also the same year of his theory of Brownian motion—truly a remarkable performance.

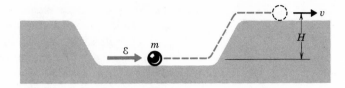

Fig. 12.10 *A mechanical analog of the photoelectric effect. Sufficient energy must be supplied to the ball for it to "escape" from the trough and proceed away with a velocity* v.

$$KE = hv - \phi \tag{12.7}$$

where ϕ is the work function for the particular material.

As a mechanical analog to this explanation of the photoelectric effect, consider a ball of mass m that is at rest in a trough, as in Fig. 12.10. If a sufficient amount of energy $\mathcal{E}$ is supplied to the ball, it will roll up the side of the trough of height H and "escape" with a final velocity v. The energy equation for the process is

$$\tfrac{1}{2}mv^2 = \mathcal{E} - mgH$$

In this expression, mgH is the "work function" (that is, the potential energy barrier that must be overcome) and $\mathcal{E}$ is equivalent to the photon energy.

Example **12.2**

What will be the maximum kinetic energy of the photoelectrons ejected from magnesium (for which $\phi = 3.7$ eV) when irradiated by UV light of frequency 1.5×10^{15} sec^{-1}?

The energy of a photon with frequency 1.5×10^{15} sec^{-1} is

$$hv = (6.6 \times 10^{-27} \text{ erg-sec}) \times (1.5 \times 10^{15} \text{ sec}^{-1})$$

$$= (9.9 \times 10^{-12} \text{ erg}) \times \left(\frac{1}{1.6 \times 10^{-12} \text{ erg/eV}}\right)$$

$$= 6.2 \text{ eV}$$

Therefore, the maximum kinetic energy is

$$KE = hv - \phi$$

$$= 6.2 \text{ eV} - 3.7 \text{ eV}$$

$$= 2.5 \text{ eV}$$

The electrons ejected from a photoemissive surface under irradiation by ultraviolet light will actually have a range of kinetic energies because those electrons that are released beneath the surface will suffer collisions and will lose energy before emerging.[4] Therefore, in the retardation experiments

---

[4] Of more fundamental importance is the fact that the deeper-lying electrons in an atom require more energy for their removal than do the outer electrons. This aspect of atomic structure is discussed in Chapters 13 and 14.

Table **12.3** *Some Properties of Different Types of Electromagnetic Radiation*

Type of Radiation	Typical Wavelength	Frequency (sec^{-1})	Photon Energy
Gamma rays	10^{-11} cm	3×10^{21}	12 MeV
X rays	10^{-9} cm	3×10^{19}	120 keV
UV radiation	10^{-5} cm = 1000 Å	3×10^{15}	12 eV
Visible light (yellow)	6×10^{-5} cm = 6000 Å	5×10^{14}	2 eV
Infrared radiation (heat rays)	10^{-3} cm	3×10^{13}	0.12 eV
Microwaves	1 cm	3×10^{10}	1.2×10^{-4} eV
Radio waves	3×10^4 cm = 300 m	10^6	4×10^{-9} eV[a]

[a] This wavelength is so long and the energy is so small that the quantum character of the radiation is unimportant. That is, the energy of a single photon of this radiation is so small that it can have no measurable effect in any experiment we might consider here.

described above, the potential V_r that just reduces the electron current to zero corresponds to the kinetic energy of the most energetic photoelectrons. This kinetic energy is $KE = eV_r$, so that Eq. 12.7 can be expressed as

$$eV_r = h\nu - \phi \tag{12.8}$$

or,

$$V_r = \frac{h}{e}(\nu - \nu_o) \tag{12.9}$$

where $\phi = h\nu_o$. Comparing this last equation with Eq. 12.5, we see that the proportionality constant a must, according to the Einstein hypothesis, be equal to Planck's constant divided by the electronic charge: $a = h/e$.

Careful measurements of the photoelectric effect for several elements were made by the American physicist Robert A. Millikan (1868–1953), beginning in 1916. These experiments showed that the slopes of the lines of V_r vs ν were indeed consistent with Einstein's theory, which predicted the value h/e for the slope of all such lines. When the results of Millikan's earlier experi-

Table **12.4** *Photoelectric Properties of Some Elements*

Element	Work Function, ϕ (eV)	Threshold Frequency (sec^{-1})	Threshold Wavelength (Å)	
Cesium	1.9	4.6×10^{14}	6500	Visible light
Potassium	2.2	5.3	5600	
Sodium	2.3	5.6	5400	
Calcium	2.7	6.5	4600	
Magnesium	3.7	8.9	3400	Ultraviolet
Silver	4.7	11.4	2600	
Nickel	5.0	12.1	2500	

ments on the determination of the electronic charge (see Section 8.5) were combined with his h/e measurements, this provided the best available value for Planck's constant. (More recently, different methods have been used for precise determinations of h.)

Example **12.3**

When silver is irradiated with ultraviolet light of wavelength 1000 Å, a potential of 7.7 volts is required to retard completely the photoelectrons. What is the work function of silver?

First, the energy of a 1000-Å photon is

$$h\nu = \frac{hc}{\lambda} = \frac{(6.6 \times 10^{-27} \text{ erg-sec}) \times (3 \times 10^{10} \text{ cm/sec})}{(1.0 \times 10^{-5} \text{ cm}) \times (1.6 \times 10^{-12} \text{ erg/eV})}$$

$$= 12.4 \text{ eV}$$

Then,

$$\phi = h\nu - eV_r$$
$$= 12.4 \text{ eV} - 7.7 \text{ eV}$$
$$= 4.7 \text{ eV}$$

(Note that we have used the fact that the *energy* of an electron that is completely stopped by a *potential* of 7.7 volts is just 7.7 electron volts.)

For his detailed explanation of the photoelectric effect (*not* for his relativity theory!), Einstein was awarded the Nobel Prize in 1921. Millikan received the 1923 Prize for his experimental work on the determination of e and h.

12.4   *Waves or Particles? Two Crucial Experiments*

THE WAVE-PARTICLE DILEMMA

The wave character of light was established in the early years of the 19th century when a series of interference and diffraction experiments disproved the competing particle theory of light. But Einstein's photoelectric theory revived the notion that light behaves as a particle—at least, in its interactions with atomic electrons. Does this mean that we are forced to abandon the wave theory and return to the old particle theory? Or is there some peculiar feature of light that presents a two-sided appearance, sometimes wave-like and sometimes particle-like? If so, then how do we know when to expect one feature and not the other?

These were the questions about light that were raised by the proposal of the quantum nature of electromagnetic radiation. An equally important question concerned the properties of particles: If light plays a dual role of particle and wave, can an electron—an object that we have always considered to be a *particle*—also behave as a *wave*?

These questions were finally resolved in the 1920s by a series of experiments performed in America and in England that showed unequivocally that light *and* electrons could each exhibit properties associated with both waves and particles. This wave-particle duality was then incorporated as an essential feature into the emerging *wave mechanics* or *quantum theory*. The first of these crucial experiments concerns the particle behavior of radiation and the second demonstrates the wave character of electrons.

COMPTON SCATTERING

If Einstein's interpretation of the photoelectric effect is correct and a photon interacts with a single electron to eject it, then it should be possible to observe a *scattering* process in which a photon and an electron interact in much the same way as do two colliding balls. In 1924, the American physicist Arthur H. Compton (1892–1962) demonstrated just such a process.

In describing the Compton scattering experiment, we wish to simplify the situation and treat the electrons as *free* particles. But we know from the results of photoelectric experiments that electrons are *bound* in matter with energies of several eV. In his experiment, Compton used photons (X rays) of energy 17.5 keV. Because this energy is so much greater than the binding energy of the electrons, we shall make no appreciable error by treating the electrons as free particles.

We proceed by considering a photon of energy $h\nu$ incident on an electron at rest (Fig. 12.11), and we treat the problem in a strictly classical way. (But relativistic effects must be taken into account when we deal with fast-moving electrons.) We have available three equations that represent the following statements:

1. *Conservation of energy:* The energy before collision (the photon energy $h\nu$ plus the electron rest energy) must equal the energy after collision (the energy $h\nu'$ of the scattered photon plus the total energy of the recoiling electron).
2. *Conservation of linear momentum in the direction of the incident photon:* The momentum of the incident photon, $p = \mathcal{E}/c = h\nu/c$, must equal the sum of the components of the electron momentum and the momentum of the scattered photon in the direction of the incident photon.
3. *Conservation of linear momentum in the direction transverse to the direction of the incident photon:* Since there is no momentum in the transverse direction prior to collision, there can be no net transverse momentum after collision. Therefore, the transverse momentum components of the scattered photon and the recoiling electron must be equal and opposite.

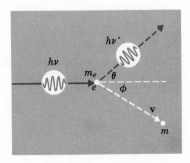

Fig. 12.11 *Schematic diagram of the Compton scattering of a photon of energy* hν *by an electron.*

By writing down these equations and performing the necessary (but tedious) algebra, we find that the frequency v' of the scattered photon is related to the frequency v of the incident photon and to the angle of scattering θ according to:

$$\frac{1}{v'} = \frac{1}{v} + \frac{h}{m_e c^2}(1 - \cos\theta) \tag{12.10}$$

or, converting to wavelengths by using $\lambda = c/v$, we have

$$\lambda' = \lambda + \frac{h}{m_e c}(1 - \cos\theta) \tag{12.11}$$

The quantity $h/m_e c$ is known as the *Compton wavelength of the electron* and has the value

$$\frac{h}{m_e c} = 2.426 \times 10^{-10} \text{ cm} \tag{12.12}$$

This is the wavelength of a photon whose energy is $m_e c^2$, the rest-mass energy of an electron.

Compton measured λ' as a function of θ for several different initial wavelengths λ and found agreement with Eq. 12.11, thus demonstrating that photons behave as *particles* not only in the photoelectric effect but in scattering processes as well. Compton received the 1927 Nobel Prize in physics for these experiments and their interpretation.

Example **12.4**

A 100-keV X ray is Compton scattered through an angle of 90°. What is the energy of the X ray after scattering?

Dividing Eq. 12.10 by h, we have

$$\frac{1}{hv'} = \frac{1}{hv} + \frac{1}{m_e c^2}(1 - \cos\theta)$$

Using the values $m_e c^2 = 511$ keV and $\cos 90° = 0$, we find

$$\frac{1}{hv'} = \frac{1}{100 \text{ keV}} + \frac{1}{511 \text{ keV}}$$

or,

$$hv' = \frac{(100 \text{ keV}) \times (511 \text{ keV})}{(100 \text{ keV}) + (511 \text{ keV})} = \frac{51100}{611} \text{ keV} = 84 \text{ keV}$$

The electron, of course, carries off the remainder of the incident energy:

$$KE = 100 \text{ keV} - 84 \text{ keV} = 16 \text{ keV}$$

THE DE BROGLIE WAVELENGTH

Einstein's explanation of the photoelectric effect and the photon scattering experiments of Compton had clearly demonstrated that electromagnetic radiation has properties that closely resemble those of material particles.

In 1924 a young Frenchman, Louis Victor de Broglie (1892–   ), proposed in his doctoral thesis that, in view of the particle-like behavior of waves, there should also be a wave-like behavior of particles.

In order to describe a wave, there must be a definable wavelength for the propagation. Now, the momentum of a photon is related to its wavelength according to (see Eq. 10.34):

$$p = \frac{\mathcal{E}}{c} = \frac{h\nu}{c} = \frac{h}{\lambda}, \quad \text{or} \quad \lambda = \frac{h}{p}$$

De Broglie argued that the wavelength of a material *particle* should follow exactly the same prescription:

de Broglie wavelength: $\boxed{\lambda = \dfrac{h}{p}}$ (12.13)

Within three years after de Broglie's ingenious proposal (which earned him the 1929 Nobel Prize), the wave properties of electrons had been demonstrated in experiments by Davisson and Germer in America and by Thomson in England which showed the diffraction of electrons. These experiments are described briefly in the following paragraph.

Example **12.5**

What is the wavelength of a 10-eV electron?

For an electron of this low energy we can use the nonrelativistic expression without appreciable error. Since $KE = \frac{1}{2}mv^2$ and $p = mv$, we have

$$p = \sqrt{2\, m_e\, KE}$$

so that

$$\lambda = \frac{h}{p} = \frac{h}{\sqrt{2\, m_e\, KE}}$$

$$= \frac{6.62 \times 10^{-27} \text{ erg-sec}}{\sqrt{2 \times (9.11 \times 10^{-28} \text{ g}) \times (10 \text{ eV}) \times (1.60 \times 10^{-12} \text{ erg/eV})}}$$

$$= 3.86 \times 10^{-8} \text{ cm}$$

$$= 3.86 \text{ Å}$$

A photon with this same wavelength would have an energy

$$\mathcal{E} = h\nu = \frac{hc}{\lambda}$$

$$= \frac{(6.62 \times 10^{-27} \text{ erg-sec}) \times (3 \times 10^{10} \text{ cm/sec})}{(3.86 \times 10^{-8} \text{ cm}) \times (1.60 \times 10^{-12} \text{ erg/eV})}$$

$$= 3.2 \text{ keV}$$

Such an energetic photon is an *X ray*.

DIFFRACTION OF X RAYS AND ELECTRONS

When electrons are allowed to fall through a potential of several thousand volts and strike a metal target, extremely short-wavelength radiation is produced. Many attempts were made in the early 1900s to measure the wavelengths of these X rays by conventional diffraction experiments using finely ruled gratings (see Section 10.5). These experiments were only marginally successful until Max von Laue (1879–1960) realized that the regular planes of atoms in crystals, with spacings of only a few Ångstroms, could be used as diffraction gratings. In 1912 von Laue succeeded in obtaining interference patterns of X rays diffracted from calcite crystals. The spacing between the planes of atoms in the crystal could be calculated approximately from a knowledge of the properties of the material. The measured diffraction patterns then showed that typical wavelengths for X rays were of the order of 1 Å, very much shorter than UV wavelengths.

The diffraction of X rays is illustrated schematically in Fig. 12.12 where the dots represent the ordered array of atoms in a simple crystalline lattice. The rows of atoms are spaced a distance d apart and therefore form a kind of grating through which the radiation can pass. In the direction specified by the angle θ with respect to the direction of the incident beam, the two scattered rays differ in their path lengths by an amount $d \sin \theta$. Therefore, on a photographic plate placed some distance away, the interference between the two rays will be *constructive* if $d \sin \theta$ is an integer number of wavelengths of the radiation; the interference will be *destructive* if $d \sin \theta$ is an odd number of half wavelengths. Consequently, there will be exposed and unexposed regions that alternate on the photographic plate. Because the crystal is a *three*-dimensional array of atoms, the directions in space in which the interference is constructive are quite limited. As a result, the exposed regions are *spots* rather than lines, as in the case of diffraction by a series of slits (which is only a *two*-dimensional array). Figure 12.13 shows such a spot pattern (called a *Laue* pattern). Measurements of the distances between the spots can be used to determine the wavelength of the incident radiation if the atomic spacing in the crystal is known. (Actually, the method

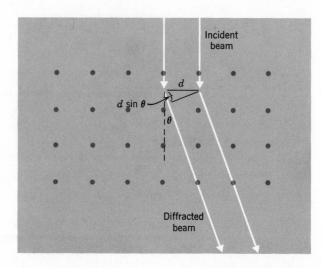

Fig. 12.12 *An incident beam of X rays (or electrons) is diffracted by the grating formed by the regular array of atoms in a crystal.*

"Fundamentals of College Physics" by W. Wallace McCormick   Wollan, Shull, & Marney, Physics Rev. *73*, 527 (1948)

Fig. 12.13 *Laue patterns for the diffraction of (a) X rays and (b) neutrons by single cubic crystals of rock salt (Na Cl). The light spots are the exposed regions of the photographic plates caused by constructive interference. Note the similarity of the interference patterns. The central bright spot in each pattern is due to the undeflected portion of the beam.*

is now used in reverse: X rays of known wavelength are employed to study the details of crystal structure by diffraction.)

Suppose that we use a *foil* of material for the diffraction experiment instead of a single crystal. Foils usually consist of large numbers of tiny crystals (*micro*crystals), which are randomly oriented. Therefore, there will be a random distribution of the angles of incidence of the X-ray beam upon the individual crystals. The interference pattern observed for the transmitted beam of X rays will then correspond to a large number of spot patterns that are located at random angles around the direction of the incident beam. That is, each spot in the single-crystal Laue pattern will produce a *circular* pattern of constructive interference. Such a circular diffraction pattern is shown in Fig. 12.14a.

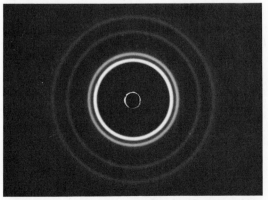

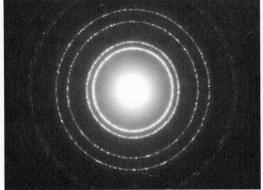

Educational Development Center   Educational Development Center

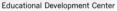

Fig. 12.14 *X-ray and electron diffraction in aluminum, showing the similarity in the diffraction patterns. The X-ray photograph (a) was made with X rays of wavelength 0.71A°. The electron diffraction photograph (b) was obtained with 600-eV electrons ($\lambda = 0.50A°$). The electron pattern has been enlarged $\times$ 1.6 in order to facilitate the comparison.*

According to de Broglie's hypothesis, electrons should exhibit wave properties and therefore should be capable of producing diffraction patterns in the same way that X rays produce these patterns. In 1927 George P. Thomson (1892–   , son of Sir J. J. Thomson) observed electron diffraction by passing electrons through foils. At almost the same time, C. J. Davisson (1881–1958) and L. H. Germer (1896–   ) directed electrons onto the surfaces of single crystals and found interference peaks in the distribution of the reflected electrons. Davisson and Thomson shared the 1937 Nobel Prize for these experiments that established the wave character of electrons.

The correspondence between X-ray diffraction and electron diffraction is strikingly illustrated in Fig. 12.14. Here are photographs of diffraction patterns obtained by passing X rays and electrons of approximately the *same wavelength* through a thin aluminum foil. Also, Fig. 12.13 shows the Laue patterns produced by X rays and by neutrons diffracted by rock salt. The close similarity between the patterns in each pair of photographs is dramatic proof of the identical wave properties of electromagnetic radiation and matter.

12.5   *The Basis of Quantum Theory*

PHOTONS

The various experiments we have just described conclusively demonstrated the dual nature of radiation and matter—an electron can propagate as a wave and light can interact as a particle. We have already learned how to describe the classical aspects of electrons and electromagnetic radiation, but how do we describe "light particles" and "electron waves"?

We have often referred to the *frequency* of an electromagnetic wave. However, we have been a bit cavalier with the terminology because, in fact, *the* frequency of a real electromagnetic wave does not exist. That is, such radiation never possesses a single, precisely defined frequency. In order to define *the* frequency of a wave, the wave must be absolutely uniform in its properties throughout space. Thus, a wave of *pure* frequency (Fig. 12.15) must have infinite extent. But all electromagnetic oscillators, be they radio antennas or atoms, oscillate only for finite periods of time. Therefore, the radiation is necessarily of finite extent and cannot have a single, precise frequency. *Real* radiation is always equivalent to a combination (or *superposition*) of oscillations with various frequencies. If these frequencies lie in a range around a central frequency, the effect is to produce *constructive* interference in one region of space and *destructive* interference everywhere else. The appearance of such a superposition of waves is shown schematically in Fig. 12.16. This localized bunch of oscillations is called a *wave packet* and, for the case of electromagnetic radiation, the packet (that is, a *photon*) propagates as a unit with the velocity of light.

Fig. 12.15 *A wave that has a pure or precise frequency must be the same everywhere; that is, it must have infinite extent.*

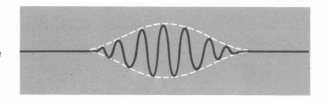

Fig. 12.16 *A wave packet (or photon) is localized in space because the superposition of waves of various frequencies produces constructive interference in one region and destructive interference everywhere else. Real photons contain 10^5–10^6 oscillations rather than the few sketched here.*

The range of frequencies that exists in typical light photons is quite small. For example, consider yellow light of nominal frequency 5×10^{14} sec^{-1}. In the emission of such light from an atom, the range of frequencies is only about 2 parts in 10^6; that is, $\Delta\nu/\nu \cong 2 \times 10^{-6}$. The corresponding range of wavelengths is only 0.01 Å; that is, the average wavelength is 6000 Å with a spread in wavelength from 5999.99 Å to 6000.01 Å. No spectral line is absolutely sharp; there is always a small natural width.

Figure 12.16 shows only 7 oscillations within the wave packet; for a typical light photon, such as that just described, the wave packet would consist of approximately 6×10^5 oscillations. This photon, containing such a large number of oscillations, retains many of its wavelike characteristics; but it still is a discrete unit and therefore interacts individually with electrons, for example, in Compton scattering or in the photoelectric effect.

ELECTRON WAVES

Let us consider a very simple type of experiment involving electrons. Suppose that we direct a beam of electrons toward a panel in which two slits have been cut. In Fig. 12.17a we show the situation for the case in which the lower slit (labeled *B*) is blocked off so that all of the electrons that penetrate the panel must go through slit *A*. At a certain distance behind the panel we place a screen for the purpose of observing the positions of impact of the various individual electrons. Such a screen can be made from a material that *scintillates;* that is, a flash of light is produced for every electron impact. Or, we could use some sort of electron detector, such as a Geiger counter, and by moving the detector up and down along the screen, record the number of electron impacts as a function of position along the screen. In any event, the essential quantity that we can measure by using one of several available techniques is the distribution of electron impacts along the screen. The curve in Fig. 12.17a represents such an *intensity* distribution; it shows that the peak of the distribution is directly in line with the slit. If the slit is *narrow,* the distribution will be *broad* (compare Fig. 10.39).

If we repeat the experiment with slit *B* open and slit *A* closed, we shall, of course, find exactly the same type of distribution centered now around the position in line with slit *B*.

What will happen if we open *both* slits? If electrons behaved as little balls, we could think of the experiment in terms of projecting a stream of marbles at the pair of slits: some marbles would go through slit *A* and, because of scattering at the slit edges, would produce a pattern on the screen like that shown in Fig. 12.17a and some would go through slit *B* and produce the pattern of Fig. 12.17b. The net result would be the *sum* of the two intensity

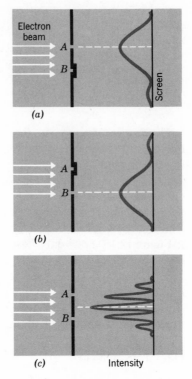

Fig. 12.17 (*a, b*) *Intensity patterns of electron impacts on a screen for a single open slit.* (*c*) *Intensity pattern for a double slit. This pattern is not the same as the sum of the two previous patterns. This pattern is drawn for the case in which the distance between the two slits is 3 times the width of each slit. Notice that increasing intensity is plotted to the left.*

patterns. But this expectation based on classical reasoning is not borne out by experiment for the case of electron beams. Actually, a much more complex pattern is observed (Fig. 12.17c). The maximum intensity is found *midway* between the two slits and there are several subsidiary maxima, falling off in intensity uniformly on each side of the central maximum. Notice also that, at some points on the screen, the intensity has actually been *decreased* by the opening of an additional slit. We can only draw the conclusion that electrons behave as waves in this experiment and produce exactly the same type of interference effects that are produced by light waves.

SELF-INTERFERENCE AND PROBABILITY

Suppose that we repeat the double-slit experiment, using either light or electrons (it does not matter which we use since both exhibit identical interference effects), but now we make the incident beam exceedingly weak—so weak, in fact, that at any one instant there can be only a single photon or electron in the vicinity of the apparatus. How will the pattern on the screen be affected? When any individual photon or electron passes through the apparatus, it will not smear itself out in conformity with the curve of Fig. 12.17c because each individual electron or photon interacts with the screen at a definite *point*. We can determine the position of the impact by observing the light flash that is produced, but *before* the impact has occurred, *we have absolutely no way in which to predict the point at which it will occur.*

After the first 10 electrons or photons have passed through the apparatus, the results might be similar to those illustrated in Fig. 12.18a where each box represents a light flash observed at the corresponding position on the screen. With this small number of events, the pattern is only a scatter of boxes. After 40 events (Fig. 12.18b), definite structure begins to emerge; the number of events in the central maximum is decidedly greater than anywhere else on the screen and there are pronounced valleys between the groups of boxes. When a large number of events has occurred (in this case, thousands), it becomes possible to represent the results with a curve that encloses the distribution of boxes (Fig. 12.18c). For any particular photon or electron, then, we can refer only to the *probability* that it will strike the screen at a given point. The height of the curve at any position along the screen is proportional to the probability that an electron or photon will interact with the screen at that position. The probability is highest for the midpoint and is least for the valleys between the various maxima. The dashed curve in Fig. 12.18c is called the *probability curve* or the *intensity curve* that describes the distribution of events. This curve is exactly the same as that predicted by the wave theory of diffraction for radiation with the same wavelength.

We are therefore led to the view that *individual* electrons or photons exhibit *interference,* but only when large numbers of particles are involved does it become possible to identify the diffraction pattern with that predicted by the wave theory. When a single electron or photon is in the apparatus at a given instant, it has only *itself* with which to interfere. This *self-interference* results from that portion of the electron wave that goes through slit *A* interfering with the portion that goes through slit *B*.

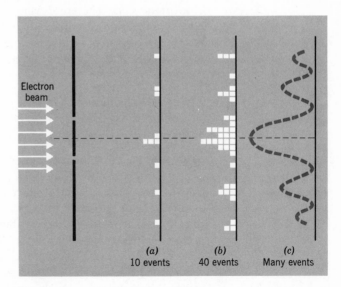

Fig. 12.18 *Distribution of light flashes on a screen after (a) 10, (b) 40, and (c) many photons or electrons have passed through the apparatus. Each box represents one light flash observed at the corresponding position on the screen. The point of occurence of any particular event cannot be predicted; only the probability (the dashed curve) can be stated.*

From these arguments we can draw two important conclusions that are crucial for the development of quantum theory.

1. Individual electrons or photons have the *wave*like property that they can interfere with themselves.
2. Individual electrons or photons have the *particle*like property that they interact with matter at discrete points but the prediction of where such interactions will take place can be made only in the probabilistic sense.

The probability (or intensity) curve that describes the interactions of electrons or photons with matter is identical with the prediction of wave theory and corresponds to the pattern of individual interactions that would occur for a very large number of events.

IS THE WAVE-PARTICLE DUALITY **REAL?**

How can we reconcile the fact that electrons and photons both appear sometimes as particles and sometimes as waves? Does each exist as part wave and part particle? Or are they both capable of transforming back and forth between these two different descriptions? The answers to these questions become clear by realizing that when we make a wave or a particle classification we have forced a *classical* description on entities that are essentially *nonclassical*. Electrons and photons do not obey the rules of classical mechanics—their behavior is described correctly only by *quantum* mechanics. It is therefore not surprising that certain ambiguities arise when we use classical ideas to describe quantum objects.

In the quantum mechanical view of Nature, an experimenter's apparatus and the object under study together constitute a system. The only meaningful way to discuss the behavior of the object studied is in terms of the results of *measurements*. Therefore, whether an electron or a photon appears as a wave or as a particle depends on the nature of the measurement that is made. The wavelike or particlelike character of an electron or a photon therefore lies only in the eye of the beholder.

12.6   *Wave Functions*

PROBABILITY AMPLITUDES AND INTENSITIES

If it is indeed true that the interaction of electrons or photons with matter can be expressed only in terms of *probabilities,* how can we give a mathematical description of such processes? For this purpose we introduce a quantity called the *wave function* of the particle or photon. This function, customarily denoted by the symbol ψ, is used in the calculation of the probability that the particle or photon will be found (through its interaction with matter) at a particular point. The *probability amplitude* for a particle at a point x_o is $\psi(x_o)$. (No multiplication is indicated by these symbols; $\psi(x_o)$ means "the value of ψ at the position $x = x_o$.")

What is the meaning of a *probability amplitude* and how do we use this quantity to calculate the probability of finding a particle at a particular point? In the case of an electromagnetic wave the field vectors, **E** and **B**,

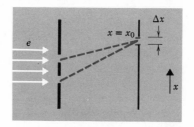

Fig. 12.19 *A beam of electrons is incident on a double slit. At the position* x = x_0 *on the screen, the wave function is* $\psi(x_0)$. *The probability that the electron will strike the screen at* x = x_0 *within a range* Δx *is proportional to* $|\psi(x_0)|^2$ Δx.

oscillate as the wave progresses; for a mechanical wave (for example, sound wave, water wave, etc.), material particles oscillate. But what oscillates in a quantum mechanical wave? No classical (that is, mechanical or electromagnetic) interpretation can be given to the wave function $\psi(x)$; nothing *waves*. In fact, the function $\psi(x)$ has no physical meaning—it is only a mathematical function that behaves in many ways as do the amplitudes of mechanical or electromagnetic waves.

In Section 10.1 we found that the energy of an oscillating mass is proportional to the *square* of the amplitude of vibration. And in Section 10.6 we found that the energy of an electromagnetic wave (that is, the *intensity* of the wave) is proportional to the *square* of the electric field vector, $|\mathbf{E}|^2 = E^2$. In quantum mechanical situations we also deal with *intensities;* in particular, the intensity of a quantum mechanical wave at a given point is just the probability of finding the particle (or photon) at that point. In calculating the intensity (or the probability), we must use the *square* of the quantum mechanical wave function whose *amplitude* is $\psi(x_0)$.

Figure 12.19 shows a beam of electrons incident on a pair of slits just as in Figs. 12.17 and 12.18. At the position $x = x_0$ on the screen, the wave function is $\psi(x_0)$. If we wish to calculate the probability that an electron will strike the screen at $x = x_0$ within a range Δx (that is, if we wish to calculate the *intensity* of electrons in this section of the screen), we must take the square of $\psi(x_0)$ multiplied by the size of the interval Δx and by a suitable proportionality constant; that is, the intensity at $x = x_0$ is

$$I(x_0) = A\,|\psi(x_0)|^2\,\Delta x \tag{12.14}$$

The probability amplitude $\psi(x)$ cannot be directly measured; only the *intensity,* which is proportional to $|\psi(x)|^2$, can be measured. Therefore, the real physical significance of the quantum mechanical wave function appears *only* in the form $|\psi(x)|^2$.

Example **12.6**

At the position $x_0 = 1$ unit on a screen, the wave function for an electron beam has the value $+1$ unit, and in an interval Δx and around $x_0 = 1$ there are observed 100 light flashes per minute. What is the intensity of flashes at $x_0 = 2, 3,$ and 4 units where ψ has the values $+4, +2,$ and -2 units, respectively?

$$I(x_0) = A\,|\psi(x_0)|^2\,\Delta x$$

Therefore,

$$I(x_o = 1) = A |\psi(x_o = 1)|^2 \, \Delta x$$
$$= A |+1|^2 \, \Delta x$$
$$= A \, \Delta x$$
$$= 100 \text{ flashes/min}$$

so that

$$A \, \Delta x = 100 \text{ flashes/min}$$

Then,

$$I(x_o = 2) = A |+4|^2 \, \Delta x$$
$$= 16A \, \Delta x$$
$$= 1600 \text{ flashes/min}$$

$$I(x_o = 3) = A |+2|^2 \, \Delta x$$
$$= 4A \, \Delta x$$
$$= 400 \text{ flashes/min}$$

$$I(x_o = 4) = A |-2|^2 \, \Delta x$$
$$= 4A \, \Delta x$$
$$= 400 \text{ flashes/min}$$

Notice that even though $\psi(x_o)$ differs in sign for the last two cases, the intensity is the same because the intensity is proportional to the *square* of the wave function.

We know from the discussion of wave phenomena (Section 10.5) that a double-slit diffraction pattern (actually *any* type of diffraction pattern) can be described in terms of interfering waves originating from the slits. Similarly, if we have two wave functions, $\psi_1(x)$ and $\psi_2(x)$, which describe electron waves from two slit sources, the probability of finding an interaction at the point $x = x_o$, that is, the *intensity* $I(x_o)$, is given by

$$I(x_o) = A |\psi_1(x_o) + \psi_2(x_o)|^2 \, \Delta x \qquad (12.15)$$

That is, in calculating the intensity, the amplitudes, ψ_1 and ψ_2, must be added *before* the squaring operation is carried out. This is a particularly important point because the wave function $\psi(x)$ carries a *sign* and therefore can be either positive or negative. Thus, if $\psi_1(x_o) = +2$ units and $\psi_2(x_o) = -2$ units, there is *destructive* interference and the probability for an interaction at $x = x_o$ is *zero*:

$$I(x_o) = A |\psi_1(x_o) + \psi_2(x_o)|^2 \, \Delta x$$
$$= A |(+2) + (-2)|^2 \, \Delta x = 0$$

However, if $\psi_1(x_o) = +2$ and $\psi_2(x_o) = +2$, there is *constructive* interference, with the result

$$I(x_o) = A\,|\psi_1(x_o) + \psi_2(x_o)|^2\,\Delta x$$

$$= A\,|(+2) + (+2)|^2\,\Delta x$$

$$= 16A\Delta x$$

Compare these results with $A\,(|\psi_1(x_o)|^2 + |\psi_2(x_o)|^2)\,\Delta x$, which corresponds to the sum of the *individual* intensities with no interference.

THE FREE PARTICLE

A quantum mechanical *free* particle can move through space without any restriction on the value of the energy that it can have. Thus, the particle wave can have any wavelength $\lambda = h/p$, and it can have any kinetic energy, which, in a nonrelativistic situation, is expressed as

$$KE = \frac{1}{2}mv^2 = \frac{1}{2}\frac{(mv)^2}{m} = \frac{p^2}{2m} \tag{12.16}$$

The relationship between the kinetic energy and the momentum, therefore, is *parabolic*, as shown in Fig. 12.20; every point on the curve represents an allowed energy and the corresponding allowed momentum.

For the allowed energies of a free particle, there is no difference between the results of classical mechanics and quantum mechanics—all energies are allowed. However, if we *confine* the particle in some way by subjecting it to forces that restrict its motion, the two theories no longer yield the same results.

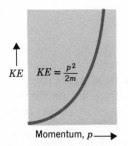

Fig. 12.20 *For a free particle, the curve that relates the kinetic energy to the momentum is a parabola, and every point on the curve refers to an allowed energy and the corresponding allowed momentum.*

PARTICLE BOUND IN A ONE-DIMENSIONAL "BOX"

Consider next a particle that is required to move along a straight line (for example, the x-axis) between the points $x = 0$ and $x = L$. Classically, we can think of this particle as bouncing between a pair of unyielding walls, always maintaining straight-line motion in the $+x$- or $-x$-direction. From the standpoint of classical theory, there is again no restriction on the energy that the particle can have. Energy and momentum are still related by Eq. 12.16 and any combination of KE and p that satisfies this condition is allowed.

Now consider a quantum particle (for example, an electron) that is required to move in the same way. We must think of *this* particle in terms of its wave character and, in particular, we must examine the conditions imposed on the wave function $\psi(x)$ by the presence of the walls. The solution to this problem is actually quite simple because it is exactly the same as that discussed in Section 10.3 for the standing waves on a string. The essential point is that the particle's wave function must be *zero* at $x = 0$ and $x = L$ because the particle is not allowed to escape from the "box." (Recall that the square of $\psi(x)$ is proportional to the probability of finding the particle at a particular point; therefore, $\psi(x)$ must be zero just outside the "box" and hence also at the walls.) That is, standing de Broglie waves must be fitted into the "box." This can occur only when an integer number of wavelengths is equal to $2L$. Thus,

$$n\lambda_n = 2L, \qquad n = 1, 2, 3, \ldots \tag{12.17}$$

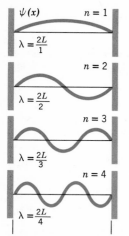

Fig. 12.21 *For a quantum particle bound in a one-dimensional "box", standing de Broglie waves (ψ-waves) must be fitted into the "box". The allowed wavelengths are therefore integer fractions of $2L$.*

The first 4 allowed ψ-waves are illustrated in Fig. 12.21.

The probability of finding the particle at a particular point within the "box" is proportional to $|\psi(x)|^2$. This quantity is shown in Fig. 12.22 for the case $n = 4$. Notice that there are 4 regions in which there is a high probability of finding the particle and that there is zero probability not only at the walls but also at certain points *within* the "box." Clearly, this result is contrary to that of classical mechanics.

ALLOWED ENERGIES IN THE "BOX"

We can now calculate the energies corresponding to the allowed wavelengths by using the de Broglie relation and Eq. 12.21. The allowed momenta are

$$p_n = \frac{h}{\lambda_n} = n\frac{h}{2L}, \qquad n = 1, 2, 3, \ldots \tag{12.18}$$

and the corresponding energies are

$$KE_n = \frac{p_n^2}{2m} = n^2\left(\frac{h^2}{8mL^2}\right), \qquad n = 1, 2, 3, \ldots \tag{12.19}$$

Therefore, we have the important result that only certain discrete energies are allowed. Instead of the classical result in which every point on the KE versus p parabola corresponds to a possible energy, the quantum result shows that only certain of these points are in fact allowed (see Fig. 12.23).

Another important point to realize is the fact that *zero* kinetic energy for the particle is *not* allowed—the particle cannot be at rest in the box. The state of rest requires zero momentum and, hence, an infinitely long wavelength. Such a de Broglie wave cannot fit into any finite "box" and therefore is not allowed. In general, no quantum system (with the exception of the ideal free particle) can possess zero kinetic energy. Even at the absolute zero of temperature, when according to classical theory all motion must cease, a quantum system still possesses a certain kinetic energy, called the *zero-point energy*.

A particle in a box (either one-dimensional or three-dimensional) can be considered to be in a *potential energy well;* that is, the interior of the box corresponds to some finite potential (for example, *zero*) whereas the walls and the exterior are at *infinite* potential. Thus, the particle can never escape the box because, in order to do so, it must have infinite energy. *Real* potential

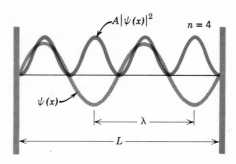

Fig. 12.22 *The probability amplitude $\psi(x)$ and the intensity $A|\psi(x)|^2$ for the case $n = 4$. The particle is most likely to be found at discrete positions within the "box."*

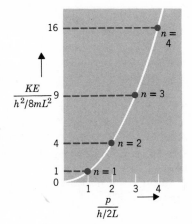

Fig. 12.23 *Only discrete energies (indicated by the horizontal lines and the dots) are allowed for a quantum particle in a box.*

wells, of course, are never infinite and a real particle can always escape from such a well if sufficient energy is supplied. For example, the electrons in a piece of metal find themselves in a certain type of potential well, but if sufficient energy is transferred to such an electron by an ultraviolet photon, the electron can escape from the metal—this is the photoelectric effect.

Different types of potential wells will give rise to different allowed energy values and spacings. If the walls are not impenetrable (as is the case for any real potential), then the wave function is not required to go to zero at the walls—there will always be some "leaking" of the wave function out of the box. But there will still be standing waves for such a potential and the energy states (for energies smaller than the minimum energy required to escape from the potential well) will still be discrete. The first three energy states and the corresponding (standing) wave functions are shown in Fig. 12.24 for the case of a one-dimensional box with *penetrable* walls; that is, the sides of the potential well are not infinitely high.

In the next chapter we shall study another important case of a finite potential—the hydrogen atom.

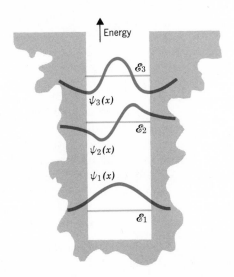

Fig. 12.24 *The wave functions for a particle in a finite one-dimensional potential well are not required to be zero—they "leak" outside the well but are rapidly damped to extremely small values.*

12.7   *The Uncertainty Principle*

WHERE DO THE ELECTRONS "REALLY" GO?

When a series of individual electrons, spaced in time, are incident on a double slit, it is necessary to treat each electron as if its "probability wave," described by $\psi(x)$, goes through *both* slits. The interference of the two portions of each wave at the screen determines the distribution of light flashes that is observed. But an electron is a *particle* and is indivisible; no one has ever observed a *part* of the mass or a *part* of the charge of an electron. Our intuition tells us that an electron cannot go through two separated slits—it must go through one or the other. Can we perform an experiment to determine where the electron "really" goes? Let us alter the double-slit experiment by placing a thin detector behind one of the slits so that the electrons will still be able to pass through but will give a signal when this occurs. Then, we allow electrons to enter the apparatus one at a time and we record only those flashes of light on the screen that are accompanied by a signal that indicates the electron went through the slit with the detector. What pattern of flashes do we find? We find exactly the *single*-slit pattern again (Fig. 12.17a)! The act of determining which slit the electron went through has destroyed the double-slit interference effect.

Perhaps the trouble with our experiment was the detector. Perhaps it was not sufficiently "thin" and actually disrupted too severely the electron trajectories passing through it. We can repeat the experiment using a source of light behind one of the slits. Whenever we detect a photon that is scattered by an electron passing through the slit, we arrange to record the position of impact of the electron on the screen. But this technique is no more successful than the first; the results of this experiment are the same as before. In fact, whatever method we devise to indicate that the electrons have gone through a particular slit, the result is always that the interference effect is destroyed.

So our intuition is wrong. Our insistence in thinking of electrons in terms of classical particles leads to inconsistencies. Because of the difficulties that arose in applying classical reasoning to individual events in the atomic domain, the German theorist Werner Heisenberg (1901–   ) concluded that there must be a general principle of Nature that places a limitation on the capabilities of all experiments. This principle, formulated in 1927, is known as the *uncertainty principle*.

According to Heisenberg's principle, it is impossible to build a detector to determine through which slit the electron passed without destroying the interference pattern. That is, there can be no device that can reveal the presence of an electron with a sufficiently delicate touch that the interference pattern will be unaffected—the act of "looking at" an electron with even a *single* photon is sufficient to change the wave function of the electron and disrupt the interference pattern.

DOES THE UNCERTAINTY PRINCIPLE EMASCULATE QUANTUM THEORY?

What good is a theory that cannot answer such a simple question as "which slit did the electron go through?" There is really no basis to fault quantum theory for its inability to answer such a question because a theory can address itself only to questions that can be settled by experiments. There is no place in a physical theory for a quantity that cannot be defined by a measurement. Since no *measurement* can decide which slit a particular electron went through in a double-slit interference experiment, we cannot expect any *theory* to provide the answer. A theory can be no better than the measurements to which it applies. The usefulness of quantum mechanics is in no way vitiated by the uncertainty principle—indeed, the uncertainty principle is the cornerstone on which quantum theory is constructed.

STATEMENT OF THE UNCERTAINTY PRINCIPLE

The concept of a *particle* is something localized in space. According to classical theory a particle has, at a given instant, both a well-defined position and a well-defined velocity. Let us attempt to apply this reasoning to an elementary particle, such as an electron.

Consider the problem of attempting to localize an electron in only one dimension. To do this we send a beam of electrons (one at a time if we wish) through a narrow slit of width d, as in Fig. 12.25. In this way the electrons are localized in the x-direction (but *only* in the x-direction) to within a distance d. The initial momentum of the electrons p is in the z-direction. Upon passing through the slit, the electron waves are diffracted and form an intensity pattern on a screen that is a distance L away from the slit. The diffraction pattern is exactly the same as would be obtained with light waves of the same wavelength (see Section 10.5). The "width" of the diffraction pattern can, somewhat arbitrarily, be defined as the distance x_1 across the central maximum to the first minimum on either side.

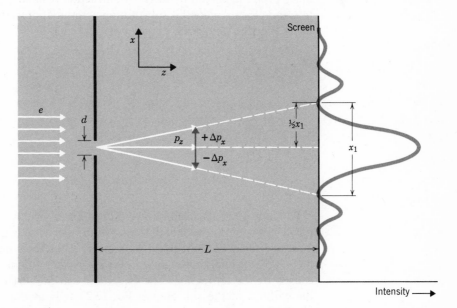

Fig. 12.25 *By confining an electron beam to a region with a dimension* d *(the slit), a transverse momentum of approximately* Δp_x *is introduced because of diffraction effects. The product of the uncertainty in position and the uncertainty in momentum is always at least as large as Planck's constant.*

That is, if an electron acquires through diffraction a transverse momentum between $+\Delta p_x$ and $-\Delta p_x$, as indicated in Fig. 12.25, the electron will strike the screen within a range x_1 around the central maximum. (We neglect the small probability that the electron will strike the screen outside the central maximum.)

Now, according to the results shown in Fig. 10.39, the distance from the central maximum of a single-slit interference pattern to the position of the first minimum is

$$\frac{1}{2}x_1 = \frac{L\lambda}{d} \tag{12.20}$$

where λ is the wavelength of the incident radiation and, for electrons, is related to the momentum by the de Broglie relation:

$$p = \frac{h}{\lambda} \tag{12.21}$$

From the two similar triangles in Fig. 12.25, we can write

$$\frac{\frac{1}{2}x_1}{L} = \frac{\Delta p_x}{p_z}$$

or,

$$\frac{\frac{1}{2}x_1}{\Delta p_x} = \frac{L}{p_z} \tag{12.22}$$

where p_z is the momentum of the electrons in the z-direction after diffraction. Since even the maximum angle of diffraction is small, p_z is essentially equal to p and we can make this substitution without appreciable error. Therefore, using the expressions for $\frac{1}{2}x_1$ and $p(=p_z)$ in Eq. 12.22, we have

$$\frac{L\lambda/d}{\Delta p_x} = \frac{L}{h/\lambda}$$

or,

$$\frac{L\lambda}{\Delta p_x d} = \frac{L\lambda}{h}$$

so that

$$\Delta p_x d = h \tag{12.23}$$

For any particular electron we do not know exactly where on the screen it will strike; we know only that most of the electrons will fall within a region of size x_1 around the central maximum. Therefore, the quantity Δp_x can be regarded as the uncertainty in the momentum of any particular electron caused by the diffraction. The uncertainty in the initial position Δx is just the width d of the slit. Hence, Eq. 12.23 tells us that the product of the uncertainty in the x-position and the uncertainty in the x-momentum is equal to Planck's constant. Actually, our calculation has been quite crude so we can state only that $\Delta x\,\Delta p_x$ is very *approximately* equal to h:

$$\boxed{\Delta x \, \Delta p_x \approx h} \tag{12.24}$$

If we wish to locate the electron more precisely, we can do so by making the slit smaller. But if Δx decreases, then Δp_x must increase in order to maintain a constant product as required by the uncertainty principle. That is, if we locate the electron more precisely, we pay for this increase in knowledge of position by a decrease in knowledge of momentum. On the other hand, if we determine the momentum (or velocity) of an electron with great precision, the act of making this measurement will lose for us the capability of knowing where the electron is after the measurement.

What are the restrictions on determining the location of a particle by means of a microscope? With such an instrument we can locate a particle to within about one wavelength of the radiation used, that is, $\Delta x \approx \lambda$. But in the act of observing the position of the particle, at least one photon must have been scattered (or absorbed) by the particle. Thus, the uncertainty in the momentum of the particle will be equal to the momentum imparted to it by the photon, which on the average will be $\Delta p_x \approx h/\lambda$. The product of Δx and Δp_x is, therefore, just $\lambda \times h/\lambda$ or h. Hence, the uncertainty principle governs this case as well. Indeed, no case has ever been found (even in *thought experiments*) in which the uncertainty principle is violated.

It is important to realize that the uncertainty principle refers to the *predictability* of events. When an electron goes through the slit of Fig. 12.25, we know only that it will strike the screen within a range x_1 of the position of the central maximum and will therefore have an uncertainty of $\pm\Delta p_x$ in its transverse momentum. *After* the electron strikes the screen, we know where it did so from the position of the light flash, but before the event takes place, we can only give the *probability* that a light flash will be observed at a particular point. Quantum theory cannot predict the result of any single event, but the *average* of a large number of events can be predicted with precision. This is the essential meaning of the uncertainty principle.

The uncertainty principle is not to be thought of as some mysterious device conceived by Nature to prevent man from probing too deeply into her methods of making atoms behave properly. Rather, the uncertainty principle is just one manifestation of the wave-particle duality of radiation and matter. Waves cannot be localized in space and so any measurement of the position of a wave-like object must be subject to uncertainty. The Heisenberg principle gives a quantitative description of this uncertainty.

Example **12.7**

The position of a free electron is determined by some optical means to within an uncertainty of 10,000 Å or 10^{-4} cm. What is the uncertainty in its velocity? After 1 sec how well will we know its position?

Nonrelativistically, we can express the uncertainty relation as

$$\Delta p_x = m_e \times \Delta v_x \approx \frac{h}{\Delta x}$$

so that

$$\Delta v_x \approx \frac{h}{m_e \, \Delta x}$$

$$= \frac{6.6 \times 10^{-27} \text{ erg-sec}}{(9.1 \times 10^{-28} \text{ g}) \times (10^{-4} \text{ cm})}$$

$$\cong 7 \times 10^5 \text{ cm/sec}$$

Therefore, after 1 sec the electron could be anywhere within a distance of 7×10^5 cm or 7 km! The act of locating the electron at one instant to within a distance as small as 10^{-4} cm stringently limits our knowledge of where the electron is at future times.

Example **12.8**

A 1-gram block rests on a frictionless surface and we measure the position of the block to a precision of 0.1 mm. What velocity have we imparted to the block by the act of measuring its position?

$$\Delta v_x \approx \frac{h}{m \, \Delta x}$$

$$= \frac{6.6 \times 10^{-27} \text{ erg-sec}}{(1 \text{ g}) \times (10^{-2} \text{ cm})}$$

$$= 6.6 \times 10^{-25} \text{ cm/sec}$$

This velocity is so small that we must consider the block to be still "at rest." The implications of the uncertainty principle for macroscopic objects are unimportant, but for microscopic objects they are crucial. That there is any uncertainty at all in the measurement arises from the fact that we can *see* the block only by virtue of the scattering of photons from the block and this process imparts momentum to the block.

THE UNCERTAINTY PRINCIPLE AND DETERMINISM

After the development of Newtonian theory, there grew up a deterministic philosophy of Nature. According to this philosophy, if, at one instant of time, the positions and velocities of all of the particles in a closed system could be determined and if the forces of interaction between the particles were known, then the complete future behavior of the system could be calculated—that is, the future of the system was *predetermined*. Of course, in a practical sense, such a calculation is impossible. If the position of only one particle in the system is inaccurate to the slightest degree, then because of the interactions of this particle with the other particles, the inaccuracy will be propagated and magnified to the extent that (given a sufficiently long time) the behavior of the system will be substantially different from that predicted by Newtonian rules. But in addition to this practical difficulty, there is the fundamental limitation on predictability imposed by quantum theory and the uncertainty principle.

Quantum theory does not prove that the philosophy of determinism is wrong. Quantum theory says only that there is no observational way ever to prove that the deterministic view is correct. Philosophers must look elsewhere for a proof or a disproof of determinism. The physicist can only accept the limitations on observations and measurements imposed by quantum theory (unless and until it is shown to be in error). He must work with *probability* instead of *certainty*.

ANOTHER FORM OF THE UNCERTAINTY PRINCIPLE

Suppose that we try to measure the frequency of a photon by counting the number of cycles that are observed in a time interval Δt. Because the photon is not of infinite extent, it does not have a pure frequency; instead, it is a mixture of frequencies (see Fig. 12.16). This fact will render our count uncertain by about one cycle. That is, the uncertainty in the frequency measurement is

$$\Delta \nu \approx \frac{1 \text{ cycle}}{\Delta t} = \frac{1}{\Delta t} \text{ Hz}$$

The corresponding uncertainty in the photon energy will be h times $\Delta \nu$:

$$\Delta \mathcal{E} = h \, \Delta \nu \approx \frac{h}{\Delta t}$$

so that

$$\boxed{\Delta \mathcal{E} \, \Delta t \approx h} \tag{12.25}$$

This statement of the uncertainty principle has exactly the same content as our previous statement, $\Delta x \, \Delta p_x \approx h$. In 1928, Niels Bohr summarized the conclusions that had been reached concerning indeterminism in quantum theory by stating that *if an experiment allows us to observe one aspect of a physical phenomenon, it simultaneously prevents us from observing a complementary aspect of the phenomenon.* This statement is known as Bohr's *principle of complementarity.* The complementary features to which the principle applies may be the position and momentum of a particle, the wave and particle character of matter or radiation, or the energy and time interval for an event. Quantum mechanical indeterminacy as stated in the Heisenberg uncertainty principle is included in the more general principle of complementarity.

Example **12.9**

In emitting a photon, an atom radiates for approximately 10^{-9} sec. What is the uncertainty in the energy of the photon?

$$\Delta \mathcal{E} \approx \frac{h}{\Delta t}$$

$$\cong \frac{6.6 \times 10^{-27} \text{ erg-sec}}{10^{-9} \text{ sec}} \times \frac{1}{1.6 \times 10^{-12} \text{ erg/eV}}$$

$$\cong 4 \times 10^{-6} \text{ eV}$$

If the photon has a nominal wavelength of 6000 Å, the energy is 2 eV (see Table 12.3), so that the *relative* uncertainty in the energy is

$$\frac{\Delta \mathcal{E}}{\mathcal{E}} \approx \frac{4 \times 10^{-6} \text{ eV}}{2 \text{ eV}} = 2 \times 10^{-6}$$

THE PION MASS

The energy-time statement of the uncertainty principle has the following implication: conservation of energy can be violated by an amount $\Delta \mathcal{E}$ for a period of time Δt if the system returns to its original state before the product $\Delta \mathcal{E} \, \Delta t$ exceeds Planck's constant. In Section 8.6 we remarked that the exchange of a pion between a proton and a neutron was allowed by the uncertainty principle even though the mass of the pion was too large to be consistent with a rigid interpretation of energy conservation.

The pion exchange force is effective over a distance approximately equal to the size of a nucleon, $R \cong 1.5 \times 10^{-13}$ cm. If a pion makes a round trip between a pair of nucleons, traveling a distance $2R$ at a typical nuclear velocity of, for example, $\frac{1}{3}$ the speed of light,[5] a time interval $\Delta t = 2R/\frac{1}{3}c = (3 \times 10^{-13} \text{ cm})/(10^{10} \text{ cm/sec}) = 3 \times 10^{-23}$ sec is required. The energy increment that the uncertainty principle allows to be gained during this time interval corresponds to the rest-energy of the pion:

$$\Delta \mathcal{E} = m_\pi c^2 \approx \frac{h}{\Delta t}$$

Therefore,

$$m_\pi c^2 \approx \frac{6.6 \times 10^{-27} \text{ erg-sec}}{3 \times 10^{-23} \text{ sec}} \times \frac{1}{1.6 \times 10^{-12} \text{ erg/eV}}$$

$$\cong 135 \text{ MeV}$$

That is, the pion mass is approximately 270 times larger than the mass of an electron (since $m_e c^2 \cong 0.5$ MeV). We shall continue the discussion of the interesting properties of the pion in Chapter 16.

Summary of Important Ideas

The details of *blackbody radiation* and the *photoelectric effect* can be explained only if electromagnetic radiation occurs in discrete packets or *photons.*

The fundamental atomic constants are: the velocity of light c, the mass of the electron m_e, the charge of the electron e, and Planck's constant h.

Depending on the type of measurement that is made, electrons and photons can exhibit properties of either *waves* or *particles.*

A massive particle has associated with it a *de Broglie wavelength* $\lambda = h/p$. Radiation has associated with it an equivalent mass $m = \mathcal{E}/c^2$.

---

[5] We know that we must choose $v \cong \frac{1}{3}c$ for this crude calculation only after the correct result has been obtained by a rigorous derivation!

In a double-slit experiment, *no* measurement can be made to determine through which slit the photon or electron went without destroying the double-slit interference pattern.

Because of the wave nature of radiation and particles, we can never predict the exact behavior of any particular photon or particle; we can only predict the *average* behavior of large numbers of photons or particles. Individual events can be discussed only in terms of *probabilities*.

The quantum mechanical wave function $\psi(x)$ that describes a particle or a photon is a *probability amplitude;* only the *square* of the wave function, which is proportional to the *intensity,* can be measured and therefore has physical meaning.

Except for the (ideal) free particle, all quantum mechanical systems are constrained to have only certain discrete energies and momenta.

The *Heisenberg uncertainty principle* expresses the fact that we cannot simultaneously measure with arbitrarily high precision *complementary* aspects of a particle or a photon (such as *momentum* and *position* or the *energy* of an event and the *time* interval during which it took place).

Questions

12.1 Herman von Helmholtz once said that he was puzzled to explain what an electric charge (that is, an electron) was, except the recipient of a symbol. Comment on the view that an *electron* is just a name that is convenient for describing various observations rather than a *thing*.

12.2 Describe an experiment to distinguish between an X ray whose wavelength is $\lambda_x = 10^{-8}$ cm and an electron whose de Broglie wavelength is $\lambda_e = 10^{-8}$ cm. What experiments would *not* be suitable?

12.3 Discuss some of the changes in everyday events that would result if Planck's constant were suddenly increased to 1 erg-sec.

12.4 In a double-slit experiment, 25 light flashes per second are observed at a certain position on the screen when slit A is open and slit B is closed. When A is closed and B is open, 16 flashes/sec are observed. Can you calculate the intensity of flashes when *both* slits are open? Explain carefully.

12.5 Refer to Fig. 12.22 where it is shown that the probability of finding a particle at a given position in a "box" is a maximum at certain positions and is *zero* at other positions. How is it possible for the particle to "move" from one position of maximum probability to another since in order to do so it must pass through a position that it is not allowed to occupy? Explain the situation carefully in terms of measurements that can be made.

12.6 Refer to Fig. 12.24. What are the *probabilities* of finding the particles, whose wave functions are shown, at various positions within the potential well? Sketch the probability functions.

12.7 According to John A. Wheeler, the following two items are complementary in the sense of Bohr's principle of complementarity: (a) the use of a word to convey *information,* (b) the analysis of the *meaning* of the word. Discuss complementarity in this case.

12.8   Are *angular position* and *angular momentum* complementary quantities in the sense of Bohr's complementarity principle? If so, state an "uncertainty principle" for these quantities.

12.9   One of the reasons cited to justify the construction of expensive high-energy accelerators is that particles (and photons) of high energy are needed to probe the detailed structure of nuclei and nucleons. Why is this so?

12.10 Choose a quantity that was once thought to be continuous and discuss how it was shown to be discrete. Choose a quantity that was once thought to be discrete and discuss how it was shown to be "fuzzy."

12.11 Discuss the proposition that we shall eventually be able to overcome the limitations now set by the uncertainty principle and shall then be able to discuss microscopic phenomena with the same kind of deterministic approach that is appropriate for macroscopic mechanical phenomena (that is, the approach of Newtonian dynamics).

Problems

12.1   What must be the temperature of a blackbody for the maximum of the wavelength distribution to occur for yellow light ($\lambda = 6000$ Å)?

12.2   The effective surface temperature of the Sun is approximately $5800°K$. If the Sun radiates as a blackbody (this is approximately the case), at what wavelength do we expect the maximum in the emitted light?

12.3* A beam of UV radiation ($\lambda = 1000$ Å) delivers 10^{-6} watt to a certain photoemissive surface. How many photons strike the surface each second? If 1 percent of the photons eject photoelectrons, what will be the resulting electron current?

12.4   The threshold wavelength for the photoelectric effect on a certain material is 3000 Å. What is the work function of the material?

12.5   A potential of 2.7 volts is required to stop completely the photoelectrons from a certain material when the electrons are ejected by 2100-Å radiation. What is the work function of the material?

12.6   The work function for platinum is 5.32 eV. What is the longest wavelength photon that can eject a photoelectron from platinum?

12.7   The work function of barium is 2.48 eV. What is the maximum kinetic energy of a photoelectron ejected from barium by photons of wavelength 2000 Å?

12.8* A mercury arc lamp is used as the source of UV radiation to study the photoelectric effect on lithium. By using various filters, discrete wavelengths can be isolated from the spectrum. The following wavelengths were used and the corresponding voltages were found necessary to stop completely the photoelectrons:

λ (Å)	V_r (volts)
2536	2.4
3132	1.5
3663	0.9
4358	0.35
5770	(no photoelectrons)

Plot the data on an appropriate graph. (Be certain to plot V_r versus v—not versus λ.) Find the work function of lithium. Calculate the experimental value of h/e and compare with the value computed from the constants listed in Table 12.2.

12.9 Under certain conditions the retina of the human eye can detect as few as five photons of blue-green light ($\lambda = 5 \times 10^{-5}$ cm). What is the corresponding amount of energy received by the retina in ergs and in eV? If five such photons strike the eye and are absorbed each second, what is the rate of energy transfer in watts?

12.10 What is the mass equivalent of a photon of wavelength 6×10^{-5} cm (yellow light)? How many such photons would be required to make up the rest-mass energy of one electron?

12.11 A 20-keV X ray is Compton scattered by a free electron. What is the wavelength of the incident X ray? If the X ray is scattered through an angle of 90°, what is the final wavelength? What is the final energy of the X ray? What is the kinetic energy of the recoil electron?

12.12 A photon whose wavelength is 0.8 Å is Compton scattered through 90°. What is the energy of the scattered photon and how much energy is imparted to the electron?

12.13 What is your wavelength if you are running at a speed of 10 m/sec? What is the significance of such a wavelength?

12.14 What is the frequency of radiation whose wavelength is 1 Å? What energy is carried by such a photon?

12.15 Through what potential difference must an electron fall (starting at rest) in order that its wavelength be 1.6 Å?

12.16* Complete the following table:

	Energy (eV)	λ (cm)
	1	—
Electron	—	1×10^{-8}
	1000	—
	1	—
Photon	—	1×10^{-5}
	1000	—

12.17 A proton, an electron, and a photon all have de Broglie wavelengths of 1 Å. If they all leave a given point a $t = 0$, what are the arrival times at a point 10 m away?

12.18 What is the velocity of a helium atom that has a wavelength of 1 Å?

12.19 When a neutron is in thermal equilibrium with objects at room temperature it has an energy of 0.025 eV; such neutrons are called *thermal* neutrons. What is the wavelength of a thermal neutron?

12.20 What is the energy of a photon whose wavelength is (a) the size of an atom (10^{-8} cm), (b) the size of a nucleus (5×10^{-13} cm)?

12.21 In producing a photon of green light ($\lambda = 5000$ Å), an atom radiates for 10^{-9} sec. How many oscillations are there in the photon? What is the "length" of the photon?

12.22 At a certain point x_0 on a screen the value of the wave function from source A is $\psi_A(x_0) = +3$; this source alone produces 18 flashes per second in a narrow region around x_0. The value of the wave function from source B is $\psi_B(x_0) = -5$. How many flashes per second will be observed from source B alone and how many from sources A and B combined?

12.23* The resolution of an optical microscope is limited by diffraction effects and the ultimate resolution is approximately equal to the wavelength of the light used to view the specimen. By what percentage does the resolution improve by using blue light instead of red light? What is the limit of resolution of an *electron* microscope that uses 20-keV electrons rather than light? What type of microscope would you use to study the internal structure of a cell (diam. ≈ 0.1 mm)? What type would be suitable for the study of a virus (diam. ≈ 100 Å)?

12.24 A baseball ($m = 150$ g) is thrown through a 20-cm wide slot in a fence. What transverse velocity could the baseball acquire by passing through the slot?

12.25 A beam of electrons ($v = 10^8$ cm/sec) passes through a 0.01 mm slit. What is the width of the central diffraction maximum on the screen of 1 m away?

12.26 An electron diffraction experiment is performed with a crystal whose atomic planes are spaced 1 Å apart (see Fig. 12.12). The first diffraction minimum is found at an angle of 30° with respect to the direction of the incident beam. What is the electron wavelength and energy?

12.27 An electron is localized in the x-direction to within 1 mm. How precisely can its x-momentum be known?

12.28 A proton in a nucleus is localized to within a distance approximately equal to the nuclear radius. What is the approximate uncertainty in the velocity of a proton in an iron nucleus ($R \cong 6 \times 10^{-13}$ cm)? What is the corresponding uncertainty in energy? (A nonrelativistic calculation is adequate for the accuracy desired.)

12.29 The electron in a hydrogen atom may be considered to be confined to a region of radius 5×10^{-8} cm around the nucleus. Use the uncertainty principle to estimate the momentum of the electron and, from this, its kinetic energy. (For purposes of estimating these quantities, make only a one-dimensional calculation; that is, use $\Delta x \, \Delta p_x \approx h$.) Why does the electron remain attached to the nucleus?

12.30 A certain atomic energy level has a mean lifetime of 10^{-8} sec. (That is, after the level is excited by some means, it requires, on the average, a time of 10^{-8} sec before it spontaneously radiates a photon.) What is the uncertainty in the energy of the emitted radiation?

Laser Sol Mednick

13 *Atoms and Quanta*

In the preceding chapter we used the results of several crucial experiments concerning electrons and photons to build the basic framework of quantum theory. The actual development of quantum theory followed a much more tortuous path and utilized the results of many more experiments. In this chapter we sketch a parallel line of development based on the interpretation of experiments with more complicated systems, namely, atoms. This phase of the campaign was directed toward understanding the optical spectra of atoms. Niels Bohr's early work on this problem set the stage for the tremendous outpouring of theoretical and experimental results that, in the short space of four years from 1924 to 1928, firmly established quantum theory as a proper description of atomic processes.

We can give in this chapter only a brief account of some of the more important lines of reasoning that so rapidly produced the new theory of how Nature behaves in the atomic domain. This quantum theory is still with us, improved by the inclusion of relativistic effects and aided by sophisticated methods of computation; it has proved to be the most precise description of natural phenomena that science has ever known. At the end of this chapter we shall discuss some of the important and interesting ways in which quantum theory has been applied to modern atomic problems.

13.1 Atomic Models

THOMSON'S MODEL

By 1902 a sufficient number of experiments had been performed to provide convincing evidence that the electron is one of the universal and fundamental constituents of all matter. Sir J. J. Thomson showed, on the basis of classical electromagnetic theory, that the size[1] of the electron must be about 10^{-13} cm. Furthermore, kinetic theory of the 19th century had shown that atoms have sizes of a few Ångstroms (that is, a few times 10^{-8} cm). Thomson therefore reasoned that the positive electrical charge of an atom, which was required to balance the negative charge of the electrons and thereby render the atom electrically neutral, should be contained in the vast regions of space in the atom unoccupied by the tiny electrons. In 1906 Thomson proposed a model in which an atom was considered to contain electrons, equal in number to the chemical atomic number of the element, in which the total charge of the electrons was neutralized by a positively-charged medium that contained most of the mass of the atom (Fig. 13.1). Because in this model the electrons were embedded in a positive sea in much the same way as raisins in a plum pudding, the irreverent appellation of "plum-pudding model" was applied to Thomson's scheme.

Although Thomson's model contained attractive features from the standpoint of the involvement of electrons in atomic structure, it survived only

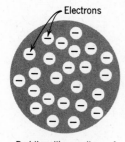

Electrons

Pudding-like medium of positive charge

Fig. 13.1 *Thomson's "plum-pudding" model of the atom. The electron "raisins" are embedded in the positively-charged "plum pudding." The diameter of the system is a few Angstroms.*

[1] We cannot really speak of the *size* or the *radius* of an electron because we have no way to define rigorously such a dimension except for a rigid body and the electron is certainly not a rigid body.

until 1911 when Ernest (later Lord) Rutherford succeeded in demonstrating that the positively-charged portion of an atom is not distributed throughout the atom but is concentrated in a massive core of exceedingly small size—the *nucleus* of the atom.

RUTHERFORD'S MODEL

Just as the atomic physics of the last decade of the 19th century was dominated by J. J. Thomson, the first decade of the 20th century belonged to Ernest Rutherford (1871–1937). Working first at McGill University in Canada and then moving to Manchester in 1907, Rutherford exhaustively studied the newly discovered radiations from radioactive substances. He was particularly interested in the positively-charged radiations called α *rays* or α *particles*. By 1908 he had shown conclusively that the α rays are helium atoms carrying a charge of $+2e$.

Almost immediately on arriving in Manchester, Rutherford began a systematic investigation of the scattering of α particles in matter. He had learned that a *single* α-particle, when it strikes a zinc sulfide screen, produces a visible flash of light. Therefore, the apparatus to study α-particle scattering was arranged as shown schematically in Fig. 13.3. Alpha particles from a radioactive source were confined to a narrow cone by a lead collimator. After scattering by the gold foil, the α particles struck a zinc sulfide screen and were detected by observing the light flashes with a small microscope. The detector could be rotated in order to measure the relative number of α particles scattered at various angles θ.

According to Thomson's atomic model, α particles should be able to pass freely through the gold atoms; only occasionally should an α particle be slightly deflected by the Coulomb field of the electrons (Fig. 13.4). It was therefore expected that a beam of α particles would be somewhat "smeared out" in passing through the thin foil and that average scattering angles of

Fig. 13.2 *Lord Rutherford, the key figure in unraveling the mysteries of radioactivity and in establishing the nuclear model of the atom. For his work on radioactivity, Rutherford was awarded the 1908 Nobel Prize (in chemistry).*

William Numeroff

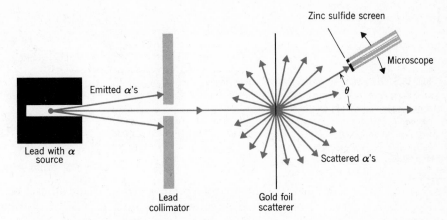

Fig. 13.3 *Schematic representation of the apparatus used by Rutherford to study the scattering of α particles. The entire apparatus was contained within an evacuated chamber in order to prevent the absorption of the α particles in air. Working under Rutherford's direction, Geiger and Marsden performed the experiments.*

a few degrees would result. Indeed, this small-angle scattering was observed, but, most unexpectedly, it was found that about 1 α particle in 20,000 of those incident was turned completely around by a sheet of gold only 4×10^{-5} cm thick and emerged from the side facing the source. Rutherford commented: "It was quite the most incredible event that has ever happened to me in my life. It was almost as incredible as if you had fired a 15-inch shell at a piece of tissue paper and it came back and hit you."

It took several years (until 1911) for Rutherford to convince himself that he understood completely the meaning of the unexpected α-particle scattering that was observed at large angles of deflection. He concluded that the only way in which it is possible to account for the experimental results

Fig. 13.4 *According to Thomson's model of the atom, incident α particles should suffer small-angle deflections but none should be scattered through large angles by the gold atoms in a foil. The experiments of Rutherford and his co-workers showed this view of the scattering of α particles by atoms to be incorrect.*

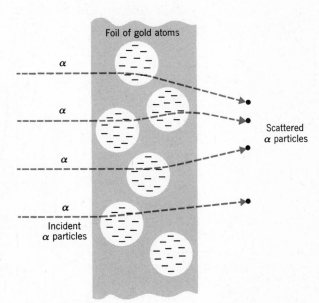

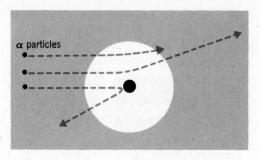

α particles

Fig. 13.5 *According to Rutherford's nuclear model of the atom, an α particle should usually pass through the atom with only little deflection, but occasionally the direction of motion of the α particle will bring it sufficiently close to the nucleus to cause a large-angle scattering.*

is to assume that the positive charge of the atom is concentrated in a small volume at the center of the atom instead of distributed throughout the atom as in Thomson's model. Thus, Rutherford proposed the *nuclear* model of the atom.

In a collision between an α particle and an atom, the α particle should suffer very little deflection from the atomic electrons (just as in the scattering from a Thomson atom), but according to Rutherford, when the trajectory brings the α particle close to the nucleus, the intense electrical repulsion can cause a considerable change in the direction of motion of the α particle. Typical encounters between α particles and atoms are shown in Fig. 13.5.

By assuming that the repulsive Coulomb force, which acts between an α particle and an atomic nucleus maintains its $1/r^2$ character even down to extremely small intra-atomic distances ($\sim 10^{-12}$ cm), Rutherford was able to derive an expression for the distribution of scattered particles in α particle-nucleus collisions. He showed that his nuclear model of the atom predicts the probability for scattering at an angle θ to be inversely proportional to the fourth power of the sine of $\theta/2$, that is, to $1/\sin^4(\theta/2)$. This function is shown in Fig. 13.6 in which each unit on the vertical scale is a factor of 10. From this figure it can be seen that the probability for scattering at angles greater than 90° (that is, into the *backward* direction) is exceedingly

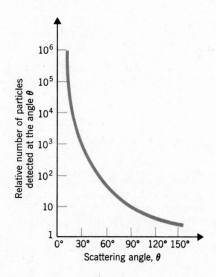

Fig. 13.6 *The relative probability (or number of light flashes expected per unit time in the detector) for Rutherford scattering at various angles (proportional to $1/\sin^4 \frac{\theta}{2}$).*

small compared to small-angle scattering. In fact, the ratio of scattering at $\theta = 120°$ compared to that at $\theta = 5°$ is approximately 10^{-5}.

The careful scattering measurements that were made by Geiger and Marsden in Rutherford's laboratory verified the nuclear model in every respect—not only was it conclusively demonstrated that atoms consist of nuclear cores of extremely small dimensions ($\sim 10^{-12}$ cm) surrounded by atomic electrons, but it was also verified that Coulomb's law is valid at these small distances.

13.2 The Hydrogen Atom

SPECTROSCOPIC RESULTS

In the middle of the 18th century it was discovered that the light from flames does not consist entirely of a continuous spectrum (such as the blackbody spectrum, Fig. 12.4) but that there are certain discrete parts of the spectrum that are more intense than the background continuum and stand out clearly as *lines*. Line spectra from a variety of sources (including the Sun) were investigated extensively during the latter half of the 19th century and *atomic spectroscopy* became a highly developed field of study.

It had been discovered in 1885 by Johann Balmer (1825–1898), a Swiss music teacher with an interest in numbers, that the wavelengths of the lines in the optical spectrum of hydrogen (Fig. 13.7) could be represented by a simple mathematical expression. Although Balmer originally presented his formula in terms of the wavelength λ, it was soon appreciated that the expression was more revealing if stated in terms of $1/\lambda$. In this form, Balmer's formula for the hydrogen spectrum becomes

$$\frac{1}{\lambda} = R\left(\frac{1}{2^2} - \frac{1}{n^2}\right), \qquad n = 3, 4, 5, \ldots \tag{13.1}$$

where R is called the *Rydberg constant,* after the Swedish spectroscopist who made extensive investigations of atomic spectra. The value of R for the hydrogen spectrum is now known to be

$$R = 109,677.58 \text{ cm}^{-1} \text{ (hydrogen)} \tag{13.2}$$

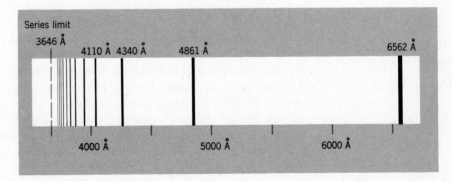

Fig. 13.7 *The Balmer series in the spectrum of hydrogen. The series limit (3646 Å) corresponds to substituting* n = ∞ *in the Balmer formula (Eq. 13.1).*

although, of course, in Balmer's time the value was not known so precisely. How supremely successful the Balmer formula is in representing the hydrogen spectrum is demonstrated in Table 13.1 where the wavelengths calculated by Balmer are compared with those obtained from measurements made by the Swedish physicist, Anders Jonas Ångstrom (1814–1874). These four lines lie in the visible part of the spectrum. On the basis of his formula, Balmer also predicted several other lines that should occur in the ultraviolet region. Indeed, in the spectra of some stars, as many as 50 lines in the Balmer series have been observed.[2]

Although Balmer's formula was exceptionally accurate in reproducing the observations, no one understood why this should be so. It was almost 30 years before Niels Bohr provided the first glimmer of understanding.

Table **13.1**   *Comparison of Hydrogen Spectral Lines with Calculations from the Balmer Formula*

Line Designation	n	λ (Computed by Balmer)	λ (Observed by Ångstrom)
H_α	3	6562.08 Å	6562.10 Å
H_β	4	4860.80 Å	4860.74 Å
H_γ	5	4340.0 Å	4340.1 Å
H_δ	6	4101.3 Å	4101.2 Å

BOHR'S INTERPRETATION OF ENERGY STATES

In 1913 Bohr took a bold and surprising step in an attempt to interpret the spectroscopic results for the hydrogen atom. He had accepted Rutherford's model of the atom with its nuclear core and outer electrons. According to classical theory, a system consisting of a massive, positively-charged core and light, negatively-charged electrons can be stable only if the electrons are in motion. Thus, an atom should be similar to a miniature solar system with a nuclear "Sun" and "planetary" electrons. The analogy would be expected to be quite good (after all, the electrical and gravitational forces both depend on $1/r^2$) were it not for the fact that classical theory also predicts that accelerating electric charges radiate energy in the form of electromagnetic waves. Therefore, the orbiting "planetary" electrons would be expected to lose their motional and electrical energy by radiation and rapidly fall toward the nucleus. A calculation based on classical electromagnetic theory shows that the electron in a hydrogen atom will radiate all of its energy in a small fraction of a second. But, of course, the atom does not do this—what is wrong with the classical model?

It was Bohr's audacious proposal that classical electromagnetic theory simply does not apply to an electron circulating in an orbit around a nucleus. At the same time he reasoned that the two terms in Balmer's formula referred to the total energies of two *allowed* orbits (or energy states) of the electron in the hydrogen atom.

[2]The tenuous atmospheres of stars provide more favorable sources of hydrogen radiation than do laboratory sources.

In order to convert the Balmer formula for $1/\lambda$ into an energy equation, we multiply by hc to obtain

$$\frac{hc}{\lambda} = hc\, R\left(\frac{1}{2^2} - \frac{1}{n^2}\right)$$

But $c/\lambda = \nu$, so that $hc/\lambda = h\nu = \mathcal{E}$, and $\mathcal{E}$ is interpreted as the photon energy. Therefore, we can write

$$\mathcal{E} = h\nu = hc\, R\left(\frac{1}{2^2} - \frac{1}{n^2}\right) = \mathcal{E}_n - \mathcal{E}_2 \tag{13.3}$$

if we identify[3]

$$\mathcal{E}_2 = -hc\frac{R}{2^2}; \quad \mathcal{E}_n = -hc\frac{R}{n^2} \tag{13.4}$$

Bohr's point was that no energy is lost by radiation while the electron is in the orbit labeled n or in the orbit labeled 2; radiation occurs only when an electron makes a *transition* between two allowed, radiationless orbits, and the energy of the emitted photon is equal to the difference in energy of the electron in these two orbits.

Although Bohr had enunciated a radical view in denying the applicability of Maxwellian electrodynamics in the atomic domain, his next step was bolder yet.

BOHR'S ANGULAR MOMENTUM HYPOTHESIS

In order to justify his interpretation of the hydrogen spectrum in terms of radiations accompanying the transitions between allowed energy states, Bohr sought to calculate the energies of these states. He was able to obtain a set of discrete allowed states only by making the drastic assumption that *angular momentum is quantized.* By specifying that the angular momentum must be an integer multiple of $h/2\pi$, Bohr was finally able to derive the Balmer formula.[4] (The combination $h/2\pi$ is used so frequently in atomic theory that it is given a special symbol:[5] $h/2\pi = \hbar$.) Bohr's condition is, therefore (see Eq. 5.13),

$$L = m_e v r = n\hbar, \quad n = 1, 2, 3 \dots \tag{13.5}$$

n is called the *principal quantum number* for the particular state. Now, we have already found (Example 6.5) that the velocity of an electron in a circular orbit in a hydrogen atom is given by

$$v = \sqrt{\frac{r}{m_e}F_E} = \sqrt{\frac{r}{m_e} \times \frac{e^2}{r^2}} = \frac{e}{\sqrt{m_e r}} \tag{13.6}$$

---

[3] The negative sign in the expressions for the energies denotes the fact that the energy states are bound; see Fig. 7.15.

[4] Although Planck's constant had first appeared in the study of blackbody radiation, there was a clue that somehow h might be connected with angular momentum, namely, h has the *dimensions* of angular momentum, erg-sec!

[5] The symbol $\hbar$ is pronounced "*h-bar*" or, less commonly, "*h-cross.*" Numerically,

$$h/2\pi = \hbar = 1.0546 \times 10^{-27} \text{ erg-sec}$$
$$= 6.583 \times 10^{-16} \text{ eV-sec}$$

Therefore, substituting for v in Eq. 13.5, we have for the quantized angular momentum,

$$L = m_e \times \frac{e}{\sqrt{m_e r}} \times r = e\sqrt{m_e r} = n\hbar$$

Squaring the last equality and solving for r, we find for the radii of the allowed orbits,

$$r_n = \frac{n^2 \hbar^2}{m_e e^2} \tag{13.7}$$

where we have attached a subscript n to the radius to indicate that it is the radius for a particular value of n.

Example **13.1**

What is the radius of the first Bohr orbit for hydrogen?

$$r_{n=1} = a_1 = \frac{\hbar^2}{m_e e^2}$$

$$= \frac{(1.05 \times 10^{-27} \text{ erg-sec})^2}{(9.1 \times 10^{-28} \text{ g}) \times (4.80 \times 10^{-10} \text{ statC})^2}$$

$$= 0.53 \times 10^{-8} \text{ cm} = 0.53 \text{ Å}$$

Next, for the total energy of the nth orbit, we write, using Eq. 13.6 for v,

$$\mathcal{E}_n = KE + PE$$

$$= \frac{1}{2} m_e v^2 - \frac{e^2}{r_n}$$

$$= \frac{1}{2} m_e \left(\frac{e^2}{m_e r_n}\right) - \frac{e^2}{r_n}$$

$$= -\frac{1}{2}\left(\frac{e^2}{r_n}\right) \tag{13.8}$$

Substituting for r_n from Eq. 13.7, we have, finally,

$$\mathcal{E}_n = -\frac{1}{2}\left(\frac{m_e e^4}{\hbar^2}\right)\left(\frac{1}{n^2}\right) \tag{13.9}$$

Thus, by using the quantization condition on the angular momentum, Bohr succeeded in obtaining a set of discrete energy states. The value of the energy for a particular state depends on the value of the principal quantum number n for that state. The difference in energy between a state with principal quantum number n and one with n' is

$$\Delta\mathcal{E}_{nn'} = \mathcal{E}_n - \mathcal{E}_{n'}$$

$$= \frac{1}{2}\left(\frac{m_e e^4}{\hbar^2}\right)\left(\frac{1}{n'^2} - \frac{1}{n^2}\right) \tag{13.10}$$

If an electron makes a transition between these states (n to n', where $n > n'$), the wavelength of the radiation will be given by

$$\frac{1}{\lambda_{nn'}} = \frac{\nu_{nn'}}{c} = \frac{h\nu_{nn'}}{hc} = \frac{\Delta\mathcal{E}_{nn'}}{hc}$$

$$= \frac{1}{2}\left(\frac{m_e e^4}{\hbar^2}\right) \times \frac{1}{hc}\left(\frac{1}{n'^2} - \frac{1}{n^2}\right) \tag{13.11}$$

Substituting numerical values for the fundamental constants in this expression, Bohr found that the value of the quantity multiplying the terms in parentheses is very close to the experimental value of the Rydberg constant determined from the hydrogen spectrum (Eq. 13.2):

$$R = \frac{m_e e^4}{4\pi\hbar^3 c} = 109,737.31 \text{ cm}^{-1} \tag{13.12}$$

Bohr was then able to duplicate almost exactly the results of measurements of the hydrogen atom spectrum.[6]

Bohr's success in explaining the hydrogen spectrum on the basis of his half-classical, half-quantum model was not exactly hailed as a triumph. In fact, he was severely criticized for tampering with centuries of classical theories; even Bohr was at a loss to explain the fundamental significance of his curious mixture of classical dynamics and quantum hypotheses. It was more than 10 years before the development of the new quantum mechanics provided the proper explanation of Bohr's remarkable results.

HYDROGEN ENERGY STATES

By substituting $n' = 2$ into Eq. 13.11, Bohr had been able to compute the wavelengths of the lines in the Balmer spectrum. By using $n' = 3$, he was able to account for a series of infrared lines discovered by Paschen in 1909. Bohr argued that other values of n' should give rise to additional spectral series. In fact, Lyman found the series of ultraviolet lines that was predicted for $n' = 1$, and in 1922 Brackett and Pfund discovered the series for $n' = 4$ and $n' = 5$.

The first 5 orbits of the Bohr hydrogen atom are shown in Fig. 13.8 along with lines that represent the transitions of the spectral series for $n' = 1, 2$, and 3. The energies associated with electrons residing in these orbits (or energy states) are shown in the energy level diagram in Fig. 13.9. Since the electron is *bound* to the nucleus, according to our standard convention all energies are *negative*. The amount of energy required to free an electron from an atom in a particular state is called the *binding energy* for that state.

---

[6]Actually, Bohr did not use exactly this expression. The calculation made here assumes that the electron executes circular orbits around a hydrogen nucleus which is at rest, whereas, in fact, we should consider the electron and the nucleus to execute orbits around their common center of mass. The correction for this effect, which Bohr took into account, amounts to 1 part in 1836, the ratio of m_e to the mass of the nuclear proton, m_p. This correction converts the numerical value of R given in Eq. 13.12 into the value obtained from hydrogen spectral measurements (Eq. 13.2). For atoms other than hydrogen, the correction is even smaller. There are, in addition, relativistic corrections to the formula which were not made until considerably later.

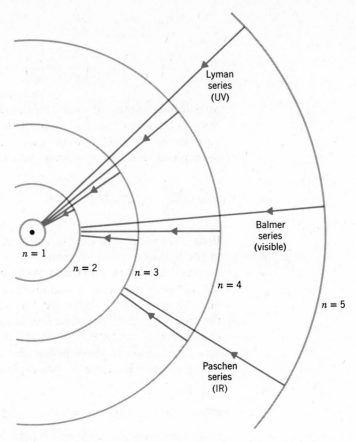

Fig. 13.8 *Orbits in the Bohr model of the hydrogen atom. Portions of three of the spectral series are shown.*

Example **13.2**

Calculate the binding energy of the hydrogen atom in its ground state. Using Eq. 13.9 with $n = 1$, we have

$$E_1 = -\frac{m_e e^4}{2\hbar^2}$$

$$= -\frac{(9.1 \times 10^{-28}\ \text{g}) \times (4.80 \times 10^{-10}\ \text{statC})^4}{2 \times (1.05 \times 10^{-27}\ \text{erg-sec})^2}$$

$$= -2.18 \times 10^{-11}\ \text{erg} \times \frac{1}{1.60 \times 10^{-12}\ \text{erg/eV}}$$

$$= -13.6\ \text{eV}$$

This (negative) energy is the *total* energy of the state, and so, the *binding energy* (that energy that must be supplied to raise the total energy to *zero* and thereby release the electron) is 13.6 eV.

DE BROGLIE WAVES IN THE HYDROGEN ATOM

Bohr's explanation of the hydrogen spectrum in terms of electrons orbiting with quantized angular momenta was tremendously successful in accounting for the observed spectral lines. But the quantization rule remained as an *ad hoc* hypothesis, not based on any deeper theory, until de Broglie made his famous proposal of the wave character of matter. As soon as it was appreciated that an electron with a momentum p had associated with it a wave of wavelength $\lambda = h/p$, it became clear why only certain orbits are available to the electron in a hydrogen atom.

Consider the first Bohr orbit of hydrogen (that is, the orbit with $n = 1$). According to Eq. 13.7, the radius of this orbit is

$$a_1 = \frac{\hbar^2}{m_e e^2} \qquad \text{(radius of first Bohr orbit)} \tag{13.13}$$

and the circumference of the orbit is

$$2\pi a_1 = \frac{2\pi\hbar^2}{m_e e^2}$$

The wavelength of the electron moving in this orbit is

$$\lambda_1 = \frac{h}{p} = \frac{h}{m_e v} = \frac{2\pi\hbar}{m_e v} \tag{13.14}$$

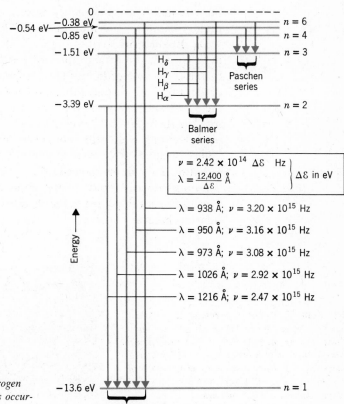

Fig. 13.9 *Energy level diagram of the hydrogen atom. The vertical lines represent transitions occurring in three of the spectral series.*

Now, for the case $n = 1$, Eq. 13.5 becomes $m_e vr = \hbar$; but for $n = 1$, r is a_1, so that

$$v = \frac{\hbar}{m_e a_1}$$

Then, Eq. 13.14 simplifies to

$$\lambda_1 = \frac{2\pi\hbar}{m_e \times (\hbar/m_e a_1)} = 2\pi a_1 \qquad (13.15)$$

Thus, the circumference of the electron orbit is exactly equal to the de Broglie wavelength of the associated electron wave. For other values of n we find, in general,

$$2\pi r_n = n\lambda_n \qquad (13.16)$$

That is, the circumference of the nth orbit is just n wavelengths of the electron wave.

Why is it so important that an allowed orbit contain exactly an integer number of wavelengths? The answer lies simply in the interference property of waves. If a wave requires a distance equal to the orbit circumference to exactly complete an integer number of cycles, this means that the wave will join smoothly onto itself and constructive interference or self-reinforcement will result (see Fig. 13.10a). On the other hand, if we try to fit into a certain orbit a wave with a wavelength that is not an integer fraction of the circumference, *destructive* interference will result and the wave will rapidly damp to zero amplitude (see Fig. 13.10b).

De Broglie's proposal of the wave character of matter therefore explains in a simple and straightforward way the puzzling angular momentum quantization rule of Bohr. (This, in fact, was the first successful application of de Broglie's wavelength—momentum condition.) But in spite of the increased understanding brought about by incorporating de Broglie's idea into the theory, Bohr's atomic model was still basically classical in character—electrons were still thought to move in orbits around the nuclear "Suns"—and no one understood why the electrons were prevented from radiating while in an allowed orbit. The merging of Bohr's and de Broglie's concepts provided much of the impetus that led within a short period of time to the development of the modern theory of atomic structure in quantum mechanical terms.

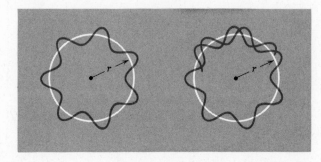

Fig. 13.10 (*a*) *Constructive interference results when an integer number of wavelengths are just fitted onto the circumference of an orbit so that this state of the system is maintained by self-reinforcement.* (*b*) *Destructive interference results when the condition* $n\lambda_n = 2\pi r_n$ *is not satisfied and the state is rapidly damped to zero amplitude.*

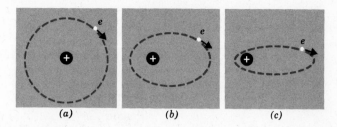

Fig. 13.11 *Three possible elliptical orbits for an electron. For a given value of the energy of a state, the circular orbit (a), which is just a special case of an ellipse, has the largest possible angular momentum. The most elongated elliptical orbit (c) has the least angular momentum. The nucleus is always at the focal point of the ellipse (neglecting the mass of the electron compared to that of the nucleus).*

13.3 *Angular Momentum and Spin*

SOMMERFELD'S ELLIPTICAL ORBITS

In order to overcome a difficulty in the Bohr theory when applied to the spectra of alkali metals (lithium and sodium, for instance), in 1915 the German theorist Arnold Sommerfeld (1868–1951) made a logical extension of the Bohr theory. In his model of the hydrogen atom, Bohr had considered only circular orbits. But a particle that executes a classical orbit and has a definite energy can be in any of an infinite number of *elliptical* orbits as well as in the unique circular orbit. Although the *energy* for each such orbit is the same,[7] the *angular momentum* is different (Fig. 13.11). Sommerfeld allowed the possibility of elliptical orbits and in so doing he removed from the principle quantum number n the role of specifying the angular momentum.

In the Sommerfeld elliptical orbit model, the principal quantum number n is retained as a measure of the energy of a state, and a new quantum number l is introduced to specify the orbital angular momentum. The orbital angular momentum is still quantized, taking on only values that are integer multiples of $\hbar$, given by $l\hbar$:[8]

$$L = l\hbar, \qquad l = 0, 1, 2 \ldots \tag{13.17}$$

It is a result of the details of the model that l can have only positive values from zero to $n - 1$. Thus, in the Bohr-Sommerfeld model (as in the complete quantum mechanical analysis), there is *zero* angular momentum associated with the state whose principal quantum number is 1 (that is, $l_{max} = n - 1 = 0$ for $n = 1$), whereas for $n = 2$, l can have the values 0 or 1, and for $n = 3$, $l = 0$, 1, or 2, etc.:

[7] As we shall see, in the correct quantum mechanical description, the energy of a particular state depends to some extent on the angular momentum of that state.

[8] Quantum theory, as developed after the Bohr-Sommerfeld model, shows that the magnitude of the angular momentum vector is actually $L = \sqrt{l(l + 1)}\hbar$. However, this distinction is unimportant for the arguments presented here.

$$\boxed{l = 0, 1, 2, \ldots, n - 1} \tag{13.18}$$

In describing electron orbital angular momentum states it is customary to use a letter to denote the value of l. The convention is as follows:

$l = 0$ S state

$l = 1$ P state

$l = 2$ D state

$l = 3$ F state

$l = 4$ G state

This curious code is a holdover from the early days of spectroscopy when S, P, D, and F meant, respectively, that the spectral lines which are associated with transitions to these angular momentum states were classified as *sharp, principal, diffuse,* or *fundamental.* For higher values of l, the letter designations continue alphabetically.

Notice the difference between the original Bohr theory, which gives to the ground state of the hydrogen atom (that is, the $n = 1$ state) one unit of angular momentum ($L = \hbar$), and the Sommerfeld modification which gives $L = 0$. But how can an orbiting electron have *zero* angular momentum? According to classical reasoning, this condition can be satisfied only if the electron moves in a straight line that *passes through* the nucleus. Such a situation is clearly unphysical and we can only conclude that this simple theory is not correct in detail. This is not surprising since the model leans heavily on classical concepts. The approach taken in quantum theory foregoes completely any discussion of electron "orbits," and so the prediction of zero angular momentum for a state is in no way bothersome. In fact, by direct experimentation, we find $L = 0$ for the hydrogen ground state.

In spite of its basic shortcomings, the Bohr-Sommerfeld theory was applied

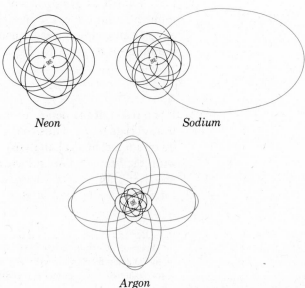

Neon   *Sodium*

Argon

Fig. 13.12 *The electronic orbits of neon, sodium, and argon according to the Bohr-Sommerfeld scheme. See also the model of the radium atom shown in Fig. 1.6.*

with considerable success to *one-electron systems* (such as the hydrogen atom, once-ionized helium, twice-ionized lithium, etc.). But with more complicated electronic systems, insurmountable difficulties were encountered. Even with their always great ingenuity, neither Bohr nor Sommerfeld could juggle the circular and elliptical orbits of complicated atoms (Fig. 13.12) to give the proper results. Only the application of the full power of quantum theory was eventually able to solve this enormous problem.

THE ORBITAL MAGNETIC MOMENT[9]

Since angular momentum is a *vector* quantity it has direction as well as magnitude. Ordinarily, there is no physical quantity that specifies a preferred direction in space, and so the *direction* of **L** has no significance. However, if there is present a *magnetic field*, for example, then a particular space direction is indeed specified. The relevance of a magnetic field to the direction of the angular momentum is due to the fact that an orbiting electron is equivalent to a current loop which acts as a tiny magnet and therefore has an interaction with the magnetic field. Figure 13.13 shows the angular momentum vector **L** of an orbiting electron. Because the circulating charge is *negative,* the *magnetic moment* vector **μ** due to the orbital motion of the electron is directed *opposite* to the angular momentum vector.

The orbital magnetic moment of an electron is a direct consequence of the electron's charge and orbital angular momentum. Since L is quantized, so is μ. In fact, μ can take on only values that are discrete multiples of a basic unit called the *Bohr magneton:*

$$\mu_0 = \frac{e\hbar}{2m_e c} = 9.27 \times 10^{-21} \text{ erg/gauss} \tag{13.19}$$

When $L = 0$ (that is, $l = 0$), then $\mu = 0$.

THE ZEEMAN EFFECT

If an atom is placed in a magnetic field **B,** electrons in the atom will experience a magnetic force by virtue of their motion in the field. The energy of the atomic state is thereby changed. For a large number of atoms in the field, we would expect a completely random distribution of the atomic magnetic moment vectors **μ** relative to the field vector **B.** Consequently, the energy of the atoms in a given atomic state could have any value within a range that corresponds to the range of the angle θ (see Fig. 13.14) between **μ** and **B** (that is, for θ between 0° and 180°). If an atom can make a transition, starting with any energy in this range, there will not be a unique energy (or frequency) for the transition. Therefore, we would expect atomic radiations that occur in magnetic fields to exhibit *broadened* spectral lines, as in Fig. 13.15b. However, experimentally this is found not to be the case. In 1896 Pieter Zeeman (1865–1943) discovered that a number of spectral lines that appeared as *singlets* in the absence of a magnetic field became *multiplets* when the emission occurred in a magnetic field, each member of the multiplet being a sharp line.

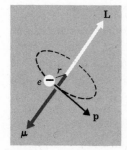

Fig. 13.13 *The angular momentum vector* **L** *and the magnetic moment vector* **μ** *for a circulating electron.*

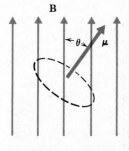

Fig. 13.14 *The energy of interaction between a circulating electron and a magnetic field depends on the angle θ.*

[9]Review Section 9.5.

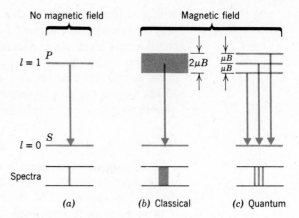

Fig. 13.15 (a) *The spectral line due to a transition between two states in the absence of a magnetic field is* sharp. (b) *According to the classical view, the presence of a magnetic field will broaden the line.* (*Only the P state is broadened; the S state has no orbital angular momentum and therefore no orbital magnetic moment and so remains unbroadened.*) (c) *In the quantum picture, only discrete states* (2l + 1 *in number*) *are allowed in a magnetic field and the P state is broken into three sub-states. The radiation consists of a triplet of spectral lines.* (*Spin effects are not included here.*)

The *Zeeman effect* can be interpreted within the framework of the Bohr theory but only by making an additional quantum hypothesis and by introducing an additional quantum number.

Since discrete spectral lines, instead of broadened lines, are observed for atoms that emit radiation in a magnetic field, there must be a restriction on the directions that the magnetic moment vector (and, hence, also the angular momentum vector) can assume relative to the field direction. The Zeeman effect can be explained with the assumption that *the component of* **L** *in the direction of the field is quantized.* That is, not only is the magnitude of **L** restricted to discrete multiples of $\hbar$, but so is the component of **L** that lies along the field direction (see Fig. 13.16). The quantum number m_l is used to specify the magnitude of the angular momentum component in the direction of the field. If this direction is called the z-axis, the quantization

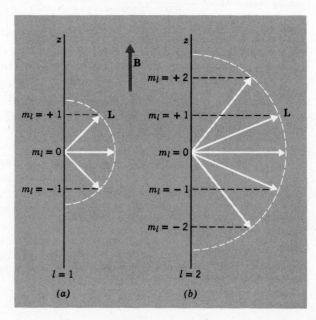

Fig. 13.16 *The angular momentum vector* **L** *can assume only those orientations in a magnetic field for which the component along* **B** *is an integer multiple of* $\hbar$: $L_z = m_l \hbar$. *The vector* **L** *never aligns with* **B** *since the magnitude of* **L** *is* $\sqrt{l(l + 1)}\,\hbar$ *and so is always greater than* $l\hbar$.

Table **13.2** *Magnetic Substates*

State	l	m_l	$2l + 1$
S	0	0	1
P	1	$+1, 0, -1$	3
D	2	$+2, +1, 0, -1, -2$	5
F	3	$+3, +2, +1, 0, -1, -2, -3$	7

rule can be expressed as

$$L_z = m_l \hbar, \qquad m_l = l, l - 1, \ldots, 0, \ldots, -l + 1, -l \qquad (13.20)$$

where L_z is the component of **L** along the z-axis.

Since the component of **L** can be either *along* the direction of **B** or *opposite* to it, m_l can have both positive and negative values; $m_l = 0$ is always possible (**L** perpendicular to **B**) and the maximum and minimum values of m_l are $+l$ and $-l$, respectively. Thus, there are always $2l + 1$ allowed values of m_l (see Table 13.2).

It is now easy to understand the origin of the Zeeman multiplets. Refer to Fig. 13.15c. In the presence of a magnetic field, the P state splits into three levels, corresponding to the three allowed orientations of **L** with respect to **B**: $m_l = +1, 0, -1$; these are called the *magnetic substates*. The lower state has $l = 0$ (and, hence, $\mu = 0$) and therefore is not split. Each of the three P substates can radiate to the S state and for each transition there is a characteristic frequency. Hence, the spectrum shows a triplet of lines.

It is not required that a magnetic field be used to specify a particular direction in space. Other natural ways of accomplishing this are to use the direction of an electric field or the direction of motion of a moving atom. It is important to realize that if a measurement is made to determine the component of the angular momentum in *any* particular direction, *no matter how that direction is selected,* the result will *always* be an integer multiple of $\hbar$.[10] The special significance of an electric or a magnetic field as the direction-determining quantity is that the *energy* of the system depends on its orientation with respect to the field and the frequency of the emitted radiation depends on the energy.

SPIN

In spite of the obvious great success of the Bohr-Sommerfeld picture of atomic structure, there were still many unexplained facts even for relatively simple systems. Foremost among these in the matter of the interpretation of atomic spectra were two observations: (a) certain spectral lines that were expected to be single were found, on close examination, actually to be *doublets,* and (b) certain Zeeman patterns were found to be *anomalous,* for example, P to S transitions consisted of more lines than the expected triplet.

---

[10] If an arbitrary direction in space is chosen (for example, *east*), the act of making a measurement of the component of angular momentum in that direction *produces* a quantization of **L** in that direction.

In 1925, Goudsmit and Uhlenbeck argued that these two effects indicate that an electron possesses angular momentum and a magnetic moment quite apart from the angular momentum and magnetic moment that arise from orbital motion. In classical terms, we can picture an electron as a spinning, charged ball—the mechanical spin produces an angular momentum and the spinning charge is equivalent to a tiny current loop and hence has associated with it a magnetic moment. This classical model has no meaning within the framework of quantum theory (where we speak only of the *intrinsic* angular momentum and *intrinsic* magnetic moment of the electron[11]), but nevertheless it is a convenient picture and is often used; indeed, it is customary to refer to intrinsic angular momentum simply as *spin*.

Spin is a vector quantity and is denoted by the symbol **S.** Just as in the case of orbital angular momentum, where $L = l\hbar$, the magnitude of **S** is given by[12]

$$S = s\hbar \tag{13.21}$$

but, unlike the orbital angular momentum quantum number l that can take on values $0, 1, 2, \ldots n - 1$, the *spin quantum number s* has a *single* value: $s = \frac{1}{2}$. Therefore, there are only *two* allowed projections of **S** on a preferred axis; that is (see Fig. 13.17),

$$S_z = m_s\hbar, \qquad m_s = +\tfrac{1}{2}, -\tfrac{1}{2} \tag{13.22}$$

Only these two projections are allowed ($S_z = 0$, for example, is not allowed) because the projections of **L** or **S** can differ only by *integer* multiples of $\hbar$.

TOTAL ANGULAR MOMENTUM

The spin and the orbital angular momentum of an electron combine to produce the *total* angular momentum. Since L and S are quantized, the total angular momentum J is also quantized:[12]

$$J = j\hbar \tag{13.23}$$

where j is the *total angular momentum quantum number* of the electron and can have only two possible values for a given l:

$$\left.\begin{array}{l} j = l + s = l + \tfrac{1}{2} \\ \text{or} \quad j = l - s = l - \tfrac{1}{2} \end{array}\right\} \tag{13.24}$$

except that for $l = 0$ only $j = \frac{1}{2}$ is possible.

Therefore, for every *orbital* angular momentum, two *total* angular momenta are possible. For $l = 1$, we have $j = \frac{1}{2}, \frac{3}{2}$, for $l = 2$, we have $j = \frac{3}{2}, \frac{5}{2}$, and so forth. The value of j for a state is given as a subscript following the letter designation for the orbital angular momentum. Thus, a state with $l = 1, j = \frac{3}{2}$ is denoted by $P_{3/2}$ and $P_{1/2}$ means $l = 1, j = \frac{1}{2}$. Some of the states of the hydrogen atom are shown in Table 13.3; the numbers preceding the letter designations refer to the values of n.

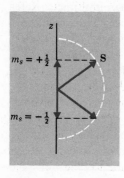

Fig. 13.17 *The projection of the spin vector on a preferred axis can take on only two values, $+\frac{1}{2}\hbar$ and $-\frac{1}{2}\hbar$.*

---

[11]That is, these quantities are *intrinsic* properties of an electron, just as charge and mass are *intrinsic* properties.

[12]Actually, $S = \sqrt{s(s + 1)}\hbar$ and $J = \sqrt{j(j + 1)}\hbar$, but these refinements are not crucial for the arguments here.

Table **13.3**   *Some of the States of the*
Hydrogen Atom

n	l	Designation
1	0	$1 S_{1/2}$
2	0	$2 S_{1/2}$
	1	$2 P_{1/2}, 2 P_{3/2}$
3	0	$3 S_{1/2}$
	1	$3 P_{1/2}, 3 P_{3/2}$
	2	$3 D_{3/2}, 3 D_{5/2}$
4	0	$4 S_{1/2}$
	1	$4 P_{1/2}, 4 P_{3/2}$
	2	$4 D_{3/2}, 4 D_{5/2}$
	3	$4 F_{5/2}, 4 F_{7/2}$

The importance of electron spin in the matter of atomic spectra lies in the fact that the associated magnetic moment (equal to one Bohr magneton, μ_0) produces a difference in the energy of the state of the electron depending on the orientation of **S** with respect to **L**. The reason is that, from the standpoint of the electron's reference frame, the positively-charged nucleus rotates around the electron and constitutes a current loop that produces a magnetic field at the position of the electron. The energy of interaction of the electron's intrinsic magnetic moment with this field depends on the orientation of the magnetic moment vector (and, hence, of the spin vector) with respect to the field. Thus, there is a kind of "internal Zeeman effect" that is present even though there is no externally applied magnetic field. Figure 13.18 shows a P ($l = 1$) state that is split into two substates, $P_{3/2}$ and $P_{1/2}$, because of the magnetic moment interaction.[13] Therefore, instead of a single spectral line, corresponding to a P → S transition, there are actually two closely spaced lines, corresponding to the $P_{3/2} \to S_{1/2}$ and $P_{1/2} \to S_{1/2}$ transitions. It is just such a situation that produces the famous doublet of yellow lines in the spectrum of sodium at 5890 and 5896 Å.

[13] This interaction is also called the *spin-orbit* interaction because it depends on whether the spin vector is parallel or antiparallel to the orbital angular momentum vector.

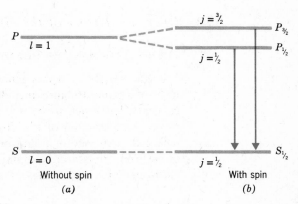

Fig. 13.18 *The existence of electron spin causes all states with nonzero orbital angular momenta to be split into two states with* $j = l + \frac{1}{2}$ *and* $j = l - \frac{1}{2}$. *The transitions from these two states to the* $S_{1/2}$ *state produce a close doublet in the spectrum. This is the situation in the sodium atom which produces the pair of yellow lines at 5890 and 5896 Å. The splitting of the* $P_{1/2}$ *and* $P_{3/2}$ *states compared to the splitting of the P and S states is greatly exaggerated in the figure.*

FERMIONS AND BOSONS

Thus far we have mentioned only electrons in the discussion of spin. In fact, all elementary particles, as well as aggregates of such particles (for example, nuclei) have measurable intrinsic angular momentum properties. Some elementary particles, such as protons and neutrons, have the same spin as electrons ($s = \frac{1}{2}$). Pions have no intrinsic angular momentum ($s = 0$) and photons carry a single unit ($s = 1$). One group of elementary particles (the omega hyperons) has $s = \frac{3}{2}$. The spin of a nucleus is determined by the way the individual spins of the constituent protons and neutrons add together. If a nucleus consists of an even number of nucleons, the spin can be 0, 1, 2, . . . ; if the number is odd, the spin can be $\frac{1}{2}, \frac{3}{2}, \frac{5}{2}, \ldots$. Nuclear states with as much as 20 units of angular momentum have been identified.

In certain types of quantum mechanical situations, particles with half-integer spins ($s = \frac{1}{2}, \frac{3}{2}, \ldots$) have a fundamentally different behavior compared to those with integer spins ($s = 0, 1, 2, \ldots$). For some discussions, therefore, it is convenient to group together the particles in each of these classes. We refer to particles with half-integer spins as *Fermi particles* (or *fermions*), after Enrico Fermi (1901–1954), the great Italian-American physicist who was instrumental in establishing the theory of the emission of electrons from radioactive nuclei. Particles with integer spins are called *Bose particles* (or *bosons*), after the Indian physicist S. N. Bose (1894–   ) who first investigated the statistical properties of such particles (actually, photons).

Table **13.4**   *Spins of Some Elementary Particles*

Classification	Particle	Spin
Fermions	Electron	$\frac{1}{2}$
	Neutrino	$\frac{1}{2}$
	Muon	$\frac{1}{2}$
	Proton	$\frac{1}{2}$
	Neutron	$\frac{1}{2}$
	Omega hyperons	$\frac{3}{2}$
Bosons	Pion (π meson)	0
	Kaon (K meson)	0
	Photon	1

13.4   *Quantum Theory of the Hydrogen Atom*

THE DEVELOPMENT OF QUANTUM THEORY

By the mid-1920s it was generally recognized that the Bohr-Sommerfeld ideas of atomic structure, including as they did both classical and quantum concepts, left much to be desired in terms of a complete and satisfying physical explanation of the properties of atoms. In 1925 and 1926 there emerged a new view of atomic processes that was based, not on a description of electron orbits and "jumping" electrons, but on the *wave* properties of electrons. The classical idea of orbits was abandoned and in its place came the *wave mechanics* or *quantum theory* of elementary processes. In 1925 Werner Heisenberg and Erwin Schrödinger (1887-1961) produced equivalent mathematical descriptions of electron behavior, and Goudsmit and Uhlenbeck introduced the concept of electron spin. In the following year Max Born (1882-   ) contributed the probability interpretation of wave functions. Immediately on the introduction of these fundamental new ideas, there began a furious outpouring of results using the new theory. By 1928 Pauli had formulated the basic principle that explains why atomic electrons are arranged in shells, Heisenberg had developed the uncertainty principle, and P.A.M. Dirac (1902-   ) had introduced relativity into quantum theory. With these advances, all of the necessary fundamental ideas concerning atomic structure were available and the answer to any specific question involved mainly a computational problem. (Of course, some of these "computational problems" were—and still are—extremely formidable.) The development of quantum theory was a gigantic step forward in our understanding of Nature; it is even more remarkable when it is realized that all of the crucial developments took place in such a short period of time. Quantum theory is certainly an equal, if not a greater tribute to the powers of the human intellect than was Newton's formulation of the law of universal gravitation and his explanation of planetary motion.

HYDROGEN ATOM WAVE FUNCTIONS

If we surrender the classical idea of electron orbits, how are we to understand the various energy states of the hydrogen atom? Schrödinger approached this problem in the same way that was discussed in Section 12.6 for the calculation of the wave functions and energy states for a particle in a potential well. He set down an equation that was quite similar to the equation that describes the propagation of mechanical waves and in which he included a term to represent the effect of the electrostatic potential energy of the electron in the field of the nuclear proton. The solution of this equation showed that there are certain discrete energies allowed for the system. These energies, which emerge in a natural way from the Schrödinger equation, correspond precisely to the energies in the Bohr theory but there are no "orbits" for the electron. Instead, each energy state has associated with it a wave function which represents the amplitude of the electron wave at any point in the vicinity of the nucleus. The square of this amplitude is proportional to the probability for finding the electron at any particular position.

The hydrogen atom problem and Schrödinger's solution is illustrated in Fig. 13.19. Figure 13.19a shows the potential, given by $-e^2/r$, and the positions of the first three energy states (compare Fig. 12.24 for the finite "box"). The radii a_1, a_2, and a_3 correspond to the Bohr orbit radii. Figure 13.19b shows the radial wave functions (or *probability amplitudes*) as functions of the radial distance from the nucleus for these first three states. Notice that in each case the wave function extends appreciably beyond the limit of the Bohr radius; that is, the wave function "leaks" out because the potential is finite. Notice also that the S-state wave functions have nonzero amplitudes near the nucleus whereas the wave functions for $l > 0$ all vanish at $r = 0$. We can understand these results of the quantum mechanical

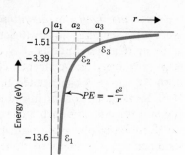

Fig. 13.19 (a) *The potential for the hydrogen atom and the first three energy states in this potential. (Compare the energy level diagram in Fig. 13.9.) The radii* a_1, a_2, *and* a_3 *correspond to the Bohr orbit radii.*

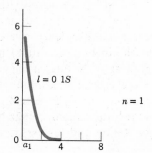

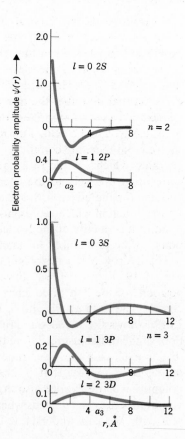

Fig. 13.19 (b) *The radial wave functions for the first three values of the principal quantum number in the hydrogen atom problem. All of the wave functions "leak" out beyond the Bohr radii. Notice that the S-state wave functions cross the axis* n-1 *times and that for a given* n, *the wave function for each higher value of* l *crosses the axis one less time. The probability for finding the electron at a distance* r *from the nucleus and within a radial interval* Δr *is proportional to* $r^2|\psi(r)|^2 \Delta r$.

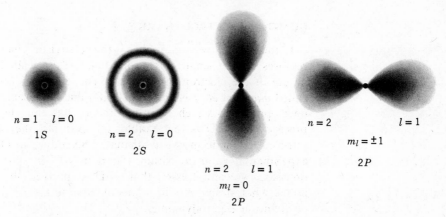

$n = 1$ $l = 0$
$1S$

$n = 2$ $l = 0$
$2S$

$n = 2$ $l = 1$
$m_l = 0$
$2P$

$n = 2$ $l = 1$
$m_l = \pm 1$
$2P$

Fig. 13.20   *Electron probability density clouds for the* $n = 1$ *and* $n = 2$ *states of the hydrogen atom. The 1S and 2S states are spherically symmetric. The density clouds for the 2P substates are drawn for the case in which the field direction is vertical. The complete three-dimensional distributions can be obtained by revolving the clouds shown about a vertical axis. The central dot represents the nucleus.*

calculation in a general way by appealing to the semi-classical model of Bohr and Sommerfeld.[14] The high angular momentum states have the most nearly circular orbits (see Fig. 13.11) and therefore are at the largest average distances from the nucleus. The zero angular momentum state must be an ellipse that has degenerated into a straight line so that the electron "passes through" the nucleus; in the classical picture this is the only way to achieve zero angular momentum. Of course, in the quantum picture, the electron never "passes through" the nucleus; there is only an *electron probability density cloud* that symmetrically surrounds the nucleus.

The wave function curves in Fig. 13.19b show that the electron does not exist at any well-defined distance from the nucleus. There are no "orbits"; instead, the electron distribution is "fuzzy" and one can only specify the probability of finding the electron at a given distance from the nucleus. It is instructive to examine the "fuzziness" of the probability density clouds in Fig. 13.20. Since the 1S and 2S states have zero orbital angular momentum, they must appear the same no matter from what direction they are viewed; that is, these states must have *spherically symmetric* probability density distributions. In the 1S state there is only one maximum in the probability density (at $r = 0$), while in the 2S state there are two maxima, a central one and another at approximately 2 Å from the nucleus. The 2P state (indeed, *any* state that has $l > 0$), on the other hand, is not spherically symmetric. If a preferred direction in space is defined (as with a magnetic field), the various magnetic substates have different probability densities (that is, different wave functions), as shown in Fig. 13.20.

[14]Although the Bohr-Sommerfeld model is not, of course, quantum mechanically correct, if the orbits are compared with the *averages* of the proper quantum wave functions, the agreement between the results of the Bohr-Sommerfeld model and those of quantum theory is quite good.

QUANTUM ELECTRODYNAMICS

In the 1950s it was realized that the relativistic quantum theory of Dirac requires certain small corrections that are associated with the quantum nature of the electromagnetic field. The extension of the Dirac theory is called *quantum electrodynamics* (QED) and represents the highest state of development to which quantum theory has thus far been carried. The precision of QED is remarkable; for example, the value of the electron intrinsic magnetic moment calculated according to QED agrees with the experimental value to within 1 part in 10^6. In spite of this astonishing precision, we cannot assert that QED is "perfect." It remains to be determined what, if any, modifications may be necessary in our present version of quantum electrodynamics.

13.5   *The Exclusion Principle and Atomic Shell Structure*

SYSTEMATICS OF THE PROPERTIES OF THE ELEMENTS

The Schrödinger-Heisenberg quantum mechanics of 1925 proved to be enormously successful in explaining the spectra of hydrogen and other one-electron systems. The introduction of spin permitted an understanding of some of the results for more complicated atoms. But at this stage it could not be claimed that the structure of atoms containing many electrons was understood in detail.

It was known, from the systematic study of atomic radiations, that there was a regular progression in the number of atomic electrons as one passed from element to element and that this electronic charge was balanced by an equal positive charge on the nucleus. Thus, hydrogen has one orbital electron and a charge of $+e$ on the nucleus; helium has two orbital electrons and a charge of $+2e$ on the nucleus. The number of electrons in the neutral atom (or, equivalently, the number of nuclear charges) is called the *atomic number* of the element and is denoted by Z.

It was also known that many of the physical and chemical properties of the elements could be organized in a systematic way and presented in the form of the *periodic table of the elements* (Fig. 13.21). In this table the elements are arranged according to *groups* and *periods,* with the members of each group having similar properties. Thus, the elements Li, Na, K, Rb, Cs, and Fr are all similar to hydrogen in that they participate in chemical reactions as if they have only a single effective electron (called a *valence* electron). For example, the Group I elements combine readily in a one-to-one fashion with the Group VII elements to form such compounds as HF, HBr, NaCl, NaF, KCl, KBr. The Group 0 elements, the so-called *noble* or *inert gases,* do not readily combine with other elements;[15] these gases have no valence electrons.

The periodic table represents the cyclic behavior of many chemical and physical properties of the elements. Each of these cycles ends with a noble gas. Thus, the various periods terminate at the atomic numbers $Z = 2, 10,$

---

[15] A limited number of inert gas compounds have been formed under special conditions.

PERIODIC TABLE OF THE ELEMENTS

Period	Group I	II	(Transition Elements)										III	IV	V	VI	VII	O
1	1 H 1.00797																	2 He 4.0026
2	3 Li 6.939	4 Be 9.0122											5 B 10.811	6 C 12.01115	7 N 14.0067	8 O 15.9994	9 F 18.9984	10 Ne 20.183
3	11 Na 22.9998	12 Mg 24.312											13 Al 26.9815	14 Si 28.086	15 P 30.9738	16 S 32.064	17 Cl 35.453	18 Ar 39.948
4	19 K 39.102	20 Ca 40.08	21 Sc 44.956	22 Ti 47.90	23 V 50.942	24 Cr 51.996	25 Mn 54.9380	26 Fe 55.847	27 Co 58.9332	28 Ni 58.71	29 Cu 63.54	30 Zn 65.37	31 Ga 69.72	32 Ge 72.59	33 As 74.9216	34 Se 78.96	35 Br 79.909	36 Kr 83.80
5	37 Rb 85.47	38 Sr 87.62	39 Y 88.905	40 Zr 91.22	41 Nb 92.906	42 Mo 95.94	43 Tc (99)	44 Ru 101.07	45 Rh 102.905	46 Pd 106.4	47 Ag 107.870	48 Cd 112.40	49 In 114.82	50 Sn 118.69	51 Sb 121.75	52 Te 127.60	53 I 126.9044	54 Xe 131.30
6	55 Cs 132.905	56 Ba 137.34	57–71 *	72 Hf 178.49	73 Ta 180.948	74 W 183.85	75 Re 186.2	76 Os 190.2	77 Ir 192.2	78 Pt 195.09	79 Au 196.967	80 Hg 200.59	81 Tl 204.37	82 Pb 207.19	83 Bi 208.980	84 Po (210)	85 At (210)	86 Rn (222)
7	87 Fr (223)	88 Ra (227)	(89–103) †	(104)	(106)													

*Lanthanide rare–earth elements

57 La 138.91	58 Ce 140.12	59 Pr 140.907	60 Nd 144.24	61 Pm (145)	62 Sm 150.35	63 Eu 151.96	64 Gd 157.25	65 Tb 158.924	66 Dy 162.50	67 Ho 164.930	68 Er 167.26	69 Tm 168.934	70 Yb 173.04	71 Lu 174.97

† Actinide rare–earth elements

89 Ac (227)	90 Th 232.038	91 Pa (231)	92 U 238.03	93 Np (237)	94 Pu (242)	95 Am (243)	96 Cm (245)	97 Bk (249)	98 Cf (249)	99 Es (254)	100 Fm (252)	101 Md (256)	102 No (254)	103 Lw (257)

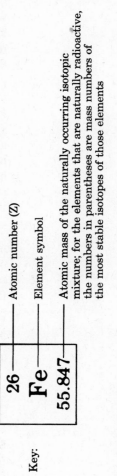

Key:

26 —— Atomic number (Z)

Fe —— Element symbol

55.847 —— Atomic mass of the naturally occurring isotopic mixture; for the elements that are naturally radioactive, the numbers in parentheses are mass numbers of the most stable isotopes of those elements

Fig. 13.21 Periodic table of the elements.

18, 36, 54, and 86. This periodicity is revealed in a striking way by the *ionization energies* of the elements. (This *ionization energy* is the minimum energy required to remove an electron from an atom and convert it into a singly-charged ion.) Figure 13.22 shows that the ionization energy tends to be quite large for the noble gases and to be quite low for the element with the next higher atomic number (a Group I element with an easily removed valence electron). There is a more-or-less uniform increase in the ionization energy as we proceed across any given period from Group I to Group 0.

In spite of the obvious importance of the systematic behavior of elements with regard to both physical and chemical properties, there was no clue in either the old Bohr-Sommerfeld model or in the early quantum theory as to the reason. All that could be said was that electrons seemed to exist in layers or *shells,* with each successive shell ending or closing with a noble gas so that there are no electrons (*valence* electrons) available to participate in chemical reactions. The elements at the beginning of each shell (the Group I elements) have one valence electron; the Group II elements have two valence electrons; and so on. The significance of this electronic shell structure and the meaning of the shell-closure numbers 2, 10, 18, 36, 54, and 86 remained a mystery until it solution was given in a simple and elegant way by Wolfgang Pauli.

THE EXCLUSION PRINCIPLE

The key to the problem of atomic shell structure was discovered by Pauli in 1925. The closing of atomic shells implies that an arbitrarily large number of electrons cannot be placed in a given shell. Pauli realized that such a restrictive effect must have a truly fundamental cause, and his solution to the problem was the formulation of the following principle, known as the *exclusion principle:*

No two electrons in an atom can have identical sets of quantum numbers.

That is, if one atomic electron is in a certain quantum state defined by a

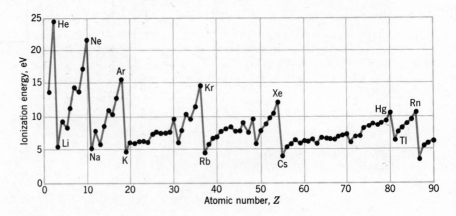

Fig. 13.22 *Ionization energies of the elements as a function of the atomic number.*

Table **13.5**   *Electron States in the First Two Atomic Shells*

Shell	n	l	m_l	m_s	No. of Electrons
K	1	0	0	$+\frac{1}{2}$	} 2 (1S)
		0	0	$-\frac{1}{2}$	
L	2	0	0	$+\frac{1}{2}$	} 2 (2S)
		0	0	$-\frac{1}{2}$	
		1	0	$+\frac{1}{2}$	
		1	0	$-\frac{1}{2}$	
		1	$+1$	$+\frac{1}{2}$	} 6 (2P)
		1	$+1$	$-\frac{1}{2}$	
		1	-1	$+\frac{1}{2}$	
		1	-1	$-\frac{1}{2}$	

(L shell total: 8)

set of quantum numbers, n, l, m_l, and m_s, then other electrons in that atom are excluded from that particular quantum state.[16] It is most remarkable that the details of atomic structure can follow from a principle so simply stated. But such is the beauty of Nature's way.

ATOMIC SHELL STRUCTURE

How many states are available to an electron in an atom? We shall limit our considerations now to the *ground states* of neutral atoms; that is, the Z electrons in the atom are arranged in the way that produces the *minimum* total energy for the system. For $n = 1$, only $l = 0$ is possible and, therefore, only $m_l = 0$ is possible. But there are two possible spin states, $m_s = +\frac{1}{2}$ and $m_s = -\frac{1}{2}$. Therefore, two electrons exhaust the $n = 1$ states and the first shell is filled for $Z = 2$ (helium), as shown in Fig. 13.23. The first shell, which contains only the two $n = 1$, S electrons is called the *K shell*.[17]

In order to form lithium ($Z = 3$), we must add the third electron in a state with $n = 2$ and with lithium we begin the L shell. For $n = 2$ we have available two 2S states ($l = 0, m_l = 0, m_s = \pm\frac{1}{2}$) and six 2P states (two states from $l = 1, m_l = +1, m_s = \pm\frac{1}{2}$, two states from $l = 1, m_l = 0, m_s = \pm\frac{1}{2}$, and two states from $l = 1, m_l = -1, m_s = \pm\frac{1}{2}$). Therefore, the L shell has a total of 8 possible electron states and this shell consists of the 8 elements from $Z = 3$ (lithium) through $Z = 10$ (the noble gas, neon), as indicated in Fig. 13.23. All of the possible states for $n = 1$ and $n = 2$ are listed in Table 13.5.

[16]The exclusion principle is not limited to electrons (actually, all *fermions* obey the restrictions) nor to atomic phenomena. In the following chapter we shall discuss phenomena of matter in bulk that require the exclusion principle for their explanation. *Bosons* do *not* obey the exclusion principle.

[17]It is customary to give letter designations to the sets of electrons with the same principal quantum number. Electrons with $n = 1$ are called K electrons; $n = 2$, L electrons; $n = 3$, M electrons; etc. The first two electronic shells contain only K electrons and L electrons, respectively. But as we shall see, the higher shells contain electrons with more than a single value of n.

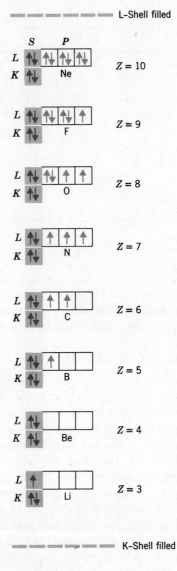

Fig. 13.23 *Filling of the first two shells of atomic electrons. In the first column are the S states (1S and 2S) and in the next three columns are the P states (l = 1, m$_l$ = +1, 0, −1). The boxes represent the magnetic sub-states; each substate contains two spin states, m$_s$ = ±$\frac{1}{2}$, which are represented by the arrows. The K shell is filled at helium and the next 8 electrons must be placed in the L shell which is completed at Z = 10, neon.*

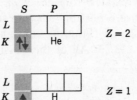

HIGHER SHELLS

We would expect the third shell to consist of the entire complement of M states with $n = 3$: two 3S states, six 3P states, and ten 3D states, giving a total of 18 possible states. However, in the third shell an additional effect comes into play. We know from our previous discussion that electrons in states with high angular momentum find themselves, on the average, much farther from the nucleus than the electrons in low angular momentum states. These distant electrons therefore do not react to a nucleus of charge $+Ze$; instead, they move under the influence of a lesser charge that results from the partial cancellation or shielding of the nuclear charge by the inner, low angular momentum electrons. The outer electrons experience a reduced force and are therefore only loosely bound to the atom. For the $n = 3$ (and higher) states this shielding effect produces an important change in the energetics of an atomic system. The energy of the 4S state is actually *lower* than that of the 3D state (that is, the 4S state is more tightly bound). Consequently, the 4S state fills *before* the 3D state. Similarly, the 5S state fills before the 4D state. This distortion of the "normal" energy scheme is shown schematically in Fig. 13.24 where the third shell is seen to close at the 3P state and contains only 8 electrons. The fourth shell consists of the 4S, 3D, and 4P states and contains a total of 18 electrons. Similarly, the fifth shell contains 18 electrons.

The order of filling of the electron subshells is:

1S, 2S, 2P, 3S, 3P, 4S, 3D, 4P, 5S, 4D, 5P, 6S, 4F, 5D, 6P, 7S, 5F, 6D

The *transition elements* in the periodic table correspond to those atoms in which the shielding effect has displaced the subshells and a subshell with the "wrong" value of n is being filled. For example, with argon ($Z = 18$), the first 18 electron states (up to and including the 3P states) have been filled in the normal manner (see Fig. 13.24). But the 19th and 20th electrons, because of the shielding effect, go into the 4S state instead of the 3D state. With the addition of more electrons, we must "back up" and fill the 10 available 3D states before going on to the 4P states. These 10 elements, formed by "back filling" the 3D states, are the transition elements of period 4. The other transition elements of higher periods occur because of similar effects. When we reach $Z = 57$ (and also $Z = 89$), the "back filling" occurs in such a way that an entire series of elements (the *lanthanides,* beginning at $Z = 57$, and the *actinides,* beginning at $Z = 89$) fits into a single box in the first column of the transition elements. These series are therefore listed in separate rows at the bottom of the periodic table (Fig. 13.21).

The combination of the Pauli exclusion principle and the effect of shielding by the inner electrons accounts completely for the observed shell structure of atomic electrons.

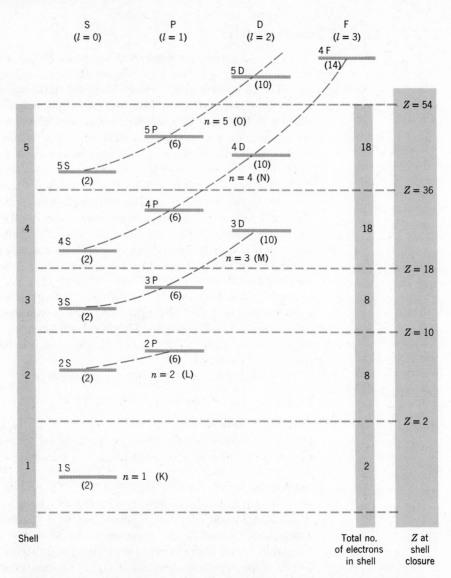

Fig. 13.24 *Distortion of the "normal" sequence of filling subshells because of the shielding of the nuclear charge by the inner, low angular momentum electrons. As a result of shielding, the high angular momentum electrons are raised in energy (i.e., are less tightly bound) and the 3D electrons actually go into the fourth shell and the 4D electrons go into the fifth shell. (Energies are not to scale.)*

13.6 *Atomic Radiations*

X RADIATION

Most of the properties of atoms—chemical, electrical, magnetic, optical, etc.—depend upon the configurations of the outermost electrons. Only in the event of a very energetic disturbance are the tightly bound inner electrons involved in the process. The reason is easy to see on the basis of energetics: it requires, for example, only 7.4 eV to remove the outermost electron from a lead atom ($Z = 82$), but an energy of 75 keV or 75,000 eV is necessary to remove one of the K electrons. The difference is smaller, of course, for atoms with lower atomic numbers but it is always much easier to remove an outer electron than an inner electron.

If sufficient energy is supplied to an atom by collision with a fast electron (as in an X-ray tube) or by irradiation with an energetic photon, then it is indeed possible to remove one of the inner K electrons (see Fig. 13.25a). The atom will not remain long in this condition with a vacancy in its K shell. It is energetically more favorable for an electron in a higher shell to make a transition and occupy the K-shell vacancy (Fig. 13.25b). It is most likely that an L electron will make this transition, emitting an energetic photon (called a K_α X ray) in the process. But then there is a vacancy in the L shell which is filled by an electron from one of the higher shells. Eventually, after this cascading of electrons and the emission of a series of X-ray photons, a free electron from the surroundings will be captured into the outer shell and return the atom to an electrically neutral condition.

Table 13.6 gives the K-shell ionization energy, the K_α X-ray energy, and the minimum ionization energy (that is, for removal of an outer electron) for several elements. The K_α X ray, corresponding to a transition between the L and K shells, is always the most prominent feature of an atom's X-ray spectrum even though the transitions $M \to K$ (yielding a K_β X ray), $N \to K$ (yielding a K_γ X ray), and so on, are more energetic.

Extensive studies of atomic X-ray spectra were made in 1913–1914 by H. G. J. Moseley (1887–1915), a brilliant student of Rutherford who met

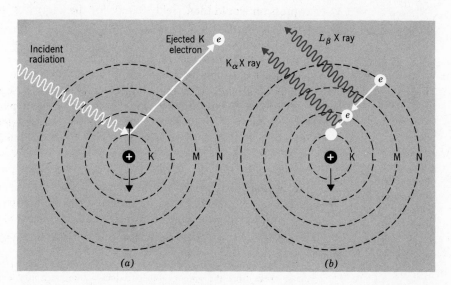

Fig. 13.25 *Schematic representation of the production of X rays. (a) A high-energy photon is incident on an atom; it penetrates to the innermost shell and ejects one of the two K electrons. (b) One of the L electrons then makes a transition, filling the vacancy in the K shell and emitting a K_α X ray in the process. Subsequently, an N electron makes a transition, filling the new vacancy in the L shell and emitting an L_β X ray. Finally, the vacancy in the outermost shell is filled by the capture of a free electron from the surroundings and the atom is again electrically neutral.*

Table **13.6**   *Ionization and* K_α *X-Ray Energies for Some Elements*

Element	Z	K-shell Ionization Energy (keV)	K_α X-Ray Energy (keV)	Minimum Ionization Energy (eV)
Al	13	1.56	1.49	6.0
Cu	29	8.99	8.06	7.7
Mo	42	20.0	17.5	7.4
Ag	47	25.5	22.1	7.6
W	74	69.6	59.3	8.1
Pb	82	88.1	75.0	7.4

an untimely death in World War I. Moseley showed from the systematics of his X-ray spectra that there is a direct connection between the X-ray energies and the atomic numbers of the emitting atoms. He found that the energy of the K_α X ray from an atom of atomic number Z is very closely given by

$$\mathcal{E}_{K_\alpha} = 10.3 \, (Z - 1)^2 \text{ eV} \tag{13.25}$$

The occurrence of the factor $(Z - 1)^2$ in this equation, instead of Z^2 as expected from the simple theory,[18] is due to the fact that when *one* of the K electrons is removed, the *other* remains and shields the nuclear charge, thereby reducing the "effective atomic number" by one. Of course, for hydrogen, there is no second electron to screen the nucleus and Moseley's expression would then yield 10.3 eV for the energy of the radiation. In fact, the L → K (or $n = 2$ to $n = 1$) transition in hydrogen has a wavelength of 1216 Å (see Fig. 13.9), which corresponds to an energy of 10.2 eV. Equation 13.25 provides an extremely good representation of the K_α X-ray energies for atomic numbers up to about 50; for higher values of Z, the values calculated from the simple formula tend to be slightly small.

Prior to Moseley's work, the elements had been placed in the periodic table according to increasing atomic *mass,* but this led to inconsistencies in certain cases. From the systematics of his X-ray spectra, Moseley established for the first time the atomic *numbers* of several elements. It then became clear that the inconsistencies occurred only in those cases for which the normal increase in atomic mass with Z was reversed[19] and that the problem could be resolved by ordering the elements in the periodic table according to atomic *number* instead of atomic *mass.* Moseley was also able to show that there were three elements then missing from the table of

[18] In Eq. 13.9 the energy of a hydrogen atom in a given state is proportional to e^4; this factor arises from the *square* of the product of the electron and nuclear charges, that is, $[(Ze) \times e]^2$ where $Z = 1$ for hydrogen. Therefore, in general, the energy is proportional to Z^2. (The e^4 factor, as well as the other constants, are incorporated into the overall multiplicative constant, 10.3 eV, which appears in Eq. 13.25.)

[19] Although the atomic mass is usually (except for hydrogen) approximately twice the atomic number, the order of increase of the two numbers is not always the same. For example, the atomic mass of cobalt ($Z = 27$) is greater than that of nickel ($Z = 28$); see Fig. 13.21 for other examples.

elements ($Z = 43, 61$, and 75). All three were subsequently discovered after Moseley had given the clue to their existence.

MUON-ATOMIC X RAYS

The π meson or pion—the particle responsible for the strong nuclear force—is unstable when in the free state and decays within 10^{-8} sec into an electron and a *muon* (symbol: μ). The muon is also unstable, but it lives sufficiently long (2×10^{-6} sec) that it can be captured by an atom and replace one of the electrons. (This is true, of course, only for *negatively* charged muons; the positively-charged variety cannot participate in atomic capture processes.)

How will an atom that contains a muon differ from the normal atom? Equation 13.7 shows that the radius of an electron orbit in the Bohr model of the hydrogen atom is inversely proportional to m_e, the electron mass. Now, if the electron is replaced by a muon, which has a mass of $207\, m_e$, the corresponding orbit will be 207 times *smaller*. Thus, the ground-state orbit, instead of having a radius of 0.53×10^{-8} cm, will have a radius of 2.6×10^{-11} cm in the muon atom. Furthermore, the transition energy between any two states, which, according to Eq. 13.10, is proportional to m_e, will be *increased* by a factor of 207 if the electron is replaced by a muon. Therefore, the L $\rightarrow$ K transition, which normally has an energy of 10.2 eV in hydrogen, will have an energy of 2.1 keV in the muon atom.

Although the above argument makes use of the Bohr model, the general results are still valid in quantum theory. K_α X-ray energies, for example, are approximately (although not exactly) 207 times greater when a muon makes an L $\rightarrow$ K transition than when an electron makes the same transition. And the mean distance of a K muon from the nucleus of an atom is approximately 207 times smaller than the corresponding distance for a K electron. These facts make the study of μ-atomic X rays of considerable interest because, for elements with high values of Z, the K muon "orbit" actually lies *inside* the nucleus! The radiation from a muon atom is sensitive to the distribution of nuclear charge encountered by the "orbiting" muon. Therefore, measurements of the radiations from muon atoms provide information regarding the detailed structure of the nuclear surface.

21-cm HYDROGEN RADIATION

Nuclei, as well as electrons, have spins. Therefore, the total angular momentum J of an *atom* is not simply the total angular momentum of the electrons in the atom, but J is the vector sum of the electron total angular momentum and the spin of the nucleus. There are some small differences in atomic radiations that depend on whether the electron total angular momentum vector is aligned in the same or in the opposite direction compared with the nuclear spin vector. One of the interesting cases is that of the hydrogen atom. In the $n = 1$ state of the atom there is no electron orbital angular momentum, so the total angular momentum of the atom is just the sum of the electron and proton spins. The ground state of the hydrogen atom is that $n = 1$ state in which the spin of the electron and the proton are aligned in *opposite* directions so that the *total* angular momentum of

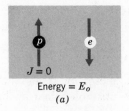

Energy = E_o

(a)

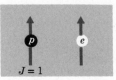

Energy = E_o + 5.9 × 10^{-6} eV

(b)

Fig. 13.26   (a) the normal or ground state of the hydrogen atom has the spin vectors for the electron and the proton in opposite directions. (b) By adding 5.9 × 10^{-6} eV to the atom, the relative spin directions can be "flipped." Conversely, when state (b) makes a transition to state (a), a 5.9 × 10^{-6} eV photon ($\lambda = 21.1$ cm) is emitted.

the atom is $J = 0$ (Fig. 13.26a). It requires only a small amount of energy, 5.9 × 10^{-6} eV, to "flip" one of the spins and produce a state in which the spins align so that $J = 1$ (Fig. 13.26b). Because the energy difference between these two spin states is so small, the "spin-flip" effect is generally unimportant. But it is just this spin transition that has provided a wealth of information about the distribution of hydrogen in our Galaxy.

Most of the matter in the Universe is in the form of hydrogen; hydrogen is the basic "stuff" out of which the heavier elements are made in the nuclear furnaces of stars. After the primordial hydrogen gas condensed to form proto-stars, a considerable amount of this hydrogen remained in the interstellar space. In fact, in our region of the local Galaxy, there is even more hydrogen in the interstellar medium than there is in stars. Now, the average density of interstellar hydrogen is about 1 atom per cm^3. Even though this is an extremely low density, the atoms sometimes (once every 25 years or so) collide with one another. These collisions can raise an atom that was originally in its $J = 0$ ground state into the $J = 1$ state. Most of these excited atoms will be returned to their ground states in further collisions, but a small fraction of the atoms will spontaneously radiate a photon of 5.9 × 10^{-6} eV. This infrequent radiation process is rendered observable because of the truly enormous amount of hydrogen in space.

As a result of this continual raising of interstellar hydrogen atoms into their $J = 1$ states and the subsequent deexcitation of a portion of these atoms by radiation, we have a convenient way of detecting the presence and the amount of hydrogen in the space between stars. The 5.9 × 10^{-6} eV photon that is emitted in the "spin flip" deexcitation process has a wavelength of 21.1 cm ($\nu = 1.42 \times 10^9$ sec^{-1}). This 21-cm radiation is in the *microwave* region of the electromagnetic spectrum (see Fig. 10.47). By using radio-telescopes tuned to the 21-cm radiation, detailed maps have been made of the distribution of hydrogen in our Galaxy (see Section 17.2).

STIMULATED EMISSION OF RADIATION—LASERS

The probability that a transition between two states of an atom will take place within a certain specified time interval depends on the product of the wave functions describing those states and a quantity characteristic of the transition. Thus, from the quantum mechanical viewpoint, there is no distinction between the transition from state A to state B ($A \rightarrow B$) and the transition $B \rightarrow A$ because the same product of wave functions and transition quantity is involved in the description of each transition. The two equivalent processes, excitation and deexcitation, are shown schematically in Fig. 13.27; in each case the photon has an energy $h\nu = \mathcal{E}_B - \mathcal{E}_A$.

It follows, therefore, that if we can arrange to have an atom in state B (the excited state) when a photon with an energy $h\nu = \mathcal{E}_B - \mathcal{E}_A$ is incident on that atom, then this photon will stimulate the deexcitation process to occur. (The photon cannot *excite* the atom because it is already excited, so it does the equivalent—it *deexcites* the atom.) This process is called *stimulated emission* and is illustrated schematically in Fig. 13.28.

The essential feature of the stimulated emission process that renders it both interesting and useful is that the incident photon and the stimulated photon proceed away from the atom *in phase*. That is, the photons travel in the same direction with their amplitudes oscillating together. The two photons therefore reinforce one another. If we had a sample of atoms, some fraction of which were in the same excited state, then a single incident photon could begin triggering the deexcitation of these atoms by stimulated emission. Each stimulated photon could, in turn, cause other atoms to emit photons and the entire system would radiate its excitation energy almost

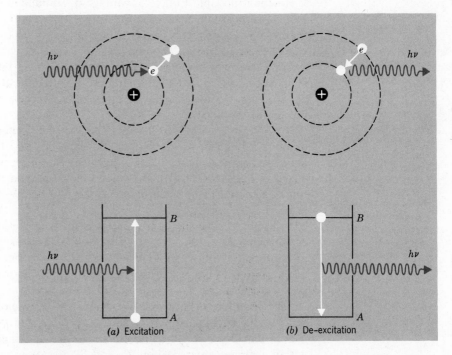

Fig. 13.27 (a) *An incident photon of energy* hν *excites an atom by raising an electron to a higher energy state.* (b) *Deexcitation occurs when the electron returns to its ground-state configuration and a photon of energy* hν *is emitted. According to the rules of quantum mechanics, the two processes are mathematically equivalent. The only difference between the two situations is that an energy* hν *is absorbed in excitation and an energy* hν *is emitted in deexcitation.*

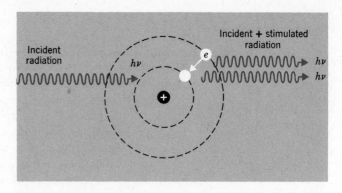

Fig. 13.28 *Stimulated emission of radiation. The incident photon of energy* hν *finds the atom in an excited state and stimulates its decay. The two photons, each with energy* hν, *proceed away from the atom in phase.*

at once with a single bundle of photons all in phase. What have we gained in such a process? Since all of the excited atoms would eventually have radiated away their excitation energy by spontaneous emission, we have done no more by stimulating the deexcitation than would have happened anyway. The difference is that in the spontaneous emission process, the photons are radiated in random directions and not in phase. Stimulated emission produces the photons essentially simultaneously and in phase.

How is it possible to take advantage of stimulated emission to produce an intense beam of in-phase radiation? If the radiation is light, the device that accomplishes this is called a *laser,* an acronym for *l*ight *a*mplification by *s*timulated *e*mission of *r*adiation.[20]

There are two main problems: First, how do we pump energy into the system of atoms so that a sufficient number of atoms are in the upper state? Second, how do we arrange for most of the photons to be emitted along the same direction?

If the high-energy state has a sharply defined energy (such as state B in Fig. 13.27), the pumping radiation must consist of photons with well-defined energy. A source of *white* light would not be suitable because such a source emits photons with a wide range of photon energies and so only a few of these can have the proper energy to be effective in pumping the atoms to the upper state. In 1960, Charles Townes and Arthur Shawlow of Columbia University called attention to an interesting property of ruby crystals that appeared to offer a solution to this problem. Ruby consists of aluminum oxide, a colorless substance, which contains a small amount of chromium as an impurity. The chromium impurity gives to ruby its characteristic red color. Figure 13.29 shows some of the energy states of the chromium atoms in ruby. The distinctive feature of this diagram is the fact that the energy states at $\mathcal{E}_2$ and $\mathcal{E}_3$ are actually *bands;* that is, the atom is not limited to a single well-defined energy but can exist with any energy within a range centered about $\mathcal{E}_2$ and $\mathcal{E}_3$. Because these bands are broad,

[20]The first practical devices to be constructed utilizing this principle operated with microwaves and were called *masers* (*m*icrowave *a*mplification by *s*timulated *e*mission of *r*adiation).

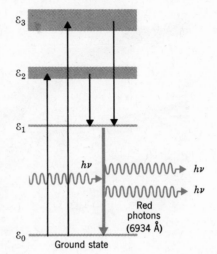

Fig. 13.29 *Some of the energy states of chromium atoms in a ruby crystal. The pumping radiation (upward arrows) excite the two energy bands, $\mathcal{E}_2$ and $\mathcal{E}_3$, which subsequently radiate to form the state $\mathcal{E}_1$ which exhibits laser action. The laser radiation (red arrow) consists of red photons ($\lambda = 6934$ Å).*

the white light from the pumping source includes large numbers of photons whose energies fall within the range that permits the pumping of these energy bands. Each of the two bands radiate primarily to the state at $\mathcal{E}_1$. Hence, the laser transition is $\mathcal{E}_1 \rightarrow \mathcal{E}_0$ and the corresponding radiation is in the red part of the spectrum at 6934 Å.

The problem of directionality can be solved in the following way. A crystal of ruby is formed into a cylinder with the end surfaces accurately parallel (Fig. 13.30). One end is silvered to form a mirror while the other end is given only a partial coating of silver so that some of the radiation can escape from this end. The pumping is provided by a high-intensity discharge lamp that spirals around the cylindrical crystal. As soon as one photon is produced in the spontaneous transition $\mathcal{E}_1 \rightarrow \mathcal{E}_0$, this triggers the laser action. Those photons that move parallel to the cylinder axis are reflected at the ends

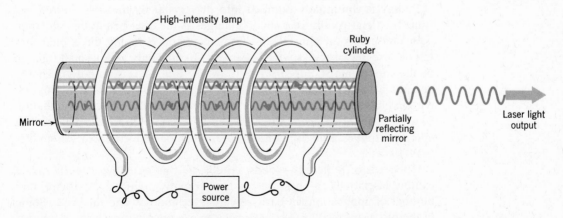

Fig. 13.30 *Schematic of a ruby laser system. The pumping radiation is furnished by a high-intensity source of white light. The stimulated photons are reflected back and forth between the parallel mirrors and build up the intensity of the radiation. The beam is formed by photons escaping through the partially reflecting surface.*

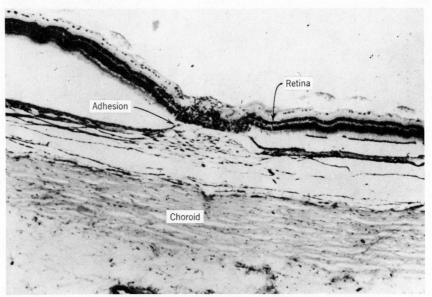

Fig. 13.31 *An adhesion between a detached retina (upper portion) and the choroid surface (lower portion) produced by the beam from a ruby laser. This technique was first used in the treatment of human patients in 1964, and since then several thousand successful treatments have been made.*

and again transverse the crystal, stimulating the emission of additional photons. A fraction of this radiation escapes through the partially-reflecting surface and constitutes the laser beam. Most of the spontaneously emitted photons are not emitted parallel to the axis; these photons are reflected in the crystal and eventually escape through the sides. These spontaneous photons do not contribute to the beam, but a sufficient number of photons *are* reflected back and forth to sustain the laser action.

Energy is continually pumped into the crystal by the light source and some fraction (usually very small) emerges as the laser beam; this radiation is in phase, has an almost pure frequency, and is highly directional. But in no sense is a laser a "source" of energy. In fact, only a very small fraction of the input energy appears in the beam. But *all* of this output energy appears in a tiny beam of small cross sectional area that is highly monochromatic. Some recently constructed lasers can produce bursts of radiation in which tens of joules of energy is released within 10^{-12} sec. This rate of energy output exceeds 10^{13} watts (!)[21] and can be delivered to areas smaller than 1 mm^2.

Hundreds of materials—solids, liquids, and gases—have been found to exhibit laser action. These lasers have rapidly found an extraordinary number of applications in basic research, technology, and medicine. Some of the more spectacular uses have been in eye operations where a laser has been found ideal for depositing just the right amount of energy to "weld"

[21]This is approximately 100 times the total electrical generating capacity of the U.S., but the laser pulse lasts for only ~10^{-12} sec.

a detached retina onto the choroid surface that lies beneath it (see Fig. 13.31). Micro-holes can be drilled in hard substances by laser beams, and the welding of materials that resist other methods can be accomplished with these devices. Modulated laser beams can carry an incredible number of communications channels and it is clear that the impact on the communications industry of the development of such devices will be enormous. We have already mentioned that a laser is being used in a precision measurement of *g* (Fig. 4.17). Also, by reflecting a laser beam from a mirror placed on the moon we are now obtaining information regarding the fluctuations of the Earth-moon distance—information that will give important clues as to the geophysics of the Earth and of the moon. Recently it has been found possible to "chop" a laser beam to give pulses of radiation with a duration of only 10^{-12} sec. The use of such light pulses will provide new information on the interaction of radiation and matter.

Summary of Important Ideas

Rutherford's analysis of α-particle scattering experiments showed conclusively that most of the mass of an atom is concentrated in a tiny, positively-charged *nuclear* core.

In order to account for the lines in the hydrogen spectrum, Bohr found it necessary to postulate that each line corresponds to a transition between two allowed *discrete energy states* and that the angular momentum of the atom is limited to *discrete multiples of $\hbar$*. Bohr departed from classical electromagnetic theory by postulating that no radiation occurs except during the transition process.

According to the Bohr model, an integer number of de Broglie electron waves must exactly fit into every allowed electron orbit.

The explanation of the Zeeman effect requires that the component of the angular momentum of an atom along the direction of the magnetic field (or any prescribed direction in space) be *quantized*. The occurrence of spectral doublets requires the introduction of the concept of electron *spin*.

The specification of the quantum mechanical state of an electron in an atom requires *four* quantum numbers: n, l, m_l, and m_s, which specify, respectively, the (gross) *energy,* the *angular momentum,* the *component of the angular momentum* in a particular direction, and the *orientation of the spin vector* relative to the angular momentum vector.

The *Pauli exclusion principle* states that no two electrons in an atom can have exactly the same set of four quantum numbers. Only *fermions* (particles with half-integer spins) obey the exclusion principle.

The occurrence of *electron shells* in atoms can be accounted for in terms of the *exclusion principle* and the effect of *shielding* by the inner electrons.

When a photon stimulates the emission from an atom of a photon with the same frequency, the two photons propagate away from the atom *in phase* and reinforce one another. The operation of *lasers* is based on this fact.

Questions

13.1   What is the significance of the limiting frequency $(n \rightarrow \infty)$ of the Balmer series for hydrogen? Does a line with this limiting frequency actually exist in the spectrum? How could such a line originate?

13.2   List the total number of spectral lines that can result from the deexcitation of a hydrogen atom in the following states (a) $n = 3$, (b) $n = 4$, (c) $n = 5$. (Use the Bohr model.)

13.3   An *absorption spectrum* is one that results when "white" light (that is, light consisting of all frequencies) is passed through a substance. The absorption lines are then *dark* lines on a background of "white" light. What lines are found in the absorption spectrum of hydrogen?

13.4   An omega hyperon is in a state with $l = 2$. What are the possible values for the total angular momentum quantum number j?

13.5   A corollary to the exclusion principle is the principle of *indistinguishability* of elementary particles. This principle states, for example, that there is no way to distinguish any one electron from another electron. Contrast the situation in which one billiard ball collides with another billiard ball to that in which one electron collides with another electron. Can one measure the angle through which the *incident* object was scattered in both situations? (The billiard balls are *numbered*, but what about the electrons?)

13.6   What difference in atomic shell structure would there be if electrons were bosons instead of fermions?

13.7   What are the maximum values for the projections of the total angular momentum along the z-axis for L, M, and N electrons?

13.8   In what positions in the periodic table do you expect to find elements that have *low* photoelectric work functions? Compare your answer with the elements in Table 12.4.

13.9   What are the quantum numbers for the outermost or *valence* electron in the ground state of sodium? Potassium?

13.10   The proton magnetic moment vector has the same direction as its spin vector, whereas the directions of the two vectors for the electron are opposite. (Why?) From the way in which the two magnetic fields interact, argue that there is an *attractive* effect in the $J = 0$ situation for a proton and an electron in the hydrogen atom (Fig. 13.26a) and a *repulsive* effect in the $J = 1$ situation (Fig. 13.26b). Show, therefore, that the $J = 0$ case is the lower in energy and corresponds to the ground state.

13.11   Examine Fig. 13.21 and find three cases in which the order of increase of atomic number does not follow the order of increase of atomic mass.

Problems

13.1   An α particle of energy 5.3 MeV from a radioactive source of Po^{210} approaches a gold nucleus "head on." How close to the nucleus can the α

particle penetrate before being stopped and deflected backward? (That is, at what distance will the electrostatic potential energy equal the initial kinetic energy of the incident α particle?) It was from such a calculation that Rutherford was able to show that nuclei are much smaller than atoms.

13.2 What is the longest wavelength photon that can induce a transition in a hydrogen atom in its ground state? When that atom deexcites, in what series will the radiation be?

13.3 What is the longest wavelength photon that can ionize a hydrogen atom in its ground state? How would you classify this photon—visible, infrared, or ultraviolet?

13.4 What frequency must a photon have in order to raise a hydrogen atom from its ground state to the state with $n = 4$? Is this a "visible" photon?

13.5 What is the velocity of an electron in the second Bohr orbit in hydrogen (radius $= 2.12 \times 10^{-8}$ cm)? What is the de Broglie wavelength of such an electron?

13.6 What is the wavelength of the series limit for the Lyman series of hydrogen lines?

13.7 Plot on a *wavelength* scale four or five lines from each of the first three series of hydrogen spectral lines (that is, the Lyman, Balmer, and Paschen series).

13.8 The next series of hydrogen lines after the Paschen series is the *Brackett series*. What are the longest and shortest wavelengths in the Brackett series?

13.9 A beam of 12.5-eV electrons bombards a quantity of hydrogen gas and excites some of the atoms. What wavelengths of radiation will be observed?

13.10 Show that Eq. 13.9 can be expressed as $E_n = -13.6/n^2$ eV.

13.11 Add together the masses of a free proton and a free electron. By what fraction does this mass change if the two particles are combined to form a hydrogen atom in its ground state?

13.12* Extend the Bohr theory to the case of the He^+ ion (once-ionized helium). What is the radius of the first Bohr orbit? How much energy is required to remove the remaining electron? What are the wavelengths of the transitions $n = 2$ to $n = 1$ and $n = 3$ to $n = 2$?

13.13 An electron is attracted to a neutron only by the gravitational force (at least for distances greater than $\sim 10^{-13}$ cm). Construct a derivation parallel to that for the Bohr model of the hydrogen atom and obtain the radius of the smallest allowed orbit. Can such "atoms" play any important role in Nature? Explain. (Neglect magnetic effects.)

13.14 Sketch de Broglie wave pictures (similar to Fig. 13.10a) for the hydrogen atom in the states $n = 2, 3$, and 4.

13.15 Continue Table 13.3 for $n = 5$ and $n = 6$.

13.16 Sketch the probability density distribution for the 3S state of the hydrogen atom.

13.17   How many times will the 4S wave function for the hydrogen atom cross the axis? Using this information, sketch the 4S wave function. Can you also sketch the 4P, 4D, and 4F wave functions?

13.18   Extend Table 13.5 to include the M electrons. How many are there? Which are not found in the third atomic shell?

13.19   Construct a diagram similar to Fig. 13.23 for the hypothetical situation in which electrons have spin $\frac{3}{2}$. What elements will be in the K shell? With what element will the L shell close?

13.20   The energy of the K_α X ray from an unknown sample is found to be 6.45 keV. What is the element?

13.21   What is the energy of the K_α X ray from silver $(Z = 47)$?

13.22   Show that the $n = 1$ orbit of a muon around a lead nucleus $(Z = 82)$ actually lies inside the nuclear radius of 1.6×10^{-12} cm. (Since the electrons are so much farther away from the nucleus than is the muon, neglect their presence and consider only a "bare" lead nucleus and a circulating muon; that is, treat the problem in exactly the same way as the Bohr hydrogen atom but with $Z = 82$).

13.23   What is the binding energy of a muon that is in the $n = 1$ state around a proton?

13.24*   The angular divergence of the photon beam from a certain laser is 10^{-4} radians. Consider such a laser that emits 1 mW of radiation directed toward a target that is 1 km away. What is the diameter of the light spot on the target? What would be the power output of a source that emits isotropically (that is, equally in all directions) and delivers the same power per unit area to the target?

13.25   Consider an atom with states of the following energies: -13.2 eV (ground state), -11.1 eV, -10.6 eV, -9.8 eV. Only the state at -11.1 eV exhibits laser action. The state at -10.6 eV radiates primarily to the state at -11.1 eV. The state at -9.8 eV radiates primarily to the ground state. What wavelength radiation would you use to pump the laser? What is the wavelength of the laser radiation?

13.26   What is the spatial length of a laser pulse that lasts 10^{-12} sec? If red light is produced, how many oscillations occur within the pulse?

Diamond

14 *The Structure of Matter*

In order to understand the details of the microscopic world, the physicist first attacks the problems of the simplest atomic structures. In the preceding chapter we saw how the solutions to these problems have contributed to our understanding of the complicated electronic systems of atoms and the spectra that they produce—how the periodic table of elements has been explained—how such fundamental concepts as electron spin and the exclusion principle have emerged—and how our belief in the vast range of validity of quantum theory has been established.

But in our everyday experience we do not deal with *atoms*. The world around us is composed of aggregates of enormous numbers of atoms in the form of solids, liquids, and gases. Therefore our next step is to understand the way in which atoms interact with one another to form molecules and then bulk matter. In this transition from the microscopic to the macroscopic world we find that we do not leave behind the effects of quantum mechanics. Indeed, the quantum nature of matter profoundly influences a wide range of properties of bulk matter. The fact that copper is a good electrical conductor whereas quartz and Teflon are extremely poor conductors is a quantum effect. The existence of the interesting *superconducting* materials, which are of increasing practical importance, is a beautiful example of the effects of quantum mechanics on bulk material. Even the individuality of human beings (indeed, *all* living things) is the result of differences in the structures of the giant molecules that carry genetic information and is therefore basically a quantum effect.

In this chapter we shall study some of the important aspects of molecules and the structure of bulk matter. We have already sketched the theory of gases (kinetic theory; see Chapter 7), and the theory of liquids is still in its embryonic stages. Therefore, in our discussion of bulk matter, we shall (except for the special case of liquid helium) confine our attention to *solids*.

14.1 Bonds Between Atoms

IONIC BINDING

We are all familiar with the chemical compound *sodium chloride* (NaCl) —it is just ordinary *salt*. The molecule of sodium chloride consists of two atoms, one of the metal sodium and one of the gas chlorine, and is typical of a large class of simple molecules. How are these two dissimilar atoms—a metal and a gas—bound together to form a stable substance such as NaCl? If we refer to the periodic table (Fig. 13.21), we see that sodium ($Z = 11$) is the first element in Period 3 and thus has one electron outside the closed L shell (that is, Na is a Group I element). Chlorine ($Z = 17$) is the Period 3, Group VII element and therefore has 7 electrons outside the closed L shell or, equivalently, lacks one electron to fill the third shell.

The single M electron of sodium is relatively easy to remove; it requires only 5.1 eV of energy to detach this electron and form a positively-charged sodium ion, Na$^+$ (see Fig. 14.1a). An atom of chlorine, on the other hand, has an affinity for electrons and, if provided with a free electron, will absorb

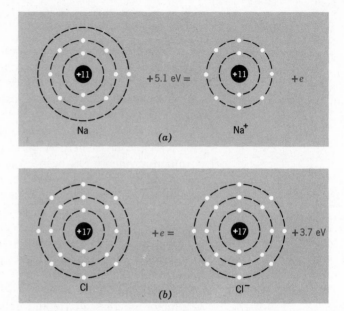

Fig. 14.1   (a) *It requires 5.1 eV of energy to detach the single M electron of sodium and form the ion, Na⁺. (b) A chlorine atom will acquire an electron from its environment to fill the third shell and form the ion, Cl⁻; in the process, 3.7 eV of energy is released. The two ions, Na⁺ and Cl⁻, attract one another and form the ionic salt Na⁺Cl⁻, (or, simply, NaCl).*

this electron into its outer shell, thus completely filling this shell.[1] This process forms a negatively-charged chlorine ion, Cl⁻, and *releases* 3.7 eV of energy (see Fig. 14.1b).

The removal of an electron from sodium and the acquisition of an electron by chlorine are complementary situations. Sodium and chlorine can therefore exist together as a chemical compound, NaCl, in which the sodium electron is used to complete the outer electron shell of chlorine. But how can such a compound be stable since it requires 5.1 eV to remove the sodium electron and only 3.7 eV is gained by forming Cl⁻? The answer lies in the fact that the electron transfer produces the ions Na⁺ and Cl⁻ and these ions are then attracted toward one another by the electrostatic force. When the centers of the two ions are separated by a distance of about 11 Å, the electrostatic potential energy of the system has contributed the requisite 1.4 eV to effect the electron transfer. In fact, the electrostatic attraction pulls the ions even closer together and binds them more tightly. But the ions cannot approach more closely than a certain small distance (which turns out to be 2.4 Å) with all of the electrons in the lowest possible energy state; if they did, two electrons with the same set of quantum numbers would be occupying the same region of space and this is prohibited by the exclusion principle. The only alternative would be for one or more electrons to be raised into a higher energy state, and under such conditions the molecule

[1] Ordinary chlorine gas is *molecular* chlorine, Cl_2, and does not have this affinity for electrons. Only *atomic* chlorine prefers to exist as the ion, Cl⁻. In those situations in which NaCl is formed, chlorine is in the atomic (ionic) state.

would no longer be a bound system. At the equilibrium separation of 2.4 Å imposed by the exclusion principle, the binding energy of the ionic molecule is 5.5 eV.

The binding together of atoms by virtue of the electrostatic attraction between ions formed by the *transfer* of an electron from one atom to the other is called *ionic binding*. In addition to NaCl, many other molecular compounds are bound in this way, for example, NaBr, KCl, RbI, and LiF. The compound MgO is produced by the transfer of *two* electrons from magnesium to oxygen, forming Mg^{++} and O^{--}. The compound Na_2S is produced by the transfer of one electron from each of the two sodium atoms to the sulfur atom, forming 2 Na^+ and S^{--}.

COVALENT BINDING

Ionic binding results from the transfer of one or more electrons from one atom to another; *covalent binding* results from the *sharing* of one or more electrons by the atoms. The simplest case of electron sharing is that of the hydrogen molecular ion H_2^+, that is, two protons sharing a single electron. A *classical* picture of the H_2^+ ion is shown in Fig. 14.2, where the circular Bohr orbits for each atom have been distorted into a single figure-8 orbit with the electron passing around both protons. In executing its orbit, the electron spends most of its time *between* the two protons. When in this region, there is an attractive force between the electron and each of the protons which more than offsets the repulsive force between the two protons. As a result, the system is stable with a proton-proton separation of 1.06 Å; it requires 2.65 eV to remove one of the protons and leave a normal hydrogen atom. Of course, according to quantum theory, there is no electron "orbit"; we can only say that there is a high probability density for the electron wave in the central region of the ion.

The description of the hydrogen molecule H_2 is essentially the same as that for H_2^+ except that *two* electrons are present and the molecule is electrically neutral. In this case the attractive force is greater and the equilibrium separation of the protons is smaller, 0.74 Å; the binding energy (that is, the energy required to produce two hydrogen atoms) is 4.5 eV, also greater than that for H_2^+.

There is one essential difference between the H_2^+ ion and the H_2 molecule. In the H_2 molecule there are *two* electrons and therefore the exclusion principle is important. If the molecule is to be stable, as in the case of the H_2^+ ion, the wave functions that describe the electrons must be concentrated in the central region of the molecule so that the attractive electrical forces can bind the system together (Fig. 14.3b). When two electrons occupy the

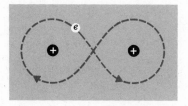

Fig. 14.2 *Classical picture of an H_2^+ ion. The single electron executes a figure-eight orbit around the two protons.*

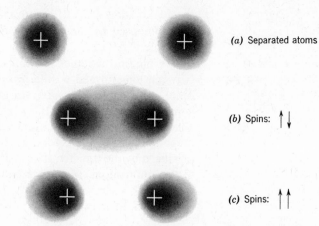

(a) Separated atoms

(b) Spins: ↑↓

(c) Spins: ↑↑

Fig. 14.3 *The effect of the exclusion principle on the H_2 molecule. (a) Separated atoms with the normal electron probability density clouds. (b) Bound H_2 molecule with spins opposite; the probability density is concentrated in the central region. (c) H_2 "molecule" with spins parallel; the electron clouds lie primarily at the extremities of the system and there is no molecular binding.*

same region of space, they must have essentially the same wave function; that is, they are in the same quantum mechanical state and have the same quantum numbers, n, l, and m_l. This can be true only if the electron spins are oppositely directed. If the spins are parallel, this energetically favorable state (that is, the state with the lowest total energy) is denied to the electrons by the exclusion principle and the state of lowest energy that is permitted is one in which the wave functions have maxima at the extremities of the molecule. In this case, the attractive force between the electrons and protons is not sufficient to overcome the proton-proton repulsion and no molecular binding occurs (Fig. 14.3c).

Among the large number of other molecules that are formed by covalent binding are water (H_2O) and ammonia (NH_3). Figure 14.4 illustrates schematically how these molecules employ covalent bonds. The box diagrams for oxygen and nitrogen, showing the electron spin states in the K and L shells are the same as those in Fig. 13.26. Oxygen has two unpaired electrons in the L shell while nitrogen has three such electrons. Each of these electrons can effect a covalent bond with a hydrogen atom, forming H_2O and NH_3. The spatial structures of these molecules are illustrated in Figs. 14.5 and

Fig. 14.4 *Schematic representations of the covalent bonding in (a) water and (b) ammonia. The double lines indicate the bonds formed by the sharing of two electrons.*

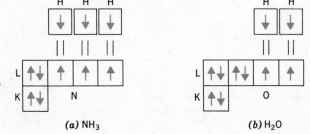

(a) NH_3

(b) H_2O

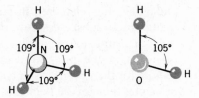

Fig. 14.5 *Representation of the water (H_2O) molecule. The straight lines indicate covalent bonds.*

Fig. 14.6 *Representation of the ammonia (NH_3) molecule.*

14.6. Of course, the molecules are not rigid as suggested in the diagrams, which are only schematic, but the *average* orientations of the electron clouds can be measured and have the directions shown. The peculiar values indicated for the bond angles are the result of the combination of the quantum mechanical requirements on the electron wave functions and the mutual electrical repulsion of the hydrogen nuclei. (Each of the angles would be 90° were it not for the repulsive effect.)

14.2 *Organic Molecules*

CARBON BONDS

The electronic configuration of the carbon atom in its ground state, which is shown in Fig. 13.23 (and reproduced in Fig. 14.7a), indicates that there are two unpaired P electrons in the L shell. We would expect, therefore, that carbon atoms would participate in the formation of molecules by contributing two electrons toward covalent bonds. It is found, however, that carbon almost always appears in molecular structures with *four* equivalent covalent bonds. The reason that all four of the L electrons in carbon participate in molecular bonding, instead of just the two unpaired electrons of the ground-state configuration, is the following. By the addition of only a small amount of energy (about 2 eV) to the carbon atom ground state, it is possible to break the 2S electron pair and promote one of the electrons into the 2P subshell (Fig. 14.7b). Thus, there are *four* unpaired electrons available for bonding if 2 eV can be supplied to the atom. Actually, the energy is supplied in the bonding process itself because the energy gained by making four covalent bonds, instead of two, more than compensates for the energy expended in breaking the 2S pair. Three of these bonds are made by P electrons and one by an S electron. This type of equivalent four-electron bonding of the carbon atom to other atoms is called SP^3 bonding or *hybridization*.

HYDROCARBON MOLECULES

A large proportion of animal and plant material is composed of compounds that consist of carbon combined with hydrogen, oxygen, nitrogen,

Fig. 14.7 *(a) The electronic configuration of the ground state of the carbon atom. Very little energy is required to break the 2S electron pair and promote one of the electrons into the 2P sub-shell where it remains unpaired. In this excited atomic state (b) there are four unpaired electrons, all of which participate in covalent bonding.*

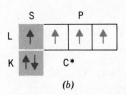

and a few other elements. Because of their association with living (organic) matter, these substances were originally classified as *organic* compounds in order to distinguish them from the ionic (inorganic) compounds such as those discussed in the preceding section. The belief that organic compounds require for their origin some *vital principle* quite different from the physical principles that govern the behavior of nonliving matter has been discredited and it is now clear that the same physical laws and the same quantum mechanical principles apply to molecules of all types, whether organic or inorganic. Indeed, many of the organic compounds that we regularly use are manufactured from such inorganic substances as water, limestone, ammonia, acids, and bases.

The simplest of the organic molecules are those that consist entirely of carbon and hydrogen—the *hydrocarbons*. Many different combinations of C and H are possible; thousands are known and they occur in gaseous, liquid, and solid forms at room temperature. The common commercial fuels, gasoline and natural or liquid (LP) gas, are mixtures of hydrocarbons. The simplest of the hydrocarbons is methane, CH_4, in which each of the four available carbon bonds is utilized to bind a hydrogen atom in the molecule. The structure of methane is shown schematically in Fig. 14.8a where the short dashes each represent a covalent bond involving two electrons. In the ethane molecule (Fig. 14.8b), one of the carbon bonds is utilized to attach another carbon atom, resulting in C_2H_6. The next in the series is propane, C_3H_8 (Fig. 14.8c).

The first four members of this series are all gases (at room temperature), the next 10 are liquids, and all of the heavier ones are solids. The solid linear-chain[2] hydrocarbon molecules are waxy substances, such as paraffin; because of this property the series of molecules is called the *paraffin series*. In principle, the process of successively detaching an H atom and adding a CH_3 group (called a *methyl radical*) onto a hydrocarbon chain can be continued indefinitely, but when the molecules become extremely long, any stresses applied to the substance (such as bending) or any heating which

---

[2]The chains are not really *linear* with all of the carbon atoms arranged in a line; instead, the chain is a staggered array (see Fig. 14.10a).

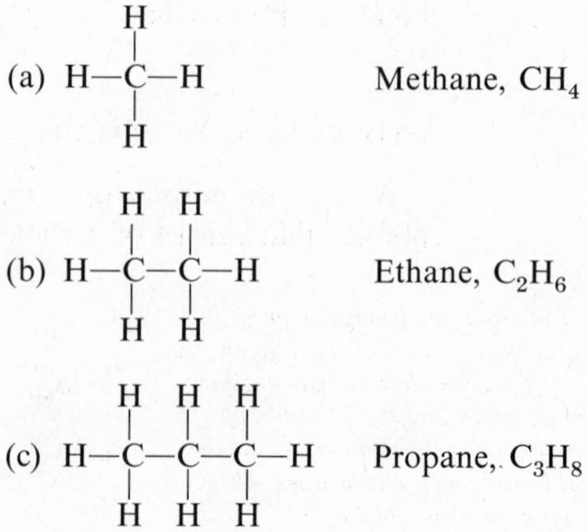

(a)   Methane, CH_4

(b)   Ethane, C_2H_6

(c)   Propane, C_3H_8

Fig. 14.8 *The first three molecules in the series of hydrocarbon chain molecules.*

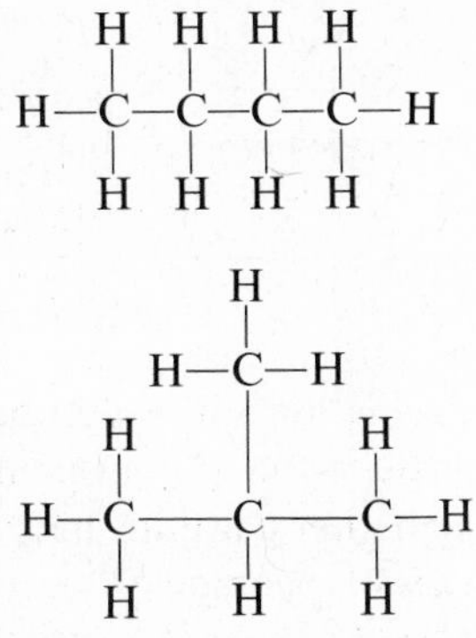

(a) $H-\overset{\overset{\displaystyle H}{|}}{\underset{\underset{\displaystyle H}{|}}{C}}-\overset{\overset{\displaystyle H}{|}}{\underset{\underset{\displaystyle H}{|}}{C}}-\overset{\overset{\displaystyle H}{|}}{\underset{\underset{\displaystyle H}{|}}{C}}-\overset{\overset{\displaystyle H}{|}}{\underset{\underset{\displaystyle H}{|}}{C}}-H$ *n*-Butane, C_4H_{10}

(b) $H-\overset{\overset{\displaystyle H}{|}}{\underset{\underset{\displaystyle H}{|}}{C}}-\overset{\overset{\displaystyle H-\overset{\displaystyle H}{|}-H}{|}}{\underset{\underset{\displaystyle H}{|}}{C}}-\overset{\overset{\displaystyle H}{|}}{\underset{\underset{\displaystyle H}{|}}{C}}-H$ Isobutane, C_4H_{10}

Fig. 14.9 *The two isomers of butane. (See also Fig. 14.10.)*

jostles the molecules tends to rupture the bonds, thereby limiting, in a practical sense, the length of the molecule. For very long hydrocarbon molecules, there are no important differences in the properties of hydrocarbons, so the bond-breaking does not affect the substance in any essential way.

ISOMERISM

The structures of the first three members of the paraffin series, methane, ethane, and propane, are *unique.* That is, there is only one way that two carbon atoms and 6 hydrogen atoms can be put together with covalent bonds to form ethane, namely, the way that is illustrated in Fig. 14.8b. For paraffin-series molecules that contain 4 or more carbon atoms, however, alternate structures are possible. The C_4H_{10} molecule, for example, has two different forms or *isomers.* The linear molecule (Figs. 14.9a and 14.10a) is called *normal butane* (or *n-butane*) whereas the branched structure (Figs. 14.9b and 14.10b) is called *isobutane*. These two isomers of butane, although

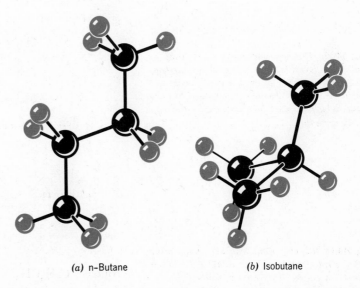

(a) n–Butane   (b) Isobutane

Fig. 4.10 *The three-dimensional structures of the two isomers of butane. (Adapted from L. Pauling and R. Hayward, The Architecture of Molecules, Freeman, San Francisco, 1954.)*

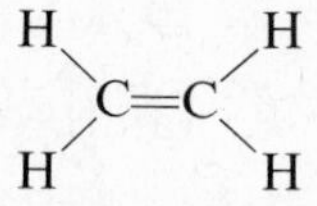

Fig. 14.11 *Two electrons from each of the carbon atoms are involved in a double covalent bond in molecules such as ethylene.*

Ethylene, C_2H_4

they have the same number and types of atoms, have slightly different properties—the boiling points, for example, differ by approximately 10°C.

The higher paraffins have increasing numbers of isomers. There are three isomers of pentane (C_5H_{12}), 5 isomers of hexane (C_6H_{14}), 9 isomers of heptane (C_7H_{16}), 75 isomers of decane ($C_{10}H_{22}$), and more than 4 million isomers of $C_{30}H_{62}$. All of the light isomers have been isolated and their properties studied.

The normal (that is, linear) hydrocarbons are not very efficient fuels, but the branched isomers, because of their greater volatility, tend to burn more easily, and are quite satisfactory fuels.

DOUBLE BONDS AND RING COMPOUNDS

Instead of sharing its outer four electrons with four other atoms, carbon can "double up" its bonds and use two electrons in a bond connecting with a single atom. Such a double bond is found in ethylene, C_2H_4, as shown in Fig. 14.11. Double bonds are also found in the so-called *ring compounds* in which 6 carbon atoms are grouped in a ring with alternating single and double bonds. The prototype ring molecule is that of benzene (Fig. 14.12).

Additional ring-type compounds (or *aromatic* compounds, so called be-

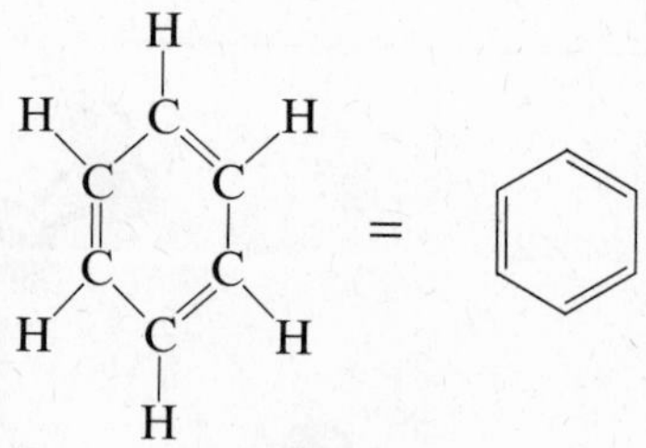

Fig. 14.12 *The ring compound benzene. Single and double bonds between carbon atoms alternate around the ring. The diamond-shape symbol, in which the explicit labeling of carbon and hydrogen atoms is suppressed, is commonly used to represent the benzene ring.*

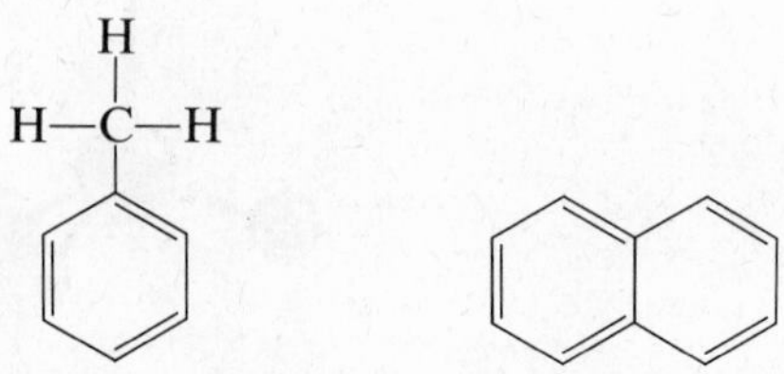

Benzene, C_6H_6

Fig. 14.13 *(a) One of the hydrogen atoms in the benzene ring is replaced by a methyl radical to form toluene (a solvent). (b) Two benzene rings are joined to form naphthalene (moth balls).*

Toluene, $C_6H_5CH_3$ Naphthalene, $C_{10}H_8$

(a)   (b)

cause the members of this class all have distinctive and pronounced odors) can be formed by replacing one or more of the hydrogen atoms with other groups of atoms, as in toluene (Fig. 14.13a), or by joining two or more rings together, as in napthalene (Fig. 14.13b).

COMPLEX ORGANIC COMPOUNDS

Many of the organic compounds that we encounter in our everyday experience contain elements other than carbon and hydrogen. Oxygen is an essential element in all *acids* (for example, *citric* acid, Fig. 14.14a, the acid common to most fruits) and in all alcohols (for example, *ethyl* alcohol or *grain* alcohol, Fig. 14.14b). Nitrogen is found in two positions in the nicotene molecule (Fig. 14.14c); one nitrogen atom replaces a CH group in a benzene ring and one occurs together with carbon in a five-sided, single-bonded ring.

The artificial fibers, such as Nylon, Acrilan, and Dacron, are organic substances. Dacron, for example, consists of a repeating series of basic units containing carbon, hydrogen, and oxygen (Fig. 14.15). The open carbon bond shown at the left-hand end of the molecular unit joins to the open oxygen bond (right-hand end) of another unit. A hundred or so of these units link together to form a chain. These long molecules (called *polymers*) can be made to lie approximately side by side and can slide past one another to some extent. Thus, an elastic fiber is produced that has excellent properties for the manufacture of clothing.

It has recently been discovered that under certain conditions water molecules will join together and form a highly branched long-chain polymer (Fig. 14.16). This interesting substance (called *polywater*) has properties quite different from ordinary water: the density is approximately 1.4 g/cm^3, solidification into a glass-like state occurs below $-40°$C, and the polymer

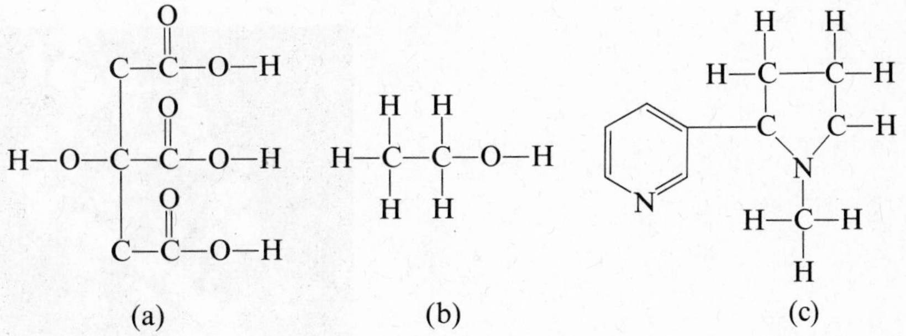

Fig. 14.14 *Three complex organic compounds: (a) citric acid, (b) ethyl alcohol, (c) nicotene.*

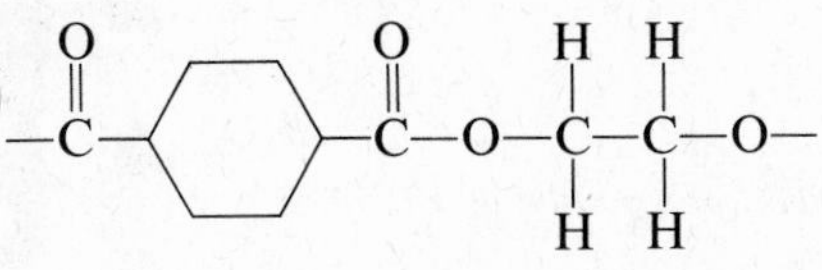

Fig. 14.15 *The basic molecular unit of Dacron. Many such units are linked together through the open bonds, shown at the ends, to form long fibers.*

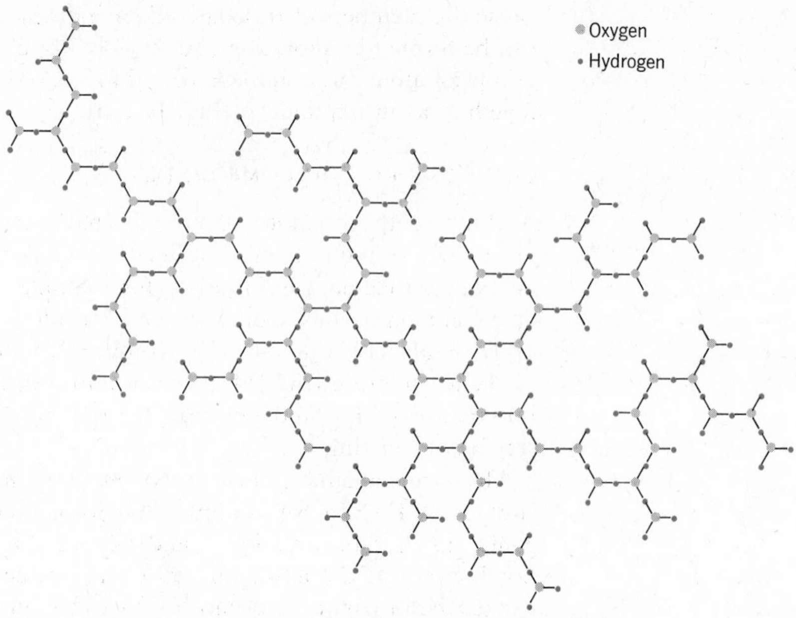

Oxygen

Hydrogen

Fig. 14.16 *A proposed scheme of the molecular structure of the polymer state of water (poly-water).*

is stable (and does not boil) at temperatures up to ~500°C. If methods can be devised to produce polywater cheaply and in large quantities, this substance may prove to be of exceptional utility in various applications (for example, as the working substance in heat transfer systems).

Many of the molecules that are found in living matter, particularly the *protein* molecules, are enormously complicated. In spite of their complex nature, much progress has been made in recent years in understanding the

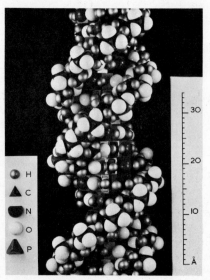

Fig. 14.17 *Scale model of a portion of a DNA molecule.*

Courtesy of Professor M. H. F. Wilkins, King's College, England composition, structure, and function of these molecules. We know a great deal, for example, about the structure of the DNA (deoxyribonucleic acid) molecule, the molecule that carries genetic information, even though these molecules can contain up to about a million atoms. A model of a portion of a DNA molecule is shown in Fig. 14.17.

14.3 *Solids*

IONIC CRYSTALS

All the compounds formed from one Group I atom and one Group VII atom (such as NaCl, LiF, etc.), as well as many other two-atom compounds, arrange themselves into cubic structures when a bulk solid is formed. The tiny cubic crystals of common salt are well known. The atomic reason for this behavior was first proposed by William Barlow who, more than 70 years ago, visualized NaCl as consisting of a tightly-packed cubic array of ball-like atoms (Fig. 14.18). Barlow's picture of the NaCl crystal structure was remarkably accurate. Modern methods of analyzing crystal structure, utilizing X-ray diffraction techniques, have shown that NaCl in solid form indeed consists of a cubic lattice of ions (Fig. 14.19). This type of X-ray crystal analysis has also been used to determine precisely the spacing between crystalline atoms; for NaCl the length of the side of a unit cube is 2.8 Å, slightly greater than the 2.4-Å separation of the ions in an isolated molecule.

A *crystal* of sodium chloride is an even more stable configuration of the ions than a simple isolated molecule. The extra stability, which amounts to 16.5 eV per ion pair, results from the fact that each Na^+ ion is surrounded by 6 Cl^- ions and *vice versa* (Fig. 14.19). The electrostatic energy of an ion in the field of its 6 neighbors provides the additional binding energy.

CRYSTAL GEOMETRIES

A crystal consists of a regularly repeating three-dimensional array of groups of atoms. There are actually only 14 different basic crystal patterns.

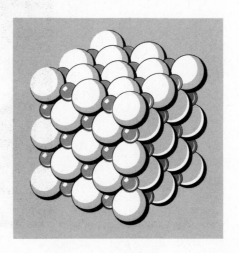

Fig. 14.18 *This arrangement of the sodium and chlorine atoms in solid NaCl was proposed by William Barlow in 1898. Barlow's scheme, which even shows the sodium ions to be smaller than the chlorine ions, has been proved correct by modern X-ray diffraction techniques.*

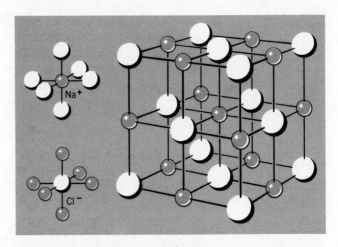

Fig. 14.19 *Solid sodium chloride is a cubic structure in which each sodium ion is immediately surrounded by 6 chlorine ions and each chlorine ion is immediately surrounded by 6 sodium ions. (The ions are shown smaller than their actual sizes relative to the cubic structure of the crystal in order to reveal more clearly the lattice structure.)*

For example, sodium chloride is of the *face-centered cubic* type, and quartz (SiO_2) has a *hexagonal* structure (see Figs. 14.20 and 14.21).

COVALENT-BONDED CRYSTALS

Elemental carbon forms two different types of crystal structures by utilizing its covalent bonds in different ways. In the graphite form, the carbon atoms are arranged in planes of interconnecting hexagons with alternating single and double bonds (Fig. 14.22a). In diamond, the basic unit contains only 4 carbon atoms with all inter-atom connections made with single bonds

Fig. 14.20 *Quartz crystals.*

Ward's Natural Science Establishment

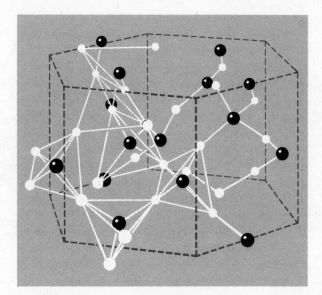

Fig. 14.21 *The hexagonal structure of the quartz crystal. The black spheres represent silicon atoms and the white spheres represent oxygen atoms. (From G. H. Wannier, Scientific American, December 1952.)*

(Fig. 14.22b). The atoms of graphite, illustrated in Fig. 14.23, lie in a series of stacked planes. Because there are only weak forces between adjacent planes, it is quite easy to cleave graphite; thin strips can be removed from bulk graphite by using a sharp instrument such as a razor blade. The diamond crystal structure, on the other hand, is not planar. The four atoms to which any given atom is bound do not all lie in the same plane. This means that there is a strong connection between the atoms in adjacent planes, as shown in Fig. 14.24. Diamond is therefore a much more rigid structure than graphite (but diamond crystals can be cleaved along planes that define the sides of the unit cubes).

METALLIC CRYSTALS

Metals form still another type of crystal structure. It is a characteristic feature of metals that, in the bulk form, the electrical fields in which the atoms find themselves are such that the outer electrons are no longer bound to particular atoms; these electrons are free to move throughout the material. (These are the *conduction electrons* to which we referred in Section 9.1 and

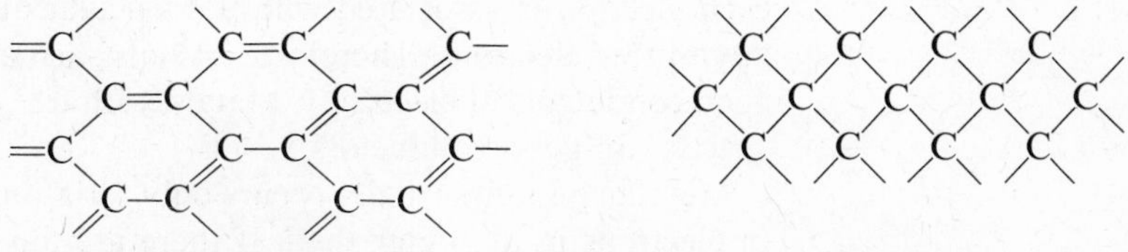

(a) Graphite   (b) Diamond

Fig. 14.22 *The crystal structures of graphite and diamond differ in that double bonds are present in graphite and the basic unit contains 6 atoms whereas diamond contains only single-bonded atoms arranged in groups of 4 atoms. The graphite atoms all lie in a plane, but the diamond atoms do not (see Figs. 14.23 and 14.24).*

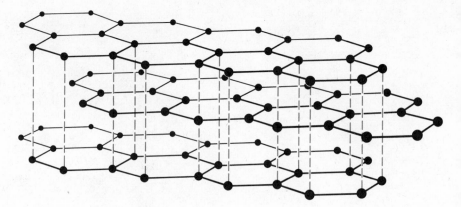

Fig. 14.23 *The three-dimensional crystal structure of graphite. The individual planes of atoms are only weakly coupled and so the graphite crystal is easy to disrupt by cleaving.*

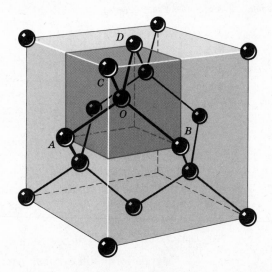

Fig. 14.24 *The three-dimensional structure of the diamond crystal. The 4 atoms (A, B, C, D) to which the atom O is connected do not all lie in the same plane. Therefore, there is a strong coupling between adjacent planes of atoms.*

which we shall discuss further in the next section.) The free electrons constitute a kind of "sea" of negative electricity in which the positively-charged metallic ions are bound, as illustrated schematically in Fig. 14.25. In ionic crystals and in crystals in which the atoms are bound by covalent bonds, each electron is associated with a particular atom or pair of atoms; there are no free electrons. Therefore, crystals such as NaCl or diamond are not good conductors of electricity. Metals, with free electrons in the inter-atomic spaces, are good conductors.

Metallic binding usually occurs only for atoms that have a small number of electrons in the outer shell. If there are too many electrons contributed by each atom, the exclusion principle forces some of these electrons into higher energy states and then the attractive forces are insufficient to cause metallic binding. Thus, the elements in Groups I and II of the periodic table, the transition elements, and some elements in Groups III and IV, form metallic crystals.

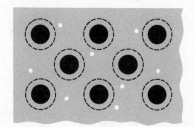

Fig. 14.25 *A typical metallic crystal. The outer electrons are detached from the metal atoms (the outer electron shell is indicated as empty in the diagram) and are free to move throughout the crystal.*

ENERGY BANDS IN SOLIDS

In the preceding chapter we discussed the behavior of electrons in *isolated* atoms. We shall now investigate how the behavior of these electrons is influenced and altered when the atoms are in a crystal of the bulk material.

Figure 14.26 shows the electrostatic potential in the vicinity of an isolated atom of lithium. The horizontal line labeled 0 indicates zero potential energy and the two lower lines indicate the 1S and 2S energy states. (Compare Fig. 13.19a for the hydrogen atom). Two electrons are located in the 1S state and one in the 2S state. The 2S electron is bound by 5.4 eV; that is, the ionization energy of an isolated lithium atom is 5.4 eV.

When lithium atoms are brought together to form a crystal, the net electrostatic field at any point in the crystal is the sum of all of the individual fields. Consequently, the potential between the atoms never rises to zero potential. In fact, the potential at a particular point between atoms in a crystal is reduced substantially below the potential at the same distance from an isolated atom (Fig. 14.27). The reduction is so pronounced in the crystal that the 2S electron, which was *bound* in the isolated atom, no longer encounters a potential barrier that is sufficient to constrain it to the vicinity of any particular atom. The 2S electrons in a lithium crystal are *free* electrons or *conduction* electrons.

The conduction electrons "belong," not to individual atoms, but to the crystal as a whole. That is, the 2S electron wave functions are not localized but extend throughout the crystal. The 2S energy state of the isolated lithium atom becomes an energy "state" of the crystal. If there are N atoms in the crystal, there are N electrons in this "state." But we know that the exclusion principle does not permit more than two electrons to exist in any single energy state. Therefore, the atomic 2S state must expand into a series of

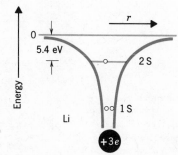

Fig. 14.26 *Schematic potential energy diagram for an isolated lithium atom. The outermost electron (i.e., the 2S electron) is bound by 5.4 eV; it requires ~75 eV to remove one of the 1S electrons and an additional ~120 eV to remove the second 1S electron.*

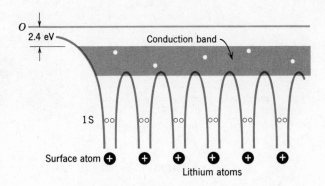

Fig. 14.27 *When lithium atoms are assembled in a crystal, the reduction of the potential between the atoms frees the 2S electrons which then partially fill the conduction band. (The energies are not to scale; a 1S electron is actually bound by 65 eV.) The electrons in the conduction band, shown schematically in the diagram, are not really localized; the electron wave functions extend throughout the crystal.*

closely spaced crystal states, each of which can accommodate two electrons. The spacing between these crystal states is so small[3] that the distribution of states is essentially continuous. As a result, the discrete atomic state becomes, in the crystal, an energy *band* (see Fig. 14.27). This band is the *conduction band* and the electrons that exist in this band are the *conduction electrons*.

The interesting physical properties of an element or compound in bulk form (for example, electrical resistance, thermal conductivity, magnetic properties) are determined in large measure by the details of the energy band structure. We shall investigate next how the crystal energy bands determine the electrical conductivity of the material.

BAND THEORY OF CONDUCTORS AND INSULATORS

In general, all the energy states of an atom appear as bands in a crystal, including all of those higher states that are empty when the atom is in its ground state. Thus, in a lithium crystal there are (unfilled) bands corresponding to the 2P, 3S, 3P, 3D, etc. states. The discrete atomic states can appear in the crystal as identifiable bands (Fig. 14.28b) or as overlapping

---

[3]These states are spread over an energy of a few eV, and a total of $\frac{1}{2}N$ states are required to accommodate N electrons. For a 1-cm^3 crystal, N is of the order of 10^{22}, so the spacing between the states is of the order of 10^{-22} eV.

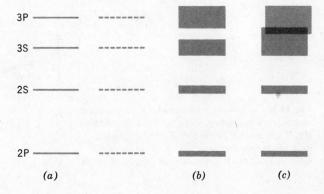

Fig. 14.28 *The discrete atomic energy states (a) correspond, in a crystal, to either individual bonds (b) or to overlapping bands (c). (The 1S state always lies at a much lower energy than the other states and so is not shown in this diagram.)*

Fig. 14.29 (*a*) *An allowed transition of an electron from a filled band into an empty band; this type of transition usually requires 5–10 eV.* (*b*) *The transition of an electron from a filled band into the forbidden region is not allowed.* (*c*) *In a partially filled band an electron can make a transition into any unoccupied state that lies within the band; such transitions ordinarily involve only very small amounts of energy.*

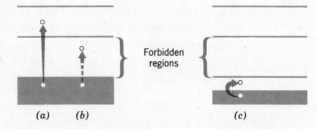

Forbidden regions

(*a*)   (*b*)   (*c*)

bands in which it is no longer possible to specify from which subshell the electrons originated (Fig. 14.28c). Overlapping always occurs for the highest energy bands but the lower states usually remain as individual bands in the crystal. The details of the crystal lattice structure determine the excitation energy at which the overlapping begins.

The filling of the energy bands by electrons follows the same prescription as does the filling of atomic energy states. The 2S atomic state, for example, can accommodate two electrons and in a crystal consisting of N atoms, the 2S band can contain 2N electrons; the 3P band (if distinct from the 3S and 3D bands) can contain 6N electrons; and so forth. If a band is completely filled, no electron in this band can be given any additional energy unless it is given a sufficient amount to raise it to an unoccupied state in a higher band (Fig. 14.29a, b). Depending on the positions of the various bands, the amount of energy required to raise an electron from one band to another may be 5–10 eV. On the other hand, if the highest energy band that contains any electrons is only partially filled (for example, the 2S band in lithium which contains N electrons and thus is half filled), there is an extremely large number of energy states *within* the band that are accessible to these electrons. Thus, an electron in a partially filled band can be given essentially *any* amount of additional energy as long as the total is less than the maximum energy allowed for the band (Fig. 14.29c).

If the highest occupied band is only partially filled, the electrons in this band can be made to drift in a particular direction by the application of an external electric field. The increase in energy brought about by this motion (since it is small) can be accommodated by the available energy states within the band. Materials that have partially filled bands can therefore conduct electricity and are called *conductors* (Fig. 14.30a).

If the highest occupied band is completely filled, increases in the energies of the electrons in that band are not allowed and the application of an

Partially filled conduction band

Empty conduction band

Forbidden regions

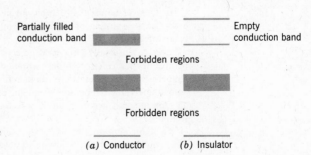

Forbidden regions

Fig. 14.30 *A conductor* (*a*) *is characterized by a partially filled band and an insulator* (*b*) *by a completely filled band above which is a forbidden region and, still higher, an empty band.*

(*a*) Conductor   (*b*) Insulator

electric field will not result in the flow of electrons. Such materials resist the flow of electricity and are called *insulators* (Fig. 14.30b).

In most insulators it requires an appreciable increase in energy (5–10 eV) to raise an electron from the filled band across the forbidden energy region and into the empty conduction band. For example, in the diamond form of carbon there is a 5-eV energy gap between the filled $2P_{1/2}$ band and the empty $2P_{3/2}$ band[4] (Fig. 14.31). What electric field strength is necessary to raise an electron from the $2P_{1/2}$ band into the $2P_{3/2}$ band? In order to answer this question we must first realize that no *real* crystal is as perfect as the ideal crystal structures we have been discussing. There are always small amounts of impurities present in real crystals and there are always small imperfections in the lattice structure. These departures from the ideal crystal form prevent the electrons from moving unimpeded through the crystal. In fact, in even the purest crystals that have been made, an electron can travel only about 10^{-6} cm before encountering one of these imperfections and being scattered with a consequent loss of kinetic energy. (The electron kinetic energy is converted into motional energy of the lattice, that is, into *heat*. This is the reason why all ordinary materials suffer a rise in temperature when they conduct an electric current.) Therefore, in order to gain 5 eV of kinetic energy in a distance of 10^{-6} cm, an electron must be accelerated by a field of 5×10^6 volts/cm! This field is about 10^{10} times that which will cause current to flow in metallic crystals of Li, Na, K, etc.[5] Diamond is therefore an extremely good insulator. Similar energy gaps in crystals such as quartz (SiO_2) and in polymers such as Mylar and Teflon[6] make these materials good insulators.

EMISSION OF ELECTRONS FROM SOLIDS

In a metallic crystal the potential barrier between the atoms is lowered to the extent that the highest energy atomic electrons are free to move throughout the crystal. A potential barrier still exists, however, on the surfaces of the crystal (see Fig. 14.27). This barrier, of course, limits the extent of the electron waves and confines them to the interior of the crystal. Because of the stacking of the electrons, two-by-two, into the closely spaced states of the conduction band, those electrons in the highest energy states find themselves bound to the crystal by an energy appreciably smaller than the atomic binding energy. In the lithium atom, for example, the 2S electron is bound by 5.4 eV, whereas in the lithium crystal, the highest energy electrons are bound by only 2.4 eV. It is therefore considerably easier to

Fig. 14.31 *The energy bands of carbon in the form of diamond (not to scale). Because of the relatively large energy gap between the filled $2P_{1/2}$ band and the empty $2P_{3/2}$ band, diamond is a good insulator.*

---

[4]In most materials that exhibit covalent bonding, the atomic fine-structure splitting (for example, the splitting of the 2P state into the $2P_{3/2}$ and $2P_{1/2}$ states) carries over into the bands of the solid.

[5]Even in an ideally pure metallic crystal there is some resistance to the flow of electrons because the lattice structure has vibrational motion (heat energy) that is characteristic of the temperature. Moving electrons can therefore lose energy in collisions with these vibrating atoms. At extremely low temperatures and for certain materials, this type of energy loss disappears and the material becomes *superconducting*. This phenomenon is discussed in the following section.

[6]The basic molecular unit of Mylar is similar to that of Dacron (Fig. 14.15); Teflon is similar to a long-chain hydrocarbon with the hydrogen atoms replaced by fluorine atoms.

eject an electron from a solid than from an isolated atom. There are three important ways in which electron emission from solids can be effected:

1. *Photoelectric emission.* Photoelectrons are produced when the energy necessary to overcome the surface potential barrier is supplied by light quanta (see Section 12.3). The photoelectric *work function* for solid lithium is equal to the binding energy of 2.4 eV and corresponds to the *minimum* energy required to eject an electron. For an electron originally in an energy state nearer the bottom of the conduction band, a correspondingly larger amount of energy is required for emission to occur. This is why there is always a range of energies for photoelectrons; a measurement of the *highest* electron energy, for a given energy of the incident photons, determines the work function for the material.

2. *Field emission.* The potential barrier at the surface of a material can be lowered by the application of an external electric field. If the field strength is sufficient ($\sim 10^6$ volts/cm), those electrons in the conduction band with the highest energies can escape.

   The electron waves that result from the field emission of electrons from crystals exhibit interference. By recording the interference patterns on photographic film, a new method of attack on problems of crystal structure is possible. This technique has been pioneered by Professor Erwin W. Müller of Pennsylvania State University. One of Professor Müller's photomicrographs of the field emission pattern from an iridium crystal is shown in Fig. 14.32. Many details of crystal structures that are not easily studied with other methods have been clarified by this new technique.

3. *Thermionic emission.* If the temperature of a solid is raised to a high value, some of the thermal energy will reside with the conduction electrons. Some of these electrons will acquire energies greater than the surface potential barrier and they will escape from the material. The heating of a wire filament coated with a material that has a low work function (for example, cesium) is the most common method of providing electrons for acceleration in electron guns, such as those found in cathode-ray oscilloscope tubes or in TV picture tubes.

SEMICONDUCTORS

Some materials that satisfy the requirement for classification as insulators, namely, a completely filled upper band, actually have a very small energy gap (an eV or less) separating the filled band from an empty conduction band. At low temperatures, essentially none of the electrons in these materials will have sufficient thermal energies to enable them to cross the energy gap into the conduction band. As the temperature is increased, however, more and more of the electrons, by virtue of thermal agitation, are found in the conduction band. Therefore, these materials—silicon and germanium are typical members of this class—exhibit a weak electrical conductivity and are called *semiconductors.* This conductivity increases markedly with temperature; between 250°K and 450°K, the number of conduction electrons in silicon, for example, increases by a factor of 10^6. Of course, the degree

Fig. 14.32 *Photomicrograph of an iridium crystal by the field emission technique. The differences in the pattern on opposite sides of the center reveal imperfections in the crystal lattice structure.*

Courtesy of Dr. Erwin Muller, Pennsylvania State University

to which a semiconductor material will conduct electricity depends on the magnitude of the energy gap; the smaller the gap, the greater the number of thermally excited electrons and, hence, the greater the electrical conductivity. Some semiconductor energy gaps are compared with those of insulators in Table 14.1.

One of the methods of increasing the conductivity of a semiconductor is to illuminate the material with light. The absorption of the light raises some of the electrons into the conduction band and thereby increase the conductivity. Materials that exhibit this property are called *photoconductors.* Germanium, for example, shows a greatly increased conductivity when illuminated by photons of energy greater than 0.7 eV.

If it is desired to endow a semiconductor with increased conduction properties that are permanent, one cannot rely on the photoconduction

Table **14.1**   *Energy Gaps for Some Semiconductors and Insulators*

Semiconductor	Energy Gap (eV)	Insulator	Energy Gap (eV)
Silicon	1.14	Diamond	5.33
Germanium	0.67	Zinc Oxide	3.2
Tellurium	0.33	Silver chloride	3.2
Indium antimonide	0.23	Cadmium sulfide	2.42

process since this phenomenon is transient and disappears when the light source is removed. Nor is it usually convenient to operate the material at an elevated temperature so that thermal excitation will provide greater conductivity. An extremely effective way to change permanently the conduction properties of semiconductors is by the introduction of minute quantities of certain impurities into the crystal structure. These impurities are called *doping agents* or *dopants* and the doping of semiconductors such as silicon and germanium is now an essential part of the process for manufacturing transistors.

The semiconductor germanium is a Group IV element and has 4 outer or valence electrons. Germanium forms a diamondlike crystal structure in which each atom is connected to 4 other atoms by double covalent bonds (Fig. 14.33a). Now, arsenic is a Group V element with 5 valence electrons and is adjacent to germanium in the periodic table. If an atom of arsenic replaces a germanium atom in a crystal, 4 of its valence electrons are used to duplicate the germanium bonds and there is one surplus electron. Thus, the arsenic atom in the crystal is actually an As^+ ion and the extra electron remains loosely bound in the vicinity of the arsenic ion (Fig. 14.33b). The binding energy of the surplus electron to the As^+ ion is extremely small, however—about 0.01 eV, an energy considerably smaller than the 0.7-eV gap energy in a pure germanium crystal. Therefore, the addition of only a small fraction of an eV of energy is sufficient to raise this electron into the conduction band. In fact, at room temperature almost all the surplus electrons contributed by the arsenic impurities are in the conduction band, whereas only a few of the germanium electrons have sufficient thermal energies to make the transition. Therefore, the presence of arsenic impurity atoms, even though their concentration may be only 1 part in 10^6 or 10^7, is the determining factor for conductivity. (In a 1-cm^3 sample of arsenic-doped germanium there will be $\sim 10^{16}$ *conduction* electrons contributed by the arsenic—this is quite a sufficient number to permit substantial conduction currents to flow.)

The energy states associated with the loosely bound electrons contributed by the arsenic ions appear in the germanium energy diagram as a set of additional states lying immediately below the lower boundary of the conduction band (see Fig. 14.34). Because the electrons that populate these states have been "donated" by the arsenic atoms, the states are called *donor states*. The semiconductor that results from doping by a donor-type impurity is

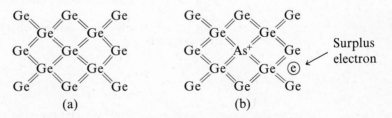

Fig. 14.33 (*a*) *The double-bonded, diamond-like crystal structure of germanium.* (*b*) *Replacing one of the germanium atoms in the crystal by an atom of arsenic produces an arsenic ion (As$^+$) and a surplus electron which is easily excited into the conduction band.*

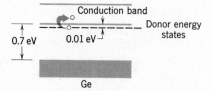

Fig. 14.34 *The energy states (donor states) contributed by the arsenic impurity atoms in a germanium crystal lie just below the conduction band. Electrons in these energy states can easily make transitions into the conduction band.*

called an *n*-type semiconductor.[7] Other Group V elements such as phosphorus or antimony can also be used to convert pure germanium into an *n*-type semiconductor.

If, instead of a Group V element, a Group III element, such as gallium, is introduced as an impurity into a germanium crystal, the effects are analogous. The difference in the two cases is that a gallium atom with three valence electrons, lacks one electron to effect the 4 double bonds that bind it to the 4 neighboring germanium atoms. In order to compensate for this deficiency, the gallium atom "steals" an electron from a germanium atom, thus producing a gallium ion (Ga^-) and a germanium ion (Ge^+). The germanium ion that lacks an electron is called an *electron hole*. The Ge^+ ion can then "steal" an electron from another germanium atom; thus, in effect the hole moves from one position to another. Thus, just as the donor electrons can migrate in the conduction band, the holes are also free to migrate in the lower energy band.

Because the gallium impurity "accepts" electrons from the germanium atoms, these impurity atoms are called *acceptors*. The energy states associated with acceptor atoms lie very close to the upper energy of the filled band (Fig. 14.35). That is, it requires very little energy for the acceptor atom to "steal" an electron from a neighboring germanium atom and thereby create a hole.

Other Group III elements that serve as acceptors in germanium crystals are boron, aluminum, and indium. A semiconductor that is doped with acceptor atoms is called a *p*-type semiconductor.[8]

ELECTRICAL CONDUCTIVITY IN SEMICONDUCTORS

An *n*-type semiconductor at ordinary temperatures has most of its donor electrons in the conduction band. Therefore, if a potential difference is placed across the semiconductor, the electrons will move under the influence of the applied field and current will flow (Fig. 14.36).

[7] So called because the result is the contribution of mobile *negatively*-charged carriers of electricity.

[8] So called because the result is the contribution of mobile *positively* charged carriers of electricity (namely, *holes*).

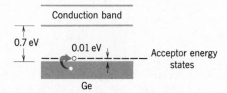

Fig. 14.35 *The energy states associated with acceptor atoms lie only slightly above the filled band. Thermally excited electrons can populate these states and leave holes in the filled band.*

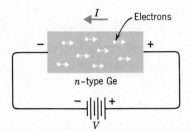

Fig. 14.36 *The conduction electrons in an* n-*type semiconductor can be made to move through the material by the application of a potential difference* V. *Notice that, as always, the direction of current flow is opposite to the direction of electron motion.*

What is the mechanism of current flow in a *p*-type semiconductor? In any crystal the atoms are bound in the lattice and do not move from these sites. Therefore, *all* electrical current in solids flows by virtue of electron movement. But in *p*-type semiconductors this movement is quite different from the motion of conduction electrons. As shown in Fig. 14.37, the acceptor atom receives an electron from a germanium atom, which then becomes a positive ion. The Ge$^+$ ion next receives an electron from another germanium atom, and so on. No single electron moves a distance more than an inter-atom spacing or two and yet the positive charge (that is, the hole) is propagated along the line of atoms. Notice that the direction of motion of the hole is the same as the direction of current flow and that both are opposite to the direction in which the electrons move. If a potential is placed across a *p*-type semiconductor, the motion of the holes constitutes a current, as illustrated in Fig. 14.38.

SEMICONDUCTOR DEVICES

From a practical standpoint, the interest in doped semiconductors is due to the fact that a number of basic elements of electronic circuits can be constructed from these materials. All semiconductor devices make use of the directional feature of current flow across a boundary between *p*-type

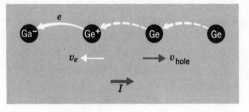

Fig. 14.37 *A hole is propagated along a line of atoms by electrons successively moving from neutral germanium atoms to Ge$^+$ ions.*

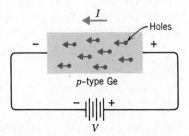

Fig. 14.38 *In* p-*type germanium the current is carried by the holes.* (*Compare Fig. 14.36.*)

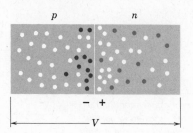

Fig. 14.39 *If* p-*type and* n-*type semiconductor materials are placed in contact (forming a* p-n *junction), the diffusion of electrons and holes across the boundary establishes a potential difference* V *across the junction.*

and *n*-type semiconductors that are placed in contact,[9] as in Fig. 14.39. There will be, even at ordinary temperatures, a diffusion of the electrons across the *p-n* junction from the *n* region into the *p* region as well as a diffusion of the holes from the *p* region into the *n* region. This migration of electrons and holes tends to build up a positive potential in the *n* region and a negative potential in the *p* region. But the magnitude of the potential is limited by the fact that the holes and electrons tend to *recombine*. The number of holes and electrons does not really diminish, however, because the thermal excitation of germanium electrons continually produces new holes and electrons. At equilibrium, the rate of thermal excitation is equal to the rate of recombination. The net result is that the flow of electron current into the *p* region is just equal to the flow of hole current into the *n* region and a potential difference exists across the *p-n* junction.

If a *p-n* junction is connected to an external voltage source with the positive lead at the *p* end of the junction, as in Fig. 14.40a, the internal potential that is built up at the *p-n* boundary (the *boundary* potential) *adds* to the external potential and *aids* the current flow (I_1) as indicated. On the other hand, if the polarity of the external voltage source is reversed, as in Fig. 14.40b, the boundary potential *opposes* the external potential and *impedes* the current flow; that is, $I_2 \ll I_1$. Thus, the *p-n* junction passes current preferentially in only one direction and acts as a *rectifier* or *diode*.

A *transistor* is a device that consists of three semiconductor elements in contact, either in the order *n-p-n* or in the order *p-n-p* (Fig. 14.41). These devices are used in electronic circuits to amplify and control electrical signals. For example, the voltage applied to the *p* region of an *n-p-n* transistor can be used to alter the potential in this region; therefore, this applied voltage can control the current flow through the device from one *n* region to the other.

The development of semiconductor devices, begun in 1949,[10] has completely revolutionized the electronics industry. The communications and computer industries now rely almost completely on semiconductor components. Recently, a tremendous advance has been made in the miniaturization of electronic devices through the introduction of *integrated circuits*. These tiny modules (called *IC chips*) are actually complete electronic subassem-

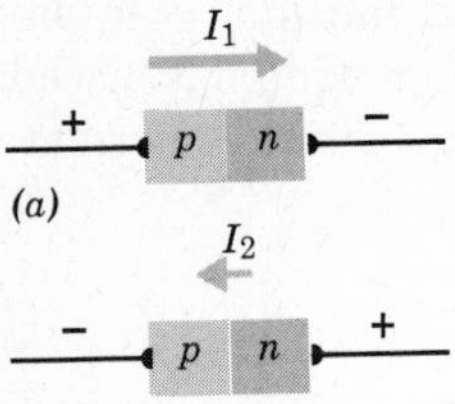

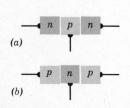

Fig. 14.40 *The internal boundary potential (see Fig. 14.39) aids current flow from the* p *region to the* n *region but impedes current flow in the opposite direction. For the same magnitude of potential difference in (a) as in (b), the current* I_1 *is considerably greater than the current* I_2.

Fig. 14.41 (*a*) *An* n-p-n *transistor.* (*b*) *A* p-n-p *transistor.*

[9] One cannot simply press together the two materials because the irregularities of the crystal structures at the surfaces would prevent the desired effects from taking place. Instead, the dopants are introduced in the proper places while the crystal is being grown or are diffused into a grown crystal from opposite sides.

[10] John Bardeen, Walter Brattain, and William Shockley shared the 1956 Nobel Prize for the discovery and the development of the transistor.

Fig. 14.42 *Two modern integrated circuit chips, compared with a length of 22-gauge insulated wire, a size commonly used for wiring present-day computers. Each chip is 1.8 mm on an edge and contains 55 complete circuits utilizing 213 transistors. One of these units serves as a complete logic element in a high-speed computer. The etching of the semiconductor wafers to form the circuits is accomplished by using an electron micro-beam. Even more sophisticated integrated circuits have been made, with densities of ~200,000 transistors per square inch.*

IBM

blies, consisting of large numbers of circuit elements (such as diodes, transistors, resistors, and capacitors) placed together on a single wafer of semiconductor material. Two such IC chips are shown in Fig. 14.42. The development of integrated circuits has produced not only savings in space and cost but has also increased the reliability of circuits and made the construction of devices considerably easier. (The reduction of circuit size is clearly of importance in devices intended for use in spacecraft where small size and mass are highly desirable. But small size is of *crucial* importance in the construction of super high-speed computers where the limitation on speed is set by how fast it is possible to transfer a signal from one point to another; since the maximum signal transfer velocity is just c, small size means faster computation.)

14.4 *Low Temperature Phenomena*

THE ATTAINMENT OF LOW TEMPERATURES

Most devices that provide low temperatures (including the household refrigerator) operate by driving a liquifiable gas through a cycle of compressions and expansions. If we do work on a gas by compressing it, the gas temperature rises. The additional thermal energy that is given to the gas can be removed by allowing it to increase the temperature of its surroundings. (In a common refrigerator, circulating air takes up the heat.) We

Table **14.2**   *Liquification
(Boiling) Points of Some Gases
at Atmospheric Pressure*

Gas	Temperature, °K
Chlorine	238.6
Xenon	166.1
Krypton	120.3
Oxygen	90.2
Argon	87.5
Nitrogen	77.4
Hydrogen	20.4
Helium	4.2

now have a gas at high pressure but at (or near) room temperature. If we next expand the gas rapidly the temperature decreases and, if the decrease is sufficient, the gas will liquify. High pressure steam, for example, if released from a boiler, will condense into water by virtue of the expansion process. A system whose function is to provide low temperatures to other objects does so by bringing the liquified gas into thermal contact with the objects; the gas is then recycled through the system. On the other hand, if the purpose is to produce liquid gas for other uses, arrangements are made to remove the liquid gas from the system.

Compression-expansion systems can be used to liquify all gases. Of course, if the gas becomes liquid only at a very low temperature (as do hydrogen and helium; see Table 14.2), then a particularly elaborate system is required. Helium (which becomes a liquid at 4.2°K) was first liquified in 1908 by Kammerlingh Onnes of Leiden.

Once temperatures as low as those of liquid helium are available, it becomes possible to use other, less efficient methods to reduce the temperature still further. It is only because helium remains in liquid form and does not solidify[11] (at atmospheric pressure) that extremely low temperatures (down to $\sim 10^{-3}$°K) can be attained.

LIQUID HELIUM—THE SUPERFLUID

The liquid form of helium is a peculiar substance. In fact, liquid helium exists in *two* completely different forms. In the temperature range from 4.2°K down to a temperature of 2.18°K (called the λ *point*), helium behaves as an ordinary, classical fluid; in this region the liquid is referred to as *helium I*. Below the λ point, however, the properties of liquid helium (called *helium II* in this temperature region) take on an extraordinary character. For $T^* < 2.18$°K, helium consists of a mixture of two fluids, one of which retains the classical properties exhibited by helium I whereas the other possesses new and remarkable properties that are responsible for the unusual behavior of the liquid at these low temperatures. The ordinary portion of the liquid is called the *normal component* and the extra-ordinary portion is termed the *superfluid component*.

---

[11] Helium is unique in this regard.

One of the properties of superfluid helium is its ability to conduct heat with *absolutely no resistance*. That is, heat that is supplied at one point in the superfluid is conducted throughout the liquid *without any losses whatsoever*. Thus, we say that the heat conductivity of the superfluid is infinite—not just "very large" but *infinite!* We can verify this statement in a qualitative way (precise measurements can do so quantitatively) by observing the boiling action in liquid helium as the temperature is lowered through the λ point. If we supply a small quantity of heat at one point in a certain volume of liquid helium that is at a temperature above the λ point, the liquid boils at this point and vigorous bubbling is seen to take place. The reason for the localized boiling is that the liquid has a finite heat conductivity and is incapable of distributing the heat sufficiently rapidly so that the temperature of the entire liquid is raised uniformly. Instead, the temperature is increased only in the region near the point at which the heat is introduced and when the local temperature exceeds 4.2°K, the liquid boils and gas bubbles are formed. Now, let us lower the temperature of the system while continuing the introduction of heat at one point. When the λ point is reached, the boiling action suddenly ceases—no more bubbles are formed. The reason for the cessation of boiling is that below the λ point a portion of the liquid is in the superfluid form with infinite heat conductivity. As rapidly as the heat is introduced, it is distributed to all parts of the liquid and therefore the entire liquid is at exactly the same temperature. Under these conditions no gas bubbles can be formed.

Other experiments show that superfluid helium experiences absolutely no resistance to flow; that is, it has *zero* viscosity. Superfluid helium flows unimpeded through narrow channels or extremely fine capillary tubes that would block the passage of any ordinary fluid, including helium I.

Liquid helium II consists of both superfluid and normal components. The relative amount of the superfluid component as a function of the temperature can be determined in experiments that measure the resisting force on an object moved through the fluid. Only the normal component resists such motion; the object passes frictionlessly through the superfluid component. Such experiments have shown that the superfluid concentration increases with decreasing temperature from zero exactly at the λ point to essentially 100 percent for temperatures below about 1°K. Thus, for $T^* \lesssim 1°K$, helium is truly a superfluid.

Liquid helium possesses a number of spectacular properties, all of which are compatible with the two-fluid model in which the liquid consists of normal and superfluid components.

How can we explain the seemingly impossible properties of the superfluid component of helium? If there are to be viscous losses in the liquid or if there is to be a resistance to the conduction of heat, the motion of one part of the liquid must be different from that of the rest of the liquid; that is, the quantum mechanical state of some of the atoms must be different from that of others. An energy loss by one atom necessarily means an energy gain by another atom and it is precisely such energy transfers (and changes of quantum mechanical states) that are responsible for viscosity and for resistance to heat conduction.

Now, the helium atom has zero spin and so is a *boson;* bosons do not obey the exclusion principle. Therefore, a collection of helium atoms does not behave at all the way a collection of fermions would behave. The conduction electrons in a metal, for example, are forced, two-by-two, into discrete energy states and thus form the conduction *band* of states. As the temperature of a collection of helium atoms is lowered, the atoms have smaller and smaller energies. Eventually, at a sufficiently low temperature, all the atoms will be in the lowest possible energy state. *All* of the atoms can be in this state; the exclusion principle does not apply to helium atoms and force them into separate states. If all of the atoms are in the same quantum mechanical state, they all have the same wave function. Thus, the atoms in superfluid helium all act together in concert.[12] Consequently, one cannot supply heat to only one region of the superfluid—all atoms are influenced in the same way. Also, since there can be no energy transfers between atoms (because all of the atoms are in the *lowest* energy state), the viscosity is zero.

The explanation of the remarkable properties of superfluid helium is a striking example of the operation of quantum mechanical principles on a macroscopic scale. Although the basic ideas of the theory of superfluid helium have been firmly established, many details of the theory are still being investigated. It is not unlikely that liquid helium still holds some surprises for us.

SUPERCONDUCTORS

When an electric field is applied to a conductor, the free electrons are set into motion and thereby produce a current. Resistance to the flow of current in a metallic crystal is caused in part by collisions of the electrons with impurities or with points of imperfection in the crystal lattice structure. However, even in an ideal crystal that is both pure and perfect, so that these sources of resistance are absent, the electrons cannot flow unimpeded because the thermal vibrations of the atoms provide sites from which the electrons can be scattered and with which they can exchange energy. As the temperature is decreased, the thermal vibrations are lessened and the motion of the electrons is less violently affected. Thus, the resistance to current flow decreases as the temperature decreases. In a pure, ideal crystal the resistance will approach zero only as $T^* \to 0$.

In 1911 Kammerlingh Onnes discovered that lead has the remarkable property that at a temperature of $7.2°K$ the electrical resistance suddenly becomes zero—not just "very small" but zero! At temperatures of $7.2°K$ or lower, lead is a *superconductor*. In one experiment, a current of several hundred amperes was induced to flow in a highly refined sample of lead shaped into a ring, and this current was found to be still flowing, apparently undiminished, after a period of a year! The resistance of superconducting lead has been measured to be at least 10^{11} times smaller than the resistance of normal lead. There is every reason to believe that in pure samples of superconducting materials, the electrical resistance is indeed *zero*. Several elements and many alloys (over 1000 are known) have now been found to be superconductors at low temperatures.

[12] There is no "disorder" in superfluid helium; that is, the entropy is zero.

The phenomenon of superconductivity is similar to that of superfluidity in that it is the result of macroscopic quantum effects. A great stride forward in the understanding of superconductivity was made in 1957 by John Bardeen, Leon Cooper, and Robert Schrieffer, whose theory is now referred to simply as the *BCS theory.* The basic idea of the BCS theory is concerned with the fact that bosons do not obey the exclusion principle. In superconducting materials, the interaction of the conduction electrons with the vibrations of the atoms in the lattice overcomes the repulsive Coulomb force and results in a small net *attraction* between the electrons. Consequently, the electrons tend to group into *pairs* and a pair of electrons, with spins opposite, behaves as a *boson.* The electron pairs are not well localized—indeed, the spacing between pairs is less than the spacing between electrons in any given pair—but nevertheless the pairs are well defined in the quantum mechanical sense and they *do* act as bosons.

The net attraction between the electron pairs—the *pairing energy*—is very small and it does not require much agitation to break the pairs. Therefore, it is only at very low temperatures that the pairs can exist. Because they behave as bosons, the electron pairs all tend to collect in the lowest possible energy state as the temperature is reduced. When the critical temperature is reached ($7.2\,^{\circ}$K for the case of lead), all of the pairs are in the lowest state and all have the same wave function that extends throughout the material. None of the pairs can change its energy state, and therefore the electrons all flow together and there is no dissipation of energy and no electrical resistance.

Certain aspects of superconductors are still not well understood and are currently under investigation, but the crucial point in the explanation of the phenomenon is contained in the BCS theory and we no longer consider superconductivity to be the great mystery that it once was.

Superconducting materials are beginning to be widely used in the construction of magnets for both research and technological applications. Electromagnets that produce strong magnetic fields are expensive to operate because of the substantial losses due to resistance effects in the windings. A conventional electromagnet that produces a field of 10^5 gauss (about the largest field that can be achieved with this type of magnet) may require a megawatt of power to maintain the field. Furthermore, such a magnet requires a cooling system that uses thousands of gallons of water per minute to prevent the windings from melting because of the generation of heat by resistance effects. Magnets are now being used in which the windings are made from various superconducting materials operated at temperatures below the critical temperature. Once the current is established in the windings of such a magnet, it continues to flow without resistance loses. Therefore, the only expenditure of power in a superconducting magnet occurs when the current is first started; in order to maintain the current it is necessary only to insure that the temperature remains below the critical temperature for the material.[13]

---

[13] Of course, no practical superconductor can be absolutely pure and so some energy losses do occur. But only very small amounts of input power are required to maintain the superconducting field.

Courtesy of Argonne National Laboratories

Fig. 14.43 *A superconducting magnet at the Argonne National Laboratory during final phase of construction. The magnet is 18 ft in diameter and contains 25 miles of specially fabricated superconducting ribbon. This magnet will be used in conjunction with a bubble chamber for elementary particle research.*

Metallic alloys and compounds have been found to be more useful than pure elements in the construction of windings for superconducting magnets. A widely used material is Nb_3Sn, which allows the production of fields up to 88 kilogauss.[14] By using V_3Ga, it is expected that fields as large as 500 kilogauss can be achieved. Figure 14.43 shows an extremely large superconducting magnet that is used in elementary particle research at the Argonne National Laboratory. The windings of this magnet are made of Nb_3Ti and a field of 20 kilogauss can be obtained.

If the resistance losses in the transport of electrical power could be eliminated or substantially reduced, enormous savings in cost would be realized. Therefore, the possibility of using superconducting materials for the construction of electrical transmission lines is of great economic importance. Perhaps within the near future we shall begin to replace the huge steel towers that now carry our electrical power with underground superconducting electrical lines.

Summary of Important Ideas

The *ionic binding* of two (or more) atoms to form a molecule results when an electron is transferred from one atom to another so that attractive

---

[14] An upper limit to the field that can be produced by a superconductor exists because the electron pairing can be broken by high fields as well as by high temperatures.

electrostatic forces bind the atoms together. The binding is *strong* when the removal or the addition of only one or two electrons leaves a *closed* electron shell.

The *covalent binding* of atoms to form molecules results when two electrons are shared between the atoms. The binding is strong when the two electrons form a "spin up, spin down" pair. The *exclusion principle* prevents more than two electrons ($m_s = \pm\frac{1}{2}$) from occupying a given energy state. In bulk material, these states are associated with the entire structure rather than with a single atom. Therefore, the atomic energy states are distributed over a certain energy range and become *energy bands*.

In a *conductor*, a portion of the highest energy band that contains electrons is available for electrons that acquire additional energy. (That is, transitions *within* the band are possible; such transitions require very little energy.)

In an *insulator*, the highest energy band that contains electrons is *completely filled*. Therefore, no transitions *within* the band are allowed and excitations are possible only when an electron is carried into the next empty band; such excitations require considerably more energy than excitations in conductors.

In a *semiconductor*, the highest energy band that contains electrons is completely filled, but the energy gap between this band and the next empty band is small thus allowing transitions to occur with thermal energies.

The conductivity of a *semiconductor* can be permanently enhanced by the addition of certain types of impurities (dopants) that produce additional energy states in the gap between the filled band and the conduction band.

In *n*-type semiconductors, the impurities provide for electrical conductions by electrons, and in *p*-type semiconductors, the conduction is by *holes*.

The strange properties of *liquid helium* II are the result of the fact that helium atoms are *bosons* and therefore do not obey the exclusion principle. Therefore, at low temperatures, the atoms all collect in the *same* energy state (the lowest possible) and have the same wave function. The atoms then all move together with no exchanges of energy and thus no energy losses occur.

The phenomenon of *superconductivity* is also the result of *bosons* (*electron pairs*) collecting in the lowest possible energy state and moving together without energy losses.

Questions

14.1 Why are neutral atoms of lithium more chemically active than Li^+ ions?

14.2 Sodium and chlorine combine to form NaCl. Two chlorine atoms combine to form Cl_2. Why does sodium not form an Na_2 molecule?

14.3 Describe the way magnesium and chlorine combine to form a molecule. (Refer to Fig. 13.21 and decide how many electrons magnesium can contribute.)

14.4   Use diagrams similar to those in Fig. 14.8 and show schematically how the N_2 molecule is formed with three covalent bonds and how the O_2 molecule is formed with two covalent bonds.

14.5   Sketch the arrangement of the electrons in molecules of MgO and Na_2S.

14.6   When a hydrocarbon is completely burned (*complete oxidation*), the carbon and hydrogen atoms combine with oxygen gas to form water and carbon dioxide. How many molecules of water and carbon dioxide are formed in the complete oxidation of butane? If the substance is incompletely oxidized, some of the carbon atoms each combine with only one atom of oxygen. Why is this an undesirable effect in the burning of fuels?

14.7   Construct a schematic diagram to show the simplest way in which chlorine can combine with carbon. How does this molecule compare with methane?

14.8   Two carbon atoms can be connected in a molecule by a *triple* covalent bond. Use this fact and draw a diagram of the structure of *acetylene* (C_2H_2). The next member of this group, containing one triple bond, is *ethyl acetylene* (C_3H_4); what is its structure? Find another possible structure for C_3H_4.

14.9   Argue why it is not possible to form molecules consisting of two helium atoms (He_2). (Consider covalent bonding.)

14.10 Use the exclusion principle to show that three hydrogen atoms cannot be bound together by covalent bonding. (However, the H_3^+ *ion* can exist. Why?)

14.11 Xylene (an aromatic solvent) is similar to toluene (Fig. 14.13a) except that there are two methyl (CH_3) groups attached to the benzene ring. Draw a diagram of the structure of xylene in each of its three isomeric forms. (That is, show that there are three possible ways for attaching two methyl groups to the ring.)

14.12 Form a new compound by adding a third benzene ring to naphthalene (Fig. 14.13b). Show that there are only two ways in which this third ring can be added. If the ring is added in line with the first two, the compound is called *anthracene;* its isomer is *phenanthrene.* (The phenanthrene skeletal structure is especially interesting because it is contained in a wide variety of complex natural products, including the opium alkaloids, bile acids, and sex and adrenal cortex hormones.)

14.13 Pentane has the chemical formula C_5H_{12}. Show the structural formula for the three different isomeric forms of pentane. (One of these forms is just the linear molecule.)

14.14 What would happen to the widths of the energy bands for a certain solid if sufficient pressure were applied to the solid to compress the atoms into a volume smaller than the original volume? (Use the uncertainty and exclusion principles.)

14.15 Do you expect the noble gases (in solid crystalline form) to be good electrical conductors? Explain.

14.16 Could an imaginary *superfish* (one that never becomes frozen) manage to swim in liquid helium at 0.5°K? At 2°K? Explain.

14.17 In natural helium, about one atom in 10^6 is the light isotope, He^3, which has spin $\frac{1}{2}$. Do you expect pure liquid He^3 to be a superfluid? Explain.

14.18   Suppose that you have a ring of superconducting material and that the temperature is below the critical temperature. No current is flowing in the ring but the ring is in a certain magnetic field (produced by a conventional electromagnet). How would you start a superconducting current to flowing in the ring?

14.19   Selenium is a semiconductor. What doping agent would you use to convert selenium into a *n*-type semiconductor? Into a *p*-type?

Problems

14.1   The ionization energy of potassium is 4.3 eV and the electron affinity energy of chlorine is 3.7 eV. In KCl the separation between the K^+ and Cl^- ions is approximately 3 Å. What is the molecular binging energy of KCl? (Begin by computing $PE_E = e^2/R$; consider the ions to be point charges.)

14.2*   A diatomic molecule can be pictured schematically as a "dumbell," two massive atoms connected by a weightless rod. Such molecules can execute rotational motion around an axis that is perpendicular to the connecting "rod." If each atom has a mass M and if the separation of the atoms is d, the rotational kinetic energy is $KE = L^2/Md^2$, where L is the angular momentum of the molecule and is quantized according to the rule $L = l\hbar$. In the hydrogen molecule, $d = 0.742$ Å. Calculate the energy in eV of the photon that is emitted when the molecule deexcites from the $l = 2$ state to the $l = 1$ state. Sketch an energy level diagram showing the rotational levels of the hydrogen molecule. Compare these molecular levels with the hydrogen atomic energy levels.

14.3*   Use the fact that the density of KCl is 1.984 g/cm^3 to compute the spacing between the K^+ and Cl^- ions in the crystalline solid. Why is the separation much larger than that between the Na^+ and Cl^- ions in NaCl?

14.4   The length of the side of the unit cube of a diamond crystal is a (see Fig. 14.24). Show that the distance between nearest neighbor carbon atoms in this crystal is $(a/2)\sqrt{3}$. (Use the Pythagorean theorem successively).

14.5   Sulfur crystals are pale yellow and transparent. Sulfur is one of the better insulators. From this information alone, estimate the magnitude of the energy gap between the conduction band and the highest filled band in sulfur crystals.

14.6   The human nervous system (indeed, *any* nervous system) is composed of discrete units called *neurons*. There are approximately 10^{10} neurons in the human central nervous system. These neurons receive, process, and store or pass along the electrical signals that relate to all bodily functions. Electronic circuit models of neurons have been devised in which about 10 transistors together with other circuit elements such as resistors and capacitors perform the basic functions of a single neuron. Use the information in Fig. 14.42 and estimate the size of a computer constructed from integrated circuit modules that has the same capacity as the human nervous system. Would it be feasible to build such a computer? (Make some crude estimates of construction cost and time; assume that all of the components are in the form of IC chips such as those shown in Fig. 14.42 and that, because of mass-production economies, the cost per chip can be reduced—not unreasonably—to $1.)

Tandem Van de Graaff Accelerator

15 *Nuclei*

Although Rutherford had established the nuclear model of the atom in 1911 and by 1919 he had observed a nuclear reaction, the science of *nuclear physics* really dates only from the 1930s. During the period from 1932 (when the neutron was first detected and recognized) to 1939 (when fission was discovered), a rapid series of advances propelled nuclear physics to the forefront of physics research and, within a few years, onto the worldwide political and social scene as well.

The subject of nuclear physics is the study of the properties and the interactions of nuclei. But the results of the discoveries in this field have had and are continuing to have a profound influence on everyday events and on research in many different areas of science. Radioactivity, for example, finds applications in such diverse areas as medicine (where radioisotopes are routinely used for diagnostics and for therapeutics, such as cancer treatments) and manufacturing (where thickness and density measurements of items on production lines are made with nuclear radiations). The use of radioactive isotopes as tracers in biochemical investigations of complicated molecular structures has contributed enormously to the rate of advancement of knowledge in this field.

Nuclear reaction studies have shown how energy is generated in stars and how the elements are produced in a variety of stellar processes. Laboratory experiments have provided nuclear information that is now being used to determine the temperature of the Sun's core. Nuclear reactions are being used in the search for petroleum and for determinations of the mineral content of ores. Legal questions have been answered by using nuclear reactions to analyze items of physical evidence. Even historians have benefited from the use of these methods—it has recently been established by reaction techniques that the cause of Napoleon's death was arsenic poisoning (but no suspects have been identified).

Nuclear reactors, operating on fission reactions, are now producing substantial amounts of commercial electrical power and will provide an increasing fraction of the world's power in the future. But uranium ores, along with our fossil fuels, will eventually be exhausted. However, *fusion* reactors (if they can be developed) may one day solve the world's energy problems by making possible the inexpensive generation of huge amounts of electrical power from nuclear reactions that use the almost inexhaustible supply of deuterium in ocean waters.

Probably the most widely known (and feared) result of nuclear research has been the development of atomic weapons. World politics has been profoundly affected by the existence of these devices. But even weapons of such awesome destructiveness can be employed constructively. The earth-moving that is necessary to build harbors and canals can probably be most economically accomplished by using atomic explosions, and underground firings have actually been used to carve out huge caverns for the storage of gases.

Even though the by-products of nuclear research have long been used in many scientific and technological areas, we are still far from reaching the goal of this research—a comprehensive understanding of nuclear structure and nuclear processes.

In this chapter we draw on and expand our previous discussions of nuclear topics: Section 3.4 (protons and neutrons, nuclear sizes, radioactivity), Section 5.11 (β decay and the neutrino), Sections 6.5 and 8.5 (nuclear forces), and Sections 7.17 and 11.4 (mass and energy).

15.1   *Properties of Nuclei*

NUCLEAR RADII

By 1911 Rutherford's analysis of α-particle scattering experiments (Section 13.1) had shown that nuclear radii are of the order of 10^{-12} cm. But it was not until the neutron had been identified and techniques developed for its use in scattering experiments that more precise values for nuclear radii were obtained. As described in Section 3.4, these experiments showed the regular increase of nuclear radii with increasing mass number (Eq. 3.12):

$$R \cong 1.4A^{1/3} \times 10^{-13} \text{ cm} \tag{15.1}$$

More recently, high energy electrons (with wavelengths that are small compared to nuclear dimensions) have been used to examine in detail the distribution of matter within nuclei. Similar studies have been made by measuring the radiation produced when a muon is captured by an atom and replaces one of the electrons (Section 13.6). Both of these types of experiments (as well as others) have demonstrated that nuclei do not have sharp boundaries (Fig. 15.1a); instead, there is a certain density of nuclear matter at the center of a nucleus, and the density decreases gradually to zero as the radial distance is increased. Because there is no well-defined surface, the nominal "radius" of a nucleus is defined as the distance at which the nuclear matter density has decreased to half its central value (Fig. 15.1b).

NUCLEAR SHAPES

Nuclei are not simply tiny spheres of nuclear matter. Nor do nuclei have the appearance of collections of little balls (representing protons and neutrons) that are held together in some fixed orientation by a nuclear "glue."

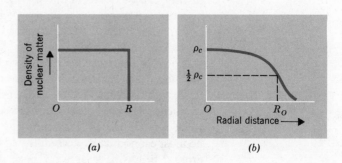

(a)   (b)

Fig. 15.1 *Several types of experiments have demonstrated that nuclei do not have sharp boundaries, as shown in (a). Rather there is a gradual decrease in the density of nuclear matter in the outer regions of nuclei (b). The nuclear "radius" R_0 is usually defined to be the radial distance at which the density of nuclear matter has decreased to half the central value, ρ_c.*

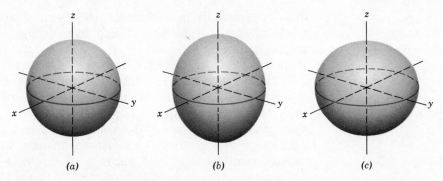

Fig. 15.2 (*a*) *A spherical nucleus.* (*b*) *A prolate ellipsoidal (i.e., football-shaped) nucleus.* (*c*) *An oblate ellipsoidal (i.e., pancake-shaped) nucleus. In each diagram the cross section in the* x-y *plane is a circle.*

Thus, the sketches that we showed in our first discussion of nuclei (Figs. 3.26 and 3.27) are poor representations of nuclei. Just as the dynamics of atomic electrons are properly described in terms of probability density distributions instead of mechanical orbits, nuclear particles must also be treated in quantum rather than in classical terms. Nuclei are just as "fuzzy" as atoms (see Fig. 3.30).

In the same way that we can describe the average positions of electrons in atoms or atoms in molecules, we can also discuss the *shapes* of nuclei. Even in this proper quantum mechanical context, the average matter density distributions for most nuclei are not simple spheres (Fig. 15.2a). In fact, the vast majority of nuclei have *deformed* shapes. The degree of deformation (that is, the amount by which the shape deviates from a sphere) varies from one nuclear species to another. Some nuclei are *prolate* ellipsoids[1] (foot-ball-shaped, Fig. 15.2b) and others are oblate ellipsoids (pancake-shaped, Fig. 15.2c). Even more complicated deformations are found for certain nuclei.

The shape of a nucleus determines the spatial distribution of both matter and charge within the nucleus. A nonspherical charge distribution produces a variety of effects that can be detected experimentally. By measuring nuclear properties that depend on the charge distribution, nuclear shapes can be determined.

NUCLEAR ENERGY STATES

The rules of quantum theory severely limit the energies available to atomic electron systems—only certain discrete energy states are allowed for any particular species of atom. Since the constituents of nuclei also obey the rules of quantum mechanics, a collection of protons and neutrons in a nucleus can exist only in certain discrete energy states that are characteristic of the particular nuclear isotope.

It is to be expected that the nuclear energy states of carbon-12 (C^{12}) are quite different from those of nitrogen-14 (N^{14}), just as the atomic energy

---

[1]The three-dimensional shape produced when an ellipse is rotated around either the major or the minor axis (see Fig. 6.6) is called an *ellipsoid.*

states of carbon are quite different from those of nitrogen. But it is also true that isotopes of the same element have different sets of energy states. The energy states of C^{12} and C^{13}, for example, differ considerably because of the additional interactions of the extra neutron in C^{13}. On the other hand, if we consider two nuclei that differ from one another only by the exchange of a proton for a neutron, we find a remarkable similarity in the nuclear energy states. For example, boron-11 (B^{11}) contains 5 protons and 6 neutrons; if we exchange one of the neutrons for a proton, we will then have a nucleus with 6 protons and 5 neutrons, namely, carbon-11 (C^{11}). The nucleon complements of these nuclei are symmetrical—5–6 and 6–5—and the term *mirror nuclei* is applied to such pairs. Figure 15.3 shows the energy states[2] of B^{11} and C^{11}; it is clear that, apart from slight shifts in the energies, there is an exact correspondence of one set of energy states to the other.

The occurrence of closely similar structures in mirror nuclei is one of the strongest pieces of evidence for believing that there is a single basic nucleon-nucleon force, which is independent of whether the nucleon pair is proton-proton, proton-neutron, or neutron-neutron (see Section 6.5). That is, in B^{11} and C^{11}, the structures of the nuclei depend primarily on the fact that there are 11 nucleons in the nucleus and only secondarily on the fact that the nucleons are grouped as $5p + 6n$ in one case and as $6p + 5n$ in the other.

NUCLEAR TRANSITIONS

When an atom makes a transition from a higher energy state to a lower one, the energy difference is radiated as a photon; typical photon energies

[2]It is customary to measure *nuclear* energies starting with *zero* for the ground state. This convention is different from that used for *atomic* energy states; for example, Fig. 13.9 shows that the unbound state of a proton and an electron corresponds to *zero* energy and that the ground state of the hydrogen atom has an energy of -13.6 eV.

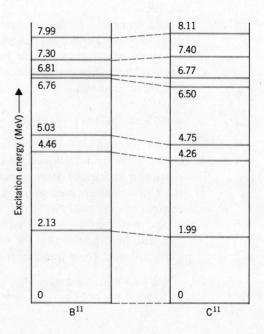

Fig. 15.3 *The energy states of B^{11} and C^{11}. Each state is labeled with its energy in MeV. There is a one-to-one correspondence in the energy states of such mirror nuclei.*

are a few eV. *Nuclear* energy states, on the other hand, have energies in the range ~1–10 MeV (see Fig. 15.3), and when transitions between these energy states take place, extremely energetic photons (called γ *rays*) are emitted.

Figure 15.4 shows the first 5 energy states of B^{10}. The vertical arrows indicate the γ-ray transitions that have been observed; for example, a 0.72-MeV γ ray is emitted when the nucleus makes a transition from the first excited state to the ground state, and a 1.43-MeV γ ray is emitted when a transition from the 3.58-MeV state to the 2.15-MeV state takes place.

15.2 *Nuclear Masses*

NUCLEAR BINDING ENERGIES

At the very small distances that are characteristic of nuclear sizes, the strong nuclear force (Section 6.5) acts between pairs of nucleons to bind these particles into nuclei. An appreciation of the effects of this nuclear interaction can be gained by examining the *binding energies* of nuclei.

Consider, first, the simplest nucleus that depends on the nuclear force for its existence—the *deuteron*, which consists of one proton and one neutron. The mass of the deuteron (m_d) is slightly *less* than the combined masses of a free proton (m_p) and a free neutron (m_n). Using the *atomic* masses[3] of hydrogen (H), deuterium (D), and the neutron (n), we have

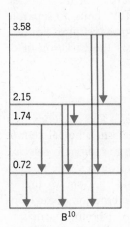

Fig. 15.4 *The energy states of B^{10}. The energy of each state is given in MeV. The vertical arrows show the observed γ-ray transitions.*

$$m_{\mathrm{H}} = 1.007\ 825 \text{ AMU} \qquad m_{\mathrm{D}} = 2.014\ 102 \text{ AMU}$$
$$\underline{m_n = 1.008\ 665 \text{ AMU}}$$
$$m_{\mathrm{H}} + m_n = 2.016\ 490 \text{ AMU}$$

That is,

$$(m_{\mathrm{H}} + m_n) - m_{\mathrm{D}} = 0.002\ 388 \text{ AMU}$$

This result is, in fact, quite general: all nuclei have masses that are *less* than the masses of the constituent protons and neutrons in the free state. The magnitude of this mass difference for a particular nucleus is indicative of the degree of *binding* of the protons and neutrons in that nucleus. According to the Einstein mass-energy relation, a *mass* difference corresponds to an *energy* difference. For the deuteron, this energy difference is

$$\mathcal{E}_b = [(m_{\mathrm{H}} + m_n) - m_{\mathrm{D}}] \times c^2$$
$$= (0.002\ 388 \text{ AMU}) \times (931.481 \text{ MeV/AMU})$$
$$= 2.224 \text{ MeV}$$

where we have used the conversion factor, $c^2 = 931.481$ MeV/AMU, from Example 11.8. This result means that it is necessary to supply 2.224 MeV

[3] Nuclear physics calculations are almost always performed using *atomic* masses instead of *nuclear* masses. The reason is that this procedure allows one to take account automatically of the atomic electrons in the mass-energy equations. In comparing the deuteron mass with the $n + p$ mass, there is no difficulty with atomic electrons because hydrogen and deuterium each have *one* atomic electron and so the electron mass cancels when the mass *difference* is computed. The use of atomic masses in other cases, however, simplifies the calculations (particularly in the case of radioactive β decay). Therefore, we adopt the procedure of *always* using *atomic* masses.

of energy to a deuteron in order to separate the nucleus into a free proton and a free neutron.[4] That is, the deuteron is *bound* by 2.224 MeV, and we refer to $\mathcal{E}_b$ as the *binding energy* of the nucleus. For complex nuclei, the term *binding energy* is used to mean the energy required to decompose the nucleus into free protons and free neutrons. The term *separation energy* means the energy required to remove from the nucleus a proton or a neutron or some specified group of nucleons (such as an α particle). For the deuteron, which consists of only two nucleons, the binding energy and the separation energy are the same.

The nuclear binding energy can also be interpreted in the following way. If, for example, a slowly moving neutron (that is, a neutron with negligible kinetic energy) is captured by a proton to form a deuteron, the initial mass-energy of the system, $(m_H + m_n) c^2$, is greater than the final mass-energy $m_D c^2$, and so the energy difference $\mathcal{E}_b$ is radiated in the form of a γ ray, as shown in Fig. 15.5.

The deuteron has an exceptionally low binding energy; for most nuclei, the binding energy *per nucleon* is approximately 8 MeV. Thus, to separate a nucleus of Ne^{20} into 10 free protons and 10 free neutrons requires approximately $20 \times (8\ \text{MeV}) = 160$ MeV. Figure 15.6 shows the variation of $\mathcal{E}_b$ (in MeV per nucleon) with mass number.

NUCLEAR MASSES AND STABILITY

Most of the nuclei that we find in Nature are *stable*. The nucleons in a stable nucleus are in the lowest possible energy state; such nuclei cannot spontaneously discharge a nucleon nor can they undergo radioactive decay. In order to separate a nucleon or a group of nucleons from a stable nucleus requires the *addition* of energy. For example, Table 15.1 shows that it requires at least 19.8 MeV to separate He^4 into any combination of nucleons but that only 7.55 MeV is required to separate the least bound particle (a proton) from N^{14}. The He^4 nucleus (that is, the α particle) is therefore a particularly stable nucleus. (Compare Fig. 15.6, which shows that the binding energy per nucleon of He^4 is substantially greater than that of any of its neighbors.) On the other hand, the He^5 nucleus is *unstable* (that is, *unbound*):

$$m(He^5) = 5.012\ 297\ \text{AMU}$$

$$\begin{aligned} m(He^4) &= 4.002\ 603\ \text{AMU} \\ m_n &= 1.008\ 665\ \text{AMU} \\ \hline m(He^4) + m_n &= 5.011\ 268\ \text{AMU} \end{aligned}$$

[4] If more than 2.224 MeV is supplied, the excess will appear in the form of kinetic energy of the proton and neutron.

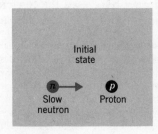
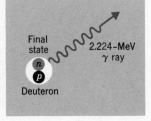

Fig. 15.5 *Schematic representation of the capture of a slow neutron (i.e., a neutron with negligible kinetic energy) by a proton to form a deuteron. The deuteron binding energy is radiated in the form of a γ ray.*

Table **15.1** *Separation and Binding Energies for He^4 and N^{14}*

	Particle Group	Separation Energy (MeV)
He$^4 \longrightarrow$	H^3 + H^1	19.814
	He^3 + n	20.578
	H^2 + H^2	23.847

Total binding energy ($\longrightarrow$ 2H^1 + 2n) = 28.296 MeV

Binding energy per nucleon = $\dfrac{28.296}{4}$ = 7.074 MeV

	Particle Group	Separation Energy (MeV)
	C^{13} + H^1	7.550
	C^{12} + H^2	10.272
	N^{13} + n	10.553
N$^{14} \longrightarrow$	B^{10} + He^4	11.613
	B^{11} + He^3	20.736
	C^{11} + H^3	22.736

Total binding energy ($\longrightarrow$ 7H^1 + 7n) = 104.659 MeV

Binding energy per nucleon = $\dfrac{104.659}{14}$ = 7.475 MeV

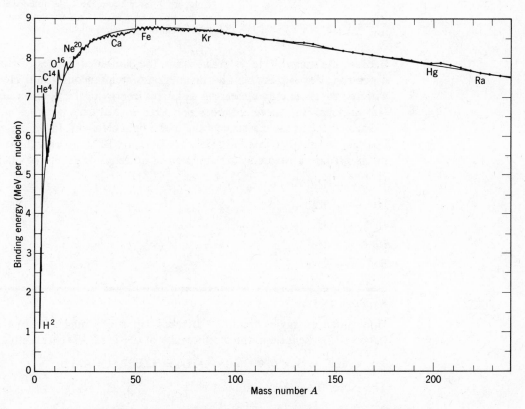

Fig. 15.6 *The binding energy per nucleon as a function of mass number. Some representative measured binding energies are shown as dots in the figure. The point for He^4 is far above the curve that represents the average behavior of nuclei because the He^4 nucleus (the α particle) is an exceptionally tightly bound group of nucleons.*

Table **15.2**   *Atomic Masses of Some Light Elements ($Z \leq 4$)*[a]

Nucleus	Z	Mass (AMU)	
n	0	1.008 665	
H^1	1	1.007 825	(stable)
H^2 or D^2 (deuterium)	1	2.014 102	(stable)
H^3 or T^3 (tritium)	1	3.016 050	
He^3	2	3.016 030	(stable)
He^4	2	4.002 603	(stable)
He^5	2	5.012 297	
He^6	2	6.018 893	
He^7	2	7.028 031	
Li^5	3	5.012 538	
Li^6	3	6.015 125	(stable)
Li^7	3	7.016 004	(stable)
Li^8	3	8.022 487	
Be^6	4	6.019 717	
Be^7	4	7.016 929	
Be^8	4	8.005 308	
Be^9	4	9.012 186	(stable)
Be^{10}	4	10.013 534	

[a]Compare Table 3.2 and the table inside the front cover that give the masses of the naturally occurring isotopic mixtures.

Because the mass of He^5 is *greater* than the combined masses of He^4 and a neutron, He^5 is unstable and disintegrates spontaneously into He^4 + n, with the excess energy appearing as kinetic energy of the neutron and the He^4 nucleus. He^5, therefore, does not occur in Nature.

Table 15.2 lists the masses of some of the light elements. From these data, it is easy to calculate that He^5, He^7, Li^5, Be^6, and Be^8 (and *only* these nuclei in the list) are unstable to the emission of nucleons or groups of nucleons:

$$\text{He}^5 \longrightarrow \text{He}^4 + n$$
$$\text{He}^7 \longrightarrow \text{He}^6 + n$$
$$\text{Li}^5 \longrightarrow \text{He}^4 + p$$
$$\text{Be}^6 \longrightarrow \text{He}^4 + 2p$$
$$\text{Be}^8 \longrightarrow 2\text{He}^4$$

Example **15.1**

How much energy is required to break up a C^{12} nucleus into three α particles? By definition, the atomic mass of C^{12} is 12 AMU (exactly). And,

$$3 \times m(\text{He}^4) = 3 \times (4.002\ 603\ \text{AMU}) = 12.007\ 809\ \text{AMU}$$

Therefore, the energy required is

$$\mathcal{E} = [3 \times m(\text{He}^4) - m(\text{C}^{12})] \times c^2$$
$$= (12.007\ 809\ \text{AMU} - 12\ \text{AMU}) \times c^2$$
$$= (0.007\ 809\ \text{AMU}) \times (931.481\ \text{MeV/AMU})$$
$$= 7.274\ \text{MeV}$$

Conversely, we can conclude that when 3 α particles combine to form a C^{12} nucleus, 7.274 MeV of energy is released. Just such a process is, in fact, important in the formation of carbon in stars (see Section 17.1).

THE LIMITS OF STABILITY

The decrease in binding energy with increasing mass number for heavy elements and the termination of nuclear stability near $A = 210$ are consequences of two effects: the *short-range* character of the nuclear force and the fact that protons and neutrons separately obey the *exclusion principle*. Because the nuclear force has a short range, there are attractive forces only between a given nucleon and its immediate neighbors. Since the Coulomb force, on the other hand, has a long range, there is an electrostatic repulsive force between a given proton and *all* of the other protons in a nucleus. For sufficiently large atomic numbers, the Coulomb repulsion dominates the attractive effects of the nuclear force. Thus, there must come a point at which it is no longer possible to form a stable nucleus by the addition of more protons. In fact, there are no *stable* nuclei with $Z > 83$.

What is the effect of the exclusion principle? If a pair of protons (spin *up*, spin *down*) occupies a particular energy state within a nucleus, an added proton (because of the exclusion principle) must go into a different energy state, usually with *greater* energy. As more protons are added, they are forced into higher and higher energy states. Eventually, the energy of one of these states will exceed the separation energy and the nucleus will be unstable. Thus, the exclusion principle, as well as the Coulomb repulsion, acts to limit the number of protons in nuclei.

The exclusion principle governs the energy states that can be occupied by neutrons in the same way that it affects protons. As neutrons are added to a nucleus, they go two-by-two into higher and higher energy states, and eventually the nucleus reaches a point of instability.

The restrictions imposed by the short-range nuclear force and the exclusion principle mean that arbitrarily large stable nuclei cannot exist; in fact, *all* nuclei with $A > 210$ (and many with smaller mass numbers) are *unstable*. These instabilities take several forms; we shall discuss these in turn, beginning with β radioactivity in the following section.

15.3   *Radioactivity*

β RADIOACTIVITY

Within a nucleus, a neutron is a stable particle just as is a proton. But Table 15.2 shows that the mass of a *free* neutron is greater than that of a hydrogen atom (that is, a proton and an electron). Therefore, it is energetically possible for a free neutron to separate into a proton and an electron, and, in fact, neutrons do undergo this type of decay process with a half-life of 12.8 min.[5] The available decay energy in the disintegration

---

[5] See Section 3.4 for the definition of *half-life* and for a qualitative discussion of radioactivity.

Table **15.3**   *Properties of Nucleons*

Property	Proton	Neutron
Mass, m	$m_p = 1.007\ 276$ AMU[a]	$m_n = 1.008\ 665$ AMU
Rest-mass energy, mc^2	938.259 MeV	939.553 MeV
Charge, q	$e = 4.803 \times 10^{-10}$ statC	0
Spin, s	$\frac{1}{2}$	$\frac{1}{2}$
Half-life, $\tau_{1/2}$	∞ (stable)	12.8 min
Decay mode	none	$n \longrightarrow p + e + \bar{\nu}_e$[b]

[a] This is the mass of the *proton;* the mass of the hydrogen *atom* (see Table 15.2) is $m_H = 1.007\ 825$ AMU.

[b] The meaning of the various symbols for neutrinos is explained in Section 16.2.

of the neutron is

$$\mathcal{E}_{np} = [m_n - (m_p + m_e)] \times c^2$$
$$= (m_n - m_H) \times c^2$$
$$= (1.008\ 665\ \text{AMU} - 1.007\ 825\ \text{AMU}) \times (931.481\ \text{MeV/AMU})$$
$$= 0.782\ \text{MeV} \tag{15.2}$$

The conversion of a neutron into a proton and an electron is the prototype of a class of nuclear disintegration processes called β *radioactivity* or β *decay.* Many nuclei (some of which occur naturally and others which must be produced artificially in the laboratory) are known to undergo β decay by the emission of electrons in exactly the same way that free neutrons decay. Actually, in all such decay processes, *neutrinos*[6] as well as electrons are emitted (Section 5.11). The neutrinos share the available decay energy (and momentum) with the electron and the residual nucleus. Therefore, in the decay of the neutron, the electron does not carry away an energy of 0.782 MeV; rather, this energy is shared among the proton, the electron, and the neutrino. As a result, the electrons emitted in any type of β-decay process do not have a unique energy; instead, there is a *distribution* of electron energies, ranging from zero to the maximum allowed by the available decay energy. The measured electron energy *spectrum* for the β decay of Bi^{211} is shown in Fig. 15.7.

All nuclear radioactive decay processes in which an electron is emitted can be considered to be the result of the β decay of a neutron *within* the parent nucleus. For example, consider the case of Li^8 (Fig. 15.8). The Li^8 nucleus consists of 3 protons and 5 neutrons and undergoes β decay with a half-life of 0.85 sec. The Li^8 decay process transforms one of the 5 neutrons into a proton so that the new nucleus has 4 protons and 4 neutrons—this new nucleus is Be^8. (The β decay of a Li^8 nucleus is shown in Fig. 5.11; the resulting Be^8 nucleus breaks up into two α particles.)

Nuclear β decay leads to an increase in the nuclear charge by one unit $(Z \rightarrow Z + 1)$, but leaves unchanged the mass number A, since the total number of nucleons remains the same. The β decay of a nucleus with atomic number Z and mass number A is energetically possible only if the atom

[6] But see Section 16.2 where we shall find that the massless entities accompanying electron emission in β decay are called *anti*neutrinos $(\bar{\nu}_e)$.

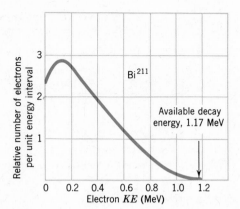

Fig. 15.7 *The measured electron energy spectrum for the β decay of Bi²¹¹. The electron energy is limited by the available decay energy, 1.17 MeV.*

with atomic number $Z + 1$ and mass number A (the *daughter* atom) has a mass smaller than that of the original (or *parent*) atom. That is, for β decay to occur, we must have

$$m(Z, A) > m(Z + 1, A)$$

If this relation is satisfied, a neutron in the nucleus (Z, A) can, in fact, transform into a proton (plus an electron and a neutrino) so that the nucleus $(Z + 1, A)$ results. The *decay energy* in such a case is the mass difference Δm multiplied by c^2 (compare Eq. 15.2):

$$\mathcal{E}_\beta = \Delta m \times c^2 = [m(Z, A) - m(Z + 1, A)] \times c^2 \tag{15.3}$$

Notice that because we use *atomic* masses here, it is unnecessary to include the electron masses in the expression for Δm. The final mass of the system consists of the mass of the *nucleus* $(Z + 1, A)$ plus the mass of the Z electrons that were the atomic electrons of the original atom (Z, A) and the mass of the ejected electron. Altogether, these masses just total to the mass of the *atom* $(Z + 1, A)$.

Typical decay energies for β-radioactive nuclei are in the range 0.01–3 MeV (see Table 15.4). Except for the small amount of kinetic energy received by the recoiling daughter nucleus when the electron and neutrino are emitted, the decay energy $\mathcal{E}_\beta$ corresponds to the *maximum* kinetic energy that the electron can possess. For example, the maximum energy permitted electrons in the β decay of Bi²¹¹ is 1.17 MeV, as indicated in Fig. 15.7. The

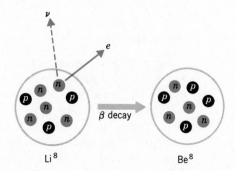

Fig. 15.8 *Schematic representation of the β decay of Li⁸. One of the Li⁸ neutrons is transformed into a proton and the new nucleus is Be⁸.*

Table **15.4**   *Some Important β-Radioactive Nuclei*

Nucleus	Half-Life	Type of Decay	Maximum Electron Kinetic Energy (MeV)	Remarks
H^3 (tritium)	12.26 yr	β$^-$	0.0186	Used in nuclear fusion devices (H-bombs)
C^{14}	5730 yr	β$^-$	0.156	Used in archeological dating; also an important tracer in biochemical studies
Na^{22}	2.60 yr	β$^+$	0.54	Useful source of positrons
Na^{24}	15.0 hr	β$^-$	1.39	Used in medical diagnostics to follow the flow of sodium in the body
K^{40}	1.3 × 10^9 yr	β$^-$	0.0118	Used in archeological dating
Co^{60}	5.24 yr	β$^-$	0.31	Accompanying γ rays used in medical therapeutics and for radiation processing of plastics, food, etc.
Sr^{90}	28.8 yr	β$^-$	0.54	Important fission product (occurs in fallout from detonation of fission bombs)
I^{131}	8.05 days	β$^-$	0.61	Used in medical diagnostics and therapeutics, particularly in thyroid ailments

average electron kinetic energy is approximately one-third the maximum energy.

Not all β-decay processes leave the residual nucleus in its ground state. In many instances, an excited state of the daughter nucleus is formed and one or more γ rays are emitted before the nucleus is left in its ground state.

POSITRON DECAY

Figure 15.9 shows the relative masses of the various nuclei with $A = 65$. For this mass number, only Cu^{65} is stable; Co^{65} has a mass greater than that of Ni^{65} and, therefore, undergoes β decay, forming Ni^{65}; furthermore, Ni^{65} has a mass greater than that of Cu^{65} and undergoes β decay, forming stable Cu^{65}. That is, two successive β decays transform a nucleus of Co^{65} into Cu^{65}. (The simultaneous emission of two electrons, which would transform Co^{65} directly into Cu^{65}, does not occur with a measurable rate.)

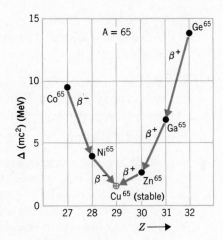

Fig. 15.9 *Nuclear β^- decay and positron (β^+) decay routes for nuclei with A = 65. There is a single stable nucleus with this mass number (Cu^{65}).*

The $A = 65$ chart also shows that there are nuclei with $Z > 29$ that are more massive than Cu^{65}. These nuclei cannot undergo normal β decay because each has a charge *greater* than that of the adjacent, less massive nucleus. A new type of β decay process is possible for these nuclei which allows them to rid themselves of one unit of the excessive positive charge that they carry. Nuclei such as Ge^{65}, Ga^{65}, and Zn^{65} emit particles that are identical to electrons in every respect except that they have a *positive* charge. These particles are called *positrons* (β^+) and the decay process is called *positron decay*. We shall learn more about these interesting particles when we discuss the properties of elementary particles in the following chapter. For now it suffices to note that there are two complementary nuclear β-decay processes, which we label β^- decay and β^+ decay; these processes can be represented in the following way:

$$\beta^- \text{ decay:} \quad n \longrightarrow p + e^- + \bar{\nu}_e \tag{15.4a}$$

$$\beta^+ \text{ decay:} \quad p \longrightarrow n + e^+ + \nu_e \tag{15.4b}$$

We understand, of course, that these processes take place *within* nuclei; the electron (or positron) and the neutrino are *created* at the instant of disintegration (they do *not* preexist in the nucleus) and are immediately ejected from the nucleus.

Example **15.2**

What is the expression that corresponds to Eq. 15.3 for the case of *positron* decay?

The positron decay of a nucleus (Z, A) forms the nucleus $(Z - 1, A)$. Originally, we have a nucleus plus Z atomic electrons with a total (that is, *atomic*) mass $m(Z, A)$. In the decay process, a positron is emitted and a nucleus with atomic number $Z - 1$ is formed. Since the nuclear charge has decreased by one unit, one of the original Z atomic electrons is superfluous and is shed. Therefore, the final system consists of a nucleus $(Z - 1, A)$ plus $Z - 1$ atomic electrons, together with the emitted positron and the excess electron. The total mass of the final system is $m(Z - 1, A) + m(\beta^+) + m(e^-)$, or $m(Z - 1, A) + 2m_e$. Hence, the total available energy for positron decay is

$$\mathcal{E}_{\beta^+} = \{m(Z, A) - [m(Z - 1, A) + 2m_e]\} \times c^2$$
$$= [m(Z, A) - m(Z - 1, A)] \times c^2 - 2m_e c^2$$
$$= \Delta m \times c^2 - 2m_e c^2$$

and the maximum positron kinetic energy is $2m_e c^2 = 1.02$ MeV *less* than the mass-energy difference between the parent and daughter *atoms*.

ELECTRON CAPTURE

A third type of electron decay process takes place in certain nuclei. Consider the case of Be[7]. Table 15.2 shows that the mass difference between Be[7] and Li[7] is $\Delta m = m(\text{Be}^7) - m(\text{Li}^7) = 0.000\ 925$ amu, so that the available decay energy is 0.862 MeV. The atomic number of Be[7] is 4, whereas Li[7] has $Z = 3$. We would therefore expect Be[7] to decay to Li[7] by positron emission, but the available decay energy (0.862 MeV) is less than that required (1.02 MeV) for this mode of decay. Consequently, as an energetically allowed alternative to β^+ decay. Be[7] captures into the nucleus one of the atomic electrons (and simultaneously emits a neutrino). This process is called *electron capture* (e.c.) and is represented in the following way:

$$\text{e.c.:} \quad p + e^- \longrightarrow n + \nu_e \tag{15.4c}$$

That is, a nuclear proton captures an atomic electron and is transformed into a neutron, thus decreasing the nuclear charge by one unit.

Electron capture is allowed if $m(Z + 1, A) > m(Z, A)$. Therefore, if β^+ decay of the nucleus $(Z + 1, A)$ is allowed, electron capture is always possible, but in such situations β^+ decay will usually be much more probable than electron capture.

RADIOACTIVE DECAY RATES

In Section 3.4 it was pointed out that the number of radioactive atoms in a sample will decay to one half the original number in a time (the *half-life*) that is characteristic of the particular material. The half-life of a radioactive substance is a *nuclear* property and is not influenced in any way by the physical condition of the bulk material (such as the temperature or the pressure or the crystalline form).[7]

The properties of any nucleus depend on the quantum mechanical wave function for that particular nucleus. If the nucleus is subject to radioactive decay, the wave function will specify the probability *per unit time* that the nucleus will undergo decay. The wave function does not depend on the time—it has the same value at one instant as it does at any other instant (as long as the decay has not taken place). Therefore, regardless of the time at which we ask the question, "What is the probability that a given nucleus will decay during the time interval τ, starting *now*?", the answer is always

---

[7] An exception to this statement is the case of *electron capture*. Clearly, if a sample of Be[7], for example, is heated until the atoms are completely ionized (as would be the case in a star), the ability of the nucleus to capture an electron would be lessened because of the decreased availability of electrons, and the half-life for the decay process would be correspondingly increased.

the same. In particular, if the time interval is the half-life $\tau_{1/2}$, the probability for decay is exactly *one-half*.

If we start with a sample of N identical radioactive nuclei, for each and every nucleus in the sample the probability is $\frac{1}{2}$ that decay will take place during the next time interval $\tau_{1/2}$. Therefore, if N is a large number,[8] at the end of the time interval, $\frac{1}{2}N$ nuclei will have decayed and $\frac{1}{2}N$ will remain. At the end of the next identical time interval, one half of the nuclei that existed at the beginning of the time interval will have decayed (that is, $\frac{1}{2} \times \frac{1}{2}N = \frac{1}{4}N$) and $\frac{1}{4}N$ nuclei will remain. The process continues with one half of the remaining sample decaying in every period $\tau_{1/2}$. The resulting *decay curve*, which shows the amount of the original sample that remains after any length of time, is shown in Fig. 3.32 for the case of radium (Ra^{226}).

α DECAY

Certain unstable nuclei, primarily those with mass numbers above 200, spontaneously emit helium nuclei (α particles). The emission of an α particle by a nucleus decreases the original nuclear charge by 2 units and decreases the original mass number by 4 units. If a nucleus identified by (Z, A) has a mass greater than the sum of the masses of the nucleus $(Z - 2, A - 4)$ and a He^4 nucleus, the nucleus (Z, A) is unstable and can decay by the emission of an α particle. The available energy for the α-decay process is

$$\mathcal{E}_\alpha = [m(Z, A) - m(Z - 2, A - 4) - m(\text{He}^4)] \times c^2 \tag{15.5}$$

Example **15.3**

What is the available energy for the α decay of Po^{210}?

Alpha-particle emission from Po^{210} leaves Pb^{206}. Therefore, the pertinent masses are:

$m(\text{Po}^{210}) = 209.98287$ AMU

$$\begin{aligned} m(\text{Pb}^{206}) &= 205.97447 \text{ AMU} \\ m(\text{He}^4) &= 4.00260 \text{ AMU} \\ \hline m(\text{Pb}^{206}) + m(\text{He}^4) &= 209.97707 \text{ AMU} \end{aligned}$$

Hence,

$$\begin{aligned} \mathcal{E}_\alpha &= (209.98287 \text{ AMU} - 209.97707 \text{ AMU}) \times c^2 \\ &= (0.00580 \text{ AMU}) \times (931.481 \text{ MeV/AMU}) \\ &= 5.40 \text{ MeV} \end{aligned}$$

Actually, this decay energy is shared by the α particle and the Pb^{206} nucleus (because the linear momenta of the two fragments must be equal and opposite). Consequently, the α particle emitted by Po^{210} has a kinetic energy of 5.30 MeV and the recoil Pb^{206} nucleus has a kinetic energy of 0.10 MeV. (See Problem 15.15.)

Essentially all nuclei with $A \gtrsim 100$ are unstable with respect to breakup by the emission of α particles, but it is only for nuclei with $A \gtrsim 200$ that

[8] Because we are dealing with *probabilities*, if N is a small number, the fluctuations can be appreciable compared to the size of the sample.

α decay is an important process. The heavier nuclei have α-decay half-lives sufficiently short that α-particle emission from a given sample occurs at a rate that permits observation, but the half-lives of the lighter nuclei tend to be so long that the α-decay process is unmeasurable, even though it is energetically allowed.

The reason for the lack of detectable α decay on the part of most nuclei with $A \lesssim 200$ can be understood in the following way. Two protons and two neutrons that cluster together within a nucleus to form an α particle find themselves in a *potential well* that is due to the attractive force exerted on them by all of the other nucleons in the nucleus. Because of the short-range nature of the nuclear force, the radial extent of the potential well is only $\sim 10^{-12}$ cm (that is, the nuclear radius R). *Within* the nucleus, the attractive nuclear force completely overwhelms the repulsive Coulomb force between the α particle and the other $Z - 2$ protons in the nucleus. At distances greater than R, however, the nuclear force vanishes, but the Coulomb force is still effective. Therefore, α particles find themselves in a potential well which consists of two parts—a *nuclear* potential (which is *negative* because the force is *attractive*) for distance $r < R$ and a *Coulomb* potential (which is *positive* because the force is *repulsive*) for distances $r > R$. This overall potential well is shown schematically in Fig. 15.10.

If the energy of an α particle in the potential well is *negative* (that is, if the total α-particle energy $\mathcal{E}_\alpha$ is less than it would be at an infinite distance from the nucleus, where $PE = 0$), the α particle is *bound* and cannot escape from the nucleus. This is the situation illustrated in Fig. 15.10a; the α-particle separation energy is $\mathcal{E}_b$. Figure 15.10b, on the other hand, shows a case in which the total α-particle energy is *positive*. If this α particle were removed an infinite distance from the nucleus, it would have a positive kinetic energy, $KE_\alpha = \mathcal{E}_\alpha$. But the α particle is prevented from escaping to infinity by the height of the Coulomb potential barrier $\mathcal{E}_C$; that is, $\mathcal{E}_\alpha$ must be greater than

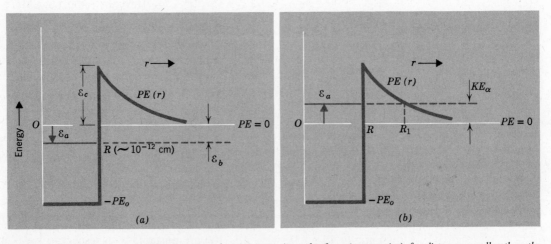

Fig. 15.10 *The nuclear potential for an α particle is negative (i.e., the force is attractive) for distances smaller than the nucleus radius R. At larger distances, the positive Coulomb potential is effective. In (a) the total energy of the α particle $\mathcal{E}_\alpha$ is negative and the particle is bound. In (b) the total energy is positive and the α particle can escape the nucleus but only by "tunneling" through the potential barrier in the region $R < r < R_1$.*

$\mathcal{E}_C$ in order for the α particle to escape. At least, this is the *classical* view of the situation. According to the *quantum mechanical* view, however, an α particle with a positive total energy *can* escape from the nucleus even though $\mathcal{E}_\alpha < \mathcal{E}_C$. The reason is that the wave function for an α particle in a finite potential well "leaks" outside the well (compare Fig. 12.24). That is, the α-particle wave function has a small, but non-zero value for distances $r > R_1$. This means that there is a small, but non-zero probability that the α particle will actually be found *outside* the nucleus. (Remember that the probability for finding a particle at a particular point is proportional to the square of the wave function at that point.) Once in this exterior region, the α particle proceeds away from the nucleus, and when outside the atom and removed from the Coulomb potential, the α-particle kinetic energy is KE_α. The emission of an α particle from a nucleus is therefore possible by virtue of the "tunneling" of the α particle through the potential barrier; this effect is due entirely to the wave character of the α particle.

Example **15.4**

What is the height of the Coulomb barrier for an α particle and a Pb^{206} nucleus?

The Coulomb barrier $\mathcal{E}_C$ is just the electrostatic potential energy between the two nuclei when the distance between their centers is equal to the sum of their radii (that is, when they are just in "contact"). The radii are found by using Eq. 15.1:

$$R_{Pb} \cong 1.4 \times (206)^{1/3} \times 10^{-13} \text{ cm} \qquad R_\alpha \cong 1.4 \times (4)^{1/3} \times 10^{-13} \text{ cm}$$
$$\cong 1.4 \times 5.91 \times 10^{-13} \text{ cm} \qquad\qquad \cong 1.4 \times 1.59 \times 10^{-13} \text{ cm}$$
$$\cong 8.2 \times 10^{-13} \text{ cm} \qquad\qquad\qquad \cong 2.2 \times 10^{-13} \text{ cm}$$

Therefore,

$$\mathcal{E}_C = \frac{(Ze)_{Pb} \times (Ze)_\alpha}{R_{Pb} + R_\alpha} \cong \frac{82e \times 2e}{(8.2 \times 10^{-13} \text{ cm}) + (2.2 \times 10^{-13} \text{ cm})}$$
$$\cong \frac{164e^2}{10.4 \times 10^{-13} \text{ cm}}$$

Now, the square of the electronic charge can be expressed as $e^2 = 1.44 \times 10^{-13}$ MeV-cm (see table inside front cover). Thus,

$$\mathcal{E}_C \cong \frac{164 \times (1.44 \times 10^{-13} \text{ MeV-cm})}{10.4 \times 10^{-13} \text{ cm}} \cong 22.7 \text{ MeV}$$

This energy is large compared to the kinetic energy of α particles emitted from radioactive nuclei.

The α particle is such a tightly bound object that the mass difference between a nucleus (Z, A) and the nucleus $(Z - 2, A - 4)$ results in a positive value for $\mathcal{E}_\alpha$ when $A \gtrsim 100$. However, the probability that an α particle can tunnel through the potential barrier depends strongly on $\mathcal{E}_\alpha$. When $\mathcal{E}_\alpha$ is large, the width of the barrier that must be tunneled through is small and the liklihood of finding the α particle outside the nucleus is large; in other words, the half-life for α-particle emission is short. Conversely, if $\mathcal{E}_\alpha$ is small, the barrier width is large, and the half-life is very long. For nuclei with $A \sim 100 - 200$, $\mathcal{E}_\alpha$ is too small to produce a half-life sufficiently short to make α decay observable.[9] Only for $A \gtrsim 200$ are the values of $\mathcal{E}_\alpha$ large enough to render α-decay processes measurable.

The strong dependence of the half-life on $\mathcal{E}_\alpha$ is shown in Fig. 15.11 for a series of thorium isotopes. The decrease in $\mathcal{E}_\alpha$ from 6.3 MeV to 4.0 MeV increases $\tau_{1/2}$ by a factor of 10^{14}.

It is significant to note that these heavy nuclei emit only α particles, not protons, or neutrons, or deuterons, or other groups of small numbers of nucleons. Only the α particle has a sufficiently large binding energy to make its total energy in a nucleus *positive*. The addition of ~ 7 MeV of energy is required to remove a nucleon from a nucleus with $A \sim 200$ and the addition of ~ 10 MeV is necessary to remove a deuteron from such a nucleus. Apart from fission into two fragments of roughly equal mass, α-particle decay is the only energetically possible process that results in the spontaneous emission of nucleons from heavy nuclei.

RADIOACTIVE DECAY CHAINS

The heaviest elements found in Nature are uranium (U, $Z = 92$), protactinium (Pa, $Z = 91$), and thorium (Th, $Z = 90$). All of the isotopes of these elements are radioactive but each element has at least one isotope with a sufficiently long half-life that the element still exists in Nature. For example, U^{238} has $\tau_{1/2} = 1.4 \times 10^9$ years. When these nuclei decay, they form new daughter elements that are also radioactive. Some of these nuclei are β radioactive and others emit α particles. A few can even decay by either

[9]There are a few nuclei in this range of mass numbers whose particular structures make $\mathcal{E}_\alpha$ large enough to produce measurable α decays; the half-lives of these nuclei are all extremely long—10^{11} to 10^{15} years. (See Table 15.5.)

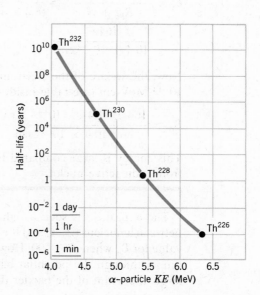

Fig. 15.11 *The variation of α-decay half-life with α-particle energy for a series of thorium isotopes. Note that small energy changes give extremely large changes in the half-life. Other series of isotopes exhibit curves of similar shape, displaced slightly from each other.*

Table **15.5** *Some Typical α-Radioactive Nuclei*

Nucleus	Half-Life	α-particle Kinetic Energy (MeV)[a]	Remarks
Ce[142]	5×10^{15} yr	1.5	Lightest naturally occurring α-radioactive nucleus
Po[210]	138 days	5.30	Much used source of α particles
Bi[214]	19.7 min	5.51	Also undergoes β decay
Po[218]	3.05 min	6.00	Formed by two successive α decays starting with Ra[226]; also known as radium-A; used in original Rutherford scattering experiment
Ra[226]	1620 yr	4.78	α particles from this source first identified as helium nuclei (Rutherford)

[a] The *decay energy* is slightly higher (see Example 15.3 and Problem 15.15).

α or β emission. A series of successive radioactive decays takes place that continues until a stable isotope of either lead (Pb, $Z = 82$) or bismuth (Bi, $Z = 83$) is formed. The stable isotopes of lead are Pb[206], Pb[207], and Pb[208]; only Bi[209] is stable. These four nuclei are the termination points for all of the radioactive decay chains that originate with the long-lived heavy elements. One such decay chain begins with U[238] and ends with Pb[206]; this series of α and β decays is shown in Fig. 15.12.

THE VALLEY OF STABILITY

A convenient way to represent a number of nuclear properties is by the type of chart shown in Fig. 15.13. In this chart, every nucleus is assigned a position according to the number of protons Z and the number of neutrons, $N = A - Z$, in the nucleus. For stable nuclei (the shaded boxes), the naturally occurring isotopic abundance is given, and for unstable nuclei, the mode of decay and the half-life is given. Nuclei that decay by particle emission (He[5], Li[5], Be[6], Be[8], and B[9], among those shown in the chart), have half-lives less than $\sim 10^{-16}$ sec and these half-lives are not shown. In such a chart, it is easy to trace the radioactive history of a particular nuclear species by following the arrows that indicate the decay modes. For example, He[8] β decays to Li[8] which, in turn, β decays to Be[8] which, finally, breaks up into two α particles.

The nuclei H[4], H[5], He[2], Li[4], etc., do not exist as configurations of nucleons that remain together for a sufficient time to be identified and therefore these nuclei are not shown in the chart.

A complete chart of the known nuclei would contain more than 1600

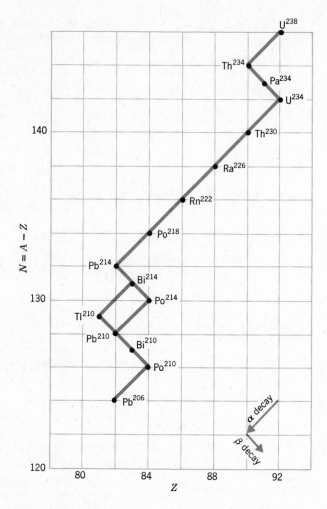

Fig. 15.12 *The radioactive decay chain that originates with U^{238} and ends with Pb^{206}. Notice that Bi^{214} can undergo either α or β decay and that after one additional decay, both branches lead to Pb^{210}. Each Bi^{214} nucleus has a certain probability for decay by β-emission and a certain probability for decay by α-particle emission.*

entries; some 330 nuclear isotopes have been found to occur naturally in the Earth (about 260 of which are stable) and almost 1300 have been produced artificially in the laboratory. A complete but schematic chart of the nuclei known at present is shown in Fig. 15.14.

A number of interesting points appear on examination of the systematics of nuclear properties when the nuclei are arranged in this fashion:

1. The stable nuclei up to $Z \cong 20$ have approximately equal numbers of protons and neutrons ($Z \cong N$).

2. For $Z \gtrsim 20$, the stable nuclei tend to have an increasing preponderance of neutrons over protons; for example, uranium has $N/Z \cong 1.6$. The reason for this effect is easy to understand when we recall that the nuclear force has a *short* range whereas the Coulomb force has a *long* range. As more nucleons are added to form heavier nuclei, the average distance between nucleons becomes greater. Therefore, the long-range Coulomb repulsion becomes more effective relative to the short-range nuclear attraction and it becomes more and more difficult to add protons to a nucleus. For this reason it becomes energetically more favorable to add neutrons to heavy nuclei; consequently, N/Z increases with Z.

3. The stable nuclei are located along a narrow band of the $N - Z$ diagram, called the *valley of stability*. Along the lines of constant A, the nuclei on either side of the valley have larger masses and undergo radioactive decay (by either β^- or β^+ emission) in order to reach a stable condition. (Of course, for the heavier nuclei, α decay is possible, and some nuclei undergo spontaneous fission in an effort to reach stability.)

4. Nuclei *above* the valley of stability are *neutron rich* and undergo β^- decay. An increase in the distance from the stable valley along a line of constant A causes: (a) the nuclei to become more massive (that is, less stable), (b) the β decay energy $\mathcal{E}_\beta$ to increase, and (c) the half-life to decrease. Sufficiently far from the valley, the neutron excess becomes very large and the instability increases to such a degree that the β^- decay process, which

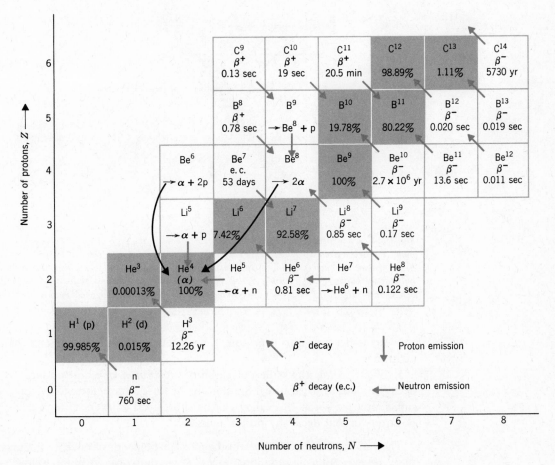

Fig. 15.13 *A portion of the chart of nuclei showing the light elements. For stable nuclei (shaded boxes), the naturally occurring isotopic abundance is given; for particle unstable nuclei, the mode of decay is given; and for β radioactive nuclei, the half-life is given.*

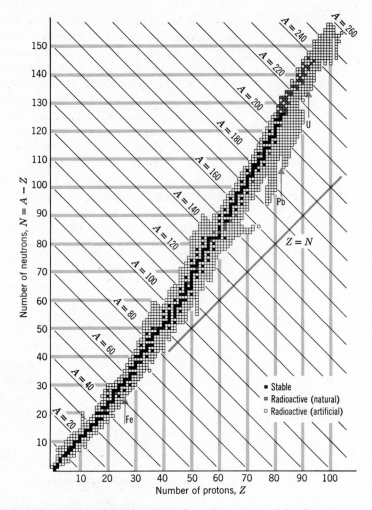

Fig. 15.14 *Chart of the known nuclei. The arrows indicate the positions of the elements uranium (the heaviest naturally occurring element), lead (the terminating element for most of the heavy radioactive nuclei), and iron (the most abundant element in the region near the maximum of the binding-energy curve). Notice that the N and Z coordinates are on opposite axes compared to Fig. 15-13. The lines of constant mass number are the diagonals marked with the values of A.*

converts a neutron into a proton, is replaced by the direct emission of a neutron.

5. Nuclei *below* the valley of stability are *proton rich* and undergo β^+ decay (or electron capture). Similar to the case above the valley, nuclei sufficiently far below the valley will be short-lived emitters of high-energy positrons or will decay by proton emission.

The statements above represent the average behavior of nuclei; of course, there are exceptions in individual cases. As we shall observe in the following section when we discuss the nuclear *shell model*, many of the deviations of nuclear properties from a smooth, average behavior have provided important clues to the understanding of nuclear structure.

15.4   *The Nuclear Shell Model*

THE "MAGIC" NUMBERS

Although the gross features of nuclei tend to vary smoothly with the numbers of protons and neutrons in the nucleus, a closer examination of detailed nuclear properties shows that there are many striking deviations from the average behavior. Certain nuclei occur with much greater natural abundances than their neighbors. Certain nuclei have much larger separation energies for the last neutron than their neighbors (see Table 15.6). Certain nuclei have much larger energies of their first excited states than their neighbors. These and other deviations from the average behavior are found to occur primarily at or near particular values of N or Z which are called "magic" numbers:

Magic numbers: N or $Z = 2, 8, 20, 28, 50, 82,$ and $N = 126$ (15.6)

For example, there are 7 stable isotopes for $N = 82$ but only one each for $N = 81$ and 83 (see Fig. 15.14). There are 10 stable isotopes for $Z = 50$ (tin), but only one for $Z = 49$ and two for $Z = 51$. There are 5 stable isotopes for $N = 20$ but *none* for either $N = 19$ or 21. Similar concentrations of exceptional stability occur for the other magic numbers. A few nuclei have magic numbers for both N and Z; these are the "doubly-magic" nuclei, $\text{He}^4 (N = Z = 2)$, $\text{O}^{16}(N = Z = 8)$, $\text{Ca}^{40}(N = Z = 20)$, and Pb^{208} $(N = 126, Z = 82)$. Doubly-magic nuclei are particularly notable for their extremely high stability.[10]

It is relatively easy to remove a neutron (or a proton) from a nucleus with N (or Z) equal to a magic number plus one (see Table 15.6). This effect is very similar to the atomic case in which the ionization energy decreases

---

[10] The number 28 is a "weak" magic number. The effect is not sufficient to make stable the nucleus Ni^{56} (N = Z = 28); however, Ca^{48} (Z = 20, N = 28) *is* stable.

Table **15.6**   *Separation Energy of the Last Neutron in Some Nuclei*

Nucleus	Z	N	Separation Energy of Last Neutron (MeV)
He^4	2	2	20.58
He^5	2	3	−1.0 (i.e., unbound)
O^{16}	8	8	15.67
O^{17}	8	9	4.14
Ca^{40}	20	20	16.62
Ca^{41}	20	21	8.36
Kr^{86}	36	50	9.76
Kr^{87}	36	51	5.51
Ce^{140}	58	82	9.06
Ce^{141}	58	83	5.53

markedly after an electron shell has closed (see Fig. 13.22). It was the realization that electrons obey the exclusion principle that was the crucial point in the explanation of atomic shell structure. Protons and neutrons are *fermions,* and therefore these particles also obey (*independently*) the exclusion principle. Can we infer, then, that protons and neutrons in nuclei are arranged in *shells?*

THE MODEL OF MAYER AND JENSEN

Although many attempts were made to construct a model that reproduced the observed shell-like structure of nuclei, all of these early efforts failed. Not until 1948 was it realized by Maria Mayer and also by J. H. D. Jensen[11] (working in collaboration with O. Haxel and H. E. Suess) that an essential ingredient in a viable shell model is a strong nuclear *spin-orbit* interaction.

As we saw in Section 13.3 (see, in particular, Fig. 13.18), the interaction of the electron *spin* with its *orbital* angular momentum results in the splitting of all states with a given value of l into two states with total angular momenta $j = l \pm s = l \pm \frac{1}{2}$. An $l = 1$ state, for example, becomes two states, one with $j = \frac{1}{2}$ ($P_{1/2}$) and one with $j = \frac{3}{2}$ ($P_{3/2}$).

In atoms the spin-orbit interaction is relatively weak and the amount of splitting of the energy states is always small compared to their energies (see Fig. 13.18). In nuclei, on the other hand, the spin-orbit interaction is *strong*

---

[11] Mayer and Jensen shared a portion of the 1963 Nobel Prize for their contribution to the theory of the nuclear shell model.

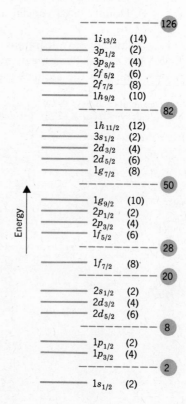

Fig. 15.15 *Schematic ordering of states in the nuclear shell model with a strong spin-orbit interaction. The dashed lines indicate the energy gaps that correspond to the closing of shells. The numbers in parentheses following each state designation indicate the number of particles that can occupy that state. Each large circled number is the total number of particles in states with lower energies—these numbers are the magic numbers. (The numbers preceding the letter designations of the orbital angular momenta do not correspond to the principal quantum number as in the designation of atomic states. In the notation used for nucleons, $1\mathrm{p}_{1/2}$ means the first $\mathrm{p}_{1/2}$ state, $2\mathrm{p}_{1/2}$ means the second $\mathrm{p}_{1/2}$ state, etc.)*

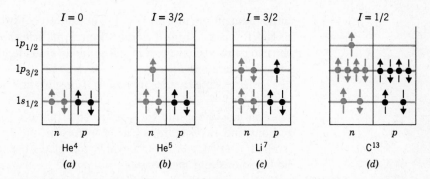

Fig. 15.16 *Schematic representation of the way in which protons and neutrons are arranged in shell-model states. The $1s_{1/2}$ state can accommodate 2 protons and 2 neutrons; the $1p_{3/2}$ state can accommodate 4 protons and 4 neutrons; the $1p_{1/2}$ state can accommodate 2 protons and 2 neutrons. For each nucleus, the configuration of lowest energy (i.e., the ground state) is shown.*

and the amount of splitting is appreciable. As a result of this large splitting, an energy state can sometimes be shifted from its "normal" position into another shell. The nuclear spin-orbit interaction therefore produces the same type of effect that electron shielding produces in atoms (see Fig. 13.24)—some energy states that would ordinarily occur together in the same shell are driven apart and actually are found in different shells. It was the full realization that a strong spin-orbit interaction would produce this effect that was one of the major contributions of Mayer and Jensen.

The inclusion of a strong spin-orbit interaction in the nuclear shell model produces the ordering of states shown in Fig. 15.15. Notice that there are energy gaps between certain adjacent pairs of states; these gaps correspond to the closing of shells (compare Fig. 13.24 for the atomic case). According to quantum theory, the total number of particles (protons *or* neutrons) that can simultaneously be in a state of given j is $2j + 1$. This prescription for counting the number of particles in a given state, combined with the ordering of states shown in Fig. 15.15, exactly reproduces the magic numbers.

The meaning of the ordering of states in the shell model is illustrated in Fig. 15.16. The lowest shell-model state, the $1s_{1/2}$ state, can accommodate two particles—one each with spin *up* and spin *down*—and, therefore, the first magic number is 2. This is exactly the same as in the case of the atomic $1S_{1/2}$ state, which can accommodate two electrons. But protons and neutrons separately obey the exclusion principle and separately fill the shell-model states. Therefore, the available $1s_{1/2}$ *states are filled* by 2 protons and 2 neutrons—He^4 (with $Z = N = 2$) is therefore a *doubly*-magic nucleus. Since the spins of all of the nucleons in He^4 are *paired* (see Fig. 15.16a), the net angular momentum[12] of the He^4 ground state is $I = 0$.

After the $1s_{1/2}$ shell is filled, the next particle must be placed in the $1p_{3/2}$ state. If this particle is a neutron, as illustrated in Fig. 15.16b, He^5 is formed and the spin of the nucleus is just the total angular momentum of the last

[12]Although the quantity to which we refer here is actually the *total* angular momentum of the nucleus (and consists of the vector combination of the individual nucleon spins and orbital angular momenta), we usually refer to I as simply the *spin* of the nucleus.

unpaired neutron, namely, $I = \frac{3}{2}$. The addition of two neutrons and a proton to He^4 forms Li^7 (Fig. 15.16c). All three particles are in the $1p_{3/2}$ state and the two neutrons have paired spins (and orbital angular momenta). The spin of Li^7 is therefore $I = \frac{3}{2}$.

The $1p_{3/2}$ state can accommodate 4 protons and 4 neutrons and this state is filled at C^{12}. If a 7th neutron is added to form C^{13} (Fig. 15.16d), this neutron must be in the $1p_{1/2}$ state. The spin of C^{13} is therefore $I = \frac{1}{2}$. Two protons and two neutrons can be placed in the $1p_{1/2}$ state and this state is filled at O^{16} ($N = Z = 8$), which completes the second nuclear shell for both protons and neutrons.

If there is both an unpaired proton and an unpaired neutron in the nucleus, a detailed quantum mechanical calculation is necessary to determine the shell-model prediction for the spin. The ability to predict correctly nuclear spins is one of the important successes of the shell model.

The development of the shell model has been one of the greatest steps forward in understanding the structure properties of nuclei. The model has been modified and improved to account for more and more experimental results. And although the shell model is not the complete answer to all nuclear questions (indeed, it has several serious shortcomings), it seems clear that any "complete" nuclear theory that is developed in the future will draw heavily on the success of the shell model.

SUPER-HEAVY NUCLEI

By extending the calculation of the ordering of states in the shell model, one can predict that the magic neutron number following 126 is 184. However, because of certain differences in the filling of neutron and proton states in heavy nuclei, the next magic proton number following 82 is believed to be 114. Therefore, the next doubly-magic nucleus following Pb^{208} ($Z = 82$, $N = 126$) should be the nucleus with $Z = 114$, $N = 184$. Such a nucleus would lie far beyond the region of the nuclear chart (Fig. 15.14) which has been studied thus far. (The element $Z = 105$, hahnium, is the heaviest experimentally identified at present.) But the prospect that element $Z = 114$ would have an isotope ($A = 298$) with a sufficiently long half-life to permit identification and study is an exciting one indeed. In fact, there may exist an entire "island" of super-heavy nuclei near ($Z = 114$, $N = 184$) that can be investigated. It appears likely that equipment will be available within a few years that will be capable of producing super-heavy nuclei—nuclei that have never before existed on Earth.

15.5   Nuclear Reactions

NEUTRON CAPTURE

A nuclear *reaction* is a process in which nucleons are added to, removed from, or rearranged within a target nucleus under bombardment by nucleons (that is, protons or neutrons), by groups of nucleons (for example, deuterons or α particles),[13] or by γ radiation. The simplest type of nuclear reaction is that in which a nucleus with mass number A captures a neutron and forms

---

[13] Other complex projectiles, such as H^3, He^3, Li^6, Li^7, C^{12}, and O^{16} ions, have also been used to initiate nuclear reactions. High energy electrons and mesons have also been used.

the nucleus $A + 1$ with the accompanying emission of a γ ray. We have already discussed just such a neutron capture process (Section 15.2), namely, the capture of a neutron by a proton to form a deuteron:

$$H^1 + n \longrightarrow D^2 + \gamma \tag{15.7}$$

It is customary to use a shorthand notation for specifying nuclear reactions. In this notation, the above reaction is written as

$$H^1(n, \gamma)D^2 \tag{15.7a}$$

where the first quantity (H^1) specifies the target nucleus and the last quantity (D^2) specifies the final (or *residual*) nucleus; within the parentheses, the first quantity (n) is the incident particle and the second quantity is the outgoing particle (in this case, a γ ray). Neutron capture by C^{12}, for example, is written as $C^{12}(n, \gamma)C^{13}$.

If the incident neutron has negligibly small kinetic energy, a neutron capture reaction always releases an amount of energy equal to the separation energy of a neutron in the final nucleus. For the case of the deuteron, we found in Section 15.2 that this energy release is 2.224 MeV.

CHARGED-PARTICLE REACTIONS

In 1919, Rutherford[14] used α particles from a radioactive source to bombard nitrogen gas in the hope that some of the α particles would reach the nitrogen nuclei and induce nuclear disintegrations. Rutherford knew how far an α particle from a given source would travel in nitrogen gas before all of its kinetic energy would be dissipated in atomic collisions. But he observed that sometimes there was a particle that had a much greater range in the gas than the α particles from the source. He deduced that these long-range particles were *protons* and that they had originated in the disintegration of nitrogen nuclei by the energetic α particles. The nuclear reaction that Rutherford had observed (the first such to be discovered) was

$$N^{14} + He^4 \longrightarrow O^{17} + H^1 \tag{15.8}$$

or, in the shorthand notation, $N^{14}(\alpha, p)O^{17}$, where we follow custom and write the bombarding and outgoing particles in *nuclear* rather than in *atomic* notation (α for He^4 and p for H^1). A cloud-chamber[15] photograph of a $N^{14}(\alpha, p)O^{17}$ reaction was first taken in 1925 by P. M. S. Blackett; this photograph is reproduced in Fig. 15.17.

In all nuclear reactions we must have a balance of protons and neutrons in the initial and final states. For the $N^{14}(\alpha, p)O^{17}$ reaction, we have

$$
\left.
\begin{array}{ccccccc}
& N^{14} & + & He^4 & \longrightarrow & O^{17} & + & H^1 \\
\text{No. protons:} & 7 & + & 2 & = & 8 & + & 1 \\
\text{No. neutrons:} & 7 & + & 2 & = & 9 & + & 0
\end{array}
\right\} \tag{15.9}
$$

Using this rule, we can always identify the fourth nucleus in a reaction if the other three nuclei are known.

---

[14] Almost all of the important early developments in nuclear physics are associated with Lord Rutherford and his students.

[15] A *cloud chamber* is a device that renders visible the track of a nuclear particle by virtue of the condensation of water droplets on the ions left in the wake of the particle. These devices have now been largely supplanted by *bubble chambers* (see Fig. 9.29).

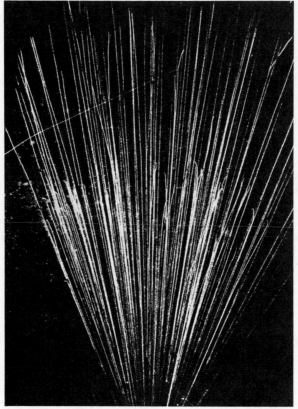

P. M. S. Blackett

Fig. 15.17 *Cloud-chamber photograph of a $N^{14}(\alpha, p)O^{17}$ reaction amidst the tracks of many α particles that do not induce reactions.*

Example **15.5**

B^{10} is bombarded with neutrons and α particles are observed to be emitted. What is the residual nucleus?

$$B^{10} + n \longrightarrow (?) + He^4$$

No. protons:	5	+ 0	=	Z	+ 2
No. neutrons:	5	+ 1	=	N	+ 2

Clearly, $Z = 3$ and $N = 4$; therefore, the residual nucleus is Li^7.

The mass-energy balance in a nuclear reaction can be calculated by using exactly the same method as that used in calculating binding energies, namely, we compare the combined mass of the target nucleus and the bombarding particle with the combined mass of the residual nucleus and the outgoing particle. For the $N^{14}(\alpha, p)O^{17}$ reaction, we have (in AMU):

$$m(N^{14}) = 14.003\ 074 \qquad\qquad m(O^{17}) = 16.999\ 133$$
$$m(He^4) = \ \ 4.002\ 603 \qquad\qquad m(H^1) = \ \ 1.007\ 825$$
$$\overline{m(N^{14}) + m(He^4) = 18.005\ 677} \qquad \overline{m(O^{17}) + m(H^1) = 18.006\ 958}$$

The *energy* difference between the initial and final pairs of particles is

$$\Delta m \times c^2 = [m(\text{N}^{14}) + m(\text{He}^4)] - [m(\text{O}^{17}) + m(\text{H}^1)]$$
$$= (-0.001\ 281\ \text{AMU}) \times (931.481\ \text{MeV/AMU})$$
$$= -1.193\ \text{MeV}$$

That is, the mass of $\text{N}^{14} + \text{He}^4$ is *less* than the mass of $\text{O}^{17} + \text{H}^1$; therefore, an energy of 1.193 MeV must be supplied to the $\text{N}^{14} + \text{He}^4$ system in order for the reaction to take place. The requisite energy can be supplied by the kinetic energy of the incident α particle.

The difference in mass between the initial and final states in a nuclear reaction is called the *Q-value* for the reaction and is usually given in energy units. A *positive* Q-value means that energy is *released* in the reaction, and a *negative* Q-value means that energy must be *supplied*. For example,

$$\left.\begin{array}{ll} \text{H}^1 + n \longrightarrow \text{D}^2 + \gamma; & Q = +2.224\ \text{MeV} \\ \text{N}^{14} + \text{H}^1 \longrightarrow \text{O}^{17} + \text{H}^1; & Q = -1.193\ \text{MeV} \end{array}\right\} \tag{15.10}$$

In reactions such as $\text{N}^{14}(\alpha, p)\text{O}^{17}$, merely supplying to the system an amount of energy sufficient to overcome the mass-energy imbalance (in this case, 1.193 MeV), does not mean that the reaction will then take place. Just as a positive-energy α particle must tunnel *out* of a nucleus in an α-decay process, an incident α particle (with $KE_\alpha < \mathcal{E}_C$) must tunnel *into* a nucleus in order to initiate a reaction. Furthermore, an outgoing charged particle, such as the proton in the $\text{N}^{14}(\alpha, p)\text{O}^{17}$ reaction, generally must also tunnel through a Coulomb barrier.

Supplying only 1.193 MeV to the $\text{N}^{14} + \text{He}^4$ system means that the $\text{O}^{17} + \text{H}^1$ system would have *zero* kinetic energy and, therefore, the proton would be unable to tunnel through the Coulomb barrier. Consequently, nuclear reactions involving charged particles occur with appreciable probability only if the energies of the incident and outgoing particles are sufficient to permit relatively easy penetrations of the Coulomb barriers. Of course, neutron absorption or neutron emission processes are not hindered by Coulomb effects; for this reason the capture of slow neutrons by nuclei takes place quite readily.

THE COMPOUND NUCLEUS—RESONANCE REACTIONS

When an α particle strikes a N^{14} nucleus, the ejection of a proton and the formation of the residual O^{17} nucleus does not occur instantaneously. In fact, the $\text{N}^{14} + \alpha$ system remains together as an *intermediate* or *compound* nucleus for a certain period of time before breaking up into a proton and an O^{17} nucleus. The compound nucleus formed by $\text{N}^{14} + \alpha$ must have $Z = 7 + 2 = 9$ and $A = 14 + 4 = 18$; hence, the nucleus is F^{18}:

$$\text{N}^{14} + \alpha \longrightarrow \text{F}^{18} \longrightarrow \text{O}^{17} + p \tag{15.11}$$

For what period of time does the F^{18} nucleus exist before it disintegrates? First let us define a "typical nuclear time." According to Eq. 15.1, the radius of a N^{14} nucleus is approximately 3.4×10^{-13} cm. A typical α particle velocity is 3×10^9 cm/sec $(=0.1c)$. Therefore, an α particle with this velocity (which corresponds to a kinetic energy of approximately 20 MeV) will

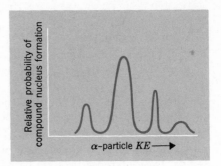

Fig. 15.18 *Nuclear reactions, such as the* N^{14} *(α, p)O^{17} reaction, exhibit resonances corresponding to excited energy states in the compound nucleus.*

remain in the vicinity of a N^{14} nucleus for a time $t \cong (3.4 \times 10^{-13}$ cm)/ $(3 \times 10^9$ cm/sec) $\cong 10^{-22}$ sec. Therefore, if an α particle and a N^{14} nucleus remain together for a time that is *long* compared to 10^{-22} sec, it is meaningful to refer to the formation of a F^{18} compound nucleus.

If we bombard N^{14} with α particles and measure the rate at which protons are emitted as we change the α-particle bombarding energy, we find that the relative probability of proton emission undergoes substantial changes with relatively small changes in the bombarding energy. In the $N^{14}(\alpha, p)O^{17}$ reaction, as in most other reactions, we find pronounced peaks (or *resonances*) in the emission probability curve at certain values of the bombarding energy (Fig. 15.18). The occurrence of these resonances indicates that there are certain discrete energies for which there is an enhanced probability for the formation of the compound nucleus. That is, these resonances correspond to *excited energy states* of the compound nucleus. Resonance peaks are typically 1–100 keV in width, and, according to the uncertainty principle, $\Delta \mathcal{E} \, \Delta t \approx h$, the lifetimes of the corresponding compound nucleus states are 10^{-18} to 10^{-20} sec. These times are much longer than the typical nuclear time of 10^{-22} sec, and so it is quite reasonable to refer to the existence of compound nucleus energy *states*. Nuclear states that lie at energies below the lowest particle separation energy are *bound* with respect to particle emission and these states cannot be formed as resonance states in particle-induced reactions; for example, all states in N^{14} below 7.55 MeV are bound (see Table 15.1). These bound states can decay *only* by γ-ray emission and have lifetimes that range from $\sim 10^{-15}$ sec to $\sim 10^{-8}$ sec or even longer in certain special cases.

The study of compound nucleus resonances has provided a wealth of information concerning nuclear energy states. Indeed, some of the most precise information available regarding the properties of nuclear excited states has been obtained from experiments of this type.

15.6 *Fission and Fusion*

THE DISCOVERY OF FISSION

Shortly before his death in 1937, Lord Rutherford stated that "the outlook for gaining useful energy from the atoms by artificial processes of transformation does not look very promising." Although Rutherford's intuition in scientific matters was almost always infallible, within a few years, a series of scientific and technological advances had shown this particular view to be incorrect—incorrect, in fact, to an astonishing degree.

In 1939, the German radio-chemist, Otto Hahn, in collaboration with Fritz Strassman, bombarded uranium with neutrons and performed very careful chemical tests on the resulting radioactive material. They found that among the products of neutron absorption by uranium there was radioactive barium ($Z = 56$) an element much less massive than the original uranium. How could such a light element be formed from uranium? The mystery was soon resolved by Lise Meitner and Otto Frisch, German physicists working then as refugees in Sweden, who suggested that neutron absorption by uranium produced a breakup (or *fission*) of the nucleus into two light fragments:

$$\text{U}(Z = 92) + n \longrightarrow \text{Ba}(Z = 56) + \text{Kr}(Z = 36)$$

This was a startling new type of nuclear reaction. Instead of exchanging only a few nucleons between the incident particle and the target nucleus, as in an (α, p) reaction, this discovery showed that it was possible to split a nucleus into two massive parts.

THE DYNAMICS OF FISSION

As shown in Fig. 15.6, the binding energy of a heavy nucleus ($A \cong 240$) is approximately 7.5 MeV per nucleon. If such a nucleus were separated into two parts, each with $A \cong 120$, the binding energy would be *increased* to approximately 8.5 MeV per nucleon. That is, the decrease in binding energy per nucleon with increasing mass number means that a heavy nucleus can break up into two light fragments with the release of a substantial amount of energy. This breakup of a heavy nucleus is somewhat analogous to the splitting apart of a vibrating drop of liquid, as shown in Fig. 15.20.

Fig. 15.19 *Otto Hahn (1879–1968), one of the chief contributors to the discovery of fission. Hahn was awarded the 1944 Nobel Prize in chemistry.*

William Numeroff

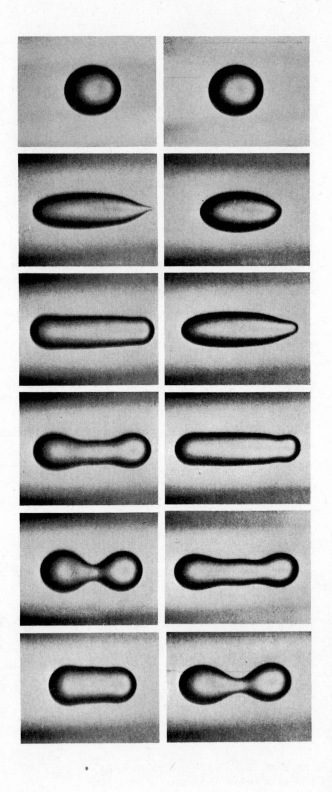

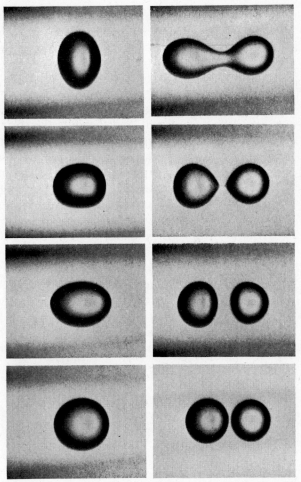

Stanley G. Thompson, Lawrence Radiation Laboratory

Fig. 15.20 *Photographs from a motion picture film showing the variation in shapes of an ordinary drop of water suspended in oil when a deformation is induced by a voltage applied across the oil. In the left-hand sequence, the drop returns to its initial spherical shape without undergoing fission. In the right-hand sequence, the initial deformation is sufficiently large that the drop fissions. In 1939, Neils Bohr and John Wheeler proposed a liquid-drop model of nuclear fission that was successful in explaining the general features of the fission process.*

Even though it is energetically favorable for a heavy nucleus to split into two parts, this process is inhibited by the strong attractive nuclear forces, and the two parts are unable to tunnel through their mutual potential barriers (as in the case of α decay for nuclei with $A \gtrsim 200$). The nucleus may become extended in an effort to fission (as in the left-hand sequence in Fig. 15.20), but it will usually return to and vibrate around its equilibrium shape. The probability of the occurrence of *spontaneous fission* is extremely small, and therefore the corresponding half-life is extremely long ($\sim 10^{17}$ years for U^{235}). If, however, some additional energy is supplied to the nucleus in the form of the binding energy of a captured neutron, this increase in energy may produce a large nuclear deformation which will be sufficient to permit the relatively easy tunneling apart of the two fragments; thus, fission can occur (as in the right-hand sequence in Fig. 15.20). For many heavy nuclei, the probability of the occurrence of *neutron-induced fission* is extremely short ($\sim 10^{-21}$ sec).[16]

The fragments that result from the fission of a heavy nucleus do not have equal masses. In fact, the mass numbers for the two fragments are usually quite different. Figure 15.21 shows the distribution of mass numbers observed for the case of the neutron-induced fission of U^{235}. The most probable pair of mass numbers is 95 + 139. (The sum does not equal 235 + 1 = 236 because, on the average, 2-3 neutrons are released during the fission process.) Two typical fission reactions involving U^{235} are

$$\left. \begin{array}{l} U^{235} + n \longrightarrow Ba^{139} + Kr^{95} + 2n \\ \phantom{U^{235} + n} \longrightarrow La^{144} + Br^{89} + 3n \end{array} \right\} \tag{15.12}$$

[16]The capture of neutron by U^{235} increases its probability for undergoing fission by a factor of $\sim 10^{45}$!

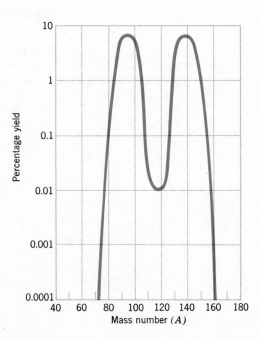

Fig. 15.21 *The distribution of fission-fragment mass numbers resulting from the neutron-induced fission of U^{235}. The most probable mass numbers are 95 and 139. The curve represents the average trend of the measured mass fractions. Of course, the curve has meaning only at the points corresponding to integer mass numbers. The reason for the highly asymmetrical shape of the curve is not yet understood in detail.*

Table **15.7** *Distribution of Energy Among Fission Products*

Product	Average Energy (MeV)
Fission fragments (kinetic energy)	168
Fission neutrons (kinetic energy)	5
γ rays (prompt)	5
β^- from fission fragments (kinetic energy)	5
γ rays from fission fragments	7
Neutrinos from β^- decays	10
	200 MeV

Since the isotopes Ba^{139}, Kr^{95}, La^{144}, and Br^{89} are all on the neutron-rich side of the valley of nuclear stability (Fig. 15.14), these fission fragments (as do almost all fission fragments) undergo radioactive β^- decay.

The way in which fission energy is distributed, on the average, among the various fission products is indicated in Table 15.7. If the body of material in which the fission event takes place is sufficiently large that all of the fission products are absorbed (except the neutrinos, which escape), a total of approximately 190 MeV will be converted into heat energy. This is truly an enormous amount of energy. As we showed in Example 7.18, if 1 kg of U^{235} undergoes fission, approximately 8×10^{20} ergs of energy is released. This amount of energy is sufficient to raise the temperature of 200,000,000 gallons of water from room temperature to the boiling point.

CHAIN REACTIONS

The fact that the fission process releases several neutrons (2.5, on the average, in the case of U^{235}) makes possible a series or chain of neutron-induced fission events that is self-sustaining. If one neutron from a fission event triggers the fission of another nucleus and one neutron from this event triggers another fission, etc., this series of fission events will sustain itself and will constitute a *chain reaction*. By controlling the environment of the fissioning nuclei it is possible to maintain a condition in which each fission event contributes, on the average, one and only one neutron that triggers another event. In this way, the rate of energy generation (the *power*) is maintained at a constant level. The controlled fission chain reaction (Fig. 15.22) is the principle of the *nuclear reactor*, now widely used in the commercial generation of electricity.

It is also possible to bring together in a small volume a sufficient amount of fissionable material so that fewer of the fission neutrons escape the system and therefore more than a single neutron from each event can trigger a new event. Figure 15.23 shows a series of fission reactions in which each event contributes *two* neutrons toward the next set of events. The rapid multiplication of the number of fissioning nuclei in this uncontrolled situation leads to the explosive release of the fission-generated energy—this is the principle of the atomic bomb (which is, of course, actually a *nuclear* bomb).

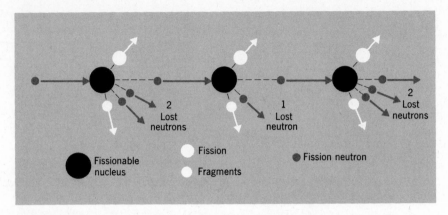

Fig. 15.22 *A controlled fission chain reaction in which one neutron from each fission event triggers another event. One or two neutrons from each fission event escape the system and are "lost."*

In order for the uncontrolled release of fission energy to take place, as many of the fission neutrons as is possible must be kept within the material. Unless the fissionable material has a mass greater than a certain value (called the *critical mass*), too many neutrons will escape the system and the rate of energy release will be too slow for an explosion to occur. The problem in constructing an atomic bomb, therefore, is to bring together into a small

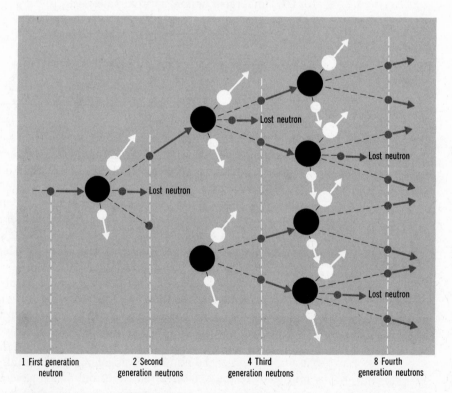

Fig. 15.23 *An uncontrolled series of fission events. The rapid release of the fission energy in such a system leads to an explosion.*

volume an amount of fissionable material at least as great as the critical mass. This assembly of the critical mass must be accomplished in an extremely short period of time ($\sim 10^{-3}$ sec), because otherwise a slow, non-explosive series of fission events will take place. One of the methods devised to overcome this problem is to drive together two or more subcritical masses by means of a conventional (chemical) explosion. The original atomic bombs of 1945 contained about 2 kg of fissionable material and were detonated in this way.

NEPTUNIUM AND PLUTONIUM

The two most important isotopes of uranium, as far as fission processes are concerned, are U^{235} and U^{238}. Only U^{235} readily undergoes fission by absorbing a slow neutron; U^{238} requires an energetic neutron to trigger a fission reaction. But the natural abundance of U^{235} is only 0.7 percent. Therefore, elaborate procedures are necessary in order to construct a bomb or to build a nuclear reactor utilizing this isotope.[17] What can be done with the abundant U^{238} isotope? If U^{238} is exposed to slow neutrons, a capture reaction takes place and U^{239} is formed. This radioactive nucleus undergoes β^- decay and forms the transuranic element *neptunium* ($Z = 93$):

$$U^{238} + n \longrightarrow U^{239} \tag{15.13}$$

$$U^{239} \xrightarrow{\ \beta^-\ } Np^{239} \quad (\tau_{1/2} = 23.5 \text{ min}) \tag{15.14}$$

Np^{239} is itself radioactive and forms the element *plutonium* ($Z = 94$) by β^- decay:

$$Np^{239} \xrightarrow{\ \beta^-\ } Pu^{239} \quad (\tau_{1/2} = 2.35 \text{ days}) \tag{15.15}$$

Pu^{239} has a sufficiently long half-life (24,360 years) that this isotope can be separated from uranium and recovered in substantial amounts. The importance of Pu^{239} lies in the fact that it readily undergoes fission induced by slow neutrons—more readily, in fact, than U^{235}. Therefore, both of the important isotopes of uranium can be used in slow-neutron fission processes: U^{235} can be used directly and U^{238} can be converted into Pu^{239}.

FUSION

Energy will be released from any group of particles that can be rearranged into a system that has a greater binding energy. Fission, of course, is one example of such a process—the total binding energy of two nuclei such as barium and krypton is greater than the total binding energy of uranium, and so the fission process releases energy. The problem of extracting energy from nuclei can also be approached from the low-mass side of the maximum in the binding energy curve. If we combine two light nuclei to form a tightly bound medium-A nucleus, energy will be released. This process is called *fusion*. For example, if two Ne^{20} nuclei (binding energy per nucleon $\cong 8$ MeV; see Fig. 15.6) are combined to form a Ca^{40} nucleus (binding energy per nucleon $\cong 8.5$ MeV), there would be a total energy release of 40×0.5

---

[17] During World War II, enormous plants were constructed at Oak Ridge, Tennessee, for the sole purpose of separating U^{235} from natural uranium.

MeV = 20 MeV. The difficulty in this particular case, of course, is that a great force would be required to overcome the Coulomb repulsion and to bring the neon nuclei into sufficiently close proximity so that the capture process would take place.

The effect of the Coulomb repulsion will be reduced if we use nuclei with small Z. If we bring two deuterons together to form an α particle, the energy release will be almost 24 MeV (see the third entry in Table 15.1) or 6 MeV per nucleon. This amount of energy release per nucleon is more than 6 times greater than for fission (200 MeV for 236 nucleons). But, in fact, when two deuterons combine, it is much more probable that a neutron or a proton, instead of a γ ray, will be emitted. That is, the probable reactions are $D^2(d, n)He^3$ and $D^2(d, p)T^3$, the Q-values for which are 3.269 MeV and 4.033 MeV, respectively. Thus, the average energy released in each D + D reaction is approximately 1 MeV per nucleon, comparable with that for the fission of a heavy element.

Since it is relatively easy to separate deuterium from normal hydrogen, there is a vast supply of deuterium available to us in the form of *water*, particularly in the oceans. How can we make use of this enormous reservoir of energy? Coulomb repulsion, which works to our advantage in the splitting of heavy nuclei in the fission process, is an obstacle that must be overcome if fusion energy is to be released. The $D^2(d, n)He^3$ and $D^2(d, p)T^3$ reactions will take place only if the two deuterons are brought to within a distance of $\sim 10^{-12}$ cm of one another. In order to approach so closely, an amount of kinetic energy must be given to the deuterons that is at least equal to the electrostatic potential energy at that distance:

$$PE_E = \frac{e^2}{r} = \frac{1.44 \times 10^{-13} \text{ MeV-cm}}{10^{-12} \text{ cm}}$$

$$\cong 0.14 \text{ MeV}$$

This amount of kinetic energy can be given to the deuterons if we raise the temperature of a deuterium gas to a sufficiently high temperature. According to Eq. 7.51,

$$\overline{KE} = \frac{3}{2} kT^*$$

so that

$$T^* = \frac{2}{3}\left(\frac{\overline{KE}}{k}\right) = \frac{2}{3} \times \frac{(0.14 \times 10^6 \text{ eV})}{(8.62 \times 10^{-5} \text{ eV/}^\circ\text{K})}$$

$$\cong 1.1 \times 10^9 \ ^\circ\text{K}$$

Actually, the D + D reactions will take place (but at a much reduced rate) even if the temperature of the gas[18] is considerably lower than $10^9 \,^\circ$K. The reaction rate is significant even at $T^* \sim 10^7 \ ^\circ$K. Reactions that require these extremely high temperatures are called *thermonuclear reactions*.

---

[18] Of course, at such a temperature, the atoms are completely ionized and we do not have a "gas" in the normal meaning of the word. A high-temperature mixture of positive ions and electrons is called a *plasma*.

One method of achieving a temperature of 10 million degrees or so[19] is by the detonation of an atomic (fission) device. In the brief fraction of a second during which the blast takes place, the temperature is sufficiently high to ignite thermonuclear reactions, which then release additional energy and maintain the elevated temperature so that all of the thermonuclear material can "burn." This is, in fact, the principle of the H-bomb.

Although uncontrolled thermonuclear reaction processes have been achieved (in the form of H-bombs), we have not yet succeeded in constructing a device in which the controlled release of fusion energy can be maintained for longer than a small fraction of a second. Experiments such as those being carried out with the *Stellarator* machine (Fig. 9.37) may eventually lead to the construction of practical fusion power plants.

Summary of Important Ideas

Nucleons are bound together in nuclei by the strong nuclear force. The mass of any nucleus is *less* than the mass of the number of free protons and free neutrons that make up that nucleus; this difference in mass-energy is the total *binding energy* of the nucleus.

Nuclei are limited in maximum size by the effects of Coulomb repulsion and the exclusion principle.

Not all groups of nucleons constitute stable nuclei; if there is a less energetic arrangement that is available to the nucleons (that is, if a configuration of smaller mass is possible), then a *radioactive decay* process will occur, which will transform the original nucleus into a nucleus of smaller mass. Radioactive decay involves one of the following possibilities: emission of an electron, emission of a positron, capture of an atomic electron by the nucleus, or emission of an α particle. (The first three of these processes also involve the emission of neutrinos.)

The stable nuclei with $A \lesssim 40$ contain approximately *equal* numbers of protons and neutrons ($Z \cong N$). For heavier nuclei, the neutron number increases more rapidly than the proton number.

For certain values of N or Z (called "magic" numbers), nuclei exhibit exceptional stability properties. The interpretation of this fact led to the development of the nuclear *shell model*.

When a nucleon or a nucleus is given a high velocity and is directed toward other nuclei, nuclear *reactions* can take place in which nuclear particles (or photons) are emitted and new nuclei are formed. In some cases the colliding nuclei form an intermediate nucleus (or *compound nucleus*) before the disintegration into the final products occurs.

Heavy nuclei (such as uranium or plutonium) can absorb a neutron and undergo *fission* by splitting into two fragments of roughly equal mass. Each

---

[19]In Section 17.1 we will discuss thermonuclear processes that take place in stellar interiors where such temperatures occur.

fission event releases approximately 200 MeV of energy. Energy is also released when two light nuclei combine to form a heavier nucleus; this process is called *fusion.*

Questions

15.1 Why must *high* energy electrons be used in scattering experiments that are designed to examine the distribution of matter within nuclei? (Consider the electron *wavelength.*)

15.2 The nucleus Be^8 spontaneously breaks up with an extremely short half-life into two α particles. Do you expect Be^8 to have a *prolate* or an *oblate* deformation? (What is the shape of the system just at the instant of breakup?)

15.3 List the energies of all of the γ rays that will be emitted when a collection of B^{10} nuclei are formed in the 3.58-MeV state. (Refer to Fig. 15.4).

15.4 For mass numbers up to $A = 209$, for only two values of A are there *no* stable nuclei. Use Fig. 15.13 and identify these A values. (This is an important point in the discussion of the formation of elements in stars; see Section 17.1.)

15.5 By what processes do you expect the following nuclei to decay: O^{14} ($Z = 8$), Ca^{50} ($Z = 20$), Cu^{67} ($Z = 29$), Sn^{111} ($Z = 50$)? (Use Fig. 15.14 to determine on which side of the valley of stability each nucleus lies.)

15.6 Discuss the decay history of C^9 (see Fig. 15.13). In what form will the original nine nucleons be at the end of the series of decay processes?

15.7 Why are there no pairs of *stable* nuclei with the same value of A but with Z differing by one unit?

15.8 If the nuclei (Z, A) and $(Z + 2, A)$ are both stable, what general statements can be made concerning the nucleus $(Z + 1, A)$? (Is this nucleus radioactive? If so, what type of decay does it undergo?)

15.9 When Be^7 captures an atomic electron to form Li^7, X rays are emitted. Why?

15.10 Refer to Fig. 15.10b. Is the α-particle wave function larger for $r < R_1$ or for $r > R_1$? What is the significance of a "large" wave function compared to a "small" wave function?

15.11 Th^{232} decays by the following series of emissions: α, β, β, α, α, α, α, β, α, β. Construct a diagram for this case similar to that shown in Fig. 15.12. Identify the intermediate (radioactive) nuclei and the final (stable) nucleus.

15.12 An early model of nuclei considered a nucleus (Z, A) to consist of A protons and $A - Z$ electrons. Argue on the basis of the spins of the particles that such a model is not consistent with the measured spin of N^{14}, namely, $I = 1$. (That is, show that the spins of 14 protons and 7 electrons cannot be combined to give $I = 1$.) This result is a powerful argument against the proton-electron model of nuclei.

15.13   Why is it not reasonable to picture the neutron as a close association of a proton and an electron?

15.14   Follow the procedure outlined in Fig. 15.16 and obtain the ground-state spins for the following nuclei: He^3, Be^8, Be^9, N^{13}, N^{15}, and O^{16}.

15.15   What are the ground-state spins of the following nuclei, according to the shell model: O^{17} ($Z = 8$), F^{17} ($Z = 9$), Ca^{40} ($Z = 20$), Ca^{41}?

15.16   In what nucleus do the protons fill all shell-model states up to and including the $2p_{3/2}$ state and the neutrons fill all states up to and including the $2p_{1/2}$ state? Is this nucleus stable?

15.17   The stable isotopes of zirconium ($Z = 40$) are Zr^{90}, Zr^{91}, Zr^{92}, Zr^{94}, and Zr^{96}. What is the highest shell-model state that is filled by the protons in zirconium? What shell-model state is being filled by the neutrons in the various zirconium isotopes?

15.18   Neutron capture by a stable target nucleus rarely leads to positron radioactivity. Why?

15.19   An experimenter bombards a natural boron target with deuterons and finds that two different radioactive species are formed, one with a half-life of 20.5 min and the other with a half-life of 0.020 sec. What reactions induced these activities? (Refer to Fig. 15.13 and remember that natural boron is a mixture of isotopes.)

15.20   What are the residual nuclei when a (p, α) reaction takes place with the following target nuclei: Be^9, B^{11}, O^{18}, and F^{19}?

15.21   Explain why a (p, n) reaction on a stable target nucleus always has a *negative* Q-value. (If the Q-value were positive, would the target be stable?)

15.22   List some stable targets and incident particles that could be used to produce nuclear reactions that yield $N^{13} + n$ in the final state.

15.23   A boron target is bombarded with a proton beam. After the beam is turned off, a β-ray detector records 100 counts/sec from radioactivity in the target. Forty minutes later, the counting rate has decreased to 25 counts/sec. What is the source of the radioactivity and what reaction has taken place? (Use Fig. 15.13.)

15.24   List some of the reactions that can take place when Be^9 is bombarded with protons.

15.25   List some nuclear reactions that can produce Be^8. (There are at least 16 that involve stable targets and employ bombarding particles with $A \leq 4$.)

15.26   What are the compound nuclei that are formed in the following reactions: (a) $B^{10}(\alpha, p)C^{13}$, (b) $Be^9(p, d)Be^8$, (c) $Be^9(p, \alpha)Li^6$, (d) $F^{19}(p, \alpha)O^{16}$, and (e) $C^{12}(He^3, \alpha)C^{11}$?

15.27   When U^{235} ($Z = 92$) absorbs a slow neutron, it undergoes fission and releases 2 or 3 neutrons. List 3 or 4 possible pairs of fission-product nuclei (different from those in Eq. 15.12) that could be formed in such a process.

15.28   Fission reactors produce substantial amounts of radioactivity. Explain why this constitutes a certain hazard. Would *fusion* reactors suffer from this same defect?

15.29   If a sample of matter containing every element were heated to thermonuclear temperatures in some nuclear cauldron, what element or group of elements would you expect to result from "cooking" this mixture until it is "done?" Explain.

Problems[a]

15.1   The 0.72-MeV state of B^{10} (see Fig. 15.4) has a half-life of 6.7×10^{-10} sec for the spontaneous emission of a γ ray. What will be approximate uncertainty in the energy of the emitted γ ray?

15.2   What group of nucleons must be separated from Li^7 to form He^4? If these nucleons are removed as a single entity, how much energy is required? (Use the data in Table 15.2.)

15.3   When He^5 decays, the range of energy which is shared by the He^4 nucleus and the neutron is approximately 0.5 MeV. How long does He^5 live on the average? (Use the uncertainty principle, $\Delta \mathcal{E} \, \Delta t \approx h$.)

15.4   How much energy is required to separate a neutron from Li^7? From Be^9? (The neutron separation energy for Be^9 is the *lowest* for any stable nucleus.)

15.5   What is the binding energy per nucleon of (a) He^3, (b) Li^6, and (c) Li^7?

15.6   Use the information in Table 15.2 and show that Be^8 is unstable and that it can break up *only* into two α particles.

15.7   If two Li^6 nuclei were brought together, what nucleus would be formed and how much energy would be released?

15.8   Use Fig. 15.6 and estimate the amount of energy that would be released if 20 protons and 20 neutrons were brought together to form Ca^{40}.

15.9   The mass of U^{238} is 238.0508 AMU. What fraction of the total mass-energy of U^{238} is its *binding energy?*

15.10   The mass of Fe^{56} ($Z = 26$) is 55.934 936 AMU and the mass of Co^{56} ($Z = 27$) is 55.939 847 AMU. Which of these nuclei is stable and which decays radioactively into the other? How much energy is available for the decay?

15.11   What is the maximum energy of electrons emitted in the β decay of tritium?

15.12   What is the value of $\mathcal{E}_\beta$ for the decay of Li^8?

15.13   Use Fig. 15.9 and estimate the maximum energy of positrons emitted in the decay of Ge^{65}.

15.14   The available energy in the α decay of Po^{210} is 5.4 MeV (see Example 15.3). Use energy and momentum conservation and show that this energy is divided between the α particle (5.3 MeV) and the residual nucleus, Pb^{206} (0.1 MeV).

15.15   The kinetic energy of α particles emitted by Ra^{226} ($Z = 88$, atomic mass = 226.02536 AMU) is 4.78 MeV and the recoil energy of the daughter nucleus, Rn^{222}, is 0.09 MeV. What is the atomic mass of Rn^{222} ($Z = 86$)?

[a]For ease of computation in the problems, use the approximate conversion factor, $c^2 \cong 930$ MeV/AMU.

15.16 How much kinetic energy will each of the α particles have when a Be^8 nucleus breaks up? (Use the data in Table 15.2.)

15.17 U^{228} emits a 6.69-MeV α particle and has a half-life of 9.3 min. U^{232} emits a 5.32-MeV α particle. Use Fig. 15.11 and estimate the half-life of U^{232}.

15.18 Calculate the Q-value for the $Li^7(d, n)Be^8$ reaction. Explain qualitatively why all (d, n) Q-values should be positive and large. The exceptions to this rule are to be found in cases in which the target nucleus is unusually stable, as in the $O^{16}(d, n)F^{17}$ reaction. (What is the Q-value for this reaction?) What is the reason for these exceptions?

15.19 What is the Q-value for the $He^4(He^3, \gamma)Be^7$ reaction?

15.20 A slow neutron is captured by Li^7 and a single γ ray is emitted. What is the energy of the γ ray?

15.21 What is the Q-value for the $Li^7(p, \alpha)He^4$ reaction? What is the Q-value for the $He^4(\alpha, p)Li^7$ reaction?

15.22 Excited states in Be^{10} can be studied by observing resonances when Be^9 is bombarded with neutrons. What is the minimum excitation energy of a Be^{10} state that can be studied in this way?

15.23 When C^{13} is bombarded with 1.75-MeV protons, a resonance with a width of 77 eV is formed. In what nucleus is the corresponding excited state? What is the approximate lifetime of this excited state?

15.24* How much energy will be released if 1 kg of deuterium is completely "burned" in fusion reactions? (Assume that the (d, n) and (d, p) reactions are equally probable so that the average Q-value is $\frac{1}{2}(3.269 + 4.033) = 3.651$ MeV.) Compare this energy with that released by the fissioning of an equal mass of U^{235}.

15.25 In an H-bomb the $T^3(d, n)He^4$ reaction, instead of the D + D reactions, is used as the primary source of thermonuclear energy. Why? (Calculate the Q-value.)

Symmetry   Courtesy of Martin-Marietta

16 *Elementary Particles*

In the days of Aristotelian science, all matter was considered to be composed of four basic substances—earth, air, fire, and water. These were the "elementary particles" of Nature. By the early 1930s, modern science had discovered a more satisfying way to describe the composition of matter in terms of four elementary particles—protons, neutrons, electrons, and photons. This was an extremely simple and an attractive scheme—with only four elementary particles its was possible, by following the rules of quantum mechanics, to account for all of the chemical elements, their compounds, and their radiations. By adding a fifth particle—the neutrino—it was possible to include a description of radioactive decay processes as well. These elementary particles appeared to be the basic and ultimate constituents of all matter.

But this apparent simplicity was not to endure for long. Within a year after the identification of the neutron, the *positron* was discovered. In 1936 the first *meson* was discovered among the products of the interactions of energetic cosmic rays with matter. In 1947 a second type of meson was found, and soon thereafter additional mesons and other unusual particles were observed. The rate of production of these particles by the action of cosmic rays was not sufficient to permit extensive detailed investigations of their properties and interactions. But techniques improved, and as accelerators capable of producing particles with higher and higher energies became available, such studies were made and, at the same time, a host of new particles was discovered.

At present, more than a hundred different mesons and other particles with strange properties are known. They range in mass from 200 times the electron mass to a few times the proton mass. All of these new particles have only a transitory existence[1]—none lives longer than a few microseconds and many decay within $\sim 10^{-23}$ sec after being formed. The end products of the decays of these particles are the familiar constituents of ordinary matter (protons, electrons, and photons) and neutrinos.

It has become customary to refer to this entire collection of objects as "elementary particles." This term does not mean that these particles are the basic ingredients of matter in the sense that they make up the atoms in our world—protons, neutrons, and electrons perform that function quite satisfactorily. But these particles do result from the fundamental interactions of ordinary pieces of matter and many of the particles participate directly or indirectly in the basic forces by which ordinary matter interacts. (For example, π mesons are the particles primarily responsible for the nucleon-nucleon force.)

In spite of the fact that many of the elementary particles enjoy only very brief careers before undergoing decay, some of these particles can, in a certain sense, be classified as "stable" particles. By way of analogy, we certainly consider an automobile to be a "stable" object, but an automobile "disintegrates" after it has traveled 80,000 miles or so during the course

---

[1] Positrons and, as we will see, *antiprotons* are stable in the free state, but they are *annihilated* when they come into contact with ordinary matter.

of several years of existence. During its lifetime, then, an automobile moves a distance $\sim 2 \times 10^7$ times its own length. A pion, on the other hand, has an average lifetime of $\sim 10^{-8}$ sec. Under typical circumstances, a pion travels with a velocity of $\sim 10^{10}$ cm/sec, and so during its lifetime a pion will move a distance of $\sim 10^2$ cm. But the "size" of a pion is only about 10^{-13} cm. Therefore, during its brief existence a pion will travel a distance that is $\sim 10^{15}$ times its own "size." On this basis, then, the pion is enormously more "stable" than an automobile. For all practical purposes in discussing the interactions of elementary particles, the pion (and any other particle that lives for a time greater than about 10^{-18} sec) can be considered to be a "stable" particle. Of course, there are elementary particles (for example, protons and electrons) that never decay—these particles are *truly* stable.

The existence of such a profusion of elementary particles has confronted physicists with perplexing questions concerning the ultimate nature of matter. What is the role played by each of these particles in the fundamental processes that take place in the subatomic domain? We know that the pion is an essential participant in the nucleon-nucleon interaction, but we have no such clear idea of the reasons for the existence of most of the other extra-ordinary elementary particles. Is there some Grand Scheme that simply and clearly specifies the relationships of the elementary particles one to another? Physicists have developed an abiding faith in the inner harmony of Nature and most physicists believe that there exists some unifying principle which, when discovered, will reveal to us the underlying reasons for the myriad individual facts that we have learned about elementary particles and elementary processes. We have already been permitted a glimpse, here and there, of powerful fundamental principles at work, but we have not yet succeeded in locating the key to the riddle of elementary particles.

16.1   *Particles and Antiparticles*

POSITRONS

The development of the theory of elementary particles really had its beginning in 1928 when P. A. M. Dirac introduced relativity into quantum theory. Prior to this time it had been necessary to introduce the value of the electron intrinsic magnetic moment into quantum theory as an assumption. With his relativistic quantum theory, Dirac was able to compute directly a value of the electron magnetic moment that agreed with experiment.[2] This was indeed a significant accomplishment, but Dirac's theory produced another important (and totally unexpected) result. Not only does Dirac's relativistic equation describe a particle that is readily identified with the electron, but it also describes a particle that is the same as the electron in every respect except that the charge is $+e$ instead of $-e$. No such *positive electron* was known in 1928 and many persons simply ignored this particular prediction of the theory as a spurious result or tried (unsuccessfully) to identify this positively-charged particle as a proton. However, while investi-

---

[2] Certain small corrections to Dirac's value for the electron magnetic moment were calculated in the 1950s by the methods of quantum electrodynamics (see Section 13.4).

gating cloud-chamber tracks of cosmic-ray particles in 1932, Carl D. Anderson (1905–   ) observed an electron track that curved the "wrong way" in the magnetic field in which his cloud chamber was located (Fig. 16.1). This was the first observation of a positive electron (or *positron*), and Dirac's prediction was confirmed.[3]

Soon after Anderson's discovery, it was established that positrons can actually be *created* by the interaction of energetic photons (γ rays) with matter. This creation process, however, always produces a *positron-electron pair* (Fig. 16.2), and therefore does not violate the general principle of charge conservation. In the creation of a positron-electron pair, electromagnetic energy is converted into mass; in order to create two electron masses, the photon energy must be at least $2m_e c^2 = 1.02$ MeV. (An energetic photon cannot produce a pair in vacuum because both energy and momentum cannot be conserved in such a process; pair production always takes place in close proximity to another particle, such as a nucleus or an electron, which absorbs the recoil momentum.)

Once a positron is created, it interacts via electromagnetic forces with the atomic electrons in its vicinity, eventually losing essentially all of its kinetic energy. As the positron drifts with very low velocity it can encounter and coalesce with an electron. The two particles then *annihilate* one another and the mass-energy of the pair appears in the form of two (or, infrequently, three) photons[4] with a total energy of $2m_e c^2$ (Fig. 16.3).[5]

[3] Anderson received a share of the 1936 Nobel Prize in physics for his discovery of the positron. Dirac and Schrödinger shared the 1933 Prize for their contributions to the development of quantum theory.

[4] An annihilation process involving a positron and a free electron in which only *one* photon is produced cannot conserve momentum.

[5] Although not, of course, referring to creation and annihilation processes, Newton nevertheless offered a charming description of these events in his *Opticks* (1704): "The changing of bodies into light and light into bodies, is very comfortable to the course of Nature, which seems delighted with transmutations."

Fig. 16.1 *Anderson's first cloud-chamber photograph of a positron. The track of the particle begins in the lower portion of the chamber where its energy is 63 MeV (deduced from the curvature in the known magnetic field—B = 15 kG). After penetrating the 6-mm lead plate, the particle has an energy of 23 MeV. This decrease in energy upon passing through the plate definitely establishes the direction of motion of the particle that produced the track. This information, when combined with the knowledge of the direction of the magnetic field and the direction of curvature of the track, shows that the particle must carry a positive charge. The density of droplets along the track shows that the mass of the particle is that of an electron (and, therefore, that the track cannot be due to a proton).*

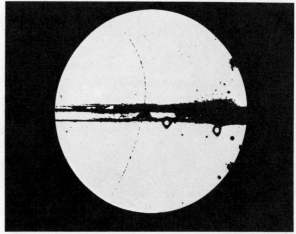

Carl D. Anderson, California Institute of Technology, 1932

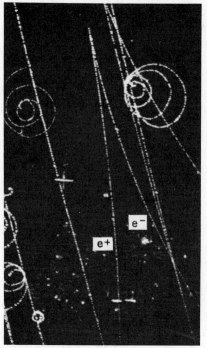

Fig. 16.2 *Bubble-chamber photograph of the creation of a positron-electron pair. An energetic γ ray enters the chamber from above and interacts with one of the (hydrogen) nuclei in the chamber to produce the pair. The chamber is located in a magnetic field, and so the tracks of the two particles curve in different directions.*

Lawrence Radiation Laboratory

ANTIPARTICLES

Electrons and positrons are said to be *antiparticles* of one another. The positron is the antiparticle of the electron and *vice versa,* but since the electron is the natural member of the pair in our world, we usually refer to the electron as the "particle" and to the positron as the "antiparticle."

The electron and the positron are not unique as a particle-antiparticle pair; in fact, relativistic quantum theory demands that *every* elementary particle have an antiparticle partner, and experiments have shown this to

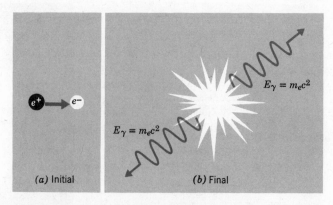

Fig. 16.3 *The annihilation of a slowly moving positron in an encounter with an electron produces two photons each with energy* $E_\gamma = m_e c^2$ *(annihilation radiation). The photons leave the annihilation site "back-to-back" in order to conserve momentum.*

be the case. A particle and its antiparticle have exactly the same mass, the same half-life and decay modes[6] (if not stable), and the same spin quantum number. However, the members of a particle-antiparticle pair have *opposite* electromagnetic properties. If the *particle* carries an electric charge, the *antiparticle* carries the *opposite* electrical charge. If the spin vector **S** and the intrinsic magnetic moment vector μ have one relative orientation for the particle, these vectors have the *opposite* relative orientation for the *antiparticle*. For example, the vectors, **S** and μ, are aligned in *opposite* directions for the electron but they are aligned in the *same* direction for the positron:

$$e^-: \uparrow \downarrow \qquad e^+: \uparrow \uparrow$$
$$\mathbf{S}\ \mu \qquad\qquad \mathbf{S}\ \mu$$

The electron and the positron are easily distinguished because they have opposite charges. But even if a particle is electrically neutral (for example, the neutron), it can be distinguished from its antiparticle if it has both spin and a magnetic moment—one member of the pair will have **S** and μ *parallel* and the other member will have **S** and μ *antiparallel*.

As we shall see in Section 16.2, *neutrinos* are electrically neutral and have spin, but they have no magnetic moment. However, since neutrinos have no existence unless they are traveling with the velocity of light, we can always (in any reference frame) define a linear momentum vector **p** for these particles which has a *unique* direction. Then, the relative orientation of **S** and **p** distinguishes *neutrinos* from *antineutrinos*. The vectors, **S** and **p**, are always antiparallel (opposite) for neutrinos and are always parallel for antineutrinos (see Fig. 16.5).

If a particle has an intrinsic mass, it can never travel with the velocity of light (as does the neutrino), and, therefore, a reference frame can always be found in which the particle is at rest. Consequently, the direction of the linear momentum vector for such a particle is not unique. If, in addition, the particle has no spin (and, hence, no magnetic moment), it is not possible to distinguish the *particle* from the *antiparticle*. Any particle of this type is therefore its *own* antiparticle.[7] The neutral pion (π^0) and the neutral eta meson (η^0) as well as the photon[8] are in this special category of particles that have antiparticles identical to themselves.

Fermions (particles with half-integer spin such as electrons and nucleons) are *always* created or annihilated in *pairs;* no exception to this rule has ever been found. On the other hand, *bosons* (particles with integer spin such as photons, pions, and K mesons) can be created or absorbed *singly* or *multiply.*

[6] The peculiar neutral *K* mesons, which will be discussed in Section 16.3, are the only known exceptions.

[7] Again, the neutral K mesons are exceptions; although these particles have no electrical charge, no spin, and no magnetic moment, nevertheless, there exists a distinction between the particle and the antiparticle because the *decay modes* are different.

[8] The photon actually carries one unit of spin angular momentum; however, unlike the neutrino, the photon can have **S** oriented either along or opposite to its direction of motion. Therefore, there is no difference between a photon and an "antiphoton." Quantum theory shows, moreover, that it is not *zero* spin that leads to the indistinguishability of a neutral particle and its antiparticle but, rather, the fact that the particles are *bosons;* photons, pions, and eta mesons are all bosons.

Of course, energy, momentum, and charge must always be conserved in any creation or annihilation process.

ANTIPROTONS AND ANTINEUTRONS

After the theoretical prediction of the existence of the positron was confirmed by experiment, it was natural to wonder whether *antiprotons* and *antineutrons* exist. As is the case for positrons, antiprotons and antineutrons are fermions and can be produced only in pairs with their antiparticles, namely protons and neutrons. It requires an energy $2m_e c^2 = 1.02$ MeV to produce an electron-positron pair and; by the same token, it requires an energy $2m_p c^2 = 1876$ MeV to produce a proton-antiproton pair and an energy $2m_n c^2 = 1879$ MeV to produce a neutron-antineutron pair. The concentration of such huge amounts of energy in a single elementary particle that initiates the creation event can be achieved only in the largest accelerators, and such accelerators were not available until the 1950s. In 1955, however, a group working with the 6-GeV accelerator that had recently been constructed at the University of California was successful in producing and identifying antiprotons (symbol:[9] $\bar{p}$); in the following year, the antineutron ($\bar{n}$) was discovered.

Electrons and positrons interact via the electromagnetic force and when an electron-positron annihilation event takes place, the products are the quanta of the electromagnetic field—photons. Nucleons interact primarily via the strong nuclear force and when a proton-antiproton or neutron-antineutron annihilation event takes place the products are *pions*,[10] the quanta of the nuclear force field. Because pions are *bosons,* any number of pions, consistent with energy and charge conservation, can be produced in a p-$\bar{p}$ or an n-$\bar{n}$ annihilation event, but at least *two* pions must be produced in order to conserve momentum also. Figure 16.4 shows the annihilation of an antiproton by a proton in a bubble chamber; in this event, 8 charged

[9] *Antiparticles* are usually denoted by a bar over the symbol for the particle. In some cases, such as in e^- and e^+, the bar notation is not used.

[10] See Section 8.5 for a discussion of the role of the pion in the nucleon-nucleon interaction.

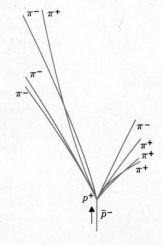

Fig. 16.4 *Annihilation of an antiproton with a proton in a hydrogen bubble chamber. Eight charged pions (and probably some neutral pions that do not leave tracks in the chamber) are produced.*

Brookhaven National Laboratory

pions (and probably several neutral pions which leave no tracks) are produced.

Antiprotons are *stable* particles, but a *free* antineutron, if it does not undergo annihilation, will eventually decay into an antiproton and a positron. Because the half-life of the antineutron is so long (presumably, the same as that of the neutron—12.8 min), annihilation will take place before decay occurs; the spontaneous decay of an antineutron has never been observed.

The most complex form of antimatter that has yet been produced and identified is the *antideuteron* ($d = \bar{p} + \bar{n}$). Conceivably, *antiatoms*, consisting of antiprotons, antineutrons, and positrons, could be produced; but because they would annihilate immediately on contact with ordinary matter, no such complete antiatom has yet been identified.

16.2 *Leptons and the Weak Interaction*

THE LEPTON FAMILY

Particles which have small or zero mass and which interact via the weak force[11] are called *leptons*.[12] Electrons (e^+ and e^-) and the neutrinos (ν_e and $\bar{\nu}_e$) associated with electrons in various weak processes constitute one branch of the family. Muons[13] (μ^+ and μ^-) and the neutrinos (ν_μ and $\bar{\nu}_\mu$) associated with muons in various weak processes constitute the other branch of the family.

ELECTRONS AND NEUTRINOS

One of the reasons that Pauli postulated the existence of the neutrino[14] was the necessity to conserve angular momentum in nuclear β decay. For example, if tritium underwent the decay $H^3 \rightarrow He^3 + \beta^-$ it would not be possible to balance the intrinsic angular momenta because the spins of H^3, He^3, and the electron are all $\frac{1}{2}$, and, according to the rules of quantum mechanics, there is no way to combine two spin-$\frac{1}{2}$ vectors to produce another spin-$\frac{1}{2}$ vector. This difficulty (as well as the problem of energy and momentum balance) disappears if an additional spin-$\frac{1}{2}$ particle is emitted along with the electron. Pauli's neutrino is therefore a *fermion*.

Just as is the case for other fermions (for example, electrons and nucleons), the neutrino has an antiparticle partner. The particle associated with nuclear β^- decay is called the *antineutrino* ($\bar{\nu}_e$) and that associated with nuclear β^+ decay is called the *neutrino* (ν_e). Nuclear β^- and β^+ decays can be represented as (see Eqs. 15.4)

$$\beta^- \text{ decay:} \quad n \longrightarrow p + e^- + \bar{\nu}_e \tag{16.1a}$$

$$\beta^+ \text{ decay:} \quad p \longrightarrow n + e^+ + \nu_e \tag{16.1b}$$

---

[11] Of course, a pair of leptons that have charge and mass will interact via the electromagnetic and gravitational force as well; but the essential point is that leptons have *no strong interactions whatsoever*.

[12] From the Greek word *leptos* (meaning "small").

[13] Recall that muons result from the decay of pions (see Section 16.3).

[14] See Section 5.11.

Now, it is possible to manipulate nuclear reaction equations in a way that is analogous to that for algebraic equations. For example, if we write

$$a + b = c + d$$

then we know that an equally valid expression is

$$a + b + (-c) = d$$

That is, the appearance of $+c$ on one side of the equation is equivalent to the appearance of $-c$ on the other side. In nuclear reaction equations, we can replace a *particle* on one side of an equation by its *antiparticle* on the other side.[15] Therefore, Eq. 16.1b can be written as

$$p + e^- \longrightarrow n + \nu_e \qquad\qquad (16.2)$$

This is just the expression for the *electron capture* process (see Eq. 15.4c). Similarly, we can also write, from Eq. 16.1b,

$$p + \overline{\nu}_e \longrightarrow n + e^+ \qquad\qquad (16.3)$$

In other words, *the emission of a neutrino is equivalent to the absorption of an antineutrino.*

In 1953 Cowan and Reines used the reaction indicated in Eq. 16.3 to demonstrate experimentally for the first time the existence of the neutrino. Because the reaction is one involving the weak interaction, the probability that a proton (which can be a *nuclear* proton) will capture an antineutrino is extremely small. In fact, the probability is so small that if 10^{12} antineutrinos were incident on the Earth, all except *one,* on the average, would pass through the entire Earth without interacting! Therefore, an extremely prolific source of antineutrinos is necessary if any $p + \overline{\nu}_e$ capture events are to take place. Cowan and Reines chose a nuclear reactor as their source because the fragments liberated in nuclear fission events undergo β^- decay and therefore, according to Eq. 16.1a, emit antineutrinos.

The detection scheme employed by Cowan and Reines made use of the unique characteristics of the history of the neutron and positron products of $p + \overline{\nu}_e$ capture. The emitted neutron undergoes a nuclear capture reaction in the detector and produces one or more high-energy γ rays; the positron annihilates with an electron and produces two 0.51-MeV γ rays. The simultaneous appearance of these two types of γ-ray events is a unique signature of the emission of a neutron and a positron; such occurrences unmistakably signify that a $p + \overline{\nu}_e$ capture has taken place. Notice that this detection scheme is sensitive only to *antineutrinos;* if a *neutrino* interacts with a nucleus, it will be absorbed by a neutron according to the reaction

$$n + \nu_e \longrightarrow p + e^-$$

and no signature such as occurs for $p + \overline{\nu}_e$ capture will result.

A series of carefully conducted experiments enabled Cowan and Reines to obtain unambiguous evidence for the detection of events induced by antineutrinos and to establish finally that neutrinos and antineutrinos are in fact *real* particles, not just inventions designed to salvage the conservation laws in nuclear β decay.

---

[15]The algebraic manipulation is possible because c + (−c) = 0. Similarly, a particle plus its antiparticle equals "zero;" that is, they can *annihilate.*

THE DISTINGUISHABILITY OF NEUTRINOS

Several experiments have shown conclusively that the massless particles that are emitted in β^- decay are *not* the same as those emitted in β^+ decay; that is, ν_e and $\overline{\nu}_e$ are *distinguishable particles*. This fact has been demonstrated, for example, by observing that the reaction

$$\text{Cl}^{37} + \overline{\nu}_e \longrightarrow \text{Ar}^{37} + e^- \quad \text{(not allowed)}$$

does *not* take place, but that the reaction *is* allowed if ν_e is substituted for $\overline{\nu}_e$:

$$\text{Cl}^{37} + \nu_e \longrightarrow \text{Ar}^{37} + e^- \quad \text{(allowed)} \tag{16.4}$$

In what way is a neutrino different from an antineutrino? Since these particles have no mass or charge and always travel with the velocity of light, the distinction between ν_e and $\overline{\nu}_e$ can depend only on the dynamic properties of the particles; these properties are specified by the intrinsic angular momentum and linear momentum vectors. Experiments have shown that the vectors, **S** and **p,** have the *same* direction for antineutrinos whereas these vectors have *opposite* directions for neutrinos (Fig. 16.5). An antineutrino always advances in the direction in which a right-hand screw advances when turned in the same direction as its "spinning motion." Therefore, we say that an antineutrino is *right-handed;* similarly, a neutrino is said to be *left-handed.* (In modern terminology, we say that an antineutrino has *positive helicity* and that a neutrino has *negative helicity.*)

MUONS AND NEUTRINOS

When pions decay to muons, neutrinos are emitted:

$$\left.\begin{aligned}\pi^+ &\longrightarrow \mu^+ + \nu_\mu \\ \pi^- &\longrightarrow \mu^- + \overline{\nu}_\mu\end{aligned}\right\} \tag{16.5}$$

The muons formed in these decays are also unstable, but muon decay differs in an important respect from pion decay. Unlike the "two-body" pion decay, muon decay is a "three-body" process in which *two* neutrinos are emitted:

$$\left.\begin{aligned}\mu^+ &\longrightarrow e^+ + \overline{\nu}_\mu + \nu_e \\ \mu^- &\longrightarrow e^- + \nu_\mu + \overline{\nu}_e\end{aligned}\right\} \tag{16.6}$$

Again, experiments concerning the production and absorption of muon neutrinos have shown that ν_μ and $\overline{\nu}_\mu$ are distinguishable particles; ν_μ is left-handed and $\overline{\nu}_\mu$ is right-handed. Of equal importance for our theories

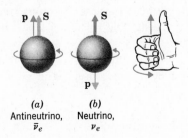

Fig. 16.5 *The spin vector* (**S**) *and the linear momentum vector* (**p**) *have the same direction for* $\overline{\nu}_e$ *and have opposite directions for* ν_e. *The arrows on the circles show the directions in which the particles are "spinning" and the direction of the spin vector is then determined by using the right-hand rule.*

(a)
Antineutrino,
$\overline{\nu}_e$

(b)
Neutrino,
ν_e

Table **16.1**   *Properties of Muons*

	$\mu^{\pm}$
Mass	206.77 m_e
Mass-energy	105.659 MeV
Spin	$\frac{1}{2}$
Half-life	1.5×10^{-6} sec
Decay modes	$\mu^+ \longrightarrow e^+ + \bar{\nu}_{\mu} + \nu_e$
	$\mu^- \longrightarrow e^- + \nu_{\mu} + \bar{\nu}_e$

of elementary particles is the fact that electron neutrinos do not participate in weak interaction processes in the same ways that muon neutrinos do. That is, the neutrino pair ν_e, $\bar{\nu}_e$ is *not* the same as the neutrino pair ν_{μ}, $\bar{\nu}_{\mu}$.

Electron neutrinos have $s = \frac{1}{2}$, zero charge, and zero mass; one is right-handed and one is left-handed. So also for muon neutrinos. In what way, then, can electron neutrinos and muon neutrinos be so different that the weak interaction can distinguish between them? No one knows.

THE CONSERVATION OF LEPTONS

In the various weak processes an extremely wide variety of interactions is possible. Neutrinos (and antineutrinos) can be emitted or absorbed, muons can be formed and decay, electrons and positrons can be emitted or can combine and annihilate. Is there any underlying order to this profusion of events? A clue to the answer is to be found in the fact that certain processes simply do not occur. For example, a muon always decays by the emission of an electron and two neutrinos (Eq. 16.6), but never by the emission of an electron and a γ ray. Why should this latter process be inhibited—in fact, apparently absolutely prohibited?

Numerous experiments with elementary particles indicate that there is a general rule (which is *believed,* but is not derived from any fundamental theory) to the effect that *any process that is not forbidden by some conservation principle will occur.* Therefore, we are led to expect that there exists a new conservation law which prevents certain processes that would be permitted if only energy, momentum, and charge conservation were effective.

The principle that we seek is actually quite a simple one. It is: *the number of leptons in an isolated system remains constant.* But we know that leptons can be created and annihilated; so if the rule is to be obeyed, we must count leptons in a particular way. To each lepton *particle* (e^-, ν_e, μ^-, ν_{μ}) we assign a *lepton number,* $N_l = +1$, and to each lepton *antiparticle* (e^+, $\bar{\nu}_e$, μ^+, $\bar{\nu}_{\mu}$) we assign a lepton number, $N_l = -1$ (see Table 16.2). All other elementary particles (such as photons, mesons, and nucleons) have lepton number $N_l = 0$. The rule then states that *the sums of the lepton numbers on each side of any reaction or decay equation must be equal.*

By counting leptons according to this scheme, it is easy to verify that leptons are conserved in all weak interaction processes. For example,

$$\left. \begin{array}{c} n \longrightarrow p + e^- + \bar{\nu}_e \\ \text{Lepton number, } N_l: \quad 0 \; = \; 0 + 1 \; + (-1) \end{array} \right\}$$

(16.7a)

$$\left.\begin{array}{l} \mu^+ \longrightarrow e^+ + \bar{\nu}_\mu + \nu_e \\ \text{Lepton number, } N_l: \quad -1 \quad = \quad -1 + (-1) + 1 \end{array}\right\} \qquad (16.7b)$$

Leptons are clearly conserved in the process, $\mu^+ \rightarrow e^+ + \gamma$, but we know that this particular decay does not occur. There is a more restrictive way of stating the principle of lepton conservation which delineates more sharply the types of processes that are allowed. In this scheme we separate the lepton family of particles into an electron branch and a muon branch, and we assign *electron numbers* (N_e) and *muon numbers* (N_μ) to each particle. For members of the electron branch, the electron numbers are the same as the lepton numbers, but the muon numbers are all *zero*. Similarly, for members of the muon branch, the muon numbers are the same as the lepton numbers, but the electron numbers are all *zero* (see Table 16.2). Thus, for every particle we have $N_l = N_e + N_\mu$. The restrictive principle of lepton conservation then states that *the electron number and the muon number are individually conserved in any weak interaction process.* For example,

$$\left.\begin{array}{l} \qquad\qquad \mu^+ \longrightarrow e^+ + \bar{\nu}_\mu + \nu_e \\ \text{Lepton number, } N_l: \quad -1 \quad = \quad -1 + (-1) + 1 \\ \text{Electron number, } N_e: \quad 0 \quad = \quad -1 + \quad 0 \quad + 1 \\ \text{Muon number, } N_\mu: \quad -1 \quad = \quad \quad 0 + (-1) + 0 \end{array}\right\} \qquad (16.8)$$

This principle explicitly forbids the decay, $\mu^+ \rightarrow e^+ + \gamma$, because neither electron number nor muon number is conserved in the process.

The restrictive principle of lepton conservation is a powerful physical principle. Just as for the other conservation laws (linear momentum, angular momentum, mass-energy, and electric charge), there is no case known for which the lepton principle is violated. What is the deeper meaning of this rule that is so strictly obeyed by the weak interaction? No one yet knows. We can only remark that the lepton principle appears to be the simplest way of summarizing the results of a large number of experiments. This is, at least, the first step toward the construction of a fundamental theory of weak interactions.

WHAT IS THE QUANTUM OF THE WEAK INTERACTION?

The pion is the mediator of the strong interaction between a pair of nucleons. Does there exist a cousin of the pion that is responsible for

Table **16.2** *Lepton Quantum Numbers*

Particle or Anti-Particle	Lepton Number, N_l	Electron Number, N_e	Muon Number, N_μ
e^-	+1	+1	0
e^+	-1	-1	0
ν_e	+1	+1	0
$\bar{\nu}_e$	-1	-1	0
μ^-	+1	0	+1
μ^+	-1	0	-1
ν_μ	+1	0	+1
$\bar{\nu}_\mu$	-1	0	-1

mediating the weak interaction between a pair of leptons? The answer to this question is not known. (But see Problem 16.10.) It is even possible that the field theory approach to the weak interaction (which necessarily involves the introduction of a mediating particle) is not applicable. However, field theory has been so successful in other areas of physics that there is a natural reluctance to forego this approach to the problem of the weak interaction. There continues to be great theoretical and experimental interest in the weak interaction and only time will reveal whether our present lines of attack will be successful and will lead to a fundamental understanding of the weakly interacting aspects of matter.

16.3   *Strongly Interacting Particles*

MESONS

The term *meson* is applied to any boson that participates directly or indirectly in the propagation of the strong interaction.[16] The best known meson is the π meson or *pion*. Later in this section we mention other types of mesons (the K and η mesons) and their relationship to the strong interaction.

All mesons decay eventually into electrons and neutrinos; the half-lives for these decay processes range from $\sim 10^{-8}$ sec to $\sim 10^{-23}$ sec. All of the mesons with masses less than about 0.6 of the proton mass have "long" half-lives ($\gtrsim 10^{-18}$ sec) and are therefore "stable" from the standpoint of elementary processes (see the discussion in the introductory section of this chapter). The 8 "stable" mesons are listed in Table 16.3. The heavy mesons, $m \gtrsim 0.6 m_p$, are all "unstable" and have half-lives $\lesssim 10^{-21}$ sec. We shall soon see the reason for the "stability" of certain mesons and the "instability" of others.

PIONS

By far, the most important and best understood member of the meson family is the *pion*. The reason for this status is, of course, the role played by the pion as the quantum of the strong nuclear force. As pointed out in Section 8.5, Yukawa realized the necessity for the existence of a particle of intermediate mass[17] that would propagate the attractive force between nucleons. Although Yukawa's prediction was made in 1935, the pion was not discovered until a 1947 experiment revealed the presence of these mesons in cosmic rays. In the following year, pions were first produced artificially in an accelerator and since that time we have had available beams of pions for use in detailed studies of their properties and interactions.

There are two types of charged pions, π^+ and π^- (which are the antiparticles of one another), and a neutral pion, π^0 (which is its own antiparti-

---

[16] The *muon,* which is the primary decay product of the pion, is sometimes called the μ *meson.* But muons are *leptons* (particles that participate in the *weak* interaction) and are not properly classified as mesons.

[17] See Section 12.7 where the uncertainty principle is used to estimate the pion mass from the range of the nuclear force.

Table **16.3** *"Stable" Mesons*[a]

Particle	Symbol	Approximate Mass	Approximate Half-Life	Anti-Particle[b]
π meson (pion)	π^+	$270m_e$	10^{-8} sec	π^-
	π^0	$260m_e$	10^{-16} sec	(π^0)
K meson (kaon)	K^+	$965m_e$	10^{-8} sec	K^-
	K^0	$975m_e$	(c)	$\overline{K^0}$
η meson	η^0	$1080m_e$	10^{-18} sec	(η^0)

[a] These mesons are "stable" in the sense of elementary processes; that is, $\tau_{1/2} \gtrsim 10^{-18}$ sec. The spin of each of these mesons is *zero* (they are all *bosons*).

[b] Notice that the π^0 and η^0 mesons are their own anti-particles.

[c] K mesons are extremely complicated particles; some of their unusual properties are described later in this section and in Section 16.4. The neutral kaons, K^0 and $\overline{K^0}$, do not even have a unique half-life because they are approximately 50-50 mixtures of two other kaons, K_1^0 and K_2^0. The half-life of K_1^0 is $\sim 10^{-10}$ sec and the half-life of K_2^0 is $\sim 10^{-8}$ sec.

cle). The masses of the charged pions are approximately 9 electron masses greater than the mass of the neutral pion. Pions have zero spin and are therefore *bosons*. A summary of the properties of pions is given in Table 16.4.

If the bombarding energy is sufficient, pions are produced copiously in nucleon-nucleon collisions. Single pion and multiple pion production is possible (the latter, of course, requires greater input energy); for example,

$$\left.\begin{aligned}
p + p &\longrightarrow p + n + \pi^+ \\
n + p &\longrightarrow n + p + \pi^0 \\
p + p &\longrightarrow p + p + \pi^+ + \pi^- + \pi^0
\end{aligned}\right\} \tag{16.9}$$

Pion production by γ-ray irradiation of nucleons (*photoproduction*) also occurs:

$$\left.\begin{aligned}
\gamma + p &\longrightarrow p + \pi^0 \\
&\longrightarrow n + \pi^+
\end{aligned}\right\} \tag{16.10}$$

The photoproduction of pions is much less probable than pion production in nucleon-nucleon collisions because the strength of the electromagnetic

Table **16.4** *Properties of Pions*

	π^+	$\pi^\pm$	π^-	π^0
Mass		$273.14m_e$		$264.14m_e$
Mass-energy		139.58 MeV		134.98 MeV
Charge	$+e$		$-e$	0
Spin		0		0
Half-life		1.8×10^{-8} sec		1.4×10^{-16} sec
Decay products	$\mu^+ + \nu_\mu$	($\sim 100\%$)	$\mu^- + \bar{\nu}_\mu$	$\gamma + \gamma$ (98.8%)
	$e^+ + \nu_e$	($\sim 0.01\%$)	$e^- + \bar{\nu}_e$	$e^+ + e^- + \gamma$ (1.2%)
	$\mu^+ + \nu_\mu + \gamma$	($\sim 0.01\%$)	$\mu^- + \bar{\nu}_\mu + \gamma$	$e^+ + e^- + e^+ + e^-$ (0.0035%)

interaction (which is responsible for photoproduction processes) is $\sim \frac{1}{100}$ of that of the strong interaction (which is responsible for pion production in nucleon-nucleon collisions).

Pions are also produced in nucleon-antinucleon annihilation events; for example,

$$\bar{p} + p \longrightarrow \pi^+ + \pi^- + \pi^0 \tag{16.11}$$

Indeed, the nucleon-nucleon force is such a tremendously strong interaction, that pions (the quanta of that interaction) are produced with relative ease in any type of energetic nucleon process.

THE DECAY OF PIONS

The first pion decays to be observed took place in photographic emulsions exposed to cosmic rays (see Fig. 16.6). When a π^+ meson comes to rest in

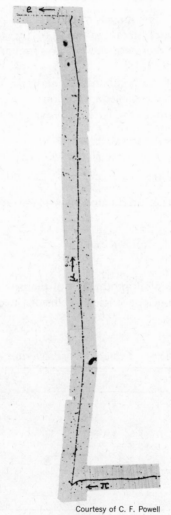

Fig. 16.6 *Decay of a π meson in a photographic emulsion exposed to cosmic rays. The pion comes to rest in the emulsion (at the bottom of the picture). In the decay process, a muon (μ) is emitted (along with a neutrino which leaves no track in the emulsion). The muon eventually comes to rest in the emulsion (at the top of the picture) and decays with the emission of an electron.*

Courtesy of C. F. Powell

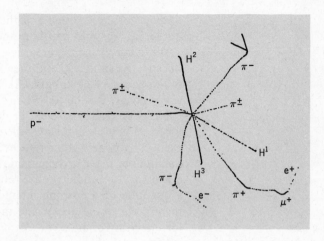

Fig. 16.7 *This tracing of a nuclear emulsion photograph shows several different types of events involving pions. An antiproton is incident from the left and is annihilated by a proton within a nucleus in the emulsion. The annihilation process disintegrates the nucleus and produces 5 charged mesons. Nucleons are also emitted in the forms H^1, H^2, and H^3. In the lower portion of the diagram, one pion decays in the normal way by the emission of a muon, and one pion undergoes a rare electron decay. In the upper right-hand portion of the diagram a π^- meson is absorbed by a nucleus and emits two heavy, charged fragments. To find all of these various processes occurring in a single event is extremely rare.*

an emulsion,[18] it emits a muon which always has the same range in the emulsion. That is, in the π-μ decay, the muon always has the same energy. This implies (momentum conservation!) that the decay is a two-body process, because if the decay resulted in three or more particles, the muon would not always have a unique energy. (Compare the case of nuclear β decay, discussed in Sections 5.11 and 15.2.) The particle emitted along with the muon in pion decay is a muon neutrino:

$$\left. \begin{aligned} \pi^+ &\longrightarrow \mu^+ + \nu_\mu \\ \pi^- &\longrightarrow \mu^- + \bar{\nu}_\mu \end{aligned} \right\} \tag{16.12}$$

In a small fraction of events ($\sim$0.01 percent), a charged pion decays directly into an electron and a neutrino:

$$\left. \begin{aligned} \pi^+ &\longrightarrow e^+ + \nu_e \\ \pi^- &\longrightarrow e^- + \bar{\nu}_e \end{aligned} \right\} \tag{16.13}$$

The neutral pion, on the other hand, decays predominantly into a pair of γ rays:

$$\pi^0 \longrightarrow \gamma + \gamma \tag{16.14}$$

Charged pions live, on the average, about 10^{-8} sec, whereas neutral pions exist for only about 10^{-16} sec (see Table 16.4). Why do these pions live as long as they do, and why is there such a tremendous disparity between the

---

[18]When a π^- meson comes to rest, it is quickly attracted to and absorbed by a nucleus, thereby causing a nuclear disintegration (see Fig. 16.7). Negative pions therefore rarely have an opportunity to decay when in the presence of matter.

lifetimes of the charged and the neutral variety? In order to answer these questions, we must examine the manner in which these particles decay.

When an elementary particle decays, it always does so predominantly through the strongest (and therefore the *fastest*) interaction that is available to the particle (see Table 16.5). If it is allowed, decay via the strong interaction always dominates. When a proton is supplied with 300 MeV of energy, it forms an "excited state" in much the same way that a hydrogen atom forms its first excited state when it is supplied with 10.2 eV of energy. An excited hydrogen atom returns to its ground state by the emission of a quantum of the electromagnetic field that binds the atom together, namely, a *photon*. An excited proton returns to its ground state by the emission of a quantum of the strong interaction field, namely, a *pion*. This process is therefore a *strong* decay, and measurements have shown that the half-life for the decay of an excited proton is $\sim 10^{-23}$ sec. *This time is typical of all strong interaction processes.*

Although pions interact strongly with nucleons and are created in strong interaction processes, pions cannot *decay* via the strong interaction because there are no less massive particles that interact strongly. Therefore, pion decay must involve either the electromagnetic or the weak interaction. Charged pions cannot undergo a purely electromagnetic decay because the electric charge must be carried off by some kind of *particle*. The decay of charged pions is therefore restricted to the weak interaction, and the products of π^+ and π^- decays are muons, electrons, and neutrinos—all weakly interacting leptons. Since decays via the weak interaction are slow processes, we expect charged pions to be relatively long-lived particles, and, indeed, the half-life is $\sim 10^{-8}$ sec. Neutral pions decay through the stronger electromagnetic interaction and have a half-life $\sim 10^{-16}$ sec.

Table **16.5** *Approximate Relative Strengths of the Basic Forces*[a]

Strong	1
Electromagnetic	10^{-2}
Weak	10^{-13}
Gravitational	10^{-38}

[a] For greater detail, see Table 6.2.

KAONS

During the 10-year period following the discovery of the pion in 1947, studies of cosmic-ray interactions, as well as investigations of reactions produced by high-energy particles in the newly constructed accelerators, revealed more than 30 elementary particles. Many of the newly discovered particles exhibited unusual properties and these particles became collectively known as *strange particles.*

The first of the strange particles to be discovered were the K mesons or *kaons*. We now know that there are *four* types of kaons—a positively charged kaon (K^+), a neutral kaon (K^0), and the two corresponding antiparticles

(K^- and $\overline{K^0}$). Kaons are formed in high-energy collisions between strongly interacting particles, for example, in proton-proton or pion-proton collisions. The charged kaons have been observed to decay by a variety of modes involving leptons and pions; for example:

$$
\left.
\begin{aligned}
K^+ &\longrightarrow \mu^+ + \nu_\mu && (64\%) \\
&\longrightarrow \pi^+ + \pi^0 && (21\%) \\
&\longrightarrow \pi^+ + \pi^+ + \pi^- && (\ 6\%) \\
&\longrightarrow e^+ + \nu_e + \pi^0 && (\ 5\%)
\end{aligned}
\right\}
\qquad (16.15)
$$

In addition, there are several other decay modes that have been observed to occur with low probabilities. The corresponding K^- decays are obtained by converting each particle into its antiparticle. The decay of a charged kaon into three charged pions is shown in Fig. 16.8.

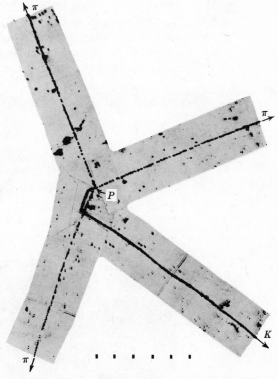

Fig. 16.8 *The decay of a charged K meson (the sign of the charge cannot be determined) into three charged pions. The kaon enters from the lower right and undergoes a scattering through almost 90° before coming to rest at P where it decays. The event occurred in a photographic emulsion exposed to cosmic rays.*

Courtesy of C. F. Powell

The neutral kaons, K^0 and $\overline{K^0}$, are unique among all elementary particles. Many experiments have shown that neither K^0 nor $\overline{K^0}$ is a well-defined particle; rather, each neutral kaon is an almost equal mixture of two other particles, K_1^0 and K_2^0. (But notice that although K^0 and $\overline{K^0}$ are antiparticles of one another, K_1^0 and K_2^0 do *not* have this property.) The K_1^0 and K_2^0 particles have different half-lives and are identified experimentally by virtue of their different decay modes:

$$K_1^0 \longrightarrow \pi^+ + \pi^- \qquad K_2^0 \longrightarrow \pi^+ + e^- + \overline{\nu}_e$$
$$\longrightarrow \pi^0 + \pi^0 \qquad\qquad \longrightarrow \pi^- + e^+ + \nu_e$$
$$\longrightarrow \pi^+ + \mu^- + \overline{\nu}_\mu \tag{16.16}$$
$$\longrightarrow \pi^- + \mu^+ + \nu_\mu$$
$$\longrightarrow \pi^+ + \pi^- + \pi^0$$
$$\longrightarrow \pi^0 + \pi^0 + \pi^0$$

The K_1^0 half-life is approximately 6×10^{-11} sec, whereas the K_2^0 is considerably longer-lived ($\tau_{1/2} \cong 4 \times 10^{-8}$ sec).

Kaons exhibit the most complicated behavior of all the mesons. Why do kaons undergo both leptonic and nonleptonic (pion) decays? Why is there such a variety of decay modes? Why do the neutral kaons consist of combinations of still other particles. No one really knows. Do kaons play a significant role in the strong interaction? The answer to this question seems to be "yes," but no one knows the manner nor the extent to which kaons mediate the strong force. If kaons did not exist (indeed, if there were *no* strange particles at all!), we would probably feel quite secure in the belief that pions are the sole mediators of the strong interaction and that no other particles are required to "understand" the strong force. But such a statement merely reflects the inadequacy of our information at present and the difficulty that we now experience in understanding the significance of the strange particles.

HYPERONS

The first elementary particles to be found with masses greater than that of the proton were the *lambda particles*. Although these strange particles (Λ^0 and $\overline{\Lambda^0}$) are electrically neutral, they are readily identified by the V-shaped tracks that the charged decay products leave in emulsions or bubble chambers. Figure 16.9 shows a bubble-chamber photograph of a proton-antiproton collision that results in the production of a $\Lambda^0 - \overline{\Lambda^0}$ pair.

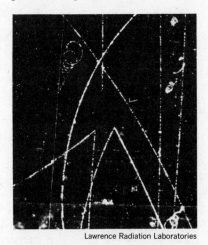

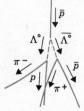

Lawrence Radiation Laboratories

Fig. 16.9 *Production of a $\Lambda^0 - \Lambda$ pair by a* p − p̄ *collision in a bubble chamber. The Λ^0 decays into a proton and a π^- meson, and the $\overline{\Lambda^0}$ decays into an antiproton and a π^+ meson.*

Each lambda particle travels only a short distance in the chamber before undergoing decay:

$$\Lambda^0 \longrightarrow \pi^- + p; \qquad \overline{\Lambda^0} \longrightarrow \pi^+ + \bar{p} \tag{16.17}$$

Following the discovery of the lambda particles, several additional heavy strange particles were found; these particles bear the labels Σ (sigma particles), Ξ (xi particles), and Ω (omega particles). This group of strange particles together with the lambda particles are collectively called *hyperons*.[19]

All of the hyperons undergo decays that lead to *nucleons*. Because of this fact, hyperons and nucleons are grouped together and called *baryons*.[20] Some of the properties of baryons are given in Table 16.6.

Table **16.6**   *Baryons*

	Particle	Mass-Energy (MeV)	Half-Life (sec)	Spin	Anti-Particle[a]
Nucleons	p	938.2	(stable)	$\frac{1}{2}$	$\bar{p}$
	n	939.6	760	$\frac{1}{2}$	$\bar{n}$
Hyperons	Λ^0	1115.5	1.7×10^{-10}	$\frac{1}{2}$	$\overline{\Lambda^0}$
	Σ^+	1189.5	5.6×10^{-11}	$\frac{1}{2}$	$\overline{\Sigma^+}$
	Σ^0	1192.5	$<10^{-14}$	$\frac{1}{2}$	$\overline{\Sigma^0}$
	Σ^-	1197.4	1.1×10^{-10}	$\frac{1}{2}$	$\overline{\Sigma^-}$
	Ξ^0	1314.9	2.0×10^{-10}	$\frac{1}{2}$	$\overline{\Xi^0}$
	Ξ^-	1321.3	1.2×10^{-10}	$\frac{1}{2}$	$\overline{\Xi^-}$
	Ω^-	1672	7.6×10^{-11}	$\frac{3}{2}$	$\overline{\Omega^-}$

[a] Note, for example, that $\overline{\Sigma^+}$ (the anti-particle of the positively-charged Σ^+ hyperon) actually carries a *negative* charge. However, the $\overline{\Sigma^+}$ hyperon is *not* the same as the Σ^- hyperon—these particles are quite distinct (they have different masses and half-lives and they are formed and decay in different ways).

Some typical decays of the hyperons are

$$\left. \begin{array}{l} \Sigma^+ \longrightarrow p + \pi^0 \\ \Sigma^0 \longrightarrow \Lambda^0 + \gamma \\ \Xi^- \longrightarrow \Lambda^0 + \pi^- \\ \Omega^- \longrightarrow \Lambda^0 + K^- \end{array} \right\} \tag{16.18}$$

The Ω^- hyperon has a special significance because its existence was predicted before its discovery. In 1961, the theorists M. Gell-Mann and Y. Ne'eman, working independently, devised a scheme for the classification of strongly interacting elementary particles. All the known particles fitted neatly into this scheme, but there was one unfilled position. The missing particle was expected to exist only as a negatively charged particle labeled Ω^- (and its positively charged antiparticle, $\overline{\Omega^-}$). The mass of the Ω^- hyperon was predicted to be approximately 3284 m_e. An experiment was designed specifically to search for the Ω^- hyperon and in 1964 it was found. The mass

---

[19] From the Greek word *hyper* (meaing "over" or "above") because all these particles are more massive than the proton.

[20] From the Greek word *barys* (meaning "heavy").

has recently been determined (by using the bubble chamber photograph shown in Fig. 16.10) to be 3272 m_e with an uncertainty of only 2 m_e. Thus, the Gell-Mann-Ne'eman theory has been confirmed in a striking fashion.[21]

THE CLASSIFICATION OF ELEMENTARY PARTICLES

Because of the large number of elementary particles that have been discovered, it proves convenient to classify these particles into various groups. Although we have already mentioned the categories, we summarize here the classification scheme for reference.

The two main groups of elementary particles are those that interact strongly (called *hadrons*) and those that interact weakly (called *leptons*). The hadrons consist of *mesons* and *baryons,* and the latter are further subdivided into *nucleons* and *hyperons.* (The *strange particles* include kaons and hyperons.) The leptons consist of electrons, muons, and neutrinos. The "stable" elementary particles can therefore be grouped as follows:

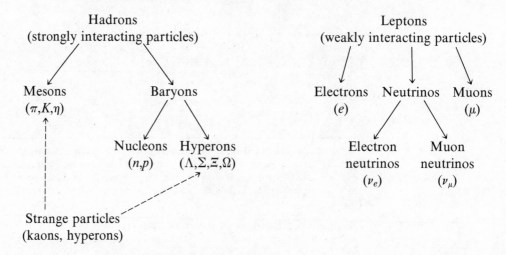

CONSERVATION OF BARYONS

In the discussions of mass-energy conservation (Section 7.17) and fission (Section 15.5), it was pointed out that energy can be extracted from a nucleus by rearranging the protons and neutrons into a configuration having a greater binding energy; but the nucleons themselves cannot be destroyed. We also know that antinucleons are created only in conjunction with nucleons. These facts can be summarized in a conservation principle for nucleons analogous to that which is obeyed by leptons. If we count each nucleon as +1 particle and each antinucleon as −1 particle, we can state that *the number of nucleons remains constant in any type of process.*

The principle of *nucleon conservation* is obeyed *exactly* in all nuclear processes (such as reactions, fission, and radioactive decay); but we can go even further. We take as our cue the fact that all of the hyperons undergo decays that lead (in one or two steps) to nucleons. Therefore, we consider

---

[21]Gell-Mann received the 1969 Nobel Prize in physics for his work on classification schemes for elementary particles, one result of which was the prediction of the Ω^- hyperon.

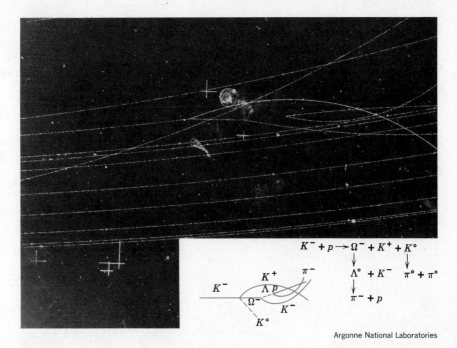

Fig. 16.10 *This bubble-chamber photograph (taken in 1968) of the production and decay of an Ω^- hyperon was used to determine the mass of the particle (3272 m_e). The decay histories of the Ω^- and the particles produced with it are shown in the inset. By 1969 only about 20 events involving Ω^- hyperons had been observed.*

the entire baryon class of particles and assign a baryon number, $B = +1$, to each *particle* and a baryon number, $B = -1$, to each *antiparticle* (all leptons and mesons have $B = 0$). Then, we can make the general statement that *the baryon number is conserved in any type of process*.

Some *decay processes* that conserve baryon number and are therefore *allowed* (and have been observed) are:

$$\left.\begin{aligned} n &\longrightarrow p + e^- + \bar{\nu}_e \\ \Sigma^- &\longrightarrow n + \pi^- \\ \overline{\Lambda^0} &\longrightarrow \bar{p} + \pi^+ \end{aligned}\right\} \tag{16.19}$$

The following *reactions* also conserve baryon number (and have been observed):

$$\left.\begin{aligned} \pi^- + p &\longrightarrow K^0 + \Lambda^0 \\ \overline{K^0} + p &\longrightarrow \Sigma^+ + \pi^0 \\ \mu^- + p &\longrightarrow n + \nu_\mu \\ K^- + d &\longrightarrow \Lambda^0 + p + \pi^- \end{aligned}\right\} \tag{16.20}$$

Two decays that do *not* conserve baryon number are:

$$\left.\begin{aligned} \Sigma^- &\longrightarrow \bar{p} + \pi^0 \\ \Lambda^0 &\longrightarrow K^+ + K^- \end{aligned}\right\} \text{ (forbidden)}$$

These decays are *forbidden* and have not been observed.

STRANGENESS

Baryon conservation alone is not sufficient to account for the fact that certain processes involving hadrons do not occur. For example, when protons are bombarded by positive pions, the reaction

$$p + \pi^+ \longrightarrow \Sigma^+ + \pi^+ \quad \text{(forbidden)}$$

never occurs even though baryon number is conserved. Just as in the case of leptons, we need a more restrictive conservation principle. For this purpose a new quantum number, called *strangeness*, was proposed by M. Gell-Mann and K. Nishijima in 1953. The classification of hadrons according to the strangeness quantum number $\mathcal{S}$ is given in Table 16.7, and the corresponding conservation principle is that *strangeness is conserved in strong interactions* (but not in weak interactions).

Table **16.7** *Strangeness Quantum Numbers for the "Stable" Hadrons*

Strangeness, $\mathcal{S}$ \ Charge, q	$+e$	0	$-e$
$+3$	$\overline{\Omega^-}$		
$+2$	$\overline{\Xi^-}$	$\overline{\Xi^0}$	
$+1$	$K^+, \overline{\Sigma^-}$	$K^0, \overline{\Lambda^0}, \overline{\Sigma^0}$	$\overline{\Sigma^+}$
0 (a)	π^+, p	$\pi^0, \eta^0, n, \overline{n}$	$\pi^-, \overline{p}$
-1	Σ^+	$\overline{K^0}, \Lambda^0, \Sigma^0$	K^-, Σ^-
-2		Ξ^0	Ξ^-
-3			Ω^-

ᵃ Pions, the η^0 meson, and nucleons are not *strange particles,* and they have the quantum number $\mathcal{S} = 0$. The strangeness quantum number is not defined for leptons.

Some reactions that conserve strangeness as well as baryon number and are therefore *allowed* (and have been observed) are:

$$
\begin{aligned}
&\text{(Fig. 16.9)} \quad
\begin{array}{lllll}
& p + & \overline{p} & \longrightarrow \Lambda^0 + & \overline{\Lambda^0} \\
\mathcal{S}: & 0 + & 0 & = (-1) + & 1 \\
B: & 1 + & (-1) = & 1 & + (-1)
\end{array}
\end{aligned}
\tag{16.21a}
$$

$$
\begin{aligned}
&\text{(Fig. 16.10)} \quad
\begin{array}{lllll}
& p + & K^- & \longrightarrow \Omega^- + K^+ + & K^0 \\
\mathcal{S}: & 0 + & (-1) & = (-3) + 1 & + 1 \\
B: & 1 + & 0 & = 1 + 0 & + 0
\end{array}
\end{aligned}
\tag{16.21b}
$$

$$
\begin{aligned}
&\text{(Fig. 16.11)} \quad
\begin{array}{llll}
& p + & \pi^- & \longrightarrow \Lambda^0 + K^0 \\
\mathcal{S}: & 0 + & 0 & = (-1) + 1 \\
B: & 1 + & 0 & = 1 + 0
\end{array}
\end{aligned}
\tag{16.21c}
$$

These reactions all take place within extremely short time intervals ($\sim 10^{-23}$ sec), indicating that they proceed via the strong interaction. A bubble-chamber photograph of the $p + \pi^- \to \Lambda^0 + K^0$ reaction is shown in Fig. 16.11.

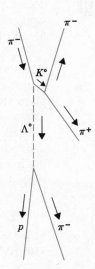

Fig. 16.11 *A negative pion, incident from the top, strikes a proton in a bubble chamber and produces two strange particles according to the reaction* $p + \pi^- \rightarrow \Lambda^0 + K^0$. *The Λ^0 and K^0 particles both decay into charged particles within the chamber.*

Some strangeness-forbidden reactions (which have not been observed) are:

$$\left. \begin{aligned} n + \pi^- &\longrightarrow K^- + \Sigma^0 \\ p + \Lambda^0 &\longrightarrow \Sigma^+ + \Xi^0 \end{aligned} \right\} \quad \text{(forbidden)}$$

(Check that baryon number is conserved in these reactions but that the strangeness quantum number does not balance.)

If we refer to Table 16.6 we see that all of the hyperons have "long" lifetimes ($\tau_{1/2} \gtrsim 10^{-18}$ sec) and are therefore "stable." But hyperons are strongly interacting particles (hadrons), and typical times for strong decays are $\sim 10^{-23}$ sec. Why, then, should the hyperon lifetimes be so long? The answer to this question can be found by examining the decay modes of the hyperons; for example (see Eqs. 16.18),

$$\left. \begin{aligned} \Sigma^+ &\longrightarrow p + \pi^0 \\ \mathcal{S}: \quad (-1) &\neq 0 + 0 \end{aligned} \right\} \tag{16.22a}$$

$$\left. \begin{aligned} \Omega^- &\longrightarrow \Lambda^0 + K^- \\ \mathcal{S}: \quad (-3) &\neq (-1) + (-1) \end{aligned} \right\} \tag{16.22b}$$

These decays *do not conserve strangeness*. On the other hand, there are no strangeness-conserving decay modes that are available to these particles. For example, the decays $\Sigma^+ \to K^0 + p$, and $\Omega^- \to K^- + \Sigma^0$, which would conserve strangeness and baryon number, are prevented from occurring by mass-energy conservation. (Verify that the combined mass of the decay products on the right-hand side exceeds the mass of the particle on the left-hand side in each case.)

We must conclude that when hyperons decay, strangeness conservation is necessarily violated. That is, hyperon decays cannot take place through the *strong* interaction (which *does* conserve strangeness) but, instead, the decays proceed through the *weak* interaction[22] (which is *not* restricted by strangeness conservation). Hyperon decays are therefore *slow* processes and, hence, the particle lifetimes are *long* (that is, $\tau_{1/2} \gtrsim 10^{-18}$ sec).

The weak decay of hyperons is not a process as simple as the relatively well understood β decay of the neutron, $n \to p + e^- + \bar{\nu}_e$, which, of course, is also a weak decay. Hyperon decays such as $\Sigma^- \to n + \pi^-$ do *not* take place via the weak interaction in the same way that β decay takes place. These *nonleptonic* weak decays of hyperons are not well understood.

There do exist certain rare decay modes of hyperons that conserve strangeness. These are the *leptonic* decays, such as $\Sigma^- \to \Lambda^0 + e^- + \bar{\nu}_e$, but the interaction is more complicated than the β-decay interaction. Greater success has been achieved in the interpretation of these strangeness-conserving leptonic decays than in the case of the strangeness-violating nonleptonic decays.

THE NEW CONSERVATION PRINCIPLES

The study of leptons and hadrons has provided us with three new conservation principles:

1. Electron number and muon number are independently conserved in *all* processes.
2. Baryon number is conserved in *all* processes.
3. Strangeness is conserved in all *strong* interactions.

The first two of these conservation principles appear never to be violated; they are *absolute* conservation laws on a par with the conservation laws of linear momentum, angular momentum, mass-energy, and electrical charge. The strangeness principle, however, is a different type of conservation principle in that it is limited—strangeness conservation holds only for *strong* interactions, but not for *weak* interactions. These facts may well be essential clues in the eventual discovery of the nature of the forces by which elementary particles interact. But at the present time the new conservation principles are little more than empirical schemes which we use to codify our experimental results. The fundamental significance of these principles is not yet known.

---

[22] There is also one purely electromagnetic decay, $\Sigma^0 \to \Lambda^0 + \gamma$, that conserves strangeness and has an intermediate lifetime of $\sim 10^{-19}$ sec.

RESONANCE "PARTICLES"

The occurrence of a resonance in a nuclear reaction indicates the existence of an excited energy state in the compound nucleus (see Section 15.4). Resonances also occur in the interactions of elementary particles. Most of the resonance states that have been observed, however, decay in times that are $\sim 10^{-23}$ sec. Because these states have the measurable properties usually associated with particles (energy, angular momentum, strangeness), it is customary to classify all resonance states as "particles," although the term seems somewhat inappropriate for the short-lived cases. By counting particles in this way, the list of elementary particles now includes more than 100 entries.

The longest-lived resonance particle is the η^0 meson whose half-life of $\sim 10^{-18}$ sec places it just within the category of "stable" particles according to our definition (see Table 16.3). The η^0 meson can be considered to be a resonance state of three pions; that is, the decay of the η^0 into particles occurs primarily according to

$$\left. \begin{aligned} \eta^0 &\longrightarrow \pi^+ + \pi^- + \pi^0 \\ &\longrightarrow \pi^0 + \pi^0 + \pi^0 \end{aligned} \right\} \tag{16.23}$$

Because the pion lifetime is short, it is not feasible to produce pion "targets." Therefore, we cannot bring together three pions in the same way that we can bring together an α particle and a N^{14} nucleus to form a F^{18} compound nucleus. Consequently, the resonance properties of a system of three pions must be studied in an entirely different way. For example, one of the reactions observed when deuterium is bombarded by positive pions is

$$D + \pi^+ \longrightarrow p + p + \pi^+ + \pi^- + \pi^0 \tag{16.24}$$

By studying the energies of the pions in a large number of such events it has been possible to conclude that the pions actually emerge as a unit (that is, as a resonant grouping). The reaction can therefore be represented as

$$D + \pi^+ \longrightarrow p + p + \eta^0 \tag{16.24a}$$

with the η^0 meson representing the group $\pi^+ + \pi^- + \pi^0$, and decaying into these particles with a half-life of $\sim 10^{-18}$ sec. Because the η^0 lifetime is so short these particles can move at most only a few Ångstroms before decaying and therefore the direct observation of η^0 emission is not possible—its existence must be inferred from the dynamics of the particles emitted in the reaction.

Many other resonance states involving various combinations of mesons, hyperons, and nucleons have been discovered and some degree of order is beginning to emerge in the classification and description of these particles.

QUARKS

One of the more interesting schemes for dealing with elementary particles is the *quark model,* another invention of M. Gell-Mann. In this model, all elementary particles are considered to be composed of specific combinations

Table **16.8**   *Properties of Quarks*

Symbol	Charge, q	Strangeness, $\mathcal{S}$	Baryon Number, B	Spin, s
Quarks	$+\frac{2}{3}e$	0	$\frac{1}{3}$	$\frac{1}{2}$
	$-\frac{1}{3}e$	0	$\frac{1}{3}$	$\frac{1}{2}$
	$-\frac{1}{3}e$	-1	$\frac{1}{3}$	$\frac{1}{2}$
Anti-Quarks	$-\frac{1}{3}e$	$+1$	$-\frac{1}{3}$	$\frac{1}{2}$
	$+\frac{1}{3}e$	0	$-\frac{1}{3}$	$\frac{1}{2}$
	$-\frac{2}{3}e$	0	$-\frac{1}{3}$	$\frac{1}{2}$

of three basic particles (called *quarks*[23]) and their antiparticles. Quarks have unusual properties: the electrical charge is $\pm\frac{1}{3}e$ or $\pm\frac{2}{3}e$ and the baryon number is $\pm\frac{1}{3}$ (see Table 16.8). Thus, quarks have fundamental properties not possessed by any other particles. But various combinations of these hypothetical particles duplicate the properties of all the known hadrons to an extraordinary degree. Baryons can be considered to consist of *three* quarks, whereas the properties of mesons can be reproduced with *two* quarks; for example,[24]

Proton: $q = +e;\ \mathcal{S} = 0;\ B = 1;\ s = \frac{1}{2}$

Λ^0: $q = 0;\quad \mathcal{S} = -1;\ B = 1;\ s = \frac{1}{2}$ (16.25)

π^+: $q = +e;\ \mathcal{S} = 0;\ B = 0;\ s = 0$

In addition, the quark model has been qualitatively successful in reproducing the known lifetimes, magnetic moments, and decay modes of elementary particles. Are quarks *real* particles or is the quark model only a convenient description of elementary particles that is devoid of real physical content? No one knows.

Because of their distinctive electrical charges, quarks (if they really exist) should be readily identifiable. Although several experiments have been designed to search for quarks, none has ever been found. If they exist, quarks must be bound together so tightly that extremely high energies are required to dislodge them from elementary particles.

Even though the quark model has registered some startling success in accounting for the properties of hadrons, the model is far from being in

[23] The name *quark* is taken from an obscure passage in James Joyce's *Finnegan's Wake:* "Three quarks for Muster Mark!"

[24] The arrows represent the directions of the spins. In combinations, a pair of oppositely directed arrows indicates a cancellation of the spins.

a satisfactory state. There is the possibility, though, that we may eventually be able to describe all strong processes in terms of only three quarks and their antiparticle partners instead of dealing with a zoo-like collection of 100 or so particles. But before such a desirable state of affairs can be attained, quarks must be detected and their properties studied. Recent experiments involving the scattering of high energy electrons from nucleons indicate the existence of a length, small compared to 10^{-14} cm, which should play an important role in the structure of nucleons. Perhaps there are some small entities within a nucleon—perhaps quarks.

16.4 *Symmetry in Elementary Processes*

SYMMETRY AND THE CLASSICAL CONSERVATION PRINCIPLES

The conservation principles of physics provide convenient means for summarizing a variety of facts concerning physical processes. But are the conservation principles only schematic devices for correlating experimental facts or is there a deeper significance to these principles? In other words, are the conservation principles the end results of our search for key physical ideas or must we seek another layer of "fundamental truth" that lies beneath even the conservation principles?

The investigation of these questions has revealed that at least some of the conservation principles can, in fact, be derived by appealing to certain natural *symmetry* principles. Because they relate to the fundamental properties of space and time, these symmetry principles appear to be of greater basic significance than the conservation principles. It is surely an important step forward in our understanding of Nature to be able to start with the *geometrical* symmetry properties of space and time and to derive *physical* conservation laws.

In physics the term "symmetry" has a broader meaning than in ordinary geometry. The best way to explain the meaning of symmetry in the context of contemporary physics is to examine the relationship between physical symmetry and the classical conservation principles. We shall not, however, give the mathematical reasoning that leads to these relationships.

Translational symmetry. The *homogeneity* of space (that is, the fact that the properties of space are the same at one point as at any other point), means that the description of an isolated physical system is independent of any translation through space. This *translational symmetry in space* leads directly to the principle of conservation of *linear momentum.*

Rotational symmetry. The *isotropy* of space (that is, the fact that the properties of space are the same in every direction from a given point), means that the description of an isolated physical system is independent of a rotation through a fixed angle around any given axis. This *rotational symmetry in space* leads directly to the principle of conservation of *angular momentum.*

Symmetry under time translation. The *homogeneity* of time (that is, the fact that the properties of time are the same at one instant as at any other instant), means that the description of an isolated physical system is independent of the time. This *translational symmetry in time* leads directly to the principle of conservation of *energy*.

Because it has been found that these classical conservation principles can be obtained by appealing to the symmetry properties of space and time, there has been a continuing search for further symmetries in the behavior of elementary particles. Some new symmetries have indeed been found, but others, which we believe should exist, have so far eluded us.

NEW SYMMETRIES

Studies of the ways in which elementary particles and antiparticles participate in reactions and decay processes has led to several new symmetry principles:

Charge-conjugation symmetry. If, in a given reaction equation, we change every particle into its antiparticle, we have a new reaction. This operation is called *charge conjugation* (or the *C*-operation). The application of such an operation to the reaction

$$p + \bar{p} \longrightarrow \Sigma^+ + \overline{\Lambda^0} + K^0 + K^- \tag{16.26a}$$

produces

$$\bar{p} + p \longrightarrow \overline{\Sigma^+} + \Lambda^0 + \overline{K^0} + K^+ \tag{16.26b}$$

which is also an allowed reaction. Processes such as these are said to be *invariant* under the charge-conjugation operation. (As used in the description of elementary processes, "invariant" does not mean that each individual particle must remain the same, but only that the *form* of the process and the forces involved are the same.)

The *C*-operation only changes a particle into its antiparticle—there is no change of linear momentum or of spin. Because of this fact, as we shall see, weak interactions are not invariant under charge conjugation; however, strong interactions (of which the reactions above are examples) and electromagnetic interactions appear to obey the rule rigorously.

Reflection symmetry. If we were to use a mirror to view a baseball game in which a right-handed batter faces a left-handed pitcher and hits a pitched ball into right field, we would see a left-handed batter facing a right-handed pitcher and hitting the ball into left field. We would not be watching the "right" game but it would certainly be a *possible* game, one that obeys all physical laws. Baseball games are therefore invariant to mirror reflections; this reflection is called the *parity* operation (or *P*-operation).

If we view various vector quantities in a mirror, what do we see? Actually, there are two classes of physical vectors and these give different results under mirror reflection. For example, the linear momentum vector **p** shows a normal mirror image under reflection (Fig. 16.12a); but the angular momentum vector **L**, the direction of which depends on the application of the right-hand rule, is *reversed* under reflection (Fig. 16.12b). Vectors that behave as in Fig. 16.12a (velocity, linear momentum, force, electric field) are called *polar vectors;* vectors that behave as in Fig. 16.12b (angular momentum, magnetic field) are called *axial vectors.* All vectors that play a role in physical processes are either *polar* or *axial.*

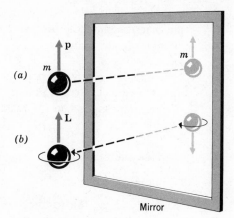

Fig. 16.12 (*a*) *A polar vector, such as the linear momentum vector* **p**, *has a "normal" mirror image.* (*b*) *An axial vector, such as the angular momentum vector* **L**, *depends on the right-hand rule for its direction and so becomes reversed upon reflection.*

The laws of classical physics (mechanics and electrodynamics) are invariant with respect to mirror reflection. That is, any process that obeys the classical laws and that occurs in Nature will have a mirror image that obeys the classical laws and may also occur in Nature (such as the reflected baseball game). This means that polar and axial vectors always occur in such combinations that the laws of classical physics are invariant under reflection. The classical laws are said to *conserve parity*. In elementary particle processes, electromagnetic and strong interactions conserve parity. But, as we shall see, weak interactions are *not* invariant under the *P*-operation and do *not* conserve parity.

Time symmetry. A ball that is dropped from a height h above a certain surface will have acquired a velocity $v = \sqrt{2gh}$ at the end of its fall. Similarly, if a ball is thrown upward with an initial velocity v, it will rise to a height h where it will (momentarily) be at rest. These two processes are symmetric in time; both are possible processes and neither violates any physical law.

If we were to film a physical event and then run the film *backwards,* we would see another possible physical event. This *time reversed* event might be one that is extremely *improbable* but it would not be one that would violate any physical law. We all have seen the backwards-running movie in which a man emerges from a swimming pool (feet first) and rises through the air to land neatly on the diving board. This event is possible—if the water molecules were to move in exactly the correct way, they could transfer to the man in the pool just the right amount of energy and momentum to force him to rise to the diving board—but the event is certainly *improbable.* The entropy principle (Section 7.16) refers to macroscopic physical systems (that is, systems containing large numbers of particles or bodies) and tells us that time progresses in the direction in which *probable* events occur. But there is no physical law that absolutely prevents an improbable event from occurring. In fact, it appears that *all* physical processes[25] are *possible* with backwards-running time. That is, all physical processes are invariant under *time reversal* (called the *T*-operation). The entropy principle

[25] With the exception of at least one particular type of elementary particle process, which we shall mention later (K mesons again!).

applies only to macroscopic systems and *not* to events that take place on the microscopic scale (that is, events involving individual particles) and so there is no way to determine the direction of the flow of time from the study of such events.

In nuclear or elementary particle reactions, time reversal invariance means that reactions can occur equally well in either direction.[26] For example, the T-operation applied to the reaction

$$\text{Li}^7 + p \longrightarrow \alpha + \alpha \qquad\qquad (16.27a)$$

leads to the reaction

$$\alpha + \alpha \longrightarrow \text{Li}^7 + p \qquad\qquad (16.27b)$$

Also, a particle that undergoes decay can be "reassembled" by the time reversal operation; symbolically, the T-operation applied to neutron decay is written as

$$T \times [n \longrightarrow p + e^- + \bar{\nu}_e] = [p + e^- + \bar{\nu}_e \longrightarrow n] \qquad (16.28)$$

Although the formation of a neutron from a proton, an electron, and a neutrino (and the requisite amount of energy) is possible, it is an extremely unlikely occurrence because it requires bringing together *three* particles at one point at one instant of time.

Symmetry of nuclear forces. It was pointed out in Section 15.1 that the nuclear force between nucleons is almost the same regardless of the type of nucleons involved. That is, the *nuclear* parts of the p-p, p-n, and n-n forces are essentially the same. This symmetry of the strong nuclear force leads to the conservation of a quantity called *I-spin* (or *isospin*). *I*-spin appears to be conserved in strong interactions but is violated in electromagnetic and weak processes.

PARITY NONCONSERVATION IN WEAK INTERACTIONS

One of the most startling discoveries in physics in the mid-1950s was the demonstration that weak interaction processes are not invariant under mirror reflections, that is, that weak interactions do not conserve parity. In 1956 it was generally believed that all physical phenomena were reflection invariant.[27] But certain inconsistencies in elementary particle processes (notably in the decay modes of K mesons) led T. D. Lee and C. N. Yang to question whether parity conservation, which had been considered a "self-evident truth," might not be violated in weak interactions. In 1957 several experiments involving nuclear β decay[28] and pion decay were carried out, and these experiments conclusively demonstrated the validity of the Lee and Yang proposal.[29]

---

[26] Recall also that the *absorption* and the *emission* of radiation by an atomic system are equivalent processes (see Section 13.6).

[27] R. P. Feynman backed his belief with a $50 (to $1) bet (he lost), and Wolfgang Pauli said that he would bet "a very high sum" on parity conservation (apparently with no takers).

[28] The β-decay experiments were first performed by C. S. Wu of Columbia University and her collaborators at the National Bureau of Standards.

[29] For their classic work on reflection invariance violations, Lee and Yang were awarded the 1957 Nobel Prize in physics, an exceptionally prompt recognition of their valuable contribution.

The nonconservation of parity in pion decay can be explained in the following way. Figure 16.13a shows the orientation of the linear momentum and spin vector for *real* π^+ decay. Notice that the diagram illustrates the following facts regarding pion decay: (1) the linear momentum vectors, $\mathbf{p}_\mu$ and $\mathbf{p}_\nu$, are equal and opposite as required by momentum conservation; (2) the spin vectors, $\mathbf{S}_\mu$ and $\mathbf{S}_\nu$, are opposite ($s = \frac{1}{2}$ for both particles) because the initial spin of the system—that of the pion—is *zero;* (3) the neutrino ν_μ is *left-handed,* that is, $\mathbf{S}$ is directed opposite to $\mathbf{p}_\nu$ (see Fig. 16.5).

If we apply the parity operation to real π^+ decay, we obtain the situation shown in Fig. 16.13b. Here, the spin vectors have been reversed because the mirror image of a *left*-handed screw is a *right*-handed screw. Therefore, $\mathbf{S}_\nu$ is in the *same* direction as $\mathbf{p}_\nu$. But a neutrino cannot have these vectors in the same direction—ν_μ neutrinos are *always* left-handed. (Only *anti*neutrinos, $\overline{\nu}_\mu$, are *right*-handed, and these particles do not result from π^+ decay.) Thus, Fig. 16.13b shows an *unphysical* situation. We must conclude that pion decay is not invariant under mirror reflection—parity is not conserved.

The dilemma brought on by the failure of reflection invariance in weak interactions can be resolved in the following way. First, consider the effect of the charge-conjugation operation on real π^+ decay. The result of $C \times [\pi^+ \rightarrow \mu^+ + \nu_\mu]$ is shown in Fig. 16.13c. Here, the linear momentum and spin vectors remain the same, but each particle is changed into its antiparticle: $\pi^+ \rightarrow \pi^-$, $\mu^+ \rightarrow \mu^-$, and $\nu_\mu \rightarrow \overline{\nu}_\mu$. However, the relative orientation of the vectors $\mathbf{S}_{\overline{\nu}}$ and $\mathbf{p}_{\overline{\nu}}$ is incorrect—antineutrinos are *right*-handed, not *left*-handed. Therefore, the *C*-operation, as well as the *P*-operation, leads to a nonphysical situation when applied to real π^+ decay. Pion decay (indeed, *every* weak interaction) is neither reflection invariant nor charge-conjugation invariant.

The solution to the problem is found in the application of the *combined*

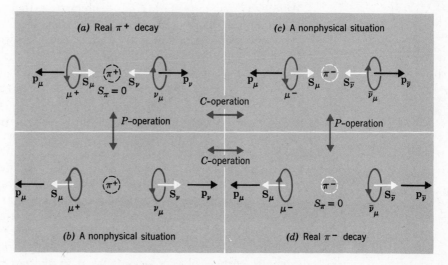

Fig. 16.13 *(a) The orientations of the momentum and spin vectors for real π^+ decay. By applying either the* P-*operation (b) or the* C-*operation (c) to real π decay, non-physical situations result. However, the combined operation,* CP, *produces real π^- decay. The π-μ decay is therefore invariant under the* CP-*operation but not under the* C- *and* P-*operations separately.*

CP-operation to real π^+ decay. That is, we bypass consideration of either the *C*-operation or the *P*-operation separately and direct our attention to the net effect of reflection *plus* charge conjugation. The result of the *CP*-operation applied to real π^+ decay is shown in Fig. 16.13d. This situation is obtained, as is easily verified, by applying *C* followed by *P* or by applying *P* followed by *C*. The *CP*-operation on real π^+ decay produces *real* π^- decay—in particular, the relative orientation of the vectors $\mathbf{S}_{\bar{\nu}}$ and $\mathbf{p}_{\bar{\nu}}$ is given correctly for the right-handed antineutrino. Pion decay (indeed, *every* weak interaction) is invariant under the combined *CP*-operation.

THE *CPT*-THEOREM

What is the significance of the fact that weak interactions exhibit *CP* invariance? We can understand this rather peculiar type of invariance by appealing to the so-called *CPT-theorem*.

It sometimes happens that the structures of our fundamental physical theories are such that it is possible to derive from them conclusions that are extremely broad and general. For example, the classical conservation principles can be obtained from considerations of the symmetry properties of space and time. Another result of great generality is the *CPT-theorem* which is based entirely on the basic concepts of quantum mechanics and relativity. This theorem states *that all interactions are invariant under the combined CPT-operation.*

For weak interactions, the *CPT*-theorem has the following implication. If weak interactions are invariant under the *CP*-operation (as shown above) and are also invariant under the *CPT*-operation, then it follows that these interactions are also invariant under the *T*-operation alone. That is, *CP*-invariance of weak interactions means that *all weak interactions are time-reversal invariant.*

Although direct tests of the *CPT*-theorem are extremely difficult to make, the theorem is believed to be rigorously true. If any violation of *CPT*-invariance were found in any process whatsoever, this would be a shattering blow to the foundations of contemporary physical theory. Therefore, if a *CP-violation* were discovered in some weak interaction, this would be interpreted as a breakdown of *T*-invariance for this process but with *CPT*-invariance still valid. In fact, a small violation of *CP* invariance was discovered in 1965 by a Brookhaven group when they carefully measured the decay rates of K^0 and $\overline{K^0}$ mesons. If *CP*-invariance were valid for the K^0 and $\overline{K^0}$ decays, the decay rates for these mesons would be exactly the same. The experiment showed, however, that there is a difference—although it is an exceedingly small difference—in the two decay rates. The conclusion is that neutral kaon decays (and, as far as we know at present, *only* these decays) violate time-reversal invariance. These decays are, therefore, the only known microscopic processes that distinguish between the two possible directions for the flow of time. One of the most serious challenges that elementary particle physics faces today is to find out what really goes on in the delicately balanced $K^0 - \overline{K^0}$ system. Perhaps the interpretation of this phenomenon (if it can be found) will lead to really profound conclusions.

UNDISCOVERED SYMMETRIES

What about the other conservation principles? What is the fundamental significance of the conservation of electrical charge, lepton number (electron and muon), baryon number, strangeness, and *I*-spin? Can these conservation laws also be related to symmetry principles in some abstract space? If we start with a symmetry principle in quantum mechanics, then it develops that it is always possible to derive a corresponding conservation law. (Whether the conservation law has any physical content is another matter.) But quantum mechanics does not specify for us the reverse route—we do not know in general how to proceed from a conservation law to an underlying symmetry principle. Whether we shall ever find symmetry principles that correspond to all of the conservation laws is a matter of conjecture.

The eminent theoretical physicist, Victor F. Weisskopf, has summarized the present status of high-energy physics in the following way:

"High-energy physics today is an experimental science. We are exploring unknown modes of behavior of matter under completely novel conditions. The field has all the excitement of new discoveries in a virgin land, full of hidden treasures, the hoped-for fundamental insights into the structure of matter. It will take some time before we can produce a rational map of that new land."

Summary of Important Ideas

All elementary particles have antiparticle partners. Photons, neutral pions, and η^0 mesons are their *own* antiparticles; all other elementary particles have distinct antiparticles.

Fermions (such as electrons and nucleons) can be produced only in particle-antiparticle *pairs*. *Bosons* (such as pions and kaons), on the other hand, can be produced in any number consistent with the conservation of charge, momentum, and energy.

Elementary particles are classified either as *leptons* (electrons, muons, and neutrinos), which are *weakly* interacting particles, or as *hadrons* (mesons and baryons), which are *strongly* interacting particles.

The *conserved* quantities in elementary processes are:

1. Linear momentum
2. Angular momentum
3. Energy (mass-energy)
4. Electrical charge
5. Electron number
6. Muon number
7. Baryon number
8. Strangeness (except in weak interactions)
9. *I*-spin (only in strong interactions)

All elementary processes (except weak interactions) are invariant with respect to the *parity* (*P*) and *charge-conjugation* (*C*) operations. Weak interactions are invariant under the combined *CP*-operation.

All elementary processes (with the exception of $K^0 - \overline{K^0}$ decays) are invariant with respect to the *time reversal* (*T*) operation.

All elementary processes are rigorously invariant under the combined *CPT*-operation.

Questions

16.1 An electron and a positron can bind together into an "atomic" system called *positronium*. What is "antipositronium?"

16.2 A beam of high energy γ rays strikes a target of He^3. Write down some of the possible photoproduction reactions that produce pions.

16.3 What reactions are possible if a negative pion is absorbed (at rest) by a proton? What reactions are possible if a negative pion is absorbed (at rest) by a deuteron?

16.4 Which of the following decays can take place and which cannot? For the latter, give the rule that would be violated if the decay were to take place.

(a) $\mu^+ \longrightarrow e^+ + e^+ + e^-$ (c) $\pi^+ \longrightarrow \pi^0 + e^+ + \nu_e$

(b) $\mu^+ \longrightarrow e^+ + \nu_\mu + \overline{\nu}_e$ (d) $\pi^0 \longrightarrow \mu^+ + \mu^- + \nu_\mu + \overline{\nu}_\mu$

16.5 Which of the following neutrino-induced reactions are forbidden and why?

(a) $\overline{\nu}_\mu + p \longrightarrow n + \mu^+$ (c) $\overline{\nu}_e + n \longrightarrow p + \mu^-$

(b) $\overline{\nu}_\mu + p \longrightarrow n + e^+$ (d) $\overline{\nu}_e + \overline{p} \longrightarrow n + e^-$

16.6 What will be the result of the absorption of a negative muon by a proton in a C^{12} nucleus?

16.7 List the ways in which a K^- meson can decay. (Refer to Eqs. 16.15.)

16.8 What is the *simplest* reaction that will produce an antiproton as the result of a *p-p* collision? (Remember conservation of baryon number.)

16.9 What are some of the reactions that can produce an antineutron as the result of a *p-p* collision?

16.10 Check the various conservation laws for the following reactions. State whether the reaction can take place at *rest* or whether energy must be supplied by the bombarding particles. Are any laws violated? Which reaction would you expect to be allowed but particularly *unlikely*?

(a) $p + \overline{n} \longrightarrow \pi^+ + \pi^0$ (c) $p + p \longrightarrow p + p + p + \overline{p}$

(b) $p + \pi^- \longrightarrow p + n + \overline{p}$ (d) $p + \overline{p} \longrightarrow \overline{n} + p + e^- + \overline{\nu}_e$

16.11 Explain why a Σ^+ hyperon cannot decay into a Λ^0 hyperon via a *strong* interaction. (The decay, $\Sigma^+ \to \Lambda^0 + e^+ + \nu_e$, can take place via the *weak* interaction. Verify that no conservation principles are violated in this process.)

16.12 Examine the following reactions from the standpoint of the various conservation principles. Which reactions are allowed? Identify the conservation principle that prevents each of the other reactions from taking place.

(a) $p + p \longrightarrow \Xi^0 + p + \pi^+$ (d) $n + K^+ \longrightarrow \Sigma^+ + \overline{\Xi^0}$

(b) $p + \pi^- \longrightarrow K^0 + \Lambda^0$ (e) $p + \gamma \longrightarrow \Sigma^- + K^0$

(c) $p + K^- \longrightarrow \Sigma^+ + \pi^-$ (f) $p + \gamma \longrightarrow \Sigma^- + \Lambda^0$

16.13 What combinations of quarks will reproduce the properties of (a) the neutron, (b) the Σ^0 hyperon, and (c) the Ω^- hyperon?

16.14 Write down the processes that result from the applications of the charge-conjugation operation to the following processes:

(a) $\pi^0 \longrightarrow 2\gamma$;   (b) $\Sigma^+ \longrightarrow p + \pi^0$;   (c) $\Omega^- \longrightarrow \Lambda^0 + K^-$

16.15 What is the time-reversed reaction corresponding to the capture of a neutron by a proton to form a deuteron with the emission of a γ ray?

16.16 What reaction results from the application of the combined *CPT*-operation to the reaction $p + \mu^- \to n + \nu_\mu$? Is this new reaction one that could occur in Nature?

16.17 What reaction results from the application of the combined *CT*-operation to the reaction $p + K^- \to \Sigma^0 + \pi^+ + \pi^-$? Is this new reaction one that could occur in Nature? Explain. If the new reaction is possible, would the observation of it be feasible? Explain.

16.18 The capture of $\bar{\nu}_\mu$ by a proton leads to a neutron and a positive muon: $p + \bar{\nu}_\mu \to n + \mu^+$. Does this result imply that the decay, $n \to p + \mu^- + \bar{\nu}_\mu$, is allowed? Why has this decay never been observed?

Problems

16.1 A positron comes to rest in matter and annihilates with an electron. If three equal-energy photons are produced in the process, what is the energy of each photon and in what relative directions do the photons leave the point of annihilation?

16.2 A p-$\bar{p}$ annihilation takes place at rest and produces 4 charged pions, all of which have the same energy. What is the kinetic energy (in MeV) of each pion?

16.3 What would be the total amount of mass-energy released if an anti-hydrogen atom annihilated with an ordinary hydrogen atom? (Neglect the atomic binding energy.) What would be the products of the annihilation?

16.4 What would be the total amount of mass-energy released if an anti-deuterium atom annihilated with an ordinary deuterium atom? (Do not forget the nuclear binding energy but neglect the atomic binding energy.)

16.5 How large would a bubble chamber have to be in order that $\frac{3}{4}$ of the charged pions produced at the center of the chamber would decay before escaping the chamber? (Assume that all of the pions travel with a velocity equal to 0.1 of the velocity of light so that a nonrelativistic calculation can be made.)

16.6 Calculate the minimum photon energy required to produce pions according to the reaction

$$p + \gamma \longrightarrow n + \pi^+$$

(Consider only energy conservation here and neglect the fact that a slightly higher photon energy will actually be required in order to conserve momentum as well.)

16.7   A negative pion is absorbed (at rest) by a helium nucleus and produces the reaction, $He^4 + \pi^- \rightarrow H^3 + n$. What is the Q-value for this reaction?

16.8   What is the Q-value for the reaction $Li^6(\pi^-, p)He^5$?

16.9   Suppose that in addition to the ordinary *pion force* there is a contribution to strong interactions from a *kaon force*. What must be the range of the kaon force compared to that of the pion force? (Refer to the calculation in Section 12.7 concerning the range of the pion force.)

16.10 If there does exist a particle (called the *W meson*) that mediates the weak interaction, it must be extremely massive. Use an argument similar to that given in Section 12.7 to estimate the mass of the pion and show that if the range of the weak interaction is less than 10^{-14} cm, the mass of the *W* meson must be at least twice the mass of the proton.

16.11 A K^+ meson decays (at rest) into two pions. What is the kinetic energy of each pion? (Neglect the mass difference between charged and neutral pions.)

16.12 How much kinetic energy is carried away by the pions in the decay (at rest) of a positive kaon, $K^+ \rightarrow \pi^+ + \pi^+ + \pi^-$?

16.13 What is the amount of mass-energy released as kinetic energy in the decay $\Sigma^0 \rightarrow p + \pi^0$?

16.14 Figure 16.11 is shown approximately "life-size." The Λ^0 and K^0 particles were emitted with velocities essentially equal to the velocity of light. Determine how long each particle lived before it decayed. Compare the Λ^0 decay time with the known half-life given in Table 16.6. The decay mode of the K^0 indicates the decay was via the K_1^0 particle. Is the decay time consistent with the known half-life of K_1^0?

The Horsehead Nebula in Orion   Mount Wilson and Palomar Observatories

17 *Astrophysics and Cosmology*

Man has gazed upon the stars for thousands of years. He has always regarded the heavens with wonder and mystery—and no less so now than in the past. For in spite of the fact that modern science has made tremendous advances in understanding the macroscopic and microscopic world around us, our progress in answering the ancient questions concerning the Universe has been slow and stumbling. How did the stars originate—or have they existed "forever?" Why is there such a variety of different types of stars? How has the Universe evolved? What is its future?

Although we have been able to give partial answers to these questions, each step forward always seems to open new areas of the unknown. Will the Universe continue to expand or will it eventually collapse? How much matter is in the Universe? Is there *antimatter* in the Universe? What is the nature of the peculiar objects that are pouring out such fantastic amounts of energy from the remote regions of space? Are there sources of energy in stars and galaxies that we have not yet discovered?

One of the great difficulties in studying astronomical phenomena is that we cannot perform controlled experiments on the objects of our study. We can only observe those events that are taking place naturally. We now believe, for example, that exploding stars (*supernovae*) perform an essential role in the manufacture of the Universe's supply of heavy elements. The study of these violent stellar events is therefore of particular interest. But supernovae occur in our Galaxy only once every few hundred years; it has been more than 350 years since a supernovae was visible to Man's unaided eye. This handicap has been alleviated somewhat by the construction of huge telescopes that gather light from galaxies billions of light years away, but we are still forced to accept for study only those events and objects that Nature provides.

Because of the severity of the limitations placed on the astrophysicist and the cosmologist, perhaps the greatest source of amazement is not the extraordinary diversity of astronomical happenings that we can see taking place, but the fact that we can analyze these events and, from them, draw conclusions regarding the course of stellar and galactic history over billions of years. Modern astronomy has indeed presented Man with some of his most interesting and unusual scientific challenges. Can we *really* predict the fate of stars and galaxies? In only 5 billion years or so we shall know the answer. While we are waiting for the conclusion of our "experiment," we must be content with the development of theories that are consistent with the facts that we continue to gather regarding the present state of the Universe.

17.1 *Nuclear Reactions in Stars*

THE FORMATION OF STARS

Most of the matter in the Universe is in the form of hydrogen and helium. Wherever it can be measured or estimated—in the Sun, in major planets, and in stars—the abundance of helium is found to be about 10 percent of the hydrogen abundance by number of atoms (or about 30 percent by mass).[1] Only a small fraction of this hydrogen and helium is contained in stars—the remainder is distributed throughout interstellar (and intergalactic) space. In stars, where the temperatures are exceedingly high, the atoms are completely ionized and constitute a *plasma*. In interstellar space hydrogen and helium are primarily in the atomic state. These interstellar atoms are the main raw materials out of which new stars are formed.

The gas in interstellar space is not distributed uniformly. In our Galaxy the average density is about 1 atom per cm^3, but the concentration varies widely from one region of galactic space to another.[2] These density fluctuations arise because of the random motions of the atoms in space. Quite by accident, the density in a certain region can become significantly higher than the average density. It sometimes happens that the amount of matter in such a high-density region exceeds a definite critical value which is ~ 1000 times the mass of the Sun. Then, the gravitational field in that region will be sufficiently high that the bulk of the gas is prevented from escaping the region and gravitational forces will tend to pull together the material into a smaller and smaller volume. But as the interstellar gas is concentrated by gravitational attraction, there will again be fluctuations in the density, and the strength of the gravitational field can become sufficiently high in a number of regions to precipitate local condensations of matter. Therefore, the original gas cloud, with a mass in excess of a thousand solar masses, can break up into a large number of small gas clouds, each of which can continue to condense under the influence of the local gravitational field. During this process some of the gas condenses into *dust grains*. These small clouds of gas and dust condense into individual stars, and, thus, a cluster of new stars is formed.

There is ample evidence to support the assertion that many stars are formed in clusters by condensation from clouds of gas and dust. The *Pleiades* (Fig. 17.1) is one of the most famous examples of an open cluster[3] of stars presumably formed by cloud condensation. This cluster contains dozens of stars in a close grouping, six of which are visible to the naked eye as a small, dipper-shaped cluster in the constellation *Taurus*. Too many stars are concentrated within too small a region of space for the clustering to be due to any cause except condensation from a common source of material. The stars of the Pleiades were surely formed together. Furthermore, a long-exposure photograph (right-hand photograph of Fig. 17.1) reveals several prominent regions of nebulous luminosity in the Pleiades. This nebulosity is caused by star light reflected from dust clouds that remain after the condensation of the stars.

STAR CLASSES

Most stars consist of homogeneous or nearly homogeneous mixtures of hydrogen (60%–90% by mass), helium (10%–40%), and heavy elements

---

[1] About 25 percent of the mass of the Sun is helium.

[2] *Outside* of galaxies the average density is lower by a factor of (perhaps) 10^6.

[3] An *open cluster* is an irregular and diffuse collection of stars, as distinct from the regular and concentrated collection of a large number of stars that constitutes a *globular cluster* (see Fig. 6.2).

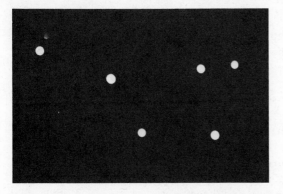

Yerkes Observatory

Fig. 17.1 *Two photographs of the Pleiades cluster. The short exposure on the left shows only the six naked-eye stars, whereas the long exposure on the right reveals many other stars as well as regions of nebulosity caused by the reflection of star light from residual dust clouds that were left after the stellar condensations.*

(0.1%–3%). Stars that have condensed from hydrogen and helium plus the heavy-element material ejected from erupting stars (*novae*) or exploding stars (*supernovae*) are called *Population I* stars. These stars are generally low in hydrogen content and are relatively rich in helium. The Sun (74% hydrogen, 24% helium, and 2% heavy elements by mass) is a normal Population I star. *Population II* stars contain much less remnant material from other stars; these stars are rich in hydrogen, poor in helium, and contain very little heavy-element material. Population II stars have apparently condensed from the primordial hydrogen and helium. The difference in initial composition between stars of the two population types leads to rather different sequences of nuclear reactions that form the elements in these stars. Moreover, even small differences in the initial heavy-element composition of Population I stars lead to different reaction histories among these stars.

THE PROTON-PROTON CHAIN

As a star condenses from a cloud of gas and dust, gravitational potential energy is released. Some of this energy is radiated away and the rest is converted into kinetic energy of the condensing atoms—that is, the *temperature* of the star is increased. This gravitational energy was long believed to be the exclusive source of energy in stars. But in the 1920s it was realized that the amount of gravitational energy possessed by a star was insufficient to account for the vast quantity of energy radiated by a typical star (such as the Sun) throughout its lifetime of billions of years. Attention was then turned to thermonuclear reactions, initiated by gravitational heating, which could account for this enormous outpouring of energy.

When a star condenses from interstellar material, the first thermonuclear reaction to take place involves only hydrogen. The capture of a proton by a proton would produce He^2, but we know that this nucleus does not exist (see Fig. 15.14). However, protons interact not only through the strong nuclear force but also through the weak force, and there exists a capture process whereby two protons can produce a stable final nucleus through a weak interaction:

$$H^1 + H^1 \longrightarrow D^2 + \beta^+ + \nu_e \tag{17.1}$$

or, in the short-hand notation, $H^1(p, \beta^+\nu_e)D^2$. Because the weak force is involved in this process, the reaction probability is extremely low—so low, in fact, that the formation of deuterium from proton-proton capture has never been observed in the laboratory. How do we know, then, that this reaction actually occurs in stars? The theory of weak interactions has been so highly developed and so thoroughly tested with reactions that *can* be produced and studied in the laboratory that we have the utmost confidence in the calculations of the $H^1(p, \beta^+\nu_e)D^2$ reaction rate even though this reaction has never been observed directly.

When a condensing star has reached the stage at which the density in the central region is ~ 100 g/cm^3 and the temperature is $\sim 10^7\,^\circ$K, the stellar protons have sufficient thermal energies for the $p + p$ capture reaction to begin to take place.[4] Once deuterium is formed in this reaction, there rapidly follow two additional reactions which result in the production of helium:

$$D^2 + H^1 \longrightarrow He^3 + \gamma \quad \text{or} \quad D^2(p, \gamma)He^3 \tag{17.2}$$

followed by a reaction involving two He^3 nuclei:

$$He^3 + He^3 \longrightarrow 2H^1 + He^4 \quad \text{or} \quad He^3(He^3, 2p)He^4 \tag{17.3}$$

The net result of this series of reactions (called the *proton-proton* or *p-p chain*) is the conversion of four hydrogen atoms into one helium atom, as shown schematically in Fig. 17.2. The total amount of energy released in this series of reactions is 26.73 MeV. The γ rays and positrons that are produced in

[4]Recall that extremely high temperatures are required in order to overcome the repulsive Coulomb effects between charged particles (see Section 15.5).

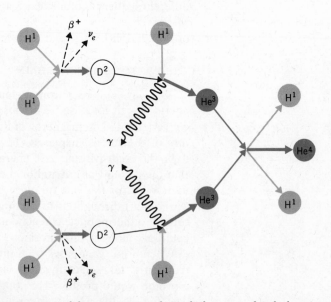

Fig. 17.2 *Schematic representation of the proton-proton chain which converts four hydrogen atoms into one helium atom. Six protons are involved in the sequence, but two are returned to the medium in the final reaction.*

these reactions are absorbed by the gas in and surrounding the thermo-nuclear core and therefore contribute to the heating of the star. The neu-trinos, on the other hand, because of their weak interaction with matter, escape from the star and carry away energy. Taking into account this energy loss, the average energy released in the star by each set of *p-p* chain reactions is approximately 26.3 MeV or about 6.5 MeV per nucleon. Each gram of hydrogen that is converted into helium releases approximately 6×10^{18} ergs of energy. In the Sun, the *p-p* reactions convert hydrogen at a rate of about 6×10^{14} g/sec, with the release of about 4×10^{26} watts of power.

In general, the conditions required for thermonuclear reactions to occur are found only in the central region of a star. Energy released in these reactions is radiated as photons from the core to the surrounding material. These photons exert a *radiation pressure* (see Section 10.6) on the outer layers of the star's gas. In the equilibrium situation, the inward gravitational force exerted on any small volume of stellar material is just equal to the outward force caused by radiation. A star does not continue to contract after the thermonuclear reactions in the core begin to produce sufficient radiation to balance the inward gravitational force.

As the radiation generated in the core streams outward, these photons are absorbed and reradiated (millions of times) by the gas atoms in the outer shell of the star until the radiation eventually escapes from the surface. Thus, the light that we see emanating from a star is characteristic of the material in the cooler surface layer of the star and not of the particular thermonuclear reactions taking place in the enormously hot interior. Nevertheless, as shall be discussed later, the *color* of the light emitted by a star's surface (when combined with a measurement of the intrinsic luminosity of the star) can still be a useful indicator of the nuclear reactions taking place in the core.

THE CARBON-NITROGEN CYCLE

If a star contains some carbon, there is another series of nuclear reactions that can convert hydrogen into helium with the release of energy. In these reactions, carbon serves as a nuclear catalyst and is returned unconsumed, to participate in additional reactions. The reaction sequence is

$$\left.\begin{array}{l} C^{12}(p,\gamma)N^{13} \\ N^{13} \longrightarrow C^{13} + \beta^+ + \nu_e \\ C^{13}(p,\gamma)N^{14} \\ N^{14}(p,\gamma)O^{15} \\ O^{15} \longrightarrow N^{15} + \beta^+ + \nu_e \\ N^{15}(p,\alpha)C^{12} \end{array}\right\} \qquad (17.4)$$

Thus, three protons are captured in a series of (p,γ) reactions and β decays; when a fourth proton is absorbed, an α particle is emitted, reforming a C^{12} nucleus. The net result is the same as that of the *p-p* chain, namely, the conversion of four hydrogen atoms into one helium atom with the release of about 6.5 MeV of energy per hydrogen atom consumed. Because this series of reactions proceeds by using and then reforming carbon and nitrogen, it is called the *CN cycle*. The cyclic nature of the reactions is illustrated in Fig. 17.3.

In a newly condensed star, the initial period of thermonuclear energy generation depends entirely on the proton-proton chain of reactions. But in stars that contain carbon, the CN cycle can compete with the *p-p* chain. The relative amounts of energy generated by the two processes in a given star depend on the temperature of the star's core. For temperatures below about $2 \times 10^7 °$K, the *p-p* chain dominates, but when $T^* \gtrsim 2 \times 10^7 °$K, the CN cycle is the primary source of energy. Thus, the more massive and brighter (and hotter) stars, such as the blue-white star *Sirius,* derive their energy from the CN cycle. The Sun's primary source of energy is the *p-p* chain.

HELIUM BURNING

The energy released in the *p-p* chain of reactions causes the central temperature of a star to increase. Consequently, the CN cycle gradually takes over as the source of energy in the star, *if* carbon is present. But how is carbon formed in Population II stars that condense primarily from hydrogen and helium? The fact that there are no stable nuclei with $A = 5$ or $A = 8$ (see Fig. 15.14) presents a serious problem in the theory of the formation of elements more massive than helium. If only hydrogen and helium are present in a star, the reactions that could form heavier elements would be the capture of a proton by a He^4 nucleus or a $He^4 + He^4$ capture reaction. However, neither Li^5, which is formed by $He^4 + p$, nor Be^8, which is formed by $He^4 + He^4$, is a stable nucleus. Therefore, element formation appears to be blocked by the instabilities at $A = 5$ and $A = 8$.

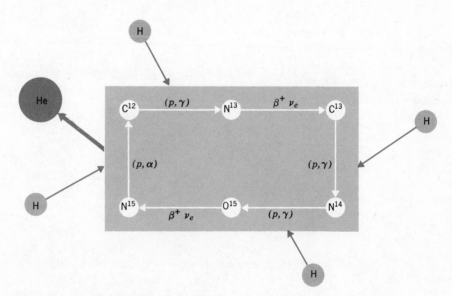

Fig. 17.3 *The carbon-nitrogen (CN) cycle of reactions that converts hydrogen into helium—four protons enter the cycle and one helium nucleus emerges. This reaction cycle was first proposed as a source of stellar energy by H. A. Bethe in 1937; Bethe was awarded the 1967 Nobel Prize in physics for his analysis of energy generation processes in stars.*

The resolution of the dilemma was proposed by E. Salpeter and E. Öpik. When a star has exhausted the hydrogen in its core through *p-p* chain reactions, the core shrinks and the remainder of the stellar material collapses upon the core. The pressure in the core then increases substantially, the temperature of the helium rises to $\sim 10^8$ °K, and the helium density increases to $\sim 10^6$ g/cm^3. Under these extreme conditions, there exists a small equilibrium concentration of Be^8 due to the coalescing of two helium nuclei even though Be^8 lives only $\sim 10^{-15}$ sec before breaking up again into helium nuclei. Because of the close proximity of helium nuclei in the high-density core, a third helium nucleus can be captured by a Be^8 nucleus during the short time interval that Be^8 exists, and C^{12} is formed. In this way the blocking at $A = 5$ and $A = 8$ is overcome and carbon is formed in a two-step helium-burning reaction, He^4 + He^4 → Be^8 followed by Be^8 + He^4 → C^{12}. The energy release in the formation of C^{12} from three α particles is approximately 7.3 MeV (see Example 15.1).

If the core of a star is relatively free of hydrogen,[5] the carbon formed in the 3He^4 → C^{12} process can participate in further helium burning through the reaction

$$C^{12} + He^4 \longrightarrow O^{16} + \gamma \tag{17.5a}$$

followed by the reactions

$$\left. \begin{array}{l} O^{16} + He^4 \longrightarrow Ne^{20} + \gamma \\ Ne^{20} + He^4 \longrightarrow Mg^{24} + \gamma \end{array} \right\} \tag{17.5b}$$
etc.

In the boundary region (which, of course, is not sharp) between the helium-burning core and the hydrogen-containing outer portion of a star, proton capture reactions can occur. Such reactions as

$$\left. \begin{array}{l} O^{16}(p, \gamma)F^{17} \\ \quad F^{17} \longrightarrow O^{17} + \beta^+ + \nu_e \end{array} \right\} \tag{17.6a}$$

$$\left. \begin{array}{l} Ne^{20}(p, \gamma)Na^{21} \\ \quad Na^{21} \longrightarrow Ne^{21} + \beta^+ + \nu_e \end{array} \right\} \tag{17.6b}$$

$$\left. \begin{array}{l} Mg^{24}(p, \gamma)Al^{25} \\ \quad Al^{25} \longrightarrow Mg^{25} + \beta^+ + \nu_e \end{array} \right\} \tag{17.6c}$$

can then produce additional nuclear isotopes.

NEUTRON CAPTURE REACTIONS

The formation of heavy elements by thermonuclear reactions with hydrogen and helium in stars is inhibited by the Coulomb repulsion between charged particles. As the atomic number of the capturing nucleus increases, it becomes more and more difficult to force a proton or an α particle through the Coulomb barrier to form a heavier nucleus. Even in the cores of ex-

---

[5]That is, if helium formation in the *p-p* chain has exhausted the hydrogen in the core. If hydrogen is present, the CN cycle reactions are much more likely to occur than further helium burning.

tremely hot stars, the thermal energies of the particles are insufficient to produce high-Z elements. However, *neutron* capture reactions are not impeded by Coulomb effects, and these reactions readily take place. It is now generally believed that neutron capture reactions in stars are responsible for the synthesis of all the heavy elements.

Neutrons can be produced in stars by reactions such as $C^{13}(\alpha, n)O^{16}$ and $Ne^{21}(\alpha, n)Mg^{24}$. Successive neutron capture reactions, together with the β decay of radioactive species, can produce heavy elements. The formation of nuclei between Fe^{56} and Zn^{68} by neutron capture is shown schematically in Fig. 17.4. Neutron capture reactions occur in red giant stars, as well as in novae and supernovae.

If neutrons are produced rapidly enough, the density of neutrons present in a star can become sufficiently large to permit the capture of neutrons by radioactive isotopes before they undergo β decay. Notice in Fig. 17.4 that Ni^{64} and Cu^{65} can be formed in this sequence of reactions only as the result of neutron capture by radioactive nuclei. And, in fact, the radioactive elements above lead ($Z = 82$) are all formed by a series of neutron captures by radioactive nuclei.

By combining laboratory data on the rates of various nuclear reactions with information concerning stellar properties based on specific star models,

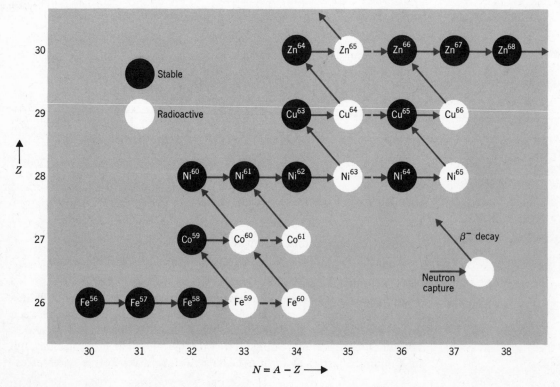

Fig. 17.4 *Successive neutron capture reactions and the β decay of radioactive isotopes produces heavy elements in stars. The chain leading from Fe^{56} to Zn^{68} is a part of this process. The dashed horizontal arrows indicate neutron capture by radioactive isotopes.*

it has been possible to account for the observed abundance of the elements[6] in a very satisfactory way. This combined effort by nuclear physicists and astrophysicists to explain the processes of element formation in stars has been one of the great triumphs of modern science.

17.2 *Non-Optical Astronomy*

OTHER WAYS TO "SEE" STARS

The visible light emitted by stars has been studied for centuries. But the atomic and nuclear processes that occur in stars also produce other types of radiations; these "non-optical" radiations include radio waves, infrared and ultraviolet radiation, X rays, γ rays, electrons, neutrinos, and heavy particles (cosmic rays). Some of these radiations (for example, infrared and X radiation) are effectively blocked from reaching our observatory instruments by the atmospheric blanket that surrounds the Earth. Neutrinos, on the other hand, are so weakly absorbed by any material that the detection of these particles is always difficult. Because of these and other problems, a variety of new techniques has been developed to study the non-optical radiations emitted by stars. In this section we describe briefly some of the important results that have been obtained by these non-optical methods.

RADIO ASTRONOMY

In 1931, K. G. Jansky, of the Bell Telephone Laboratories, discovered that there are radio waves emanating from the Milky Way. These radio waves are emitted continuously and over a broad frequency spectrum. The first instrument designed specifically to receive these cosmic radio waves—a *radio telescope*—was built in 1936. There are now hundreds of radio telescopes in operation around the world. A typical steerable "dish" type antenna is shown in Fig. 17.5 (see also Fig. 3.3). These instruments have detected radio waves emitted by many different types of astronomical objects—the Sun, the moon, some planets, certain stars, gas clouds in our Galaxy, and other galaxies.

In the late 1940s, thousands of discrete radio sources, each occupying a small region in the sky, were discovered and catalogued. One of the most intense radio sources, discovered in 1948, is located in the constellation *Cygnus* and is known as *Cygnus A*. This was the first extragalactic source to be identified (1951) with an optically visible astronomical object (Fig. 17.6).

More than 200 discrete radio sources have now been identified with optical galaxies. The "normal" galactic radio sources emit radio energy at rates $\sim 10^{38}$ ergs/sec, a small fraction of the energy output at visible wavelengths. (For example, the Sun radiates $\sim 10^{33}$ ergs/sec at visible wavelengths and our Galaxy contains $\sim 10^{11}$ stars. Therefore, the energy radiated by our Galaxy at visible wavelengths is $\sim 10^{44}$ ergs/sec.) Some "peculiar" galaxies, on the other hand, emit radio energy at rates of 10^{40}–10^{44} ergs/sec; Cygnus A, for example, emits $\sim 10^{44}$ ergs/sec of radio energy, many times the energy

---

[6]Information concerning the abundance of elements in the Universe is obtained in a variety of ways—measurements of the composition of the Earth, meteorites, and cosmic rays, and observations of the atomic line spectra from stars.

C.S.I. R.O., courtesy of K. Nash

Fig. 17.5 *The 210-ft diameter radio telescope at Parkes, Australia. This instrument has been used for much of the radio mapping of the southern sky. In the foreground is a smaller, movable radio telescope which is used in conjunction with the large instrument for interference measurements.*

output of this galaxy at visible wavelengths. Only about a half dozen other radio galaxies are known to be as energetic as Cygnus A. The reason for the enormous release of radio energy from these peculiar galaxies is not known but it is probably associated with the acceleration of charged particles in the galactic magnetic fields.

QUASARS

In the early 1960s, a discovery of great importance was made when several radio sources, for which the positions had been determined with sufficient accuracy, were identified with certain visual objects that have an unusual blue color. It has been concluded that these objects are not simply visible radiogalaxies; the reasoning is based on two important observations. First, the photographic images are sharp and star-like, not "fuzzy" as is the case for galaxies that are too distant to be resolved into individual stars. Second, the emitted radiations show variations with time that have periods of the order of a day. Because the radiation from an object cannot change significantly in a time less than the time required for light to cross the object, these unusual objects can be no larger than about one light-day (that is, $\sim 3 \times 10^{16}$ cm, or about 200 astronomical units), extremely small compared to galactic sizes. These peculiar star-like objects are called *quasi-stellar objects* or *quasars*.

Further observations have shown that the spectra of quasars exhibit extremely large red shifts. If these red shifts are interpreted in terms of the general expansion of the Universe (see Section 3.2), then some of the quasars are among the most rapidly moving and most distant objects in the Universe. For a single star to lie in the outer regions of the Universe and to have the brightness observed for the quasars means that these objects are pouring out fantastic amounts of energy—$\sim 10^{46}$–10^{47} ergs/sec, about 10^{12}–10^{13} times the output of the Sun. At this rate, a quasar would radiate an amount of energy in just one month equal to the entire mass-energy of the Sun! In order to account for such lavish expenditures of energy, quasars must have masses that are $\sim 10^{9}$ times the Sun's mass.

Although it is not absolutely certain that quasars are distant objects, it seems most probable that they are both extremely remote and extremely massive. How do such objects originate? What is their history? In what way

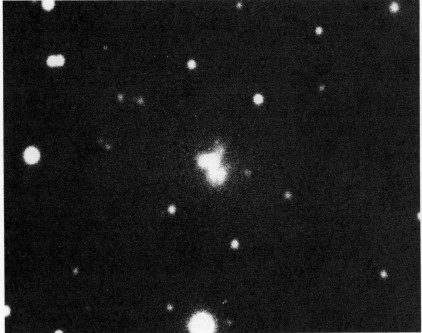

Mount Wilson and Palomar Observatories

Fig. 17.6 *Photograph taken with the 200-in. telescope of the object associated with the intense radio source, Cygnus A. This photograph led to the proposal that a pair of colliding galaxies is responsible for the emission of the radio energy emanating from Cygnus A. But subsequent measurements have shown this hypothesis to be untenable. It is now believed that the Cygnus A source is a single galaxy and that the apparent bifurcation shown in the photograph is due to the presence of a dark dust cloud that obscures the equatorial region of the galaxy. The radio energy from Cygnus A is emitted from two regions of this cloud which lie symmetrically, one on either side of the galaxy. There is no detailed understanding of the process that generates the radio energy from Cygnus A or from other peculiar galaxies.*

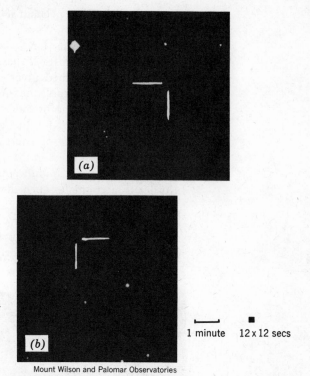

Fig. 17.7 *The intersections of the white lines indicate the positions of discrete radio sources. In (a)
the field in the vicinity of the radio source 3C49 appears empty and no identification with an optical
object can be made. In (b) the position of the radio
source 3C309.1 corresponds almost exactly with a
star-like object—a quasar.*

1 minute   12 x 12 secs

Mount Wilson and Palomar Observatories

are they related to galactic phenomena? Are quasars isolated, star-like
objects or do they lie within galaxies and outshine all of the other galactic
stars put together? At present we do not know the answers to any of these
questions. It is guaranteed, however, that further studies of quasars will lead
to some answers and to additional important questions concerning the nature
and the behavior of the Universe.

RADIO EMISSIONS FROM INTERSTELLAR HYDROGEN

As was pointed out in Section 13.6, the presence of atomic hydrogen in
space can be detected by virtue of the distinctive 21-cm radiation that is
emitted when a hydrogen atom undergoes a "spin-flip" transition. Radio
telescopes tuned especially to this radiation have been used to map the
interstellar hydrogen in our Galaxy. Not only have regions of hydrogen
concentration been detected, but the *velocities* of the atoms in these regions
(along the lines connecting the regions with the Earth) have been determined
by measuring the Doppler shifts of the 21-cm radiation. A drawing of the
Milky Way Galaxy, constructed from 21-cm measurements, is shown in Fig.
17.8. Note the similarity of the spiral structure of our Galaxy to that of
the Whirlpool galaxy (NGC 5194), shown in Fig. 3.14.

MOLECULES IN SPACE

In 1963 a new branch of radio astronomy was opened with the discovery
that there are *molecules* in interstellar space and that these molecules can

be studied through their radio-frequency emissions and absorptions. The first molecular species to be observed was the hydroxyl (OH) radical.[7] Although hydrogen is ubiquitous in our Galaxy, OH radicals are found primarily in extremely small regions of space near very hot stars. In these regions the OH abundance is $\sim 10^{-4}$ of the hydrogen abundance, a remarkably high value for a molecular unit in interstellar space.

Since the discovery of OH in space, searches for other molecular species have revealed the presence of water (H_2O), carbon monoxide (CO), cyanogen (CN), ammonia (NH_3), and formaldehyde (CH_2O), all identified on the basis of molecular radio-frequency emissions and absorptions. Much additional information must be obtained before we have any clear idea of the origin of these space molecules.

INFRARED AND X-RAY ASTRONOMY

Because electromagnetic radiation above and below the visible spectrum is strongly absorbed by the atmosphere, very little of these radiations reach the surface of the Earth. Consequently, rockets, satellites, and high-flying aircraft have been used to gather information regarding infrared and X radiation from astronomical objects.

Stars and galaxies that are in the early stages of formation are relatively cool, and so the radiation emitted by these objects tends to be concentrated

[7] A chemical *radical* is an incomplete molecule. The OH radical is chemically active and combines readily with many substances in ordinary situations (such as Na + OH → NaOH). In space, however, the density of matter is so low that the OH radical can exist independently for long periods of time.

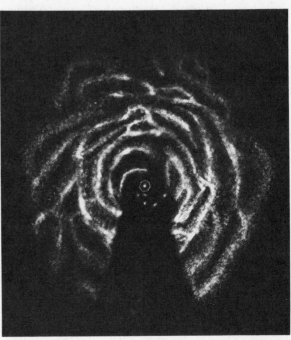

Fig. 17.8 *The spiral structure of the Milky Way as determined from 21-cm measurements of hydrogen concentrations and Doppler shifts. The cross marks the center of the Galaxy and the circled dot in the upper portion of the diagram represents the position of the Sun. The blank region at the bottom is that portion of the Galaxy which is obscured by the dense galatic center.*

Professor Westerhout, University of Maryland

in the long wavelength region. That is, proto-stars and proto-galaxies emit large amounts of infrared radiation even though the amount of emitted visible radiation is so low that these objects cannot be detected with optical telescopes. Therefore, infrared telescopes flown in aircraft, rockets, and satellites are being used to study the processes by which stars and galaxies are formed. Much valuable and interesting information is currently being obtained in these experiments.

X-ray detectors flown above the Earth's atmosphere by rockets have, thus far, revealed about 50 discrete sources of X-ray emission in the sky. Some of these sources have been identified with optically visible objects. The most plausible explanation for such radiations is similar to that proposed as the source of radio emission, namely, that high energy electrons are trapped in magnetic fields and emit high frequency radiation because they are accelerated by the fields.

By supplementing optical observations, infrared and X-ray experiments, as well as radio measurements, are beginning to provide an informative picture of the sky for a great range of wavelengths of electromagnetic radiation.

NEUTRINO ASTRONOMY

Of all the various types of radiation that are emitted by stars, neutrinos, because of their extremely weak interaction with matter, are the most difficult to detect. By the same token, only neutrinos emerge directly from stellar cores without being influenced to any appreciable extent by the outer shells of stars. Therefore, neutrinos provide us with a unique probe of stellar interiors. If we can devise sufficiently sensitive instruments for neutrino investigations, we shall eventually be able to answer many questions regarding nuclear processes in stars.

Unlike ordinary astronomical observatories, which are frequently located on high mountains in order to minimize the effects of atmospheric absorption of radiation, neutrino "observatories" are located in deep mines in order to obtain the maximum shielding of detectors from undesired (background) radiations. Several deep mines in the United States and South Africa are now being used as neutrino "observatories."

The interest in neutrino astronomy at present is centered mainly on solar neutrinos that are emitted in the β^+ decay of B^8 because if we could measure the rate of B^8 decays in the Sun, we would have a sensitive test of current theories of the Sun's interior.

In addition to the reactions in the p-p chain and in the CN cycle, there are other secondary reactions that can take place in stellar interiors. B^8 can be formed in these reactions in the following way. In the p-p chain, most of the He^3 nuclei that are produced interact with one another in the $He^3(He^3,2p)He^4$ reaction (Eq. 17.3). A small fraction of the He^3 nuclei, however, combine with He^4 nuclei and produce Be^7 in the capture reaction,

$$He^3 + He^4 \longrightarrow Be^7 + \gamma \qquad (17.7)$$

Ordinarily, Be^7 decays exclusively by electron capture (see Section 15.2), but under the extreme conditions of pressure and temperature that exist in a star's core, a certain fraction of the Be^7 nuclei will capture a proton and form B^8 before decay can occur:

$$Be^7 + H^1 \longrightarrow B^8 + \gamma \tag{17.8}$$

B^8 rapidly undergoes β^+ decay to Be^8 which, in turn, breaks up into two helium nuclei:

$$B^8 \longrightarrow Be^8 + \beta^+ + \nu_e \tag{17.9}$$
$$Be^8 \longrightarrow He^4 + He^4 \tag{17.10}$$

The reaction rates for the $He^3(\alpha, \gamma)Be^7$ and $Be^7(p, \gamma)B^8$ reactions have been measured in the laboratory. Therefore, by using this information and any stellar model for the density and temperature conditions in a star that is producing energy by the *p-p* chain, it is possible to calculate how many B^8 nuclei should be formed (and decay) each second.

The unique aspect of B^8 decay that makes possible the determination of the rate of B^8 decays in the Sun is the fact that the neutrinos from B^8 decay have a maximum energy[8] of approximately 14 MeV, whereas the neutrinos from all other solar sources have energies less than 2 MeV. Thus, a measurement of the rate at which high energy neutrinos are emitted by the Sun provides a measure of the rate of B^8 formation in the interior of the Sun. If we could devise a neutrino detector that responds *only* to neutrinos with energies greater than 2 MeV, we would be able to detect B^8 neutrinos, with complete discrimination against neutrinos from all other sources.[9]

The neutrino reaction that was used by R. Davis to prove the distinguishability of ν_e and $\bar{\nu}_e$ (see Section 16.2) provides exactly the properties needed for a B^8 neutrino detector. This reaction is

$$Cl^{37} + \nu_e \longrightarrow Ar^{37} + e^- \tag{17.11}$$

The relative masses of Cl^{37} and Ar^{37} are such that a neutrino energy of more than 5 MeV is required in order to initiate the reaction. Therefore, Cl^{37} will react only with B^8 neutrinos—no other solar neutrinos can produce Ar^{37}.[10]

Davis' neutrino detector, consisting of 100,000 gallons of C_2Cl_4, is located almost 5000 feet below ground level in the Homestake Gold Mine in South Dakota (Fig. 17.9). The detection rate for B^8 neutrinos is much smaller than expected; in fact, no net counting rate unequivocally due to B^8 neutrinos has yet been established. The present results indicate that less than 0.5 count per day can be due to B^8 neutrinos. This result is too low (by a factor of

[8] The neutrino energy spectrum is a reflection of the electron (or positron) energy spectrum in β decay—for each low energy electron (or positron) there is a high energy neutrino and *vice versa*.

[9] At least, all other sources in the *Sun*. There are other radioactive decays that yield high energy neutrinos, but the reactions that produce these radioactive nuclei are not now taking place in the Sun.

[10] Actually, the $Cl^{37}(\nu_e, e^-)Ar^{37}$ reaction has a small sensitivity to neutrinos with energies below 2 MeV, but more than 90 percent of the Cl^{37} reactions caused by solar neutrinos should be due to the high energy B^8 neutrinos even though, according to current theories, only about 0.05 percent of the Sun's neutrinos originate in B^8 decays.

Fig. 17.9   *The 100,000-gallon C_2Cl_4 neutrino detector located in the Homestake Gold Mine at Lead, South Dakota. The detector is sensitive to neutrinos from the decay of B^8 in the core of the Sun.*

R. Davis, Brookhaven National Laboratories

2) to be consistent with current theories of the thermonuclear reaction rates in the Sun's core. These measurements, when completed, may point the way toward modifications that are necessary in the theories of stellar dynamics.

17.3   *Stellar Evolution*

THE MAIN SEQUENCE

Stars condense from interstellar gases, burn their nuclear fuels, and die an explosive death as supernovae or simply "fade away" and become small, cold collections of nuclear ash. It is fortunate that we can trace stars through their evolutionary history in terms of only two quantities that are relatively easy to measure—the *intrinsic luminosity* and the *color* (which indicates the *surface temperature* of a star). Every star, at every epoch in its life, has a unique luminosity and a unique color or surface temperature and therefore can be represented by a single point in a color-luminosity diagram. During the life of a star, this representative point traces out a path in this diagram. Thus, the life and death process of a star can be shown in a simple and graphic way in terms of a curve in a color-luminosity diagram.

The dynamical behavior of a star depends on only two factors—the mass of the material from which the star has condensed and the composition of that material. In the initial phases of a star's life, only the *mass* is important in determining the stellar dynamics. The *composition* influences the sequence of nuclear reactions that take place during later phases. If we consider stars that are similar in composition to the Sun (that is, normal Population I stars), then we find that these stars spend most of their lifetimes occupying positions along a color-luminosity curve called the *main sequence* (Fig. 17.10). The *initial* position[11] of a star on this curve depends on the mass of the star—the more massive stars are hot and bright, whereas the less massive stars are cool and dim.[12]

As a star condenses, the release of gravitational potential energy increases the central temperature until thermonuclear reactions are initiated. This new source of energy causes the contraction process to become slower because of the radiation pressure exerted on the outer layers of the star. Eventually, the rate of generation of thermonuclear energy increases to the point that the outward radiation pressure on any small volume of star material is just equal to the inward gravitational pressure. The star then becomes stabilized in size and in luminosity. A star spends most of its life in this condition and, during this time, the star's representative point moves only a short distance along the main sequence.

Because of the stability of a star during most of its life, the color-luminosity diagram for any sample of stars will exhibit a distribution of representative points along the main sequence. This distribution will be indicative of the range in mass of the individual stars and also of the period of time that the stars have been evolving along the main sequence. If we plot the representative points for a number of stars (for example, the stars in the

[11]For our purposes, we shall consider the life of a star to "begin" when thermonuclear reactions start to take place in the core.

[12]Despite the fact that a star radiates huge amounts of energy, the mass of a star is relatively constant during most of its life. (See Problem 17.3.)

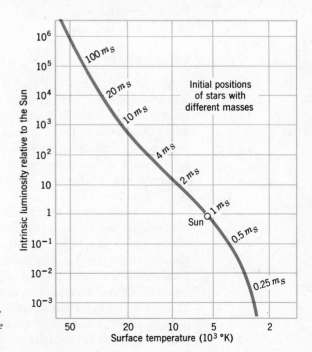

Fig. 17.10 *The main sequence for normal Population I stars (such as the Sun). The position on the main sequence at which a star begins its evolution depends on the mass of the star.* (m_S = *mass of the Sun.*) *Most stars depend ~90% of their lives on the main sequence.*

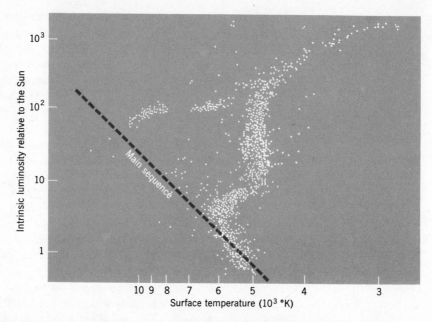

Fig. 17.11 *Color-luminosity diagram for the stars in the globular cluster M3. Such diagrams are often called H-R diagrams because the astronomers E. Hertzsprung and H. N. Russell first plotted star properties in this way. (After A. R. Sandage)*

globular cluster M3, Fig. 17.11), we find not only a group of points along the main sequence, as expected, but also a significant number of points that depart from the main sequence. These departures indicate stars that have had an opportunity to evolve into later phases of their lives and have moved off the main sequence.

EVOLUTION OFF THE MAIN SEQUENCE

A typical Population I star spends 90 percent or more of its lifetime burning hydrogen in its core by the *p-p* chain or CN cycle reactions and simultaneously moving slowly up the main sequence. The Sun has been doing this for 4.5×10^9 years and will continue to do so for another 5×10^9 years before embarking upon the last, violent stages of its life. More massive stars evolve much more rapidly—a star that begins high on the main sequence will run through its life cycle in the comparatively brief period of $\sim 10^7$ years.

When hydrogen is exhausted in a star's core, the core collapses, the temperature increases, and helium burning begins. Since a large amount of energy is released in converting helium into carbon, the luminosity of the star increases. But this energy release also means increased radiation pressure on the outer portion of the star's gases. Consequently, the outer layers of the star *expand.* This expansion causes the gases to become cooler and therefore the emitted light becomes *redder.* Thus, the star departs abruptly from the main sequence (see Fig. 17.12). This expansion and reddening continues until the star has a diameter 200–300 times the diameter it had while on the main sequence—the star becomes a *red giant*.[13] As the Sun reaches the red-giant stage, it will first toast the Earth to a cinder (because of the increased energy output) and then engulf the ashes (because of the tremendous expansion). But this catastrophic day of reckoning is ~5 billion years away.

The red-giant phase in the life of a typical star lasts ~10^7 years. Once having reached its maximum size in the red-giant stage, a star evolves rapidly and moves to the left in the color-luminosity diagram. In fact, the time required to evolve from the red-giant stage to the point at which it crosses the main sequence line (see Fig. 17.12) requires only about one percent of the star's lifetime. Thus, the Sun will make this transition in about 100 million years. At this stage in their careers, most stars become dynamically unbalanced and begin to pulsate, increasing and decreasing in both size and luminosity. Most variable stars (for example, the Cepheid variables discussed in Section 3.1) are found in this region of the color-luminosity diagram.

---

[13]The red color of the giant star *Betelgeuse* (Orion's right shoulder) is quite apparent to the unaided eye.

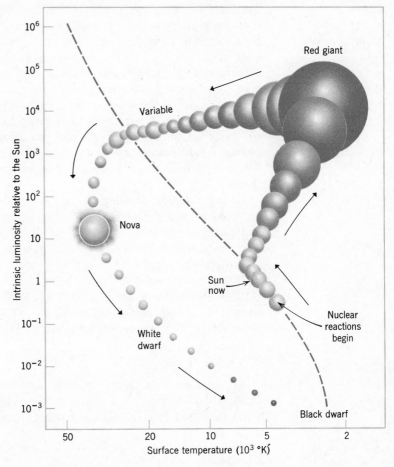

Fig. 17.12 *A simplified diagram of the evolutionary track of a typical Population I star (e.g., the Sun).*

NOVAE

Up to the point at which the red-giant stage is reached, it is believed that all stars evolve in approximately the same way. (The *rate* of evolution depends on the mass of the star.) During the red-giant and subsequent phases, however, differences in composition (and, in particular, the differences between Population I and Population II stars) become important. As a star evolves from the red-giant maximum and moves to the left in the color-luminosity diagram, the central temperature increases (as does the surface temperature). This means that new thermonuclear reactions can take place, and the composition of the star determines which reactions will occur. This is a very complicated stage in a star's life and our theories of post-red-giant evolution are still incomplete. Although we do not understand the details, it seems clear that there are at least two distinct possibilities for the behavior of a star during its terminal phase—massive stars explode and small stars just "fade away."

As a small star (mass < 1.4 times the solar mass) consumes the last of its nuclear fuel, it progresses downward on the color-luminosity diagram (see Fig. 17.12); the luminosity and energy output decrease as gravitational contraction attempts to supply sufficient energy to maintain the temperature. Before substantial cooling has taken place, the star may go through another stage of instability in which there are periodic eruptions that spew stellar material into space. Each such eruption may cause 10^{-4} to 10^{-5} of the star's mass to be ejected into space while at the same time increasing the star's luminosity. A star that exhibits such a surge of light output is called a *nova*.[14] A typical nova will increase in luminosity from about 10 times to about 10^4 times the luminosity of the Sun in a period of less than a day. This increased light output persists for a week or two and then declines. Generally, a nova will recur,[15] often many times, until a condition of stability is reached.

As a prelude to entering the region of the color-luminosity diagram occupied by *white dwarfs* (Fig. 17.12), a star may eject still more of its material, but not in as violent a way as a nova. These prewhite-dwarf ejections are probably responsible for the spectacular ring-like gas clouds (called *planetary nebulae*) that are observed to encircle some stars (Fig. 17.13).

When the last of a white dwarf's available energy is radiated away, it cools rapidly, first turning red (a *red dwarf*), and finally becoming a cold, dense cinder of a once-mighty nuclear furnace—a *black dwarf.*

SUPERNOVAE

Stars that are 1.4 or more times as massive as the Sun die a spectacular death. Instead of entering a period of relatively gentle ejections of material, as do the small stars in the nova phase, massive stars expire in a single

---

[14] So called because early observers believed these objects to be *new* stars (they were usually not visible prior to the outburst).

[15] Some novae or novae-like stars experience eruptions at intervals of months or years, while for others the average period may be thousands of years.

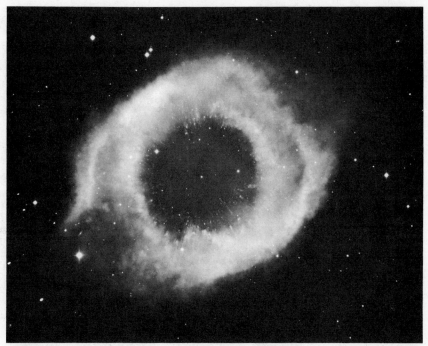

Fig. 17.13 *The planetary nebula NGC 7293 in the constellation Aquarius, photographed with the 200-in. telescope. More than 1000 examples of planetary nebulae are known. The ejection of gaseous material that forms a planetary nebula seems to be a part of the evolutionary process that leads to the formation of a white dwarf star.*

gigantic explosion. In such a *supernova,* the light intensity can be 10^4 times that of a typical nova. During the period of time in which astronomical events have been recorded, three supernovae in our Galaxy have been visible to the unaided eye.[16] The first was the explosion of 1054 which occurred in the constellation Taurus and was recorded by the Chinese (see Fig. 17.14). The second was observed by Tycho Brahe in Cassiopeia in 1572, and the most recent, the supernova of 1604 in Serpens, was described both by Kepler and Galileo. Many other supernovae in other galaxies have been observed telescopically (see Fig. 17.15). In a typical galaxy, supernovae seem to occur at a rate of one every few hundred years.

According to a current theory, supernovae originate in the following way. As helium burning proceeds in the core of a star during the red-giant phase, the central temperature increases. At temperatures of a few times 10^9 °K, fusion reactions produce nuclei with higher and higher atomic numbers until, finally, the group of elements near iron is formed. The nuclei of these elements have the largest binding energies of all nuclei (see Fig. 15.6), and therefore additional fusion reactions do not release energy. Consequently, the generation of thermonuclear energy in the core can proceed no further. Because the radiation pressure is then insufficient to maintain the stability

---

[16]Only a total of seven supernovae are known to have occurred in our Galaxy in the last ten centuries.

Mount Wilson and Palomar Observatories

Fig. 17.14   *The Crab Nebula in Taurus. This nebula consists of the gases ejected in the supernova explosion of 1054. The Crab is at a distance of 1300 pc.*

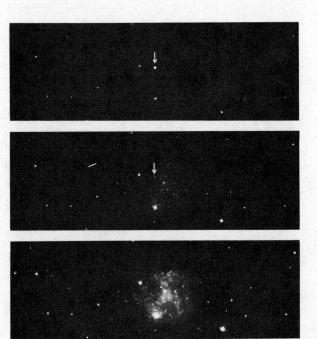

Fig. 17.15   *Three photographs of the 1937 supernova in Virgo. The first photograph (1937), a 20-min exposure, shows the bright star after the maximum intensity had been attained. The second photograph (1938), a 45-min exposure, just reveals the waning star. The last photograph (1942) fails to reveal any trace of the star even though the exposure was 85 min. (The increased exposure time for this photograph has made visible many faint stars and some nebulosity.)*

Mount Wilson and Palomar Observatories

of the outer layers of the star, a gravitational collapse of the core takes place. The implosion of the core draws in the envelope material of the star, which contains unconsumed nuclear fuels. As this new material is heated, further thermonuclear reactions occur and the material is raised to higher temperatures. This new source of energy reverses the collapse and a catastrophic explosion ensues. During the collapse and subsequent explosion (which takes place during the remarkably short period of only a few minutes), enormous numbers of neutrons are produced and heavy elements are formed by neutron capture reactions. In supernovae explosions, most of the star material, consisting of both light and heavy elements, is ejected into space.

Two distinct types of neutron capture processes occur in stars. In red giants, capture takes place on a long time scale (the so-called *slow* or *s* process) in which each radioactive species formed by neutron capture generally has time to decay before the next capture reaction takes place (see Fig. 17.4). In supernovae, on the other hand, capture takes place on a short time scale (the *rapid* or *r* process) in which a radioactive nucleus has a higher probability for capturing an additional neutron than for undergoing β^- decay. Both the *s* and the *r* process are required to account for the observed abundances of the elements in the Universe. The heavy elements formed by neutron capture reactions are ejected into space by novae eruptions and supernovae explosions; this material is then incorporated into stars of the next generation as they condense from interstellar material.

It is now believed that most of the high-energy cosmic rays that continually bombard the Earth are particles that were ejected with extremely high velocities from supernovae explosions. These particles were then trapped in the galactic magnetic field and have eventually found their way to the Earth. The study of the composition of cosmic rays has been of considerable importance in determining the abundance of the elements in the Universe.

NEUTRON STARS AND PULSARS

What remains after a supernova explosion? About 30 years ago it was proposed that when a condition of extreme pressure and temperature is attained in the core of a star, electrons can literally be squeezed into nuclei where they combine with protons to form neutrons. Electrostatic repulsion would be removed by such a process and the neutrons would collapse under gravitational attraction into a small superdense ball, so dense, in fact, that the neutrons would be prevented from undergoing the normal decay process. Stars that explode as supernovae appear to satisfy the conditions for neutron formation in their cores, and so it was thought that the remnant of a supernova explosion would be a *neutron star*. But no one knew how to identify a neutron star and the matter rested there. Recent observations, however, have revealed several unusual star-like objects that are probably the conjectured neutron stars.

In 1968 a startling new discovery was announced by a group of radio astronomers at Cambridge University. They had detected a radio source with the unique feature of pulsating at a rapid rate—approximately once per second. Soon after the Cambridge announcement, several other pulsating radio sources were discovered and these objects became known as *pulsars*.

The periods of the pulsars known at present range from 0.033 sec to just over 3 sec. The pulsar identified by the label CP 0950[17] has a period of 0.253 sec; a portion of the radiation pulse received from CP 0950 is shown in Fig. 17.16. Notice in particular that the pulse rises to its peak value in approximately 0.005 sec. Any object that can emit radiation within so short a period of time must be quite small in size—no large object could propagate information back and forth between its parts rapidly enough to maintain in-phase pulsing with such a short time spread. Since light travels only 1500 km in 0.005 sec, this distance represents the maximum diameter of CP 0950. Analyses of the radio signals from other pulsars have shown that these objects are no larger than about 30 km and are probably even smaller, perhaps only a few kilometers in diameter. (Recall that the diameter of the Sun is 1.4×10^6 km.)

Thomas Gold of Cornell University proposed a model of pulsars that appears capable of accounting for many of the remarkable properties of these objects. According to Gold, a pulsar is a rapidly rotating neutron star. The pulsed radiation arises in the following way. Electrons and protons are trapped in the enormous magnetic field of the neutron star (B $\sim 10^{12}$ gauss!). As the star spins, the magnetic field and the trapped particles spin with the star. At the outer limit of this magnetically confined plasma the particles travel with velocities approaching the velocity of light. Because the motion is circular, the particles are accelerated and therefore emit radiation. The radiation is particularly intense because of the extremely high accelerations of the particles. Another consequence of the relativistic velocities of the

[17] CP stands for *Cambridge pulsar* and 0950 indicates the approximate position of the pulsar in the sky in a certain reference scheme.

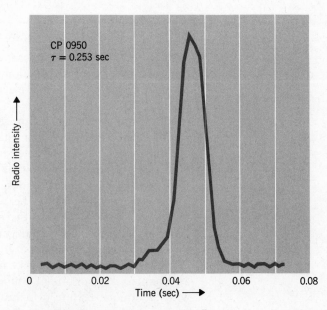

Fig. 17.16 *A portion of the record of radiation received from the pulsar CP 0950. The time between successive peaks is 0.253 sec. (Based on the data of Rickett and Lane.)*

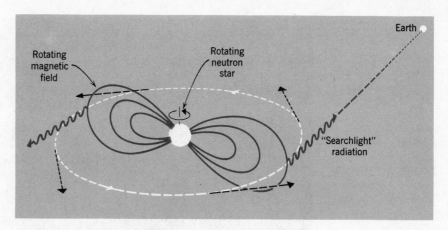

Fig. 17.17 *Gold's model of a pulsar. Charged particles are trapped in the star's magnetic field which rotates with the star. The accelerated particles emit radiation similar to that from a rotating searchlight beacon.*

particles is the fact that the radiation is emitted almost entirely along the direction of motion of the particles. Since the particles rotate with the magnetic field of the star, the radiation takes the form of a "searchlight beacon" scanning the sky (see Fig. 17.17). Once each revolution, a beam flashes toward the Earth.

In addition to providing a tractable explanation for pulsar radiation, the neutron-star model has solved a long-standing astronomical puzzle—why does the Crab Nebula (Fig. 17.14), which had its origin in the supernova of 1054, continue to shine? It is generally agreed that the Crab's luminosity is due to radiation by energetic electrons ($E \gtrsim 5 \times 10^{11}$ eV) that are accelerated in the weak magnetic field of the nebula ($B \sim 3 \times 10^{-4}$ gauss). But if these electrons initially gained their energy in the supernova explosion that ejected the gases that now give rise to the Crab's nebulosity, the energy should have been dissipated long ago. Therefore, the Crab must have some continuing source of energy to maintain its luminosity. A pulsar has now been found near the center of the Crab and this pulsar acts as a pump to keep the Crab Nebula supplied with energy. The energy output of the Crab pulsar ($\sim 10^{38}$ ergs/sec) agrees with the amount of energy that calculations show is required to maintain the luminosity of the nebula.

Pulsating *optical* radiation from the Crab pulsar was first observed in 1969. Measurements made at the University of California's Lick Observatory have shown that the Crab pulsar winks on and off with the same 0.033-sec period found for the radio signal (see Fig. 17.18).

The fact that pulsars pour out such vast amounts of energy has led to the proposal that those pulsars, which are active in our Galaxy at any time,[18] are responsible for the generation of most of the cosmic rays in the Galaxy. Thus, the unusual properties of pulsars seem to be providing answers to many puzzles that have plagued astrophysicists for years.

Although the idea of a rotating neutron star seems capable of explaining most of the present accumulation of pulsar data, it is, of course, not certain

---

[18]The lifetime of a typical pulsar is thought to be $\sim 10^8$ years.

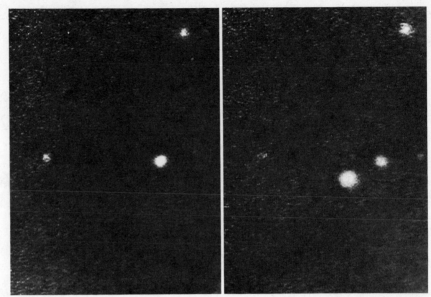

J. S. Miller & E. J. Wampler, Lick Observatory

Fig. 17.18 *Optical radiation from the Crab pulsar—(a) off and (b) on. The photographs are from a light-amplifying television camera attached to the Lick Observatory telescope. The light signals show the same pulsation property and the same period as the radio signals.*

whether this model will stand the test of further observations. But it is an interesting answer to an unusual astronomical puzzle, one that emphasizes the fact that imaginative new ideas are necessary to account for the vast range of phenomena observable in the Universe.

17.4   *Cosmology*

THEORIES OF THE UNIVERSE

The subject of *cosmology* deals with the history, structure, and evolution of the Universe as a whole. Because vast distances, high velocities, and enormous masses are involved in the Universe, cosmology and general relativity are intimately related. The first modern *cosmology*[19] was developed in 1917 by Einstein as a direct result of his original formulation of relativity theory. Einstein showed that the general theory of relativity could account for a *static* Universe—one that does not change with time—in a unique and self-consistent way. Since, at that time, the Universe did appear to be static, this accomplishment was hailed as a significant success and an undeniable vindication of the general theory.

Because the subject of cosmology attempts to account for the origin and history of the Universe, we cannot perform any experiments or make any observations that pertain directly to the heart of any given theory. Each

[19] The term *cosmology* is used both in a general sense and in a specific sense to indicate a particular theory.

theory must stand or fall on the basis of its predictions concerning observable quantities as they exist today. Therefore, when Hubble showed in 1929 that the Universe is not static but is *expanding* (see Section 3.2), Einstein's original cosmology suffered a fate that has befallen many other theories—it was shattered by observational facts.

The basic disagreement between Einstein's static cosmology and the observed expansion of the Universe was an unambiguous dichotomy—the model was clearly incorrect. Some of the later cosmologies have not been so obviously in error—failures in certain cases have rested on rather subtle points. Until the last few years, every cosmology seemed to suffer from some basic flaw. But as the observational techniques advanced and old results were improved and corrected, one cosmology (the *big bang* theory) began to emerge as the only cosmology capable of explaining all (or *almost* all) the results of observations. Of course, we can never be certain whether any theory is *true,* but at the moment, the big bang theory is generally considered to have the best chance of surviving future observational tests.

THE BIG BANG THEORY

The idea that the Universe originated in a gigantic explosion (a *big bang*) was first propounded by the Russian-American theorist, George Gamow (1904–1968) in 1948. According to the big bang theory, about 10^{10} years ago all of the matter and energy now in the Universe were concentrated in a single *fireball* in which the density was $\gtrsim 10^{25}$ g/cm^3 and the temperature was $\gtrsim 10^{16}$ °K. The radiation pressure was tremendous in this fireball and it expanded outward with explosive rapidity—the *big bang*. Those parts of the fireball that had the greatest relative velocities are now concentrated in the distant galaxies that we see (as they were $\sim 2 \times 10^9$ years ago) receding from us with high velocities. Thus, the general expansion of the Universe results in a natural way from the big bang theory.

Gamow originally proposed that all of the elements in the Universe were formed by nuclear reactions during the first few moments after the big bang. This hypothesis leads to the conclusion that the Universe should everywhere exhibit the same relative abundance of the elements. But we now know from spectral measurements that even in the Milky Way there are wide variations in element abundances among the stars. And, indeed, our current theories of stellar evolution make clear the point that element synthesis is a continuing process and that erupting and exploding stars return the products of thermonuclear reactions to the interstellar medium where they can be incorporated into newly forming stars.

Refinements of Gamow's big bang theory have shown that nuclear reactions did indeed take place in the expanding fireball but that the reactions proceeded no further than the formation of helium.[20] In the earliest moments of the fireball, the temperature and density were so great that there was thermal equilibrium between protons, neutrons, electron-positron pairs, neutrinos, and photons. As the fireball expanded, the temperature and density decreased and the neutrons were no longer prevented from under-

---

[20] Some heavier elements were formed but the amounts are negligible in comparison to the abundances found in the Universe.

going the normal decay process. But the half-life of the neutron is about 13 min, so that during the first few minutes after expansion began, the neutrons were abundant and neutron-induced reactions took place. These reactions are:

$$\left. \begin{array}{l} H^1 + n \longrightarrow D^2 + \gamma \\[4pt] D^2 + D^2 \longrightarrow \left\{ \begin{array}{l} He^3 + n \\ T^3 + p \end{array} \right. \\[12pt] He^3 + n \longrightarrow T^3 + p \\[4pt] T^3 + D^2 \longrightarrow He^4 + n \end{array} \right\} \tag{17.12}$$

Detailed calculations show that this series of reactions will build up helium to an abundance about 10 percent of that of hydrogen. Thus, the big bang theory can account for the observed universal abundance of helium (see Section 17.1).

The most spectacular confirmation of a prediction of the big bang theory has been the recent observation of *blackbody radiation,* presumably associated with the fireball. In the superdense fireball there was both matter and radiation. The radiation was in the form of enormously energetic γ rays, but as the fireball expanded and cooled, the γ radiation also "cooled" and the photon energies decreased (that is, the wavelengths increased). This radiation still persists in the Universe, but it is now in the form of radio waves, microwaves, and some infrared radiation.

In thinking about the fireball, we must remember that we cannot place ourselves at some distant position and "view" the fireball expanding toward us. The fireball *is* the Universe, and the Earth (or at least the raw material from which the Earth will eventually be formed) is immersed within the fireball. As the fireball expands, all the rest of the matter in the Universe is moving away from the Earth-to-be (or from any other piece of matter in the fireball). Therefore, the fireball radiation bombards the Earth (*then* and *now*) from all directions. Indeed, any observer in the Universe would detect this radiation arriving at his location in equal amounts from all directions.

Because the fireball has been expanding for $\sim 10^{10}$ years, the original fantastically high temperature has decreased to the point that now, according to the theory, the mean temperature of the Universe should be approximately $3°K$. The blackbody radiation from a source at $3°K$ should have a distribution of wavelengths with a maximum near 0.1 cm (see Fig. 12.4 and Eq. 12.1). If the big bang theory is correct, there should be two observable effects: (1) The radiation spectrum should have the shape characteristic of a $3°$-K blackbody and (2) this blackbody radiation should be arriving in equal amounts from all directions in space (that is, the radiation should be *isotropic*).

In 1965 the first measurements were made which detected low-energy cosmic radio waves that could be interpreted as blackbody radiation from the still-expanding-but-now-cool fireball. Subsequently, sufficient additional measurements were made to establish that the shape of the spectrum is indeed that predicted by blackbody radiation theory and corresponds to a temperature of $2.7°K$ (see Fig. 17.19). Furthermore, measurements made

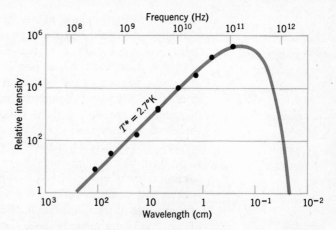

Fig. 17.19   *Measured values of the intensity of low-energy cosmic radio waves (solid circles) compared with the spectrum shape expected on the basis of blackbody radiation theory. These measurements indicate that the mean temperature of the Universe is 2.7°K. The fact that the spectrum agrees closely with the blackbody spectrum is considered to be strong evidence in favor of the big bang theory of the origin of the Universe.*

of the radiation arriving at the Earth from different directions in space have shown that the radiation is isotropic to within a few tenths of one percent.

The big bang theory of the origin of the Universe has therefore successfully passed the crucial observational tests that have so far been devised. But the theory is at present in its formative stages and sophisticated measurement techniques designed to test the theory are still being developed. We can therefore expect new confrontations between theory and experiment to take place at an increasing rate during the next few years. Whether the theory will survive these tests remains to be seen.

THE FUTURE OF THE UNIVERSE

The Universe is now expanding. Will this expansion continue into the indefinite future so that the matter in the Universe will approach a state of being infinitely dilute? Relativity theory makes a definite prediction on this point. According to the theory, there exists a *critical mass* for the Universe. If the actual mass is *less* than this critical value, the mutual gravitational attraction of all the matter in the Universe is insufficient to halt the expansion and the Universe will continue on toward infinite dilution. On the other hand, if the actual mass of the Universe *exceeds* the critical value, gravitational attraction will eventually slow the expansion, stop it entirely, and then reverse the motion. In this situation the Universe is destined for eventual collapse and the fireball will be reformed. The stage will then be set for another big bang and another expansion. That is, the Universe *oscillates* between a condensed fireball phase and a phase of maximum expansion.

Does the Universe contain sufficient mass (in the form of matter and energy) to bring about the oscillation situation? The approximate amount of matter in stars and galactic dust and gases can be estimated by several means. The amount of energy in star light, in magnetic fields in space, in the motion of gas clouds, in cosmic rays, and in neutrinos can similarly be estimated. All of this mass-energy taken together does not equal the critical mass. There is, however, a great uncertainty in this calculation for we do not know how much matter is in *intergalactic* space. The atomic hydrogen in galactic gas clouds can be detected by means of the 21-cm radiation. But intergalactic hydrogen is probably mostly *ionized* (by virtue of the absorption of radiation emanating from galaxies, radio galaxies, and quasars) and therefore will not emit the characteristic 21-cm radiation. X-ray methods are required to detect ionized hydrogen and these experiments are difficult to perform because they must be carried out above the Earth's atmosphere to eliminate absorption effects. Nevertheless, some recent measurements with rockets and satellites indicate the possible existence of appreciable amounts of ionized hydrogen in intergalactic space. If the preliminary interpretation of these measurements is indeed correct, then the amount of intergalactic matter appears to be sufficient to make the total mass of the Universe well in excess of the critical value. If this is so, then we live in an oscillating Universe that goes through a cycle from big bang to expansion to collapse every 8×10^{10} years or so. New X-ray experiments are underway and during the next few years this crucial cosmological question may be answered in a definitive way.

ANTIMATTER AND ANTIGALAXIES

The success of symmetry principles in the interpretation of elementary processes has been so striking (see Section 16.5) that physicists often seek a symmetry argument when attempting to answer a new question. Since *matter* exists in the Universe, should *antimatter* also exist? In fact, by invoking a symmetry argument we could ask: Is it not reasonable to suppose that there are *equal* amounts of matter and antimatter in the Universe? Should not there be stars and *anti*stars, galaxies and *anti*galaxies?

How would we detect an antistar if one existed? Easy. An antihydrogen atom emits an antiphoton, which we then observe and identify as having originated in antimatter. But the problem is that photons are their own antiparticles, so there is no difference between the photons emitted by an ordinary hydrogen atom and those emitted by an antihydrogen atom. Consequently, there is no way to identify antimatter by examining only the electromagnetic radiations that it emits.

What evidence can we use as a basis for a judgment as to whether antimatter exists in the Universe or in our Galaxy? We can place a limit on the amount of antimatter in our Galaxy by the following argument. If antimatter exists in the Galaxy, a portion of the interstellar gas must be antiatoms. As these antiatoms move through space they will encounter ordinary atoms and will annihilate. Each annihilation event releases a certain amount of energy, part of which will escape from the Galaxy in the form of photons and neutrinos and part of which (only about 10 percent), in the form of electrons and positrons, will remain in the Galaxy, trapped by magnetic fields. These particles will contribute to the average energy content of the interstellar gas by colliding with and transferring energy to the gas atoms (and antiatoms). Also, on a much longer time scale, the electrons and positrons will annihilate one another and produce γ radiation. The average energy density in the interstellar regions of our Galaxy can be deduced from various observations—the value obtained is 10^{-11}–10^{-12} erg/cm^3 (1–10 eV/cm^3). If we assume that *all* of this energy arises from antimatter annihilation processes, then the concentration of antimatter in the interstellar gas cannot exceed about 1 part in 10^7. Since the interstellar gas can contain at most this exceedingly small fraction of antimatter, we must also conclude that the stars in our Galaxy consist predominantly (and probably *exclusively*) of ordinary matter.[21]

Even though our Galaxy consists of ordinary matter, can other galaxies be composed of antimatter? Could some of the intense radio signals that we receive from space be due to the collision and annihilation of galaxies and antigalaxies? It is extremely difficult to understand how this could be possible. There is no firm evidence that galaxy-antigalaxy collisions are taking place or have ever taken place. The existence of antigalaxies would require some process capable of separating matter from antimatter; otherwise, the two types of matter would annihilate each other. In particular, some mechanism would have to exist to separate matter from antimatter at the time of the explosion of the big-bang fireball, thus allowing galaxies and antigalaxies to form. No such mechanism is known.[22]

In spite of the great difficulties in devising a model of a symmetric matter-antimatter Universe, some (almost desperate) attempts to salvage the idea have been made. One such speculation is that the original ultraenergetic fireball separated into or created a matter-antimatter pair of superparticles, a *cosmon* and an *anticosmon*. These two "particles" flew apart at great speed and served as the fireballs from which our Universe and an anti-Universe were formed. It seems unlikely that there will ever be observational evidence which can support (or deny) such speculation. On the other hand, if the existence of sizable amounts of antimatter in the Universe can be established (for example, by the discovery of a demonstrable case of a galaxy-antigalaxy collision), then our current ideas of cosmology will require a thorough revision.

Summary of Important Ideas

Stars condense from clouds of gas (and dust) that consist primarily of *hydrogen* and *helium*. Some clouds contain heavy-element material ejected into space by novae and supernovae.

As a star condenses, gravitational potential energy is converted into kinetic energy of the atoms and the temperature of the star increases. When $T^* \sim 10^7$ °K and $\rho \sim 100$ g/cm^3, the *proton-proton chain* reactions begin to take place. These reactions convert hydrogen into helium. The *p-p* chain is the main source of energy in the Sun.

[21] There is no evidence that high energy cosmic rays (which are believed to originate primarily in our Galaxy in supernovae explosions and in their neutron star remnants) consist of antimatter to any appreciable extent.

[22] It has been proposed that the gravitational force between matter and antimatter is *repulsive* rather than attractive. But there is no evidence that this is the case and, indeed, the concept runs counter to the foundations of relativity theory.

In stars that contain carbon, the *CN cycle* is the main source of energy if $T^* \gtrsim 2 \times 10^7$ °K.

Because there are no stable nuclei with $A = 5$ or $A = 8$, *carbon* is formed in stars by the $3\mathrm{He}^4 \rightarrow \mathrm{C}^{12}$ process.

Heavy elements are formed in stars by *neutron capture reactions.* In *supernovae* the neutron density is sufficient to produce the heavy, radioactive elements.

Certain *peculiar galaxies* and *quasi-stellar objects* (quasars) emit enormous amounts of radio energy. The process or processes by which this energy is generated is not known. Nor do we understand the roles that these objects play in the cosmological scheme of things.

All stars spend most of their lives on the *main sequence* generating energy by converting hydrogen into helium by the *p-p* chain reactions or the CN cycle reactions.

Stellar *evolution,* traced in a *color-luminosity diagram,* proceeds by a slow movement along the main sequence, followed rapidly by the red-giant, variable, and eruptive stages. Small stars ($m < 1.4\ m_{\mathrm{Sun}}$) eventually become white dwarfs, whereas massive stars ($m > 1.4\ m_{\mathrm{Sun}}$) undergo violent explosions (that is, they become *supernovae*). The more massive a star, the more rapidly the evolution takes place.

Pulsars are believed to be rapidly rotating *neutron stars* that are remnants of supernovae explosions.

The weight of observational evidence at present favors a model of the Universe which "begins" with a gigantic explosion (the *big bang* theory). Galaxies and stars condense from the gases in the expanding fireball. The Universe is now expanding from this explosion. Whether the expansion will eventually cease and the matter in the Universe will collapse on itself and form another fireball or whether the expansion will continue to a state of infinite dilution is still an open question.

The Universe seems to be composed almost exclusively of *matter;* the abundance of *antimatter* in the Universe is probably very small.

Questions

17.1 A large mass of gas, originally more or less spherical in shape, condenses into a galaxy of stars. Explain how this process can result in a galaxy that is disc-shaped, as is our Galaxy. (Consider angular momentum effects.)

17.2 What are the conditions within a star that determine whether and how fast a given thermonuclear reaction will take place?

17.3 What are the various pieces of evidence that there is matter in the interstellar regions of space?

17.4 Some stars that are still on the main sequence have binary companions that are *white dwarfs.* If the stars were formed at the same time, explain how they can now be so different.

17.5 Spectral lines of the element *technitium* have been observed in the light from certain red-giant stars. The longest-lived isotope of technitium has a half-life of 2×10^6 years. How does this fact support the argument that element formation must take place through nuclear reactions in stars?

17.6 A certain galaxy contains a large amount of hydrogen and helium (but no heavy elements) in gas clouds. Do you expect the stars in this galaxy to be relatively old or relatively young? Explain.

17.7 Discuss the history of a condensing mass of gas if electromagnetic radiation did not exert radiation pressure. Does the fact that stars exist in stable forms offer substantial proof of the reality of radiation pressure?

17.8 Consider two galaxies, one of which is very old and the other of which has been formed only recently (on a cosmological time scale!). How would the stars in these two galaxies differ?

17.9 Cosmic rays with energies up to $\sim 10^{20}$ eV have been observed to bombard the Earth from outer space. According to current theories, there should exist very few cosmic rays with energies in excess of $\sim 10^{20}$ eV; that is, the cosmic-ray energy spectrum should essentially terminate at $\sim 10^{20}$ eV. Nevertheless, detectors are now being constructed to extend the measurements into the range $10^{21} - 10^{22}$ eV. If our present ideas are correct, these detectors should record *not a single* event with an energy $\gtrsim 10^{21}$ eV. Comment on whether the construction of such detectors is sensible, according to the scientific method.

17.10 Assume that the oscillating model of the Universe is correct. Will star formation take place during all phases of a cycle of the Universe or will there be periods during which star formation ceases? Explain.

17.11 If the Universe is indeed *oscillating*, will we always observe red shifts from the distant galaxies? Explain.

17.12 What evidence is there that the entire solar system consists of ordinary matter and that antimatter is not present? (Consider meteorites, solar wind particles, and the landing of space vehicles on the moon and on Venus.)

Problems

17.1 The Sun is moving with a speed of approximately 300 km/sec around the galactic center which is at a distance of $\sim 10^4$ pc. About how many orbits of our Galaxy will the Sun make during the remainder of its expected life?

17.2 There exists another cycle of reactions, similar to the CN cycle, that converts hydrogen into helium using neon and sodium as catalysts. Construct a diagram similar to Fig. 17.3 for the Ne-Na cycle. Starting with Ne^{20}, the sequence of reactions is the same as in the CN cycle—three (p, γ) reactions and β decays followed by a (p, α) reaction.

17.3 We know that a typical star radiates enormous amounts of energy and, therefore, that the mass of the star decreases with time. Why, then, do we refer to *the* mass of a star? What fraction of the mass of the Sun has been radiated away during the last 10^9 years? (The rate of energy generation in the Sun has been essentially constant during this time.)

17.4 Cygnus A is at a distance of $\sim 7 \times 10^8$ L.Y. and emits $\sim 10^{44}$ ergs/sec of radio power. If the radio emissions from Cygnus A are propagated into space equally in all directions, how much radio energy per sec will fall on a 1-cm^2 area on Earth? How many photons per sec of hydrogen H_α radiation would be required to give the same energy flow?

17.5 The planetary nebula surrounding a star (see Fig. 17.13) is observed to increase in diameter at a rate of 0.2 arc sec per year. Doppler shift measurements show that the gases are moving with a velocity of 600 km/sec. What is the distance to the nova?

17.6 A star with a mass equal to 5 times that of the Sun passes through the supernova stage and loses 99 percent of its mass in the explosion. The residual matter condenses to form a neutron star that has a diameter of only 10 km. What is the density of matter in this neutron star? Compare this density with that in nuclei.

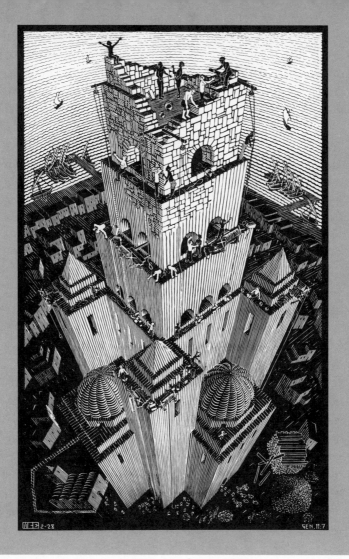

Tower of Babel

M. C. Escher

18 *Toward the Future*

Since the days of Galileo, 300 years ago, scientists have accumulated a vast store of information concerning physical processes. A great deal of this information has been put into order and from this order have emerged our physical theories. None of these theories, however, is complete—all are plagued with unanswered questions to a greater or lesser degree. Our theories of nuclei, of elementary particles, and of cosmology suffer from a profusion of ailments; even our key theories of quantum mechanics and relativity are not immune to this disease. Some of the unresolved questions of contemporary physical science are discussed briefly below. This list is by no means exhaustive (and the individual items are certainly not arranged in any order of special importance)—the list is meant to serve only as an indication of some of the important fundamental questions to which physicists are seeking the answers today.

The reader is also warned not to have the impression that *all* physicists are now working on research that is designed to attack directly these open fundamental questions. On the contrary, only a small fraction of today's physicists are engaged in research programs aimed directly at these problems. Most physicists work within their areas of special interest and competence, focusing attention on *specific* questions for which there is hope of obtaining a definite answer within a reasonable period of time. However, all physicists *worry* about these fundamental questions and each hopes that his own research will shed a bit more light on the subject and will contribute eventually to the resolution of some fundamental problem.

How firm is the foundation of quantum mechanics? In the formulation of quantum theory, the uncertainty principle is the keystone. There seems to be no way to escape the fact that individual events involving elementary particles, nuclei, and atomic systems cannot be predicted with certainty but only in a probabilistic sense. But several eminent physicists have questioned this fundamental assumption of quantum theory. (Einstein never believed it.) They ask whether there might not exist certain unknown variables ("hidden variables") that actually determine the outcome of every elementary event, but, because we have not yet discovered these "hidden variables," the only interpretation now available to us is the probabilistic one.

No one has yet been able to identify these "hidden variables" or to formulate the description of elementary events in a way that bypasses the uncertainty principle. Of all the theories of physical processes that have been developed, quantum theory seems to have the deepest and firmest roots. But it is always possible that quantum theory, in its present form, is not correct. (Only one clear-cut experimental refutation of the theory is necessary to send the theorists scurrying back to the "drawing boards.") If it can ever be proved that the uncertainty principle is a spurious concept, we shall have to look forward to an entire restructuring of physical theory.

What is the nucleon-nucleon force? We now understand a great deal about the nucleon-nucleon force but we do not yet know the fine details. It is a very complicated force indeed: there are *central* parts to the force (as in the static gravitational and electrical interactions), but there are also parts that depend on the relative velocity of the two particles; there are parts that depend on the relative orientation of the nucleon spins; and there are subtle differences in the force that depend on the relative charge states of the nucleons—that is, the nuclear parts of the *p-p*, *p-n*, and *n-n* forces are not *exactly* the same.

Will increased understanding of elementary particle interactions help us to learn more about the nucleon-nucleon force? In what way do kaons, η^0 mesons, and ρ mesons assist pions in mediating the nucleon-nucleon force? We do not know. Would complete information concerning the nucleon-nucleon force be of benefit in calculating the properties of nuclei? Perhaps, but then perhaps not, or at least not very much, because we would still be faced with the problem of calculating the properties of a large number of interacting particles. (This is the *many-body* problem, which is discussed in the following paragraph.) In spite of our ignorance of exactly where the nucleon-nucleon problem is taking us, it is a basic unsolved problem and we are continuing to attack it.

Can we solve the many-body problem? When two objects interact via a $1/r^2$ force, we can obtain a complete and general solution to the problem in terms of known mathematical functions. (This is just the problem of planetary motion—Kepler's problem.) However, if we add a third object to the system, we are no longer able to write a solution that involves known mathematical functions (except in certain special cases). In any particular case, given the initial positions and velocities of the bodies, we can always follow the subsequent motion by numerical calculations that are as precise as we care to make them. If appreciably more than three bodies are involved (as, for example, in the problem of motion of objects in the solar system), the calculations become extremely complex and even the most sophisticated electronic computers require a great deal of time to perform the computations.

Although computer calculations are quite adequate for describing the motion of small numbers of objects, the problem becomes completely intractable with this technique when very large numbers of objects are considered. Therefore, to treat such problems as the behavior of a mass of gas (where the number of particles is $\sim 10^{23}$ or greater), an entirely different approach is required. In order to cope with these problems, a statistical theory was developed (*statistical mechanics*) that can answer questions regarding the *average* properties of large numbers of particles. Statistical mechanics is an outgrowth of kinetic theory and has proved quite successful in dealing with the thermodynamic properties of matter.

If we wish to discuss, for example, the behavior of electrons in a superconductor, then the classical statistical methods are no longer adequate because the interactions of the electrons are complicated by the quantum nature of the system. We then have a quantum mechanical many-body problem. The first breakthrough in this area was the development of the Bardeen-Cooper-Schrieffer theory, which has been successful in explaining many of the features of superconductivity. But we have so far seen only the beginnings of many-body theory for matter in the solid state—many fundamental problems remain to be solved.

Many-body theory has also been applied in an effort to account for the properties of nuclei. Here we have an added complication—the nucleon-nucleon force does not obey the principle of superposition. That is, the nuclear force that binds together a collection of nucleons into a nucleus is not simply the sum of the individual nucleon-nucleon forces that act between all possible pairs of particles in the nucleus—groups of three (or four or more) nucleons exhibit additional interactions. That is, we cannot account for the properties of a nucleus such as He^3 (2 protons, 1 neutron) entirely in terms of two *n-p* interactions and one *p-p* interaction. We know that three-body forces in nuclei are much weaker than two-body forces, but we do not know with any precision the way in which they operate.

The many-body problem is an area in which there is considerable interchange of ideas among various subdisciplines of physics—the methods developed in the study of superconductivity, for example, are being exploited in the nuclear domain.

What is the nature of the weak interaction? At the present time we certainly know many more of the details of the weak interaction than we do of the strong interaction. We understand the crucial importance of relativistic considerations in the weak interaction and we appreciate the role of the four types of neutrinos and how they participate in various forms of the interaction. On the other hand, we do not completely understand the differences between the way in which the weak interaction acts in β decay and the way it acts in the leptonic decays of hyperons. We do not know the range of the weak interaction nor the way which the interaction is mediated. In the case of the strong interaction we have some knowledge of both of these quantities: the range of the strong interaction is $\sim 10^{-13}$ cm and pions are primarily responsible for mediating the interaction (just as photons are the mediators of the electromagnetic interaction). As far as we know at the moment, the weak interaction could be a *point* interaction (that is, the range could be *zero*), but this seems rather unphysical. On the other hand, if the range is finite, then we expect that there exists some type of particle (analogous to the pion) that mediates the interaction. No particle that could be the sought-for mediator of the weak interaction has been found—it remains a ghost particle.

Why is a muon? Among the vast array of elementary particles that have thus far been identified, only a few are reasonably well understood (for example, nucleons, electrons, and pions). Of the remainder, muons remain somewhat of a special mystery. The properties of muons are known, the formation modes and the decay modes are known, and it is known that muons interact via the weak interaction. But muons seems to play no role in the scheme of elementary particles as we now understand them—muons seem to have no purpose in life. When muons were first discovered, it was believed that they were the particles responsible for the nuclear force (that is, the mediators of the strong interaction). For 10 years muons filled this role (but not doing a very good job at it because the muon-nucleon interaction is hopelessly weak) until pions were discovered. It was at once recognized that pions, which interact strongly with nucleons, and not the weakly interacting muons, are the particles involved in nucleon-nucleon forces. Muons were then out of a job—and have remained unemployed ever since.

Muons seem to have all the properties of electrons except that they are heavier (and eventually decay into electrons). But electrons are quite capable of performing their functions in Nature—they do not require any assistance from muons. Perhaps muons are, after all, just fat electrons that Nature has accidentally provided.

Quarks, magnetic monopoles, and tachyons—Do they exist? Some current theoretical investigations speculate about the existence of these three objects: *Quarks* are entities with electric charges of $\frac{1}{3}e$ and $\frac{2}{3}e$ (+ and −) which, in groups of two or three, combine to produce objects with charges of 0 or $\pm e$. These resultant objects are supposed to be identified with the known elementary particles. Indeed a certain degree of success has been achieved in accounting for some of the detailed properties of elementary particles with the quark model. *Magnetic monopoles* (predicted by Dirac in 1931) are supposed to be isolated magnetic poles—that is, poles that occur singly and not in N-S pairs as we are accustomed to finding in ordinary magnets. Magnetic monopoles if discovered, would qualify as a certain type of elementary particle. (A magnetic monopole would have some strange properties; for example, it would be accelerated in a static magnetic field in the same way that an electric charge is accelerated in a static electric field.) *Tachyons*[1] are particles that move with velocities greater than the velocity of light, that is, $v > c$. Relativity theory tells us that no object can be accelerated from a velocity $v < c$ to a velocity $v = c$. Therefore, no ordinary particle can ever travel with a velocity $v > c$ because it cannot be accelerated through the forbidden velocity $v = c$. But what about particles (*tachyons*) that *always* move with velocities $v > c$? There seems to be nothing in relativity theory that absolutely precludes the existence of such particles. (Tachyons would also be forbidden ever to have velocities $v = c$ and they would have *imaginary* masses.) The velocities of ordinary particles are restricted to the range $0 \leqslant v < c$ whereas tachyons are restricted to the range $c < v \leqslant \infty$.

No one has ever detected a quark or a magnetic monopole or a tachyon. But the various theories predict definite properties for these objects and methods exist for the detection of the effects of these properties. Therefore, if any of these objects exists, it should be possible to detect its effects, assuming, of course, that our detection methods are sufficiently sensitive. Searches for these particles continue to be carried out, but there have been no positive results as yet.

How good is the general theory of relativity? Of all our basic theories, general relativity makes the fewest predictions regarding physical effects and has therefore received the least experimental confirmation. None of the experimental tests has yet been made in a definitive way—the uncertainty in each of the measurements is disconcertingly large. The prediction of the rate of advance of Mercury's perihelion agrees closely with the observed value, but a small correction in the predicted value will be necessary if a suspected

---

[1] From the Greek word *tachys* (meaning "swift").

slight oblateness in the shape of the Sun is confirmed. Theory and observation would then disagree by about 10 percent. Similarly, the prediction of the bending of light rays in a gravitational field has been checked to only about 10 percent. If the new measurements that are now in progress produce results that differ from the theoretical predictions by 10 percent, then a major overhaul of the general theory will be necessary. Such a modification has already been proposed—the suggestion is that the gravitational force actually consists of two parts, the major part being that which we now recognize as the gravitational force and the smaller part having a decidedly different character.

One of the assumptions in the general theory is that gravitational mass and inertial mass are identical (the *principle of equivalence*). The equality has been checked to a precision of 1 part in 10^{11} but a higher order of precision will be necessary before we will feel completely comfortable with this assumption.

An even more fundamental question in the general theory is whether all gravitational effects are geometrical in character. That is, is gravity just a manifestation of the curvature of space? Einstein labored long to develop this proposition but we still have no answer.

We have also made little progress in uniting the theories of general relativity and quantum mechanics. Attempts have been made to develop a quantum theory of gravitation but no one has yet succeeded.

What about cosmology? If there is any area of modern scientific thought that is beset with fundamental problems, it is certainly the subject of cosmology. The major difficulty, of course, is that we cannot perform detailed, controlled cosmological experiments in the laboratory—we must rely on observations made on objects that lie at fantastic distances from us, which we cannot influence in any way. And, because the general theory of relativity is so intimately connected with cosmology, any problem in the general theory is reflected as a difficulty in cosmological theory.

We do not know with any precision either the size or the mass of the system with which we are dealing (the *Universe*). We do not know whether the observed expansion of the Universe will continue indefinitely or whether it will eventually cease and reverse so that a collapse will ensue.

We do not know whether antimatter exists to any appreciable extent in the Universe. Are there *anti*galaxies? Probably not—we have no really convincing evidence of their existence. We do not know the nature of the *quasars* that pour out such huge amounts of energy. We do not know very much about the details of stellar evolution in the post-red-giant phase. We do not understand why molecules exist in space. We do not have any viable theory of the ultraenergetic cosmic rays.

And, of course, we do not know the origin of the Universe, although the evidence on hand at present indicates that the expansion is the result of an enormous explosion of inconceivable magnitude that took place 10 billion years ago. But where did that gigantic primordial bundle of energy come from?

Are there "biotonic" laws? The application of physical theory and physical techniques of analysis to biological systems has produced a series of startling advances. We can perform detailed analyses of the structures of the giant molecules that abound in living things; we now understand a great deal about the replication of cells through the action of DNA; and we have even been able to produce artificially some of the smaller organic molecules and even enzymes that are the building blocks of living matter. But we have not yet produced *life* by physical means. We have no reason to believe that the molecules in living matter do not obey the same physical principles that govern the behavior of inanimate objects. Quantum mechanics applies to the molecules in living cells just as it does to the atoms in a crystal. But does quantum mechanics *exclusively* govern living matter, or is there some truth to the ancient idea that there is a *vital principle* necessary to produce living matter? That is, are there any *biotonic* laws, fundamental laws that apply to biological systems but that have no counterpart in the physical world? We have no evidence to support the existence of biotonic laws, but until we approach closer to the physical basis of life we shall not know whether it is really necessary to postulate their existence.

Where do we go from here? Although we have accumulated a great store of information regarding physical phenomena, we seem to have generated more questions than answers. But in a sense, this is the meaning of progress in science. We have made an advance in our understanding of Nature if we have learned enough to ask a significant question, because then we know in what direction to turn our attention to seek an answer. We shall certainly find more answers in the future—perhaps we shall even find answers to some of the questions we have just discussed—and, just as certainly, we shall uncover new fundamental problems. This is the essence and the romance of physics.

Answers to Selected Odd-Numbered Problems

Chapter 1

1.1 $x = n^2$, $n = 1, 2, 3, \ldots$. The experimental results verify this conclusion to within a precision of about 1%.

1.3 (a) 0.0006   (c) 0.000 0039
    (b) 0.000 000 86   (d) 0.000 000 000 003

1.5 (a) 1.2×10^3 (c) 2.4×10^8
    (b) 4.8×10^5 (d) 1.8×10^8

1.7 10 ft

1.9 14 mi, East

1.11 The theory is valid.

1.13 (a) $y \cong 2.5$ (b) $y \cong 4.5$
    (c) $y \cong -5$

Chapter 2

2.1 10 sec

2.3 167 days

2.5 3 μsec

2.7 2 min

2.11 2.5×10^9

2.13 4 hr; 1.5 arc sec

2.15 5×10^{-25} cm³; $\rho(\text{atom}) \cong 3.3$ g/cm³, $\rho(\text{nucleus}) \cong 2.3 \times 10^{14}$ g/cm³

2.17 1 pint of water = 1.04 lb

2.19 1.2×10^{46} atoms; 3 m

Chapter 3

3.1 After one Venusian year, $\angle ESE' = 135°$.

3.3 2.1×10^{24} cm = 6.7×10^5 pc

3.5 0.004 arc sec

3.7 No, because the revolution of the Earth around the Sun would cause the apparent position in the sky to change.

3.9 Approximately 6 times farther away.

3.11 5×10^{-2} rad $\cong 3°$

3.13 $1.2 \times 10^3 (\text{L.Y.})^3$

3.15 1.8×10^{24}

3.17 $N_0 = 6.02 \times 10^{23}$; 1 mole

3.19 1.24 liters

3.21 3.68×10^{21} molecules

3.23 $\sim 10^{-8}$ cm

3.25 $\sim 4.2 \times 10^{-13}$ cm

3.27 $\sim 10^5$ cm

3.29 2 hr

3.31 Pb^{206} ($Z = 82$, $A = 206$)

Chapter 4

4.1 (b) 60 mi/hr, 12 mi/hr, 50 mi/hr, 80 mi/hr, 0 mi/hr, 40 mi/hr.
    (c) The driver stopped near noon (for lunch); at 1:46 P.M. he was stopped for speeding (80 mi/hr).

4.3 (a) 30 ft/sec   (c) 18 ft/sec
(b) 10 ft/sec   (d) 23.3 ft/sec

4.5 208 ft

4.7 144 ft; 96 ft/sec

4.11 600 ft; 15 sec

4.13 484 ft; 4400 ft/sec^2

4.15 5.05 g; no

4.17 $|\mathbf{A} + \mathbf{B}| = 2$ units, 30° clockwise from $\mathbf{B}$

4.19 500 m/sec

4.23 25 ft/sec

4.25 3.5×10^3 ft/sec; 155 sec; 19 mi

4.27 $R = v_0^2/g$, $h = \frac{1}{2}gt^2$, $t = v^0/\sqrt{2}g$; therefore, $h = v_0^2/4g = R/4$

4.29 210 m/sec

4.31 7.3×10^{-5} rad/sec; 0.47 km/sec; 0.33 km/sec

4.33 $a_c = 4\pi^2 r/\tau^2 = 0.1$ g

4.35 2.1×10^{10} cm/sec; no

Chapter 5

5.1 2 sec

5.3 5 N

5.5 5.6 cm/sec^2 at 27° from x-axis.

5.7 2 N

5.9 3×10^4 cm/sec

5.11 2000 N in the direction opposite to the initial motion

5.13 $\sim$6 cm/sec; possibly

5.15 2 m/sec; essentially no change in ship's velocity ($\Delta v/v = 3 \times 10^{-5}$)

5.17 $v = 1.67$ m/sec; absorbed by the Earth.

5.19 1.67 m/sec

5.21 2.7×10^{47} g-cm^2/sec

5.23 Approximately 4600 km from center of Earth ($\sim\frac{2}{3}$ of Earth radius); approximately 450 km from center of Sun ($\sim$0.07 percent of Sun radius).

Chapter 6

6.1 30.5 ft/sec^2; 20.5 ft/sec^2; 8 ft/sec^2

6.3 11.8 ft/sec^2; 0.362 N

6.5 0.04 ft; 144 ft; the ratio is 3600, which is just the square of the ratio of the Earth-moon distance to the radius of the Earth

6.7 13.6 days (i.e., $\frac{1}{2}$ actual lunar month)

6.9 77 days; 0.45 mi/sec

6.11 Kepler's third law is not satisfied

6.13 450 km; the period is the same as that for the Earth

6.15 36 dynes

6.17 6.7×10^{-2} statC

6.19 −9.8 statC

6.21 2.3×10^{-3} dynes; $a_e = 2.5 \times 10^{24}$ cm/sec^2, $a_p = 1.38 \times 10^{21}$ cm/sec^2

6.23 10^{-10} cm

Chapter 7

7.1 1st man, 115 N; 2nd man, 100 N; 10^4 J for both.

7.3 9.8×10^3 J; there is no difference in the work done in the two cases.

7.5 274.4 J; 313.6 J; the latter is greater because it includes the *PE* of each block in the initial condition when it was resting on the floor—in the first case only 7 of the blocks were moved.

7.7 9.8 cm

7.9 7.84×10^8 J; 1.18×10^9 J; no, additional *PE* is made available.

7.11 5×10^5 ergs; zero

7.13 1.47×10^7 m^3

7.15 3750 J

7.17 22.6 m/sec, 10.8 m/sec

7.19 2.25×10^5 J

7.21 140 J

7.23 1.26×10^9 J, 5×10^3 m/sec

7.25 2 ergs; 2 ergs; -6 ergs; no

7.27 2.67 ergs

7.29 zero; 1.64×10^{-11} erg

7.31 9.8×10^8 cm/sec

7.33 $v_p = -5.9 \times 10^8$ cm/sec, $v_{He} = 3.9 \times 10^8$ cm/sec

7.35 0.47°C

7.37 6 percent

7.39 1.74×10^{24} ergs/sec; 2.47×10^{41} ergs

7.41 $P_{final} = 3 P_{initial}$

7.43 4.76 atm; $\Delta T = -141.5°$

7.45 mv

7.47 0.039 eV; 1.37×10^5 cm/sec

7.49 1.57×10^8 cm/sec; 12.9 keV

7.51 1.1×10^7 kW-hr

7.53 3.6×10^5 kg/yr

Chapter 8

8.7 850; 855; the gravitational potential at the position of the moon is essentially constant.

8.11 3.22×10^{-8} dyne at an angle of 63°4 with respect to the *x*-axis

8.13 10^6 statC; 8×10^{-6} statC/cm^2

8.17 $\Phi_G = 1.11 \times 10^{-23}$ erg/g; $\Phi_E = 4.8 \times 10^{-2}$ erg/statC $= 4.8 \times 10^{-2}$ statV

8.19 4.8×10^{16} statV/cm

8.21 0.04 statV

8.23 $E = 0$, $\Phi_E = 0.6$ statV; $E = 0$, $\Phi_E = 0$

8.25 1.2×10^{-7} erg for electron; -1.2×10^{-7} erg for proton

Chapter 9

9.1 5.3×10^3 sec

9.5 $0.79

9.7 1.56×10^{21}

9.11 5.35 gauss

9.13 The particle with the greater charge will have an orbit with a radius $\frac{1}{2}$ that of the other.

9.15 $h \cong 1$ cm; up.

9.17 $(BR)_p = (BR)_\alpha$

9.19 $R = 10^{21}$ cm $\cong 1$ percent of size of our Galaxy

9.21 9.15×10^3 gauss

9.23 $B = 725$ gauss

9.25 $\mu = 1.1 \times 10^{-9}$ erg/gauss

9.27 210 MeV. For high energies, high field strengths are required; sparking limits electric field strengths to ~30 kV/cm, but there is no such limitations on magnetic field strengths ($B \approx 10^6$ gauss have been produced).

Chapter 10

10.1 $5\sqrt{5} = 11.2$ cm/sec; 25 cm/sec^2; 2.81 sec

10.3 10^3 dynes

10.5 $\sqrt{6} = 2.45$ sec

10.7 1.86 mi = 2.6 km

10.9 51.3 cm; 3.3 cm

10.13 5, 10, 15, 20 Hz; the 20-Hz tone is audible and the 15-Hz tone is near the borderline between audibility and inaudibility.

10.15 122 ft/sec = 83 mi/hr

10.21 12 cm

10.23 1.8 cm

10.25 12 m; 30 MHz

10.27 1.5×10^{21} Hz; gamma ray

Chapter 11

11.1 1.71×10^{10} cm/sec

11.3 [v (Newtonian) $- v$ (relativistic)]/v (rel.) = 0.017 percent or ~2 parts in 10^4.

11.5 $v = (\sqrt{3}/2)c = 0.866\ c = 2.59 \times 10^{10}$ cm/sec

11.7 4.52 years; 1.41 years

11.9 $v = c - 6$ cm/sec

11.11 2.6 days

11.13 1 cm; $m = 10^5\ m_0 = 54.5$ AMU $\cong m$(Fe)

11.15 $(\sqrt{3}/2c)c = 0.866\ c$

11.17 $(\sqrt{8}/3)c = 2.84 \times 10^{10}$ cm/sec

11.21 4.35×10^{21} g/sec; 1.45×10^{13} years. (The expected life of the Sun is limited by other factors and is ~10^{10} years.)

11.23 1.1×10^{-6} cm

11.25 2×10^{-6}

Chapter 12

12.1 4833 °K

12.3 6.25×10^{11} photons/sec; 10^{-9} amp

12.5 3.2 eV

12.7 3.72 eV

12.9 12.4 eV = 1.98×10^{-11} erg; 1.98×10^{-18} watt

12.11 0.62 Å; 0.64 Å; 19.3 keV; 0.7 keV

12.13 Approximately 8×10^{-35} cm

12.15 59 volts

12.17 Electron: 1.39×10^{-6} sec; proton: 2.55×10^{-3} sec; photon: 3.3×10^{-8} sec

12.19 1.8×10^{-8} cm

12.21 6×10^5 oscillations; 30 cm

12.23 Approximately $\frac{1}{3}$; 0.09 Å; optical; electron

12.25 0.14 cm

12.27 6.6×10^{-29} g-cm/sec

12.29 1.33×10^{-19} g-cm/sec; 6 eV; because the *total* energy remains constant.

Chapter 13

13.1 4.2×10^{-12} cm

13.3 910 Å

13.5 1.09×10^8 cm/sec; 6.65×10^{-8} cm

13.9 States with $n = 2$ and 3 can be excited: 1026 Å, 1216 Å, 6562 Å

13.11 $\Delta m/m = 1.45 \times 10^{-8}$; the change is ~ 1 part in 10^8.

13.13 1.22×10^{31} cm; no, the size is greater than that of the Universe!

13.17 The $4S$ wave function crosses the axis 3 times; $4P$, 2 times; $4D$, 1 time; $4F$, none.

13.19 The K shell will fill at $Z = 4$ (Be), and the L shell will fill at $Z = 20$ (Ca).

13.21 21.8 keV

13.23 2.81 keV

13.25 4770 Å; 5900 Å

Chapter 14

14.1 4.2 eV

14.3 4×10^{-8} cm; because the K^+ ion is considerably larger than the Na^+ ion.

14.5 About 2 eV (light with frequency and energy higher than for yellow light is absorbed by raising electrons into the conduction band).

Chapter 15

15.1 6×10^{-6} eV

15.3 8×10^{-21} sec

15.5 5.263 MeV; 5.332 MeV; 5.606 MeV

15.7 C^{12}, 28.18 MeV

15.9 $m(92p + 146n) = 239.9878$ AMU; $\mathcal{E}_b/mc^2 = 0.8$ percent

15.11 18.6 keV

15.13 5.4 MeV

15.15 222.02899 AMU

15.17 74 years

15.19 1.58 MeV

15.21 17.35 MeV; -17.35 MeV

15.23 N^{14}; 5×10^{-17} sec

15.25 $Q = 17.6$ MeV, which is much larger than the $D + D$ Q-values.

Chapter 16

16.1 0.34 MeV, 120° relative to one another and all in a single plane

16.3 1877.6 MeV; pions and photons

16.5 1.1 m

16.7 119 MeV

16.9 The range of the kaon force is smaller by the ratio of the pion mass to the kaon mass, or approximately 0.28.

16.11 110 MeV

16.13 119.3 MeV

Chapter 17

17.1 20

17.3 Approximately 0.01 percent

17.5 2000 L.Y.

Index

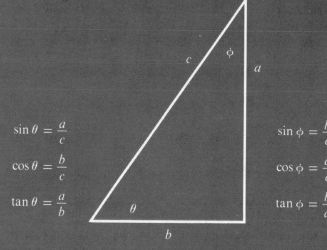

$$\sin \theta = \frac{a}{c}$$

$$\cos \theta = \frac{b}{c}$$

$$\tan \theta = \frac{a}{b}$$

$$\sin \phi = \frac{b}{c}$$

$$\cos \phi = \frac{a}{c}$$

$$\tan \phi = \frac{b}{a}$$

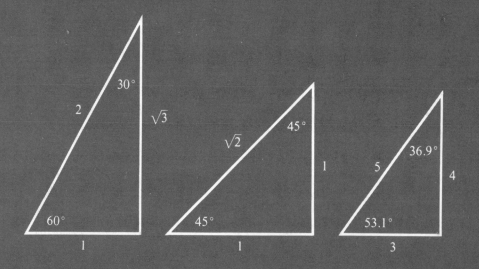

	$\theta = 0°$	$\theta = 30°$	$\theta = 36.9°$	$\theta = 45°$	$\theta = 53.1°$	$\theta = 60°$	$\theta = 90°$
$\sin \theta$	0	$\frac{1}{2}$	$\frac{3}{5}$	$\frac{1}{\sqrt{2}}$	$\frac{4}{5}$	$\frac{\sqrt{3}}{2}$	1
$\cos \theta$	1	$\frac{\sqrt{3}}{2}$	$\frac{4}{5}$	$\frac{1}{\sqrt{2}}$	$\frac{3}{5}$	$\frac{1}{2}$	0
$\tan \theta$	0	$\frac{1}{\sqrt{3}}$	$\frac{3}{4}$	1	$\frac{4}{3}$	$\sqrt{3}$	∞

$$\sqrt{2} = 1.414 \qquad \frac{1}{\sqrt{2}} = 0.707 \qquad \sqrt{3} = 1.732 \qquad \frac{\sqrt{3}}{2} = 0.866 \qquad \frac{1}{\sqrt{3}} = 0.577$$